Lectures on Infinitary Model Theory

Infinitary logic, the logic of languages with infinitely long conjunctions, plays an important role in model theory, recursion theory and descriptive set theory. This book is the first modern introduction to the subject in 40 years, and will bring students and researchers in all areas of mathematical logic up to the threshold of modern research.

The classical topics of back-and-forth systems, model existence techniques, indiscernibles and end extensions are covered before more modern topics are surveyed. Zilber's Categoricity Theorem for quasiminimal excellent classes is proved and an application is given to covers of multiplicative groups. Infinitary methods are also used to study uncountable models of counterexamples to Vaught's Conjecture, and effective aspects of infinitary model theory are reviewed, including an introduction to Montalbán's recent work on spectra of Vaught counterexamples. Self-contained introductions to effective descriptive set theory and hyperarithmetic theory are provided, as is an appendix on admissible model theory.

DAVID MARKER is LAS Distinguished Professor of Mathematics at the University of Illinois at Chicago and a Fellow of the American Mathematical Society. His main area of research is model theory and its connections to algebra, geometry and descriptive set theory. His book, *Model Theory: An Introduction*, is one of the most frequently used graduate texts in the subject and was awarded the Shoenfield prize for expository writing by the Association for Symbolic Logic.

LECTURE NOTES IN LOGIC

A Publication of The Association for Symbolic Logic

This series serves researchers, teachers, and students in the field of symbolic logic, broadly interpreted. The aim of the series is to bring publications to the logic community with the least possible delay and to provide rapid dissemination of the latest research. Scientific quality is the overriding criterion by which submissions are evaluated.

More information, including a list of the books in the series, can be found at http://www.aslonline.org/books-lnl.html

LECTURE NOTES IN LOGIC 46

Lectures on Infinitary Model Theory

DAVID MARKER
University of Illinois, Chicago

ASSOCIATION FOR SYMBOLIC LOGIC

CAMBRIDGE
UNIVERSITY PRESS

CAMBRIDGE
UNIVERSITY PRESS

University Printing House, Cambridge CB2 8BS, United Kingdom

One Liberty Plaza, 20th Floor, New York, NY 10006, USA
477 Williamstown Road, Port Melbourne, VIC 3207, Australia
4843/24, 2nd Floor, Ansari Road, Daryaganj, Delhi - 110002, India
79 Anson Road, #06-04/06, Singapore 079906

Cambridge University Press is part of the University of Cambridge.

It furthers the University's mission by disseminating knowledge in the pursuit of education, learning, and research at the highest international levels of excellence.

www.cambridge.org
Information on this title: www.cambridge.org/9781107181939
10.1017/9781316855560

Association for Symbolic Logic
Richard Shore, Publisher
Department of Mathematics, Cornell University, Ithaca, NY 14853
http://www.aslonline.org

First published 2016

A catalogue record for this publication is available from the British Library.

ISBN 978-1-107-18193-9 Hardback

To my mother Vivian and the memory of my father Charlie

CONTENTS

INTRODUCTION

This book grew out of two courses at the University of Illinois at Chicago. The first was team taught with John Baldwin, Meeri Kesälä and David Kueker in the fall of 2007 and the second I taught in Fall 2013. The main goal of my lectures in 2007 was to provide background material on the classical results on the model theory of infinitary languages needed to appreciate current work on abstract elementary classes that would be the focus of the other lecturers. When I returned to the material in 2013, this was still in my mind. One goal of the course was to cover the introductory material on the model theory of infinitary languages that would prepare someone interested in studying John Baldwin's *Categoricity* or the recent work of Zilber and his school around quasiminimal excellent classes. My second goal was to explore some of the connections between model theory and descriptive set theory, particularly effective descriptive set theory. Montálban's recent gem [59] convinced me it was timely to include this material. To make this material more accessible, I have included more-or-less self contained introductions to the needed effective descriptive set theory and hyperarithemetic theory. A dominant theme related to both goals is to study properties of counterexamples to Vaught's Conjecture.

In more detail, Part 1 focuses on foundations of the model theory of $\mathcal{L}_{\infty,\omega}$ and $\mathcal{L}_{\omega_1,\omega}$. In Chapter 1 we examine the expressive power of infinitary languages and prove the downward Löwenheim–Skolem Theorem. We also prove Chang's results that $\mathcal{L}_{\omega_1,\omega}$-axiomatizable classes can be characterized as reducts of models of a first order theory omitting a set of first order types.

In Chapter 2 we give several back-and-forth criteria characterizing $\mathcal{L}_{\infty,\omega}$-equivalence and introduce the fundamental notion of Scott analysis and prove Scott's Isomorphism Theorem. Section 2.3 is devoted to Kueker's characterization of $\mathcal{L}_{\infty,\omega}$-equivalence via countable approximations and Section 2.4 gives the briefest taste of the model theory of fragments of $\mathcal{L}_{\infty,\kappa}$ for uncountable κ.

Descriptive set theory starts to play a role in Chapter 3 when we introduce a Polish topology on the space of countable τ-structures. These ideas are the key to Morley's theorem on the number of countable models. This is the first place we start to understand potential counterexamples to Vaught's Conjecture, a theme that will recur frequently.

While the Compactness Theorem fails in $\mathcal{L}_{\omega_1,\omega}$, Henkin arguments still play a useful role. Chapter 4 is devoted to the Model Existence Theorem and its consequences. We will prove the appropriate version of Omitting Types Theorem, discuss prime and atomic models, and prove Lopez-Escobar's Interpolation Theorem and the undefinability of well order. We will show how the Interpolation Theorem can be used to show the spaces of models of an $\mathcal{L}_{\omega_1,\omega}$-sentence are exactly the invariant Borel subsets of the space of τ-structures.

Chapter 5 introduces the Erdös–Rado partition theorem and uses it to construct infinitary indiscernibles. Indiscernibles are used to prove that the Hanf number for $\mathcal{L}_{\omega_1,\omega}$ is $\beth_{\omega_1}$. In Section 5.4 we give Hjorth's proof of Knight's result that there are complete $\mathcal{L}_{\omega_1,\omega}$-sentences with models of size $\aleph_1$ but no higher cardinality. We also show how he uses these ideas to show that if there are counterexamples to Vaught's Conjecture, then there are counterexamples with no models of size $\aleph_2$. Hjorth's original proof used ideas from descriptive set theory and dynamical systems. Here we give a straightforward model theoretic proof from [4].

Part 2 is devoted to a number of constructions of uncountable models. Chapter 6 begins with Keisler's methods for constructing uncountable models by building elementary chains of end extensions. These constructions are then used to build uncountable models realizing few types. Here we closely follow Keisler's presentation from [37]. In Section 6.4 we give a nice proof by Larson of a theorem of Keisler and Morley on extending models of set theory. Section 6.5 contains Shelah's elegant proof that an $\aleph_1$-categorical sentence must have a model of size $\aleph_2$.

In Chapter 7 we consider uncountable models of Vaught couterexamples. First we prove results of Harnik and Makkai showing that a Vaught counterexample has models of size $\aleph_1$. We then prove Harrington's result that a Vaught counterexample has models of arbitrarily high Scott rank below ω_2. This is a striking contrast to Hjorth's result. Harrington's original proof relied on admissible model theory. Here we give a new proof from [4] that does not.

Much of the recent interest in infinitary languages has been generated by the work of Zilber and others on complex-like exponential fields and universal covers of commutative algebraic groups. Chapter 8 is devoted to the key model theoretic result Zilber's Categoricity Theorem for quasiminimal excellent classes. Here we closely follow the proof given by Kirby in [39]. We also give some details of the simplest interesting application of the method, the categoricity for covers of the multiplicative group $\mathbb{C}^\times$.

In Part 3 we look at some effective issues. Chapters 9 and 10 are intended to be a self contained treatment of the results we need from effective descriptive set theory and hyperarithmetic theory, though I've taken the liberty of including some of my favorite results that are not needed in the sequel. In Chapter 11 we use these tools to look at questions about bounds on Scott rank and effective constructions of models. In particular, we prove Sacks' theorem about Scott ranks of Vaught counterexamples. We also prove a compactness theorem due to Kreisel that anticipates the later work of Barwise. In Chapter 12 we prove one of Montálban's results on computability spectra of Vaught counterexamples after first proving some prerequisite results of Martin on the effect of determinacy on the Turing degrees.

When I first began this project my intention was to steer clear of admissible model theory. My view has always been that, while admissible model theory provides a fascinating bridge between set theory and recursion theory, it leads to few real model theoretic insights that are unavailable by other means. In the end, having developed the basics of hyperarithmetic theory, it seemed a shame not to tie this all together. I decided, as a compromise, to include some introductory material in an appendix.

There is of course much more that could have been included, but I have mostly restrained myself to presenting material covered in my lectures. An introduction to $\mathcal{L}(Q)$ and $\mathcal{L}_{\omega_1,\omega}(Q)$, for example, would be welcome, but I leave that for others.

Prerequisites. The course this book was based on was taught as a topics course to advanced graduate students most of whom had already taken introductory courses in model theory, set theory and descriptive set theory, but knew almost no recursion theory. While a broad preparation is ideal, much of the book can be read with a more limited background.

- *Model Theory:* The reader should be familiar with basic model theory including Henkin arguments, the Löwenheim–Skolem theorem, back-and-forth arguments, omitting types, prime and atomic models and indiscernibles. Chapters 1-5 of [56] are good preparation.
- *Set Theory:* Most of the book can be read with only a background in naive set theory. Closed unbounded and stationary sets are useful at a few points. Forcing and absoluteness arguments are used briefly in Sections 3.2 and 6.4, and heavily in Section 7.2. Familiarity with models of set theory and Gödel's constructible universe is assumed in Appendix B. Good references are [46] and [30].
- *Descriptive Set Theory:* In Chapter 3 I assume familiarity with Polish spaces and Borel sets. Effective descriptive set theory plays a major role in Part III. I introduce the effective ideas in Chapter 9, but assume familiarity with classical results from the early chapters of [34].
- *Recursion Theory:* I attempt to give a self-contained introduction to everything needed in Section 9.1 and Chapter 10.

In Section 8.2 we prove one case of Zilber's Thumbtack Lemma. This argument requires some comfortability with algebraic geometry.

Acknowledgments. This book would not have been written without the influence of my colleague and friend John Baldwin, whom I thank for countless conversations on infinitary model theory and many comments on earlier versions of these lecture notes.

I have also learned a great deal from discussing these issues over the years with a great many mathematicians. I would particularly like to thank Martin Bays, Victor Harnik, Leo Harrington, Bradd Hart, Alekos Kechris, Meeri Kesälä, Jonathan Kirby, Julia Knight, David Kuecker, Paul Larson, Chris Laskowski, Angus Macintyre, Antonio Montálban, Mark Nadel, Gerald Sacks, and Boris Zilber, as well as Alan Mekler and Greg Hjorth, both of whom are greatly missed.

I am very grateful to the students in my 2013 course who gave detailed comments on earlier drafts of these lectures: Gabe Conant, Maxwell Levine, Tori Noquez, Caroline Terry, Phil Wesolek, Sam Ziegler and Joe Zelinski.

I first learned infinitary model theory from Keisler's beautiful, but, sadly, out of print, *Model Theory of Infinitary Logic*. There are many places where I have found it difficult to improve on his exposition. Chapters 4 and 6, in particular, closely follow Keisler's presentation.

Chicago 2015

Part 1

CLASSICAL RESULTS IN INFINITARY MODEL THEORY

Chapter 1

INFINITARY LANGUAGES

Let τ be a vocabulary, i.e., a set of function symbols, relation symbols and constant symbols. In the logic $\mathcal{L}_{\infty,\omega}(\tau)$ we build formulas using the symbols from τ, equality, Boolean connectives $\neg$, $\bigwedge$ and $\bigvee$, quantifiers $\forall$ and $\exists$, variables $\{v_\alpha : \alpha \text{ an ordinal}\}$.[1]

- terms and atomic formulas are defined as in first order logic;
- if ϕ is a formula, then so is $\neg\phi$;
- if X is a set of formulas, then so are
$$\bigvee_{\phi \in X} \phi \text{ and } \bigwedge_{\phi \in X} \phi;$$
- if ϕ is a formula, so are $\exists v_\alpha\, \phi$ and $\forall v_\alpha\, \phi$.

We extend the usual definition of satisfaction by saying
$$\mathcal{M} \models \bigvee_{\phi \in X} \phi \text{ if and only if } \mathcal{M} \models \phi \text{ for some } \phi \in X$$
and
$$\mathcal{M} \models \bigwedge_{\phi \in X} \phi \text{ if and only if } \mathcal{M} \models \phi \text{ for all } \phi \in X.$$
For notational simplicity, we use the symbols $\wedge$ and $\vee$ as abbreviations for binary conjunctions and disjunctions. Similarly we will use the abreviations $\rightarrow$ and $\leftrightarrow$ when helpful.

We can inductively define the notions of *free variable*, *subformula*, *sentence*, *theory* and *satisfiability* in the usual ways.

EXERCISE 1.0.1. Suppose ϕ is an $\mathcal{L}_{\infty,\omega}$-sentence and ψ is a subformula of ϕ. Prove that ψ has only finitely many free variables.

Let κ be an infinite cardinal. In the logic $\mathcal{L}_{\kappa,\omega}(\tau)$ we form formulas in a similar way but we only use variables $\{v_\alpha : \alpha < \kappa\}$ and restrict infinite conjunctions and disjunctions to $\bigvee_{\phi \in X} \phi$ and $\bigwedge_{\phi \in X} \phi$ where $|X| < \kappa$. Thus $\mathcal{L}_{\omega,\omega}$ is just the usual first order logic. Throughout these notes we be focusing

[1] When no confusion arises we omit the τ and write $\mathcal{L}_{\infty,\omega}$ instead of $\mathcal{L}_{\infty,\omega}(\tau)$.

primarily on $\mathcal{L}_{\omega_1,\omega}$, the logic where we allow countable conjunctions and disjunctions.[2]

EXERCISE 1.0.2. Show that if κ is a regular cardinal and ϕ is a sentence of $\mathcal{L}_{\kappa,\omega}$, then ϕ has fewer than κ subformulas. Show that this fails for singular cardinals. This is one reason it is customary to restrict attention to $\mathcal{L}_{\kappa,\omega}$ for κ a regular cardinal.

DEFINITION 1.0.3. We say $\mathcal{M} \equiv_{\infty,\omega} \mathcal{N}$ if

$$\mathcal{M} \models \phi \text{ if and only if } \mathcal{N} \models \phi$$

for all $\mathcal{L}_{\infty,\omega}$ sentences ϕ. The notion $\mathcal{M} \equiv_{\kappa,\omega} \mathcal{N}$, is defined analogously.

EXERCISE 1.0.4. Show that if $\mathcal{M} \cong \mathcal{N}$, then $\mathcal{M} \equiv_{\infty,\omega} \mathcal{N}$.

When studying the model theory of infinitary logics there is one fundamental and inescapable fact:

The compactness theorem fails for infinitary languages.

EXERCISE 1.0.5. Let τ be the vocabulary with constant symbols $d, c_0, c_1, \ldots$ and let Γ be the set of sentences

$$\{d \neq c_i : i \in \omega\} \cup \left\{\forall v \bigvee_{i \in \omega} v = c_i\right\}.$$

Show that every finite subset of Γ is satisfiable, but Γ is not satisfiable. Thus the Compactness Theorem fails for $\mathcal{L}_{\infty,\omega}$ and even $\mathcal{L}_{\omega_1,\omega}$.

The failure of compactness will lead to many new phenomena and force us to find new approaches and develop new tools.[3]

EXERCISE 1.0.6. (a) Give an example of structures $\mathcal{M}_0, \mathcal{M}_1, \ldots$ and $\phi \in \mathcal{L}_{\omega_1,\omega}$ such that $\mathcal{M}_i \models \phi$ for all i, but if $\mathcal{U}$ is a nonprincipal ultrafilter on ω then $\prod M_i/\mathcal{U} \models \neg\phi$.
(b) Show that if $\mathcal{U}$ is a σ-complete ultrafilter on I, then

$$\prod_{i \in I} \mathcal{M}_i/\mathcal{U} \models \phi \Leftrightarrow \{i \in I : \mathcal{M}_i \models \phi\} \in \mathcal{U}$$

for $\phi \in \mathcal{L}_{\omega_1,\omega}$.

---

[2]Logically it would make sense to call this logic $\mathcal{L}_{\aleph_1,\aleph_0}$ but we will follow the historical precedent and refer to it as $\mathcal{L}_{\omega_1,\omega}$.

[3]Though we will not focus on it, another fruitful approach is to look for more general forms of the Compactness Theorem that hold in particular settings. One example of this is Barwise Compactness for countable admissible fragments. We will see an avatar of these results in Theorem 11.2.2 and give a quick treatment of the general result in Appendix B, but the interested reader should consult [8].

Another important example is studying compactness in languages $\mathcal{L}_{\kappa,\kappa}$ where κ is a large cardinal. See [31] or [30] for further information.

If compactness fails, why then do we study the model theory of infinitary languages? One reason is that we get new insights about first order model theory. But the simplest answer is that there are many natural classes that are axiomatized by $\mathcal{L}_{\omega_1,\omega}$-sentences.

EXERCISE 1.0.7. Show that the following classes are $\mathcal{L}_{\omega_1,\omega}$-axiomatizable for appropriate choices τ.

(a) torsion abelian groups;
(b) finitely generated groups;
(c) non-finitely generated groups;
(d) linear orders isomorphic to $(\mathbb{Z}, <)$;
(e) archimedian fields;
(f) connected graphs;
(g) recursively saturated models of PA;
(h) ω-models of ZFC (i.e., models of ZFC where the integers are standard);
(i) models of T omitting p, where T is a first order theory and p is a type.

EXERCISE 1.0.8. Show by compactness that none on these classes is axiomatizable in first order logic.

Examples (ii), (iv) and (v) immediately show the failure of the Upward Löwenheim–Skolem Theorem in infinitary languages. There are no uncountable linear orders isomorphic to $(\mathbb{Z}, <)$ and any archimedian ordered field is isomorphic to a subfield of the real numbers and hence has cardinality at most $2^{\aleph_0}$. We give another example in Exercise 1.1.13.

EXERCISE 1.0.9. Show by induction that for ordinal α there is an $\mathcal{L}_{\infty,\omega}$-sentence Φ_α describing $(\alpha, <)$ up to isomorphism.

EXERCISE 1.0.10. Let κ be a regular cardinal. Show there are $\alpha, \beta < (2^{2^\kappa})^+$ such that $(\alpha, <) \equiv_{\kappa,\omega} (\beta, <)$.

Taken together these two exercises give examples of structures $\mathcal{M}, \mathcal{N}$ with $\mathcal{M} \equiv_{\kappa,\omega} \mathcal{N}$ but $\mathcal{M} \not\equiv_{\infty,\omega} \mathcal{N}$ for all κ.[4]

We will also, from time to time, look at *pseudoelementary classes* which are simply reducts of elementary classes.

DEFINITION 1.0.11. We say that a class $\mathcal{K}$ of τ-structures is a $\mathrm{PC}_{\omega_1,\omega}$*-class* if there is a vocabulary $\tau^* \supseteq \tau$ and $\phi \in \mathcal{L}_{\omega_1,\omega}(\tau^*)$ such that

$$\mathcal{K} = \{\mathcal{M} : \text{ there is a } \tau\text{-structure } \mathcal{M}^* \text{ expanding } \mathcal{M} \text{ with } \mathcal{M}^* \models \phi\}$$

i.e., $\mathcal{K}$ is the class of τ-reducts of models of ϕ.

Similarly we say that $\mathcal{K}$ is a PC-*class* if ϕ is a first order sentence and $\mathcal{K}$ is a PC_δ-*class* if ϕ is a first order theory.

---

[4]For a specific example: suppose $\kappa < \lambda$ are uncountable cardinals. Show that $\kappa \equiv_{\omega_1,\omega} \lambda$, but $\kappa \not\equiv_{\infty,\omega} \lambda$. See [43] for details.

EXERCISE 1.0.12. Show that the following classes are $\mathrm{PC}_{\omega_1,\omega}$.

(a) orderable groups;
(b) free groups;
(c) 1-trainsitive linear orders, i.e., linear orders where for any a, b there is an order automorphism taking a to b;
(d) fields that are pure transcendental extensions of $\mathbb{Q}$;
(e) incomplete ordered fields;
(f) ordered fields with an integer part (i.e., an ordered subring with no element between 0 and 1 such that every element of the field is within distance at most one of some element of the subring).

EXERCISE 1.0.13 (Silver). Let $\tau = \{U\}$ where U is unary and let $\mathcal{K} = \{\mathcal{M} : |\mathcal{M}| \leq 2^{|U(\mathcal{M})|} \wedge |\mathcal{M} \setminus U(\mathcal{M})| = |\mathcal{M}|\}$.

(a) Show that $\mathcal{K}$ is PC-class.
(b) Show that $\mathcal{K}$ is κ-categorical if and only if $\kappa = \beth_\alpha$ for some limit ordinal α.

This example shows that the straightforward generalization of Morley's Categoricity Theorem to PC-classes fails

EXERCISE 1.0.14. Let $\mathcal{K}$ be a $\mathrm{PC}_{\omega_1,\omega}$-class. Show that

$$\mathcal{K}_0 = \{\mathcal{M} \in \mathcal{K} : \mathcal{M} \text{ countable}\}$$

is also a $\mathrm{PC}_{\omega_1,\omega}$-class.

1.1. Fragments and Downward Löwenheim–Skolem

We will prove a useful and natural version of the Downward Löwenheim–Skolem Theorem. The next exercise shows that even here we will need to be careful.

EXERCISE 1.1.1. Give an example of a countable vocabulary τ and an $\mathcal{L}_{\omega_1,\omega}$-theory T such that every model of T has cardinality at least $2^{\aleph_0}$.

We will often restrict our attention to subcollections of the set of all $\mathcal{L}_{\omega_1,\omega}$-formulas. We will look at collections of formulas with some natural closure properties. One of these will be formal negation an operation that shows how we inductively move a negation inside a quantifier or a Boolean operation. Closure under formal negation isn't needed in this proof, but will be used in later arguments.

DEFINITION 1.1.2. For each formula ϕ we define $\sim\phi$, a *formal negation* of ϕ as follows:

(i) for ϕ atomic, $\sim\phi$ is $\neg\phi$;
(ii) $\sim(\neg\phi)$ is ϕ;
(iii) $\sim \bigwedge_{\phi\in X} \phi$ is $\bigvee_{\phi\in X} \sim\phi$ and $\sim \bigvee_{\phi\in X} \phi$ is $\bigwedge_{\phi\in X} \sim\phi$;
(iv) $\sim\exists v\phi$ is $\forall v\sim\phi$ and $\sim\forall v\phi$ is $\exists v\sim\phi$.

EXERCISE 1.1.3. Show by induction that

$$\mathcal{M} \models \neg\phi \text{ if and only if } \mathcal{M} \models {\sim}\phi$$

for all $\phi \in \mathcal{L}_{\infty,\omega}$.

DEFINITION 1.1.4. We say that a set of $\mathcal{L}_{\infty,\omega}$-formulas $\mathbb{A}$ is a *fragment* if there is an infinite set of variables V such that if $\phi \in \mathbb{A}$, then all variables occurring in ϕ are in V and A satisfies the following closure properties:

(i) all atomic formulas using only the constant symbols of τ and variables from V are in $\mathbb{A}$;
(ii) if $\phi \in \mathbb{A}$ and ψ is a subformula of ϕ, then $\psi \in \mathbb{A}$;
(iii) If $\phi \in \mathbb{A}$, v is free in ϕ, and t is a term where every variable is in V, then the formula obtained by substituting t into all free occurrences of v is in $\mathbb{A}$;
(iv) $\mathbb{A}$ is closed under $\sim$;
(v) $\mathbb{A}$ is closed under $\neg$, $\wedge$, $\vee$, $\exists v$, and $\forall v$ for $v \in V$.

EXERCISE 1.1.5. (a) Suppose κ is regular (in particular this holds for $\mathcal{L}_{\omega_1,\omega}$). Prove that if T is a set of $\mathcal{L}_{\kappa,\omega}$-sentences with $|T| < \kappa$, then there is $\mathbb{A}$ a fragment of $\mathcal{L}_{\kappa,\omega}$ such that $T \subseteq \mathbb{A}$ and $|\mathbb{A}| < \kappa$.
(b) Show that there is a smallest such fragment.

EXERCISE 1.1.6. Let T and $\mathbb{A}$ be as above. Show that every formula in $\mathbb{A}$ has only finitely many free variables. Give an example showing that even though are only finitely many free variables, there may be infinitely many bound occurrences.

EXERCISE 1.1.7. We say that a formula is in *negation normal form* if the $\neg$ only occurs applied to atomic formulas. Show by induction that for any fragment $\mathbb{A}$ and any $\mathcal{L}_{\infty,\omega}$-formula $\phi \in \mathbb{A}$, there is $\psi \in \mathbb{A}$ such that ψ is equivalent to ϕ and ψ is in negation normal form.

Conjunctive and disjunctive normal form does not work as well.

EXERCISE 1.1.8. Let $\tau = \{P_{i,j} : i, j \in \omega\} \cup \{c\}$ where each $P_{i,j}$ is a unary predicate and c is a constant symbol. Consider the $\mathcal{L}_{\omega_1,\omega}$-sentence

$$\bigwedge_{i\in\omega}\bigwedge_{j\neq k}(P_{i,j}(c) \to \neg P_{i,k}(c)) \wedge \bigwedge_{i\in\omega}\bigvee_{j\in\omega} P_{i,j}(c).$$

Show that there is an equivalent sentence in disjunctive normal form in $\mathcal{L}_{(2^{\aleph_0})^+,\omega}$. (See also Exercise 3.1.6.)

We write $\mathcal{M} \equiv_{\mathbb{A}} \mathcal{N}$ and $\mathcal{M} \prec_A \mathcal{N}$ for elementary equivalence and elementary submodels with respect to formulas in $\mathbb{A}$ (where $\mathbb{A}$ could be $\mathcal{L}_{\omega_1,\omega}$ or $\mathcal{L}_{\infty,\omega}$).

THEOREM 1.1.9 (Downward Löwenheim–Skolem). *Let $\mathbb{A}$ be a fragment of $\mathcal{L}_{\infty,\omega}$ such that any formula $\phi \in \mathbb{A}$ has at most finitely many free variables.*

Let $\mathcal{M}$ be a τ-structure with $X \subseteq \mathcal{M}$. There is $\mathcal{N} \preceq_{\mathbb{A}} \mathcal{M}$ with $X \subseteq \mathcal{N}$ and $|\mathcal{N}| \leq \max(|\mathbb{A}|, |X|)$.

In particular, if τ is countable and $\mathbb{A}$ is a countable fragment of $\mathcal{L}_{\omega_1,\omega}(\tau)$, then every τ-structure has a countable $\mathbb{A}$-elementary submodel.

The proof is a simple generalization of the proof in first order logic. It is outlined in the following Exercise.

EXERCISE 1.1.10. (a) Prove there is $\tau^* \supseteq \tau$ and $\mathbb{A}^* \supseteq \mathbb{A}$ and $\mathcal{M}^*$ an τ^*-expansion of $\mathcal{M}$ such that $|\tau^*|, |\mathbb{A}^*| = |\mathbb{A}|$ and for each $\mathbb{A}^*$-formula $\phi(\overline{v}, w)$ with free variables from $v_1, \ldots, v_n, w$, there is an n-ary function symbol $f_\phi \in \tau^*$ such that

$$\mathcal{M}^* \models \forall \overline{v}\, (\exists w\, \phi(\overline{v}, w) \rightarrow \phi(\overline{v}, f_\phi(\overline{v}))).$$

(b) Prove that if $\mathcal{N}$ is a τ^*-substructure of $\mathcal{M}^*$, then $\mathcal{N} \preceq_{\mathbb{A}^*} \mathcal{M}^*$.

(c) Prove that there is a τ^*-substructure $\mathcal{N}$ of $\mathcal{M}^*$ with $X \subseteq \mathcal{N}$ and $|\mathcal{N}| \leq \max(|\mathbb{A}^*|, |X|)$.

EXERCISE 1.1.11. For τ and $\mathbb{A}$ define the appropriate notion of *built-in Skolem functions*. Prove that for any τ and $\mathbb{A}$ and $\mathbb{A}$-theory T there are $\tau^* \supseteq \tau$, $\mathbb{A}^* \supseteq \mathbb{A}$ and $T^* \supseteq T$ with $|\mathbb{A}^*| = |T^*| = |\mathbb{A}|$ where T^* has built-in Skolem functions and any model of T has and expansion to a model of T^*.

EXERCISE 1.1.12. Let $\mathbb{A}$ be a fragment of $\mathcal{L}_{\infty,\omega}$ where every formula has finitely many free variables and let $\mathcal{K}$ be a $\mathrm{PC}_{\omega_1,\omega}$-class. Show that if $\mathcal{M} \in \mathcal{K}$ is a τ-structure and $X \subset \mathcal{M}$, then there is $\mathcal{N} \prec_{\mathbb{A}} \mathcal{M}$ such that $|\mathcal{N}| = \max(|X|, |\mathbb{A}|))$ and $\mathcal{N} \in \mathcal{K}$.

EXERCISE 1.1.13. As mentioned above, in first order logic, the Upward Löwenheim-Skolem is an easy consequence of Compactness. In infinitary logics it is generally false. Let $\tau = \{U, S, E, c_1, \ldots, c_n, \ldots\}$, where U and S are unary predicates, E is a binary predicate and each c_i is a constant. Let ϕ be the conjunction of

(i) $\forall x\, U(x) \leftrightarrow \neg S(x)$,

(ii) $\forall x \forall y\, (E(x, y) \rightarrow U(x) \wedge S(y))$,

(iii) $c_i \neq c_j$, for $i \neq j$,

(iv) $U(c_i)$ for all i,

(v) $\forall y \forall z\, ([S(y) \wedge S(z) \wedge \forall x\, ((E(x, y) \leftrightarrow E(x, z))] \rightarrow y = z)$,

(vi) $\forall x (U(x) \rightarrow \bigvee_{i=1}^{\infty} x = c_i)$.

Prove that every model of ϕ has size at most $2^{\aleph_0}$.

Elementary chains behave as they do in first order logic.

EXERCISE 1.1.14. Suppose $\mathbb{A}$ is a fragment of $\mathcal{L}_{\infty,\omega}$, $(I,<)$ is a linear order and $(\mathcal{M}_i : i \in I)$ is an elementary chain of τ-structures, i.e., $\mathcal{M}_i \preceq_{\mathbb{A}} \mathcal{M}_j$ for $i<j$. Let $\mathcal{M} = \bigcup \mathcal{M}_\alpha$. Then $\mathcal{M}_i \preceq_{\mathbb{A}} \mathcal{M}$ for all $i \in I$.

EXERCISE 1.1.15. Give an example showing that PC classes need not be preserved by elementary chains.

EXERCISE 1.1.16. Let τ be countable. Suppose $(\mathcal{M}_\alpha : \alpha < \omega_1)$ is a chain of countable τ-structures such that $\mathcal{M}_\beta = \bigcup_{\alpha<\beta} \mathcal{M}_\alpha$ for β a limit ordinal. Let $\mathcal{M} = \bigcup_{\alpha<\omega_1} \mathcal{M}_\alpha$. Suppose $\phi \in \mathcal{L}_{\omega_1,\omega}$ and $\mathcal{M} \models \phi$. Show that $\{\alpha : \mathcal{M}_\alpha \models \phi\}$ is closed unbounded.

1.2. $\mathcal{L}_{\omega_1,\omega}$ and omitting first order types

In Exercise 1.0.7 we noted that the class of models of a first order theory omitting a type is expressible in $\mathcal{L}_{\omega_1,\omega}$. The next result, due to Chang, shows that any class axiomatizable by an $\mathcal{L}_{\omega_1,\omega}$-sentence is the reduct of a class of models of a first order theory omitting a set of types.

THEOREM 1.2.1. *Let τ be a countable vocabulary and let T be a countable set of $\mathcal{L}_{\omega_1,\omega}$-sentences. There is a countable vocabulary $\tau^* \supseteq \tau$, a first order τ^*-theory T^* and a set of partial types Γ such that:*

(i) *if $\mathcal{M} \models T^*$ and $\mathcal{M}$ omits all types in Γ, then the τ-reduct of $\mathcal{M}$ is a model of T;*
(ii) *every model of T has an τ^*-expansion that is a model of T^* omitting all types in Γ.*

PROOF. Let $\mathbb{A}$ be the smallest fragment containing all sentences in T. We expand τ to τ^* so for each formula ϕ in $\mathbb{A}$ with free variables from $v_1,\dots,v_n$ we have an n-ary relation symbol R_ϕ. T^* is formed by taking the following sentences:

(i) if ϕ is atomic add $\forall \overline{v}\,(R_\phi \leftrightarrow \phi)$;
(ii) if ϕ is $\neg\theta$ add $\forall \overline{v}\,(R_\phi \leftrightarrow \neg R_\theta)$;
(iii) if ϕ is $\bigwedge_{\theta\in X}\theta$ add $\forall \overline{v}\,(R_\phi \rightarrow R_\theta)$ for all $\theta \in X$ and let γ_ϕ be the type $\{\neg R_\phi\} \cup \{R_\theta : \theta \in X\}$;
(iv) if ϕ is $\bigvee_{\theta\in X}\theta$, add $\forall \overline{v}\, R_\theta \rightarrow R_\phi$ for all θ in X and let γ_ϕ be the type $\{R_\phi\} \cup \{\neg R_\theta : \theta \in X\}$;
(v) if ϕ is $\exists w\,\theta$, then add $\forall \overline{v}\,(R_\phi \leftrightarrow \exists w R_\theta)$;
(vi) if ϕ is $\forall w\,\theta$, then add $\forall \overline{v}\,(R_\phi \leftrightarrow \forall \overline{v}\forall w\, R_\theta)$;
(vii) for each sentence $\phi \in T$, add R_ϕ to T^*.

Let Γ be the collection of types γ_ϕ described above.[5]

[5] As described here for each sentence ϕ we have added a 0-ary predicate symbol. If you are not comfortable with this approach we could rephrase this by adding a single constant c. Then for each sentence ϕ we could add unary predicate R_ϕ and make assertions about $R_\phi(c)$.

EXERCISE 1.2.2. (a) Suppose $\mathcal{M} \models T^*$ and $\mathcal{M}$ omits every type in Γ. Prove that

$$\mathcal{M} \models \forall \overline{v}\, \phi \leftrightarrow R_\phi$$

for all $\phi \in \mathbb{A}$.

Conclude that the τ-reduct of a model of T^* is a model of T.

(b) Prove that every $\mathcal{M} \models T$ has an expansion that is a model of T^* omitting all types in Γ by interpreting R_ϕ as ϕ.

This completes the proof. ⊣

We will examine a refinement of this result in Theorem 2.2.18.

Chapter 2

BACK AND FORTH

2.1. Karp's Theorem

We next work on several characterizations of $\equiv_{\infty,\omega}$ due to Karp.

Definition 2.1.1. Let $\mathcal{M}$ and $\mathcal{N}$ be τ-structures. A *partial isomorphism system* between $\mathcal{M}$ and $\mathcal{N}$ is a non-empty collection P of partial τ-embeddings $f : A \to \mathcal{N}$ where $A \subseteq \mathcal{M}$ such that:

(i) for all $f \in P$ and $a \in \mathcal{M}$ there is $g \in P$ such that $g \supseteq f$ and $a \in \text{dom}(g)$;
(ii) for all $g \in P$ and $b \in \mathcal{N}$ there is $g \in P$ such that $g \supseteq f$ and $b \in \text{img}(g)$.

We write $\mathcal{M} \cong_p \mathcal{N}$ if there is a partial isomorphism system between $\mathcal{M}$ and $\mathcal{N}$.

Exercise 2.1.2. If $\mathcal{M}$ and $\mathcal{N}$ are countable and $\mathcal{M} \cong_p \mathcal{N}$, show by a back-and-forth argument that $\mathcal{M} \cong \mathcal{N}$.

Definition 2.1.3. Let $\mathcal{M}$ and $\mathcal{N}$ be τ-structures. The game $G(\mathcal{M}, \mathcal{N})$ is played as follows. At stage n, Player I plays $a_n \in \mathcal{M}$ or $b_n \in \mathcal{N}$. In the first case Player II responds with $b_n \in \mathcal{N}$ and in the later case Player II responds with $a_n \in \mathcal{M}$. Player II wins the play of the game if the map $a_n \mapsto b_n$ is a partial τ-embedding.

Theorem 2.1.4. *The following are equivalent*:

(i) $\mathcal{M} \equiv_{\infty,\omega} \mathcal{N}$;
(ii) $\mathcal{M} \cong_p \mathcal{N}$;
(iii) *there is a partial isomorphism system P between $\mathcal{M}$ and $\mathcal{N}$ where every $p \in P$ has finite domain*;
(iv) *Player II has a winning strategy in $G(\mathcal{M}, \mathcal{N})$.*

Proof. (iii) $\Rightarrow$ (ii) is clear.

(ii) $\Rightarrow$ (i). Let P be a system of partial isomorphisms. We prove that for all $\phi(v_1, \dots, v_n) \in \mathcal{L}_{\infty,\omega}$, $f \in P$ and $a_1, \dots, a_n \in \text{dom}(f)$, then

$$\mathcal{M} \models \phi(\overline{a}) \Leftrightarrow \mathcal{N} \models \phi(f(\overline{a})).$$

We prove this by induction on formulas. This is clear for atomic formulas and the induction step is obvious for $\neg$, $\bigwedge$ and $\bigvee$. Suppose $\phi(\overline{v})$ is $\exists w\, \psi(\overline{v}, w)$.

If $\mathcal{M} \models \psi(\overline{a}, c)$. There is $g \in P$ with $g \supseteq f$ and $c \in \text{dom}(P)$. By induction, $\mathcal{N} \models \psi(f((\overline{a})), f(c))$, so $\mathcal{N} \models \phi(f(\overline{a}))$.

On the other hand, if $\mathcal{N} \models \psi(f(\overline{a}), d)$. There is $g \in P$ with $g \supseteq f$ and $c \in \text{dom}(g)$ such that $g(c) = d$. By induction, $\mathcal{M} \models \psi(\overline{a}, c)$. Thus $\mathcal{M} \models \phi(\overline{a})$.

This proves the claim. In particular, if ϕ is a sentence $\mathcal{M} \models \phi \Leftrightarrow \mathcal{N} \models \phi$.

(iv) $\Rightarrow$ (iii). Let η be a winning strategy for Player II. Let P be the set of all maps $f(a_i) = b_i$ where $a_1, \ldots, a_n, b_1, \ldots, b_n$ are the results of some play of the game where at each stage Player I has played either a_n or b_n and Player II has responded using η. Since η is a winning strategy for Player II, each such f is a partial τ-embedding. Since Player I can at any stage play any element from $\mathcal{M}$ or $\mathcal{N}$, P satisfies (i) and (ii) in the definition of a partial isomorphism system.

(i) $\Rightarrow$ (iv). We need one fact.

CLAIM. *Suppose* $(\mathcal{M}, \overline{a}) \equiv_{\infty,\omega} (\mathcal{N}, \overline{b})$ *and* $c \in M$, *there is* $d \in N$ *such that*

$$(\mathcal{M}, \overline{a}, c) \equiv_{\infty,\omega} (\mathcal{N}, \overline{b}, d).$$

Suppose not. Then for all $d \in \mathcal{N}$ there is ϕ_d such that $\mathcal{M} \models \phi_d(\overline{a}, c)$ and $\mathcal{N} \models \neg\phi_d(\overline{b}, d)$. But then

$$\mathcal{M} \models \exists v \bigwedge_{d \in N} \phi_d(\overline{a}, v)$$

and

$$\mathcal{N} \not\models \exists v \bigwedge_{d \in N} \phi_d(\overline{b}, v)$$

a contradiction.

We now describe Player II's strategy. Player II always has a play to ensure $(\mathcal{M}, a_1, \ldots, a_n) \equiv_{\infty,\omega} (\mathcal{N}, b_1, \ldots, b_n)$. As long as Player II does this the resulting map will be a partial $\mathcal{L}$-embedding. $\dashv$

EXERCISE 2.1.5. If $\mathcal{M} \equiv_{\infty,\omega} \mathcal{N}$, then for any countably generated substructure $\mathcal{M}_0 \subseteq \mathcal{N}$ there is an isomorphic $\mathcal{N}_0 \subseteq \mathcal{N}$.

EXERCISE 2.1.6. We say that an abelian group is $\aleph_1$-*free* if every countable subgroup is free.

Prove that if G is free abelian and $H \equiv_{\infty,\omega} G$, then H is $\aleph_1$-free. We will show the converse in Corollary 2.2.9.

EXERCISE 2.1.7. For any abelian group A we say that a subgroup A_0 is *thin* if there is a subgroup A_1 such that $A = A_0 \oplus A_1$ and A_1 is isomorphic to A.

Let G be the free abelian group on countably many generators. Let P be the set of isomorphisms between thin subgroups of G and thin subgroups of $\mathbb{Z}^\omega$.

(a) Prove that P is a partial isomorphism system. Conclude that $G \equiv_{\infty,\omega} \mathbb{Z}^{\omega}$;
(b) The group $\mathbb{Z}^{\omega}$ is known not to be free abelian (see for example [32] Theorem 21). Conclude that the class of free abelian groups is $\mathrm{PC}_{\omega_1,\omega}$ but not $\mathcal{L}_{\infty,\omega}$-axiomatizable. In Appendix A we show that there are $2^{\aleph_1}$ non-isomorphic $\aleph_1$-free abelian groups of cardinality $\aleph_1$.

EXERCISE 2.1.8. Prove that $\mathcal{M} \equiv_{\infty,\omega} \mathcal{N}$ if and only if there is a forcing extension of the universe $\mathbb{V}[G]$ where $\mathcal{M} \cong \mathcal{N}$.

For another view, we begin by defining a deceptively simple sequence of equivalence relations. For each ordinal α, we will have a relation

$$(\mathcal{M}, \overline{a}) \sim_{\alpha} (\mathcal{N}, \overline{b})$$

where $\mathcal{M}$ and $\mathcal{N}$ are τ-structures, $\overline{a} \in \mathcal{M}^n$ and $\overline{b} \in \mathcal{N}^n$ and $n = 0, 1, 2, \ldots$.

- $(\mathcal{M}, \overline{a}) \sim_0 (\mathcal{N}, \overline{b})$ if $\mathcal{M} \models \phi(\overline{a})$ if and only if $\mathcal{N} \models \phi(\overline{b})$ for all atomic τ-formulas ϕ.
- For all ordinals α, $(\mathcal{M}, \overline{a}) \sim_{\alpha+1} (\mathcal{N}, \overline{b})$ if for all $c \in \mathcal{M}$ there is $d \in \mathcal{N}$ such that $(\mathcal{M}, \overline{a}, c) \sim_{\alpha} (\mathcal{N}, \overline{b}, d)$ and for all $d \in \mathcal{N}$ there is $c \in \mathcal{M}$ such that $(\mathcal{M}, \overline{a}, c) \sim_{\alpha} (\mathcal{N}, \overline{b}, d)$.
- For all limit ordinals β, $(\mathcal{M}, \overline{a}) \sim_{\beta} (\mathcal{N}, \overline{b})$ if and only if $(\mathcal{M}, \overline{a}) \sim_{\alpha} (\mathcal{N}, \overline{b})$ for all $\alpha < \beta$.

We show that $\sim_{\alpha}$ captures the more complicated notion of equivalence for sentences of quantifier rank at most α.

DEFINITION 2.1.9. Define $\mathrm{qr}(\phi)$ the *quantifier rank* of $\mathcal{L}_{\infty,\omega}$-formulas as follows:

- If ϕ is atomic, then $\mathrm{qr}(\phi) = 0$;
- $\mathrm{qr}(\neg\phi) = \mathrm{qr}(\phi)$;
- $\mathrm{qr}(\bigwedge_{\phi \in X} \phi) = \mathrm{qr}(\bigvee_{\phi \in X} \phi) = \sup_{\phi \in X} \mathrm{qr}(\phi)$;
- $\mathrm{qr}(\exists v\, \phi) = \mathrm{qr}(\forall v\, \phi) = \mathrm{qr}(\phi) + 1$.

DEFINITION 2.1.10. We say $\mathcal{M} \equiv_{\alpha} \mathcal{N}$ if and only

$$\mathcal{M} \models \phi \Leftrightarrow \mathcal{N} \models \phi$$

for any $\mathcal{L}_{\infty,\omega}$-sentence ϕ of quantifier rank at most α.

EXERCISE 2.1.11. If α is a limit ordinal, show that $\mathcal{M} \equiv_{\alpha} \mathcal{N}$ if and only if $\mathcal{M} \equiv_{\beta} \mathcal{N}$ for all $\beta < \alpha$.

EXERCISE 2.1.12. If α is a limit ordinal, then $\mathcal{M} \equiv_{\alpha} \mathcal{N}$ if and only if $\mathcal{M} \equiv_{\beta} \mathcal{N}$ for all $\beta < \alpha$.

THEOREM 2.1.13. $\mathcal{M} \equiv_{\alpha} \mathcal{N}$ *if and only if* $\mathcal{M} \sim_{\alpha} \mathcal{N}$.

We leave the proof as an exercise.

EXERCISE 2.1.14. Prove by induction on α that for all τ-structures $\mathcal{M},\mathcal{N}$, $\overline{a}\in\mathcal{M}$ and $\overline{b}\in\mathcal{N}$, $(\mathcal{M},\overline{a})\sim_\alpha(\mathcal{N},\overline{b})$ if and only if

$$\mathcal{M}\models\phi(\overline{a})\Leftrightarrow\mathcal{N}\models\phi(\overline{b})$$

for all formulas $\phi(\overline{v})$ of quantifier rank at most α.

2.2. Scott's Theorem

If τ is any vocabulary and $\mathcal{M}$ is an τ-structure we define a sequence of $\mathcal{L}_{\infty,\omega}$-formulas $\Phi^{\mathcal{M}}_{\overline{a},\alpha}(\overline{v})$ for $\overline{a}\in\mathcal{M}^{<\omega}$ and α is an ordinal as follows:

$$\Phi^{\mathcal{M}}_{\overline{a},0}(\overline{v})=\bigwedge_{\psi\in X}\psi(\overline{v}),$$

where $X=\{\psi:\mathcal{M}\models\psi(\overline{a})$ and ψ is atomic or the negation of an atomic τ-formula$\}$.

If α is a limit ordinal, then

$$\Phi^{\mathcal{M}}_{\overline{a},\alpha}(\overline{v})=\bigwedge_{\beta<\alpha}\Phi^{\mathcal{M}}_{\overline{a},\beta}(\overline{v}).$$

If $\alpha=\beta+1$, then

$$\Phi^{\mathcal{M}}_{\overline{a},\alpha}(\overline{v})=\bigwedge_{b\in\mathcal{M}}\exists w\ \Phi^{\mathcal{M}}_{\overline{a},b,\beta}(\overline{v},w)\wedge\forall w\bigvee_{b\in\mathcal{M}}\Phi^{\mathcal{M}}_{\overline{a},b,\beta}(\overline{v},w).$$

LEMMA 2.2.1. *Let $\mathcal{M}$ and $\mathcal{N}$ be τ-structures, $\overline{a}\in\mathcal{M}^l$, and $\overline{b}\in\mathcal{N}^l$. Then, $(\mathcal{M},\overline{a})\sim_\alpha(\mathcal{N},\overline{b})$ if and only if $\mathcal{N}\models\phi^{\mathcal{M}}_{\overline{a},\alpha}(\overline{b})$.*

PROOF. We prove this by induction on α. Because $(\mathcal{M},\overline{a})\sim_0(\mathcal{N},\overline{b})$ if and only if they satisfy the same atomic formulas, the lemma holds for $\alpha=0$.

Suppose that γ is a limit ordinal and the lemma is true for all $\alpha<\gamma$. Then

$$\begin{aligned}(\mathcal{M},\overline{a})\sim_\gamma(\mathcal{N},\overline{b})&\Leftrightarrow(\mathcal{M},\overline{a})\sim_\alpha(\mathcal{N},\overline{b})\text{ for all }\alpha<\gamma\\&\Leftrightarrow\mathcal{N}\models\Phi^{\mathcal{M}}_{\overline{a},\alpha}(\overline{b})\text{ for all }\alpha<\gamma\\&\Leftrightarrow\mathcal{N}\models\Phi^{\mathcal{M}}_{\overline{a},\gamma}(\overline{b}).\end{aligned}$$

Suppose that the lemma is true for α. First, suppose that $\mathcal{N}\models\phi^{\mathcal{M}}_{\overline{a},\alpha+1}(\overline{b})$. Let $c\in\mathcal{M}$. Because

$$\mathcal{N}\models\bigwedge_{x\in\mathcal{M}}\exists w\ \Phi^{\mathcal{M}}_{\overline{a},x,\alpha}(\overline{b},w),$$

there is $d\in N$ such that $\mathcal{N}\models\Phi^{\mathcal{M}}_{\overline{a},c,\alpha}(\overline{b},d)$. By induction, $(\mathcal{M},\overline{a},c)\sim_\alpha(\mathcal{N},\overline{b},d)$. If $d\in\mathcal{N}$, then because

$$\mathcal{N}\models\forall w\bigvee_{c\in\mathcal{M}}\Phi^{\mathcal{M}}_{\overline{a},c,\alpha}(\overline{b},w)$$

there is $c \in \mathcal{M}$ such that $\mathcal{N} \models \Phi^{\mathcal{M}}_{\overline{a},c,\alpha}(\overline{b}, d)$ and $(\mathcal{M}, \overline{a}, c) \sim_\alpha (\mathcal{N}, \overline{b}, d)$. Thus $(\mathcal{M}, \overline{a}) \sim_{\alpha+1} (\mathcal{N}, \overline{b})$.

Suppose, on the other hand, that $(\mathcal{M}, \overline{a}) \sim_{\alpha+1} (\mathcal{N}, \overline{b})$. Suppose that $c \in \mathcal{M}$, then there is $d \in \mathcal{N}$ such that $(\mathcal{M}, \overline{a}, c) \sim_\alpha (\mathcal{N}, \overline{b}, d)$ and $\mathcal{N} \models \Phi^{\mathcal{M}}_{\overline{a},c,\alpha}(\overline{b}, d)$. Similarly, if $d \in \mathcal{N}$, then there is $c \in \mathcal{M}$ such that $\mathcal{N} \models \Phi^{\mathcal{M}}_{\overline{a},c,\alpha}(\overline{b}, d)$. Thus, $\mathcal{N} \models \Phi^{\mathcal{M}}_{\overline{a},\alpha+1}(\overline{b})$, as desired. ⊣

A similar induction shows that even more is true. The formula $\Phi^{\mathcal{M}}_{\overline{a},\alpha}$ depends only on the $\sim_\alpha$-class of $(\mathcal{M}, \overline{a})$.

EXERCISE 2.2.2. Prove that if $(\mathcal{M}, \overline{a}) \sim_\alpha (\mathcal{N}, \overline{b})$ if and only if $\Phi^{\mathcal{M}}_{\overline{a},\alpha}(\overline{v}) = \Phi^{\mathcal{N}}_{\overline{b},\alpha}(\overline{v})$

LEMMA 2.2.3. *For any infinite τ-structure $\mathcal{M}$, there is an ordinal $\alpha < |M|^+$ such that if $\overline{a}, \overline{b} \in \mathcal{M}^l$ and $(\mathcal{M}, \overline{a}) \sim_\alpha (\mathcal{M}, \overline{b})$, then $(\mathcal{M}, \overline{a}) \sim_\beta (\mathcal{M}, \overline{b})$ for all β. The least such α is called the Scott rank of $\mathcal{M}$ and denoted* $\mathrm{sr}(\mathcal{M})$.

PROOF. Let $\Gamma_\alpha = \{(\overline{a}, \overline{b}) : \overline{a}, \overline{b} \in \mathcal{M}^l$ for some $l = 0, 1, \ldots$ and $(\mathcal{M}, \overline{a}) \not\sim_\alpha (\mathcal{M}, \overline{b})\}$. Clearly, $\Gamma_\alpha \subseteq \Gamma_\beta$ for $\alpha < \beta$.

CLAIM 1. *If $\Gamma_\alpha = \Gamma_{\alpha+1}$, then $\Gamma_\alpha = \Gamma_\beta$ for all $\beta > \alpha$.*

We prove this by induction on β. If β is a limit ordinal and the claim holds for all $\gamma < \beta$, then it also holds for β. Suppose that the claim is true for $\alpha \leq \beta$ and we want to show that it holds for $\beta + 1$. Suppose that $(\mathcal{M}, \overline{a}) \sim_\beta (\mathcal{M}, \overline{b})$ and $c \in \mathcal{M}$. Because $(\mathcal{M}, \overline{a}) \sim_{\alpha+1} (\mathcal{M}, \overline{b})$, there is $d \in \mathcal{N}$ such that $(\mathcal{M}, \overline{a}, c) \sim_\alpha (\mathcal{M}, \overline{b}, d)$. By our inductive assumption, $(\mathcal{M}, \overline{a}, c) \sim_\beta (\mathcal{M}, \overline{b}, d)$. Similarly, if $d \in \mathcal{M}$, then there is $c \in \mathcal{M}$ such that $(\mathcal{M}, \overline{a}, c) \sim_\beta (\mathcal{M}, \overline{b}, d)$. Thus, $(\mathcal{M}, \overline{a}) \sim_{\beta+1} (\mathcal{M}, \overline{b})$ as desired.

CLAIM 2. *There is an ordinal $\alpha < |M|^+$ such that $\Gamma_\alpha = \Gamma_{\alpha+1}$.*

Suppose not. Then, for each $\alpha < |\mathcal{M}|^+$, choose $(\overline{a}_\alpha, \overline{b}_\alpha) \in \Gamma_{\alpha+1} \setminus \Gamma_\alpha$. Because $\Gamma_\alpha \subseteq \Gamma_\beta$ for $\alpha < \beta$, the function $\alpha \mapsto (\overline{a}_\alpha, \overline{b}_\alpha)$ is one-to-one. Because there are only $|\mathcal{M}|$ finite sequences from $\mathcal{M}$ this is impossible. ⊣

A more subtle notion of Scott rank is sometimes useful

DEFINITION 2.2.4. For $\overline{a} \in \mathcal{M}$ define $r(\overline{a})$ as the least ordinal α such that for all $\overline{b}$

$$(\mathcal{M}, \overline{a}) \sim_\alpha (\mathcal{M}, \overline{b}) \Rightarrow (\mathcal{M}, \overline{a}) \sim_{\alpha+1} (\mathcal{M}, \overline{b}).$$

Then

$$sr(\mathcal{M}) = \sup\{r(\overline{a}) : \overline{a} \in \mathcal{M}\}.$$

We also define

$$\mathrm{SR}(\mathcal{M}) = \sup\{r(\overline{a}) + 1 : \overline{a} \in \mathcal{M}\}.$$

EXERCISE 2.2.5. Calculate SR for the following structures.

(a) $(\omega, <)$;
(b) $(\omega + 1, <)$;
(c) $(\omega + \omega, <)$;
(d) $(\mathbb{Z}, +, 0)$.

We will see later that at limit ordinals α there can be interesting distinctions between $\mathrm{SR}(\mathcal{M}) = \alpha$ and $\mathrm{SR}(\mathcal{M}) = \alpha + 1$ while in both cases $\mathrm{sr}(\mathcal{M}) = \alpha$.

EXERCISE 2.2.6. Formulate and prove a version of Lemma 2.2.3 for finite τ-structures.

EXERCISE 2.2.7. (a) Show that for all $\beta < \omega_1$ and $n \in \omega$ there is an $\mathcal{L}_{\omega_1,\omega}(\tau)$-formula $S^n_\beta(\overline{v}, \overline{w})$ where $|\overline{v}| = |\overline{w}| = n$ such that

$$\mathcal{M} \models S^n_\beta(\overline{a}, \overline{b}) \Leftrightarrow (\mathcal{M}, \overline{a}) \sim_\alpha (\mathcal{N}, \overline{b})$$

for any τ-structure $\mathcal{M}$.

(b) Show that for all $\alpha < \omega_1$ there is $\sigma_\alpha \in \mathcal{L}_{\omega_1,\omega}(\tau)$ such that $\mathcal{M} \models \sigma_\alpha$ if and only if $SR(\mathcal{M}) \geq \alpha$.

We conclude this section with Scott's Isomorphism Theorem

Let $\mathcal{M}$ be an infinite τ-structure of cardinality κ, and let α be the Scott rank of $\mathcal{M}$. Let $\Phi^{\mathcal{M}}$ be the sentence

$$\Phi^{\mathcal{M}}_{\emptyset,\alpha} \wedge \bigwedge_{l=0}^{\infty} \bigwedge_{\overline{a}\in M^l} \forall \overline{v}(\Phi^{\mathcal{M}}_{\overline{a},\alpha}(\overline{v}) \rightarrow \Phi^{\mathcal{M}}_{\overline{a},\alpha+1}(\overline{v})).$$

Because all of the conjunctions and disjunctions in $\Phi^{\mathcal{M}}_{\overline{a},\beta}$ are of size κ, $\Phi^{\mathcal{M}}_{\overline{a},\beta} \in \mathcal{L}_{\kappa^+,\omega}$ for all ordinals $\beta < \kappa^+$. Thus $\Phi^{\mathcal{M}}$ is an $\mathcal{L}_{\kappa^+,\omega}$-sentence. We call $\Phi^{\mathcal{M}}$ the *Scott sentence* of $\mathcal{M}$. If τ is a countable vocabulary and $\mathcal{M}$ is countable τ-structure, then $\Phi^{\mathcal{M}} \in \mathcal{L}_{\omega_1,\omega}$.

THEOREM 2.2.8 (Scott's Isomorphism Theorem). *Let $\mathcal{M}$ and $\mathcal{N}$ be τ-structures, and let $\Phi^{\mathcal{M}} \in \mathcal{L}_{\infty,\omega}$ be the Scott sentence of $\mathcal{M}$. Then, $\mathcal{N} \equiv_{\infty,\omega} \mathcal{M}$ if and only if $\mathcal{N} \models \Phi^{\mathcal{M}}$.*

In particular, if $\mathcal{M}$ and $\mathcal{N}$ are countable then $\phi \in \mathcal{L}_{\omega_1,\omega}$ and, if $\mathcal{N} \models \Phi^{\mathcal{M}}$, then $\mathcal{N} \cong \mathcal{M}$.

PROOF. Clearly if $\mathcal{N} \equiv_{\infty,\omega} \mathcal{M}$, then $\mathcal{N} \models \Phi^{\mathcal{M}}$.

For the other direction, suppose $\mathcal{N} \models \Phi^{\mathcal{M}}$. We will show that $\mathcal{N} \cong_p \mathcal{M}$. Let α be the Scott rank of $\mathcal{M}$. Let P be the set of partial functions f with finite domain from $\mathcal{M}$ to $\mathcal{N}$ such that if $\overline{a}$ is the domain of f and $f(\overline{a}) = \overline{b}$ then $\mathcal{N} \models \Phi^{\mathcal{M}}_{\overline{a},\alpha}(\overline{b})$. Since $\mathcal{N} \models \Phi^{\mathcal{M}}$, the empty map is in P so $P \neq \emptyset$.

Suppose we have f in P with domain $\overline{a}$ with $f(\overline{a}) = \overline{b}$. Let $c \in \mathcal{M}$. Since $\mathcal{N} \models \Phi^{\mathcal{M}} \wedge \Phi^{\mathcal{M}}_{\overline{a},\alpha}(\overline{b})$, $\mathcal{N} \models \Phi^{\mathcal{M}} \wedge \Phi^{\mathcal{M}}_{\overline{a},\alpha+1}(\overline{b})$. But $\Phi^{\mathcal{M}}_{\overline{a},\alpha+1}(\overline{v}) \rightarrow \exists w\, \Phi^{\mathcal{M}}_{\overline{a},c,\alpha}(\overline{v}, w)$. Thus there is $d \in \mathcal{N}$ such that $\mathcal{N}$ models $\Phi^{\mathcal{M}}_{\overline{a},c,\alpha}(\overline{b}, d)$ and we can extend f by sending c to d.

On the other hand suppose we are given $\widehat{d} \in \mathcal{N}$. Since $\mathcal{N} \models \Phi^{\mathcal{M}}_{\overline{a},\alpha+1}(\overline{b})$,

$$\mathcal{N} \models \forall w \bigvee_{m \in \mathcal{M}} \Phi^{\mathcal{M}}_{\overline{a},m,\alpha}(\overline{b}, w).$$

Thus there is $\widehat{c} \in \mathcal{N}$ such that $\mathcal{N} \models \Phi^{\mathcal{M}}_{\overline{a},\widehat{c},\alpha}(\overline{b}, d)$ and we can extend f be sending $\widehat{c}$ to $\widehat{d}$.

Thus P is a partial isomorphism system and $\mathcal{M} \cong_p \mathcal{N}$. By Theorem 2.1.4 $\mathcal{M} \equiv_\infty \mathcal{N}$ and, by Exercise 2.1.2, if $\mathcal{M}$ and $\mathcal{N}$ are countable, then $\mathcal{M} \cong \mathcal{N}$. ⊣

We use this to prove a converse to Exercise 2.1.6.

COROLLARY 2.2.9. *Let G be the free abelian group on countably many generators and let H be an $\aleph_1$-free group that is not free on a finite set of generators. Then $G \equiv_{\infty,\omega} H$.*

PROOF. Let $\Phi^G \in \mathcal{L}_{\omega_1,\omega}$ be the Scott sentence of G. Let $\mathbb{A}$ be a countable fragment of $\mathcal{L}_{\infty,\omega}$ containing Φ^G and a sentence asserting H is not free on finitely many generators. By Downward Löwenheim–Skolem we can find $H_1 \prec_{\mathbb{A}} H$ containing countably many independent elements with $|H_1| = \aleph_0$. Since H is $\aleph_1$-free, H is free abelian on countably many generators and hence isomorphic to G. But then $H_1 \models \Phi^G$ and hence $H \models \Phi^G$. Thus $H \equiv_{\infty,\omega} G$. ⊣

A stronger version of this result is proved in Exercise 2.4.9.

EXERCISE 2.2.10. Show that the class of $\aleph_1$-free abelian groups is $\mathcal{L}_{\omega_1,\omega}$-axiomatizable.

Complete sentences.

DEFINITION 2.2.11. A satisfiable sentence $\phi \in \mathcal{L}_{\infty,\omega}$ is *complete* if for all $\psi \in \mathcal{L}_{\infty,\omega}$

$$\phi \models \psi \text{ or } \phi \models \neg\psi.$$

COROLLARY 2.2.12. *For any $\mathcal{M}$ the Scott sentence $\Phi^{\mathcal{M}}$ is complete.*

EXERCISE 2.2.13. For any satisfiable sentence $\psi \in \mathcal{L}_{\omega_1,\omega}$ there is a complete sentence ϕ such that $\phi \models \psi$.

EXERCISE 2.2.14. A satisfiable sentence $\phi \in \mathcal{L}_{\omega_1,\omega}$ is complete if and only if it is $\aleph_0$-categorical.

EXERCISE 2.2.15. Suppose $\psi \in \mathcal{L}_{\omega_1,\omega}$ is κ-categorical for some infinite cardinal κ. Show that if $\mathcal{M}, \mathcal{N} \models \psi$ of cardinality at least κ, then $\mathcal{M} \equiv_{\infty,\omega} \mathcal{N}$. (Compare this to Exercise 5.2.4.)

EXERCISE 2.2.16. Give an example of an $\aleph_1$-categorical $\phi \in \mathcal{L}_{\omega_1,\omega}$ that is not complete.

EXERCISE 2.2.17. Let σ be a complete $\mathcal{L}_{\omega_1,\omega}$-sentence, and let $\mathcal{M}$ be the unique countable model. Show that the following are equivalent.

(i) There is an uncountable model of σ.
(ii) There is a non-surjective $\mathcal{L}_{\omega_1,\omega}$-elementary embedding of $\mathcal{M}$ into itself.
(iii) There is a non-surjective $\mathbb{A}$-elementary embedding of $\mathcal{M}$ into itself, where $\mathbb{A}$ is a countable fragment of $\mathcal{L}_{\omega_1,\omega}$-containing ϕ.

Gao [22] has shown these conditions are equivalent to the non-existence of a complete left-invariant metric on Aut($\mathcal{M}$) the automorphism group of $\mathcal{M}$.

For complete sentences in $\mathcal{L}_{\omega_1,\omega}$ we can give an improvement of Chang's Theorem 1.2.1.

THEOREM 2.2.18. *Suppose τ is countable and ϕ is a complete sentence of $\mathcal{L}_{\omega_1,\omega}(\tau)$. There is a countable vocabulary $\tau^* \supseteq \tau$ and T^* a complete first order τ^*-theory such that $\mathcal{M} \models \phi$ if and only $\mathcal{M}$ is the τ-reduct of an atomic model of T^*.*

PROOF. Let $\mathcal{M}$ be the unique countable model of ϕ and let α be its Scott rank. Let $\mathbb{A}$ be a countable fragment containing all formulas $\Phi^{\mathcal{M}}_{\overline{a},\alpha}$ for $\overline{a} \in \mathcal{M}$. Now follow the proof of Theorem 1.2.1 to obtain $\tau^* = \tau \cup \{R_\psi : \psi \in \mathbb{A}\}$ and Γ a collection of τ^*-types such that models of $\Phi^{\mathcal{M}}$ are exactly the reducts of models of T^* that omit all types in Γ. Let T be the complete τ^*-theory of $\mathcal{M}$. As $\mathcal{M}$ omits all of the types in Γ, no type in Γ is implied by a principal type. Thus all types in Γ are omitted in any atomic model of T^* and the reduct of an atomic model of T^* is a model of ϕ.

For any $\mathcal{N} \models \phi$ we can take its natural expansion as a τ^*-structure. For any $\overline{a} \in \mathcal{N}$ there is $\overline{b} \in \mathcal{M}$ such that $R_{\Phi^{\mathcal{M}}_{\overline{b},\alpha}}(\overline{v})$ isolates $\mathrm{tp}^{\mathcal{N}}(\overline{a})$. Thus the expansion is an atomic model of T^*. ⊣

In later sections we will be looking at uncountable structures that are models of complete sentences. We would like to give a characterization of when this happens.

DEFINITION 2.2.19. Let $\mathbb{A}$ be a fragment of $\mathcal{L}_{\omega_1,\omega}(\tau)$. If $\mathcal{M}$ is a τ-structure and $a_1, \ldots, a_n \in \mathcal{M}$ then the $\mathbb{A}$-*type* of $\overline{a}$ is

$$\mathrm{tp}^{\mathcal{M}}(\overline{a}, \mathbb{A}) = \{\phi(\overline{v}) \in \mathbb{A} : \mathcal{M} \models \phi(\overline{a})\}$$

and $S_n(\mathbb{A}, \phi)$ is the set of $\mathbb{A}$-types of n-tuples realized in some model of T.

EXERCISE 2.2.20. Let τ be a countable vocabulary and let $\mathcal{M}$ be an uncountable τ-structure.

(a) Suppose there is a countable $\mathcal{N}$ such that $\mathcal{M} \equiv_{\infty,\omega} \mathcal{N}$. Show that $\mathcal{M}$ realizes only countably many $\mathcal{L}_{\omega_1,\omega}$-types.

For the remainder of the Exercise assume $\mathcal{M}$ realizes only countably many $\mathcal{L}_{\omega_1,\omega}$-types.

(b) Prove that $\mathcal{M}$ has countable Scott rank.
(c) Prove that there is a countable $\mathcal{N} \equiv_{\infty,\omega} \mathcal{M}$.

2.3. Countable approximations

In this section we give another characterization of $\equiv_{\infty,\omega}$ due to Kueker [44]. We begin with some set theoretic preliminaries. Kueker, and independently, Jech, introduced the following generalization of the filter of closed unbounded subsets of a regular cardinal.

For any set X let $\mathcal{P}_{\omega_1}(X)$ be the collection of countable subsets of X.

DEFINITION 2.3.1. $A \subseteq \mathcal{P}_{\omega_1}(X)$ is *closed* if $\bigcup a_n \in X$ for any $a_1 \subset a_2 \subset \cdots$ in X and $A \subseteq \mathcal{P}_{\omega_1}(X)$ is *unbounded* if for all $a \subset X$ countable, there is $b \in A$ with $a \subseteq b$.

LEMMA 2.3.2. *Suppose $A_0, A_1, \ldots \in \mathcal{P}_{\omega_1}(X)$ are closed and unbounded. Then so is $\bigcap A_i$ is closed and unbounded.*

PROOF. It is clear that $\bigcap A_i$ is closed, so we need only show it is unbounded.

Let $a \in \mathcal{P}_{\omega_1}(X)$. Let $\langle \cdot, \cdot \rangle : \omega^2 \to \omega$ be a bijective pairing function such that $\langle m, i\rangle < \langle m, j\rangle$ if $i < j$. We build

$$a = a_{-1} \subset a_0 \subset a_1 \subset \cdots \in \mathcal{P}_{\omega_1}(X)$$

such that if $n = \langle i, j\rangle$ we choose $a_n \in A_i$ with $a_{n-1} \subset a_n$. This is always possible since each A_i is unbounded. Then

$$b = \bigcap_{n\in\omega} a_n = \bigcap_{j\in\omega} a_{\langle i,j\rangle} \in A_i$$

for all $i \in \omega$. Thus $b \in \bigcap A_i$ and $\bigcap A_i$ is unbounded. ⊣

Let $D(X) = \{A \subseteq \mathcal{P}_{\omega_1} : A$ has a closed, unbounded subset$\}$.

EXERCISE 2.3.3. Show that $D(X)$ is a countably complete filter on $\mathcal{P}_{\omega_1}(X)$.

We also have closure under diagonal intersections.

LEMMA 2.3.4. *Suppose $I \subseteq X$ and A_i is closed and unbounded for all $i \in I$ then so is*

$$\triangle A_i = \{a \in \mathcal{P}_{\omega_1}(X) : a \in A_i \textit{ for all } i \in a \cap I\}.$$

PROOF. Suppose $a_0 \subset a_1 \subset \cdots$ are in $\triangle A_i$ and $a = \bigcup a_i$. Let $j \in a \cap I$. Then there is n_0 such that $j \in a_n$ for $n \geq n_0$. But then $a_n \in A_j$ for $n \geq n_0$ and, since A_j is closed, $a \in A_i$. Thus $a \in \triangle A_i$ and $\triangle A_i$ is closed.

Let $a \in \mathcal{P}_{\omega_1}(X)$. Let $a_0 = a$. Given a_n, we know from Lemma 2.3.2 that $\bigcap_{i\in a_n\cap I} A_i$ is closed and unbounded. Thus there is

$$a_n \subset a_{n+1} \in \bigcap_{i\in a_n\cap I} A_i.$$

Let $a = \bigcup a_n$. Suppose $i \in a$. There is n_0 such that $i \in a_n$ for $n \geq n_0$. Thus $a_{n+1} \in A_i$ for $n \geq n_0$ and, since A_i is closed, $a \in A_i$. Thus $a \in \triangle A_i$ and $\triangle A_i$ is unbounded. ⊣

Kueker also gave a useful game theoretic formulation of membership in $D(X)$.

EXERCISE 2.3.5. For $A \subseteq \mathcal{P}_{\omega_1}(X)$ consider the game $G(A)$ where Players I and II alternate playing elements of X.

Player I	Player II
c_0	
	d_0
c_1	
	d_1
$\vdots$	$\vdots$

Player II wins if $\{c_i, d_i : i \in \omega\}$ is in A. Prove that Player II has a winning strategy in $G(A)$ if and only if $A \in D(X)$.

We use the filter $D(X)$ as a notion of "almost all". Let X be a set. Let τ be a countable vocabulary and let $\mathcal{M}$ be a τ-structure. We say that X is *large enough to approximate* $\mathcal{M}$ if the universe of $\mathcal{M}$ is contained in X.

EXERCISE 2.3.6. Suppose X is large enough to approximate $\mathcal{M}$. Show that $\{a \in \mathcal{P}_{\omega_1}(X) : \mathcal{M} \cap a \text{ is a substructure of } \mathcal{M}\} \in D(X)$.

Let $\mathcal{M}^a$ denote $\mathcal{M} \cap a$. We say that $\mathcal{M}^a$ is a substructure *almost everywhere*.

DEFINITION 2.3.7. Suppose $\phi \in \mathcal{L}_{\infty,\omega}$ and $a \in \mathcal{P}_{\omega_1}(X)$ we define the approximation ϕ^a to ϕ inductively as follows:

- $\phi^a = \phi$ if ϕ is atomic;
- $(\neg\phi)^a = \neg\phi^a$;
- $(\forall v\, \phi)^a = \forall v\, \phi^a$ and $(\exists v\, \phi)^a = \exists v\, \phi^a$;
- $\left(\bigwedge_{i\in I} \phi_i\right)^a = \bigwedge_{i\in I\cap a} \phi_i^a$ and $\left(\bigvee_{i\in I} \phi_i\right)^a = \bigvee_{i\in I\cap a} \phi_i^a$.

We say that X is *large enough to approximate* ϕ if $I \subset X$ whenever $\bigvee_{i\in I} \psi_i$ or $\bigwedge_{i\in I} \psi_i$ is in the smallest fragment of $\mathcal{L}_{\infty,\omega}$ containing ϕ.

EXERCISE 2.3.8. Show that if $\phi \in \mathcal{L}_{\infty,\omega}$ and ψ is the canonical formula in negation normal form (see Exercise 1.1.7) such that $\models \phi \leftrightarrow \psi$, then for any $\mathcal{M}$ if X is large enough to approximate ϕ and $\mathcal{M}$, then

$$\models \phi^a \leftrightarrow \psi^a \text{ a.e.}$$

We work in an X large enough to approximate $\mathcal{M}$ and ϕ. Intuitively, we think of X as some sufficiently rich substructure of the universe of sets $\mathbb{V}$ containing $\mathcal{M}$ and ϕ.

THEOREM 2.3.9. *For any $\mathcal{M}$ and $\phi \in \mathcal{L}_{\infty,\omega}$*

$$\mathcal{M} \models \phi \text{ if and only if } \mathcal{M}^a \models \phi^a \text{ a.e.}$$

i.e., $\{a \in \mathcal{P}_{\omega_1}(X) : \mathcal{M}^a \models \phi^a\} \in D(X)$.

PROOF. First note that, since $D(X)$ is a filter, it is impossible that

$$\mathcal{M}^a \models \phi^a \text{ a.e. and } \mathcal{M}^a \models \neg\phi^a \text{ a.e.}$$

Thus it suffices to prove the $\Rightarrow$ direction.

Next note that, by Exercise 2.3.8, we may assume that ϕ is in negation normal form. We prove by induction that if ψ is a subformula of ϕ, $\overline{b} \in \mathcal{M}$ and $\mathcal{M} \models \psi(\overline{b})$, then $\mathcal{M}^a \models \psi^a(\overline{b})$ a.e. Note that $\overline{b} \in \mathcal{M}^a$ a.e., so this makes sense.

- If ψ is atomic or negated atomic this is clear since $\mathcal{M}^a$ is a substructure of $\mathcal{M}$. Note that these are the only negations we need to deal with.
- If ψ is $\bigvee_{i\in I} \theta_i(\overline{b})$, then there is $j \in I$ such that $\mathcal{M} \models \theta_j(\overline{b})$. Note that $j \in a$ for almost every $a \in P_{\omega_1}(X)$. By induction $\mathcal{M}^a \models \theta_j^a(\overline{b})$ a.e. Thus $\mathcal{M}^a \models \psi^a(\overline{b})$ a.e.
- Suppose ψ is $\bigwedge_{i\in I} \theta_i(\overline{b})$ and $\mathcal{M} \models \psi(\overline{b})$. By induction,
$$A_i = \{a : \mathcal{M}^a \models \theta_i^a(\overline{b})\} \in D(X)$$
for all $i \in I$. By Lemma 2.3.4, $\triangle A_i \in D(X)$. Let $a \in \triangle D(X)$. If $i \in a \cap I$, then $a \in A_i$ so $\mathcal{M}^a \models \theta_i^a(\overline{b})$. Thus $\mathcal{M}^a \models \psi^a$ a.e.
- If ψ is $\exists v\, \theta(v, \overline{b})$, this is similar to the $\bigvee$-case.
- if ψ is $\forall v\, \theta(v, \overline{b})$ this is similar to the $\bigwedge$-case. ⊣

EXERCISE 2.3.10. Fill in the details in the $\exists v\, \theta$ and $\forall v\, \theta$ cases.

COROLLARY 2.3.11. (i) *If $\phi \in \mathcal{L}_{\omega_1,\omega}$, then $\mathcal{M} \models \phi$ if and only if $\mathcal{M}^a \models \phi$ a.e.*

(ii) *If $\mathcal{M}$ is countable and $\phi \in \mathcal{L}_{\infty,\omega}$, $\mathcal{M} \models \phi$ if and only if $\mathcal{M} \models \phi^a$ a.e.*

THEOREM 2.3.12. (i) *$\mathcal{M} \equiv_{\infty,\omega} \mathcal{N} \Leftrightarrow \mathcal{M}^a \cong \mathcal{N}^a$ a.e.*

(ii) *$\mathcal{M} \not\equiv_{\infty,\omega} \mathcal{N} \Leftrightarrow \mathcal{M}^a \not\cong \mathcal{N}^a$ a.e.*

PROOF. Note that filter $D(X)$ is not an ultrafilter, thus (ii) is not an immediate consequence of (i). Though, since it is a filter, it is enough to prove the $\Rightarrow$ directions of both (i) and (ii).

Suppose $\mathcal{M} \not\equiv_{\infty,\omega} \mathcal{N}$. Let $\phi \in \mathcal{L}_{\infty,\omega}$ such that $\mathcal{M} \models \phi$ and $\mathcal{N} \models \neg\phi$. By the previous theorem, $\mathcal{M}^a \models \phi^a$ a.e. and $\mathcal{N}^a \models \neg\phi^a$ a.e. Since $D(X)$ is a filter $\{a : \mathcal{M}^a \models \phi^a \text{ and } \mathcal{N}^a \models \neg\phi^a\} \in D(X)$ and, thus, $\{a : \mathcal{M}^a \not\cong \mathcal{N}^a\} \in D(X)$.

Suppose $\mathcal{M} \equiv_{\infty,\omega} \mathcal{N}$. By Theorem 2.1.4 there is a non-empty system P of partial isomorphism with finite domains from $\mathcal{M}$ to $\mathcal{N}$. Let $A = \{a \in \mathcal{P}(\omega_1(X))$: such that $\{f \in P : \mathrm{dom}(f) \subset \mathcal{M}^a$ and $\mathrm{img}(f) \subset \mathcal{N}^a\}$ is a nonempty system of partial isomorphism between $\mathcal{M}^a$ and $\mathcal{N}^a\}$.

EXERCISE 2.3.13. Show that $A \in D(X)$ and prove that $\mathcal{M}^a \cong \mathcal{N}^a$ for $a \in A$.

⊣

COROLLARY 2.3.14. *If $\mathcal{M}$ is countable, then $\mathcal{M} \equiv_{\infty,\omega} \mathcal{N}$ if and only if $\mathcal{M} \cong \mathcal{N}^a$ a.e.*

The corollary leads to another proof that an abelian group is $\equiv_{\infty,\omega}$ to a free abelian group if and only if every countable subgroup is free.

2.4. Larger infinitary languages

For completeness, we will briefly mention the logics $\mathcal{L}_{\kappa,\lambda}$. In these logics we allow a more general quantification rule. We will say very little in this section and refer to reader to [18] for a more detailed survey.

Suppose ϕ is a formula and $\vec{v}$ is a sequence of variables freely occuring in ϕ with $|\vec{v}| < \lambda$, then $\exists \vec{v}\, \phi$ is a formula.

Thus we are allowed existential and, by taking negations in the usual way, universal quantification over sequences of variables of cardinality less than λ.

EXERCISE 2.4.1. Show there is $\phi \in \mathcal{L}_{\omega_1,\omega_1}$ such that $\mathcal{M} \models \phi$ if and only if $|M| > \aleph_0$.

EXERCISE 2.4.2. Show there is an $\mathcal{L}_{\omega_1,\omega_1}$-sentence ϕ such that $(A,<) \models \phi$ if and only if $(A,<)$ is a well-order.

EXERCISE 2.4.3. Show that is an $\mathcal{L}_{\omega_1,\omega_1}$-sentence ϕ such that $(A,<) \models \phi$ if and only if $(A,<) \cong (\mathbb{R},<)$.

EXERCISE 2.4.4. Show that there is an $\mathcal{L}_{\omega_1,\omega_1}$ sentence ϕ such that ϕ has a model of cardinality κ if and only if the cofinality of κ is ω.

EXERCISE 2.4.5. Recall that a linear ordering $(A,<)$ is $\aleph_1$-*like* if $|A| = \aleph_1$ and $|\{b : b < a\}| \leq \aleph_0$ for all $a \in A$. Show that the class of $\aleph_1$-like linear orders is $\mathcal{L}_{\omega_1,\omega_1}$-axiomatizable.

DEFINITION 2.4.6. We say $\mathcal{M} \cong_p^\kappa \mathcal{N}$ if and only if there is a nonempty system of partial embeddings P with the property that:

(i) if $f \in P$, $A \subset \mathcal{M}$ and $|A| < \kappa$, then there is $g \in P$, $f \subseteq g$ and $A \subseteq \mathrm{dom}(g)$ and

(ii) if $f \in P$, $B \subset \mathcal{N}$ and $|B| < \kappa$, then there is $g \in P$, $f \subseteq g$ and $B \subseteq \mathrm{img}(g)$.

Our original notion $\cong_p$ is $\cong_p^{\aleph_0}$.

EXERCISE 2.4.7. (a) Prove $\mathcal{M} \cong_p^\kappa \mathcal{N} \Rightarrow \mathcal{M} \equiv_{\infty,\kappa} \mathcal{N}$ by induction on complexity of formulas.

(b) Prove $\mathcal{M} \equiv_{\infty,\kappa} \mathcal{N} \Rightarrow \mathcal{M} \cong_p^\kappa \mathcal{N}$. [Hint: Let P be the set of $f : A \to \mathcal{N}$ where $|A| < \kappa$ and f is $\mathcal{L}_{\infty,\kappa}$-elementary.]

EXERCISE 2.4.8. Let $\mathcal{M}$ and $\mathcal{N}$ be $\aleph_1$-like dense linear orders with out endpoints. Prove $\mathcal{M} \equiv_{\infty,\omega_1} \mathcal{N}$. [Hint: Consider $P = \{f : A \to B : A$ an initial segment of $\mathcal{M}$, B is an initial segment of $\mathcal{N}$, $\mathcal{M} \setminus A$ and $\mathcal{N} \setminus B$ has no least element and f is order preserving$\}$.]

Note in A.0.2 we show there are $2^{\aleph_1}$ non-isomorphic $\aleph_1$-like dense linear orders.

EXERCISE 2.4.9. We say that an abelian group G is *strongly* $\aleph_1$*-free* if for any countable $A \subset G$ there is a countable subgroup $H \subset G$ such that $A \subset H$ and H'/H is free for all countable subgroups $H \subset H' \subset G$.

Let F be an uncountable free abelian group prove that $G \equiv_{\infty,\omega_1} F$ if and only if G is strongly $\aleph_1$-free.

In Theorem A.0.4 we show that there are $2^{\aleph_1}$ non-isomorphic strongly $\aleph_1$-free abelian groups of cardinality $\aleph_1$.

Chapter 3

THE SPACE OF COUNTABLE MODELS

There are many connections between the model theory of $\mathcal{L}_{\omega_1,\omega}$ and descriptive set theory. Suppose τ is a countable vocabulary. The first observation is that the set of countably infinite τ-structures can naturally be given the structure of a Polish space. The reader should consult [34] for any unexplained notions or results from descriptive set theory.

3.1. Spaces of τ-structures

To each countable vocabulary τ we will define a Polish space on the set of τ-structures with universe ω. The notation in the general case makes this seem more mysterious then it is, so we first illustrate with a simple example. Let τ_0 be the language with one binary relation symbol R. In this case we will let $\mathbb{X}_{\tau_0} = 2^{\omega\times\omega}$. If $\mathcal{M}$ is a τ_0-structure with universe ω, define $f_{\mathcal{M}} : \omega \times \omega \to 2$ by $f_{\mathcal{M}}(i, j) = 1$ if and only if $\mathcal{M} \models R(i, j)$. Similarly, for any $f : \omega^2 \to 2$ we can define $\mathcal{M}_f$ as the τ_0-structure on ω with binary relation $R_f = \{(i, j) : f(i, j) = 1\}$. The maps $\mathcal{M} \mapsto f_{\mathcal{M}}$ and $f \mapsto \mathcal{M}_f$ are easily seen to be inverses. Thus we can identify the set τ_0-structures with universe ω with elements of $\mathbb{X}_{\tau_0}$. Of course, $\mathbb{X}_{\tau_0}$, with the product topology, is a Polish space–indeed it is homeomorphic to the Cantor space.

Suppose we instead consider τ_1 where we have one unary function symbol. Then we could identify τ_1-structures with ω^ω.

Let τ be a countable vocabulary with constant symbols $(c_i : i < \alpha_C)$, relations symbols $(R_i : i < \alpha_R)$ and function symbols $(f_i : i < \alpha_F)$ where $\alpha_C, \alpha_R, \alpha_F \leq \omega$. For each $i < \alpha_R$ and $j < \alpha_F$, let $m(i)$ and $n(j)$ be the arities of R_i and f_j respectively. Define

$$\mathbb{X}_\tau = \omega^{\alpha_C} \times \prod_{i<\alpha_R} 2^{\omega^{m(i)}} \times \prod_{j<\alpha_F} \omega^{\omega^{n(j)}}.$$

Arguing as above, it is easy to see that there is a natural bijection between $\mathbb{X}_\tau$ and the set of all τ-structures with universe ω. The ω^{α_C} part tells how to interpret the constant symbols, the $\prod_{i<\alpha_R} 2^{\omega^{m(i)}}$ part tells how to interpret the relation symbols and the $\prod_{j<\alpha_F} \omega^{\omega^{n(j)}}$ part tells how to interpret the function symbols.[6]

Suppose $\phi(v_{i_1}, \ldots, v_{i_m})$ is an $\mathcal{L}_{\omega_1,\omega}$-formula with free variables from $v_{i_1}, \ldots, v_{i_m}$ and $n_1, \ldots, n_m \in \omega$. Define

$$E_{\phi,\overline{n}} = \{\mathcal{M} \in \mathbb{X}_\tau : \mathcal{M} \models \phi(n_1, \ldots, n_m)\}.$$

LEMMA 3.1.1. *Each $E_{\phi,\overline{n}}$ is a Borel subset of $\mathbb{X}_\tau$.*

PROOF. We prove this by induction on the complexity of $\mathcal{L}_{\omega_1,\omega}$-formulas.

CLAIM. *If ϕ is atomic, then $E_{\phi,\overline{n}}$ is clopen.*

The formula ϕ is either $R(t_1, \ldots, t_l)$ or $t_1 = t_2$ where $t_1, \ldots, t_l$ are terms and R is a relation symbol. If each of the t_i is a variable, this is clear from the construction of $\mathbb{X}_\tau$.

Unwinding the terms, we can find atomic formulas $\psi_1(\overline{v}, \overline{w}), \ldots, \psi_k(\overline{v}, \overline{w})$ each of the form $R(\overline{v})$, $v_i = v_j$, $v_i = c$ or $f(\overline{v}) = v_j$ where R is a relation symbol, c is a constant symbol and f is a function symbol, and

$$\begin{aligned}\phi(\overline{v}) &\Leftrightarrow \exists \overline{w}\,(\psi_1(\overline{v}, \overline{w}) \wedge \cdots \wedge \psi_k(\overline{v}, \overline{w}))\\ &\Leftrightarrow \forall \overline{w}\,[(\psi_1(\overline{v}, \overline{w}) \wedge \cdots \wedge \psi_{k-1}(\overline{v}), \overline{w}) \rightarrow \psi_k(\overline{v}, \overline{w})].\end{aligned}$$

For example, suppose $\phi(v_1, v_2)$ is $R(v_1, g(f(v_1, v_2)))$. Then

$$\begin{aligned}\phi(v_1, v_2) &\Leftrightarrow \exists w_1 \exists w_2\,[f(v_1, v_2) = w_1 \wedge g(w_1) = w_2 \wedge R(v_1, w_2)]\\ &\Leftrightarrow \forall w_1 \forall w_2 [(f(v_1, v_2) = w_1 \wedge g(w_1) = w_2) \rightarrow R(v_1, w_2)].\end{aligned}$$

Thus we have $\phi(\overline{v}) \Leftrightarrow \exists \overline{w}\, \psi(\overline{v}, \overline{w}) \Leftrightarrow \forall \overline{w}\, \theta(\overline{v}, \overline{w})$, where each $E_{\psi,\overline{a}}$ and $E_{\theta,\overline{a}}$ is clopen. But then

$$\begin{aligned}E_{\phi,\overline{m}} &= \bigcup_{\overline{a}} E_{\psi,\overline{m},\overline{a}} \text{ an open set}\\ &= \bigcap_{\overline{a}} E_{\theta,\overline{m},\overline{a}} \text{ a closed set.}\end{aligned}$$

Thus $E_{\phi,\overline{m}}$ is clopen.

The rest of the induction is easy:

- $E_{\neg\phi,\overline{m}} = \mathbb{X}_\tau \setminus E_{\phi,\overline{m}}$;
- $E_{\bigvee \phi,\overline{m}} = \bigcup E_{\phi,\overline{m}}$;
- $E_{\bigwedge \phi,\overline{m}} = \bigcap E_{\phi,\overline{m}}$;
- $E_{\exists v\phi,\overline{m}} = \bigcup_a E_{\phi,\overline{m},a}$;
- $E_{\forall v\phi,\overline{m}} = \bigcap_a E_{\phi,\overline{m},a}$.

So, by induction on complexity, each $E_{\phi,\overline{m}}$ is Borel. ⊣

DEFINITION 3.1.2. If ϕ is a sentence of $\mathcal{L}_{\omega_1,\omega}(\tau)$, let

$$\mathrm{Mod}(\phi) = \{\mathcal{M} \in \mathbb{X}_\tau : \mathcal{M} \models \phi\}.$$

---

[6]Note that this definition still makes sense if one or several of α_C, α_R or α_F are zero. In particular if τ is the empty vocabulary, then $\mathbb{X}_\tau$ is the one element space.

Then $\text{Mod}(\phi)$ is a Borel subset of $\mathbb{X}_\tau$ and can be viewed as a standard Borel space. Not every Borel subset of $\mathbb{X}_\tau$ arises as $\text{Mod}(\phi)$ for some $\mathcal{L}_{\omega_1,\omega}$-sentence.

There is a natural action of S_∞, the set of all permutations of ω, on X_τ by homeomorphisms. If $\mathcal{M} \in \mathbb{X}_\tau$ and $\sigma \in S_\infty$ let $\sigma(\mathcal{M})$ be the τ-structure induced by σ, i.e.,

$$\sigma(\mathcal{M}) \models \phi(\overline{a}) \Leftrightarrow \mathcal{M} \models \phi(\sigma^{-1}(\overline{a})).$$

We say that $A \subseteq \mathbb{X}_\tau$ is *invariant* if $\sigma(A) = A$ for all $\sigma \in S_\infty$.

COROLLARY 3.1.3. *For any sentence $\phi \in \mathcal{L}_{\omega_1,\omega}(\tau)$, $\text{Mod}(\phi)$ is an invariant Borel subset of $\mathbb{X}_\tau$.*

PROOF. If ϕ is a sentence, then $\mathcal{M} \models \phi$ if and only if $\sigma(\mathcal{M}) \models \phi$. ⊣

In Theorem 4.3.7 we will prove the converse namely that for every invariant Borel set $A \subseteq \mathbb{X}_\tau$ there is an $\mathcal{L}_{\omega_1,\omega}(\tau)$-sentence ϕ such that $A = \text{Mod}(\phi)$.

COROLLARY 3.1.4. *For any $\mathcal{M} \in \mathbb{X}_\tau$, $\{\mathcal{N} : \mathcal{N} \cong \mathcal{M}\}$ is invariant Borel.*

PROOF. By Scott's Theorem the isomorphism class of $\mathcal{M}$ is $\text{Mod}(\Phi^{\mathcal{M}})$, where $\Phi^{\mathcal{M}}$ is the Scott sentence of $\mathcal{M}$. ⊣

EXERCISE 3.1.5. Prove that if $\mathcal{K}$ is a $\text{PC}_{\omega_1,\omega}$-class then $\mathcal{K} \cap \mathbb{X}_\tau$ is invariant Σ^1_1.

EXERCISE 3.1.6. (a) Suppose $\phi \in \mathcal{L}_{\omega_1,\omega}(\tau)$ is in disjunctive normal form, i.e., ϕ is $\bigvee_{i\in\omega} \bigwedge_{j\in\omega} \psi_{i,j}$, where each $\psi_{i,j}$ is atomic or negated atomic. Show that $\text{Mod}(\phi)$ is a Σ^0_2 subset of $\mathbb{X}_\tau$ (i.e., $\text{Mod}(\phi)$ is an F_σ-set).

(b) Find a vocabulary τ and an $\mathcal{L}_{\omega_1,\omega}(\tau)$-sentence ϕ in conjunctive normal form such that ϕ is not equivalent to any $\mathcal{L}_{\omega_1,\omega}(\tau)$-sentence in disjunctive normal form. (Hint: Find ϕ such that $\text{Mod}(\phi)$ is Π^0_2 but not Σ^0_2.)

3.2. The number of countable models

One of the nagging long standing open problems in the model theory of first order logic is *Vaught's Conjecture* that if T is a complete first order theory in a countable language, then the number of non-isomorphic models countable of T is either at most $\aleph_0$ or exactly $2^{\aleph_0}$. It is natural to extend the conjecture from first order logic to models of an $\mathcal{L}_{\omega_1,\omega}$-sentence. Thus, *Vaught's Conjecture for $\mathcal{L}_{\omega_1,\omega}$* asserts that for any sentence $\phi \in \mathcal{L}_{\omega_1,\omega}$ the number of non-isomorphic countable models is either at most $\aleph_0$ or exactly $2^{\aleph_0}$. Of course, if the Continuum Hypothesis holds, Vaught's Conjecture is a trivial consequence, but we will later give a rephrasing of Vaught's Conjecture that does not depend on the continuum hypothesis. The first progress toward Vaught's Conjecture is still the strongest general result known.

THEOREM 3.2.1 (Morley [60]). *Let ϕ be an $\mathcal{L}_{\omega_1,\omega}$-sentence. Then the number of non-isomorphic countable models of ϕ is either at most $\aleph_1$ or exactly $2^{\aleph_0}$.*

We will give Morley's original proof below, but first it worth noting that, from the point of view of modern descriptive set theory, Morley's Theorem is an easy consequence of core theorems on equivalence relations. Consider the isomorphism relation $\cong$ on $\mathbb{X}_\tau$. Then $\mathcal{M} \cong \mathcal{N}$ if and only if

$\exists f \in \omega^\omega$ f is a bijection and f preserves all of the constant, relation and function symbols of τ.

This is easily seen to be Σ^1_1. For example, let τ be the vocabulary with a single binary relation symbol R. Then $\mathcal{M} \cong \mathcal{N}$ if and only if there exists $f \in \omega^\omega$ such that f is a bijection and

$$\forall m, n\, \mathcal{M} \models R(m,n) \leftrightarrow \mathcal{M} \models R(f(m), f(n)).$$

Thus for any ϕ, isomorphism is a Σ^1_1-equivalence relation on $\mathrm{Mod}(\phi)$.

We can now apply two important theorem equivalence relations from descriptive set theory. For proofs see, for example [23] Theorems 5.35 and 9.1.5.

THEOREM 3.2.2. (i) (Silver [74]) *If E is a Π^1_1-equivalence relation on standard Borel space X with uncountably many equivalence classes, then there is a perfect set of E-inequivalent elements. In particular, this is true for any Borel equivalence relation.*

(ii) (Burgess [16]) *If E is a Σ^1_1-equivalence relation on standard Borel space X with at least $\aleph_2$ equivalence classes, then there is a perfect set of E-inequivalent elements.*

For proofs see, for example [23] 5.3.5 and 9.1.5.

Morely's Theorem is, of course, an easy consequence of Burgess' result. It is worth noting that Burgess's result can't be improved. For example define an equivalence relation on E on ω^ω such that $x \mathrel{E} y$ if and only if either neither codes a well ordering or there both code linear orders and there is an isomorphism between the orders they code. This is easily seen to be a Σ^1_1-equivalence relation with exactly $\aleph_1$-classes.

In the isomorphism equivalence relation on $\mathrm{Mod}(\phi)$ we have added information for Corollary 3.1.4 that every equivalence class is Borel. This is also not enough.

EXERCISE 3.2.3. Define an equivalence relation E on ω^ω such that $x \mathrel{E} y$ if and only if $\omega_1^x = \omega_1^y$ (see Chapter 9 for definitions). Show that E is Σ^1_1-equivalence relation with exactly $\aleph_1$ classes, all of which are Borel.

Further examples get closer to Vaught's Conjecture. In particular there are $\mathrm{PC}_{\omega_1,\omega}$ classes with exactly $\aleph_1$-countable models.

EXERCISE 3.2.4 (Friedman). (a) Suppose $\mathcal{M}$ is a countable ω-model of ZFC^- (i.e., ZFC – Power set) and the ordinals of $\mathcal{M}$ are ill-founded. Show that there is an ordinal α such that the ordinals of $\mathcal{M}$ have order type $\alpha + (\mathbb{Q} \times \alpha)$.

(b) Show that (a) gives rise to a $\mathrm{PC}_{\omega_1,\omega}$-class with exactly $\aleph_1$ non-isomorphic countable models.

EXERCISE 3.2.5 (Kunen). A linear order is *1-transitive* if for all a, b there is an order preserving automorphism taking a to b.

For $\alpha < \omega$, we say that the support of $f : \alpha \to \mathbb{Z}$ is $\{\beta < \alpha : f(\beta) \neq 0\}$ and let $\mathbb{Z}^\alpha$ be the set of finite support functions from α to $\mathbb{Z}$ ordered lexicographically. [Note $\mathbb{Z}^0 = \{\emptyset\}$.]

(a) Prove that each $\mathbb{Z}^\alpha$ is 1-transitive.
(b) Prove that every countable 1-transitive linear order is isomorphic to $\mathbb{Z}^\alpha$ or $\mathbb{Q} \times \mathbb{Z}_\alpha$ ordered lexicographically.
(c) Use (b) describe a PC-class with exactly $\aleph_1$ non-isomorphic countable models.

We also know that the isomorphism equivalence relation on $\mathrm{Mod}(\phi)$ is also the orbit equivalence relation of the action of the Polish group S_∞ on the standard Borel space $\mathrm{Mod}(\phi)$. This leads to the Topological Vaught Conjecture which is also still open.

Topological Vaught Conjecture. Suppose G is a Polish group and there is a Borel-measurable action of G on a standard Borel space $\mathbb{X}$. If G has uncountably many orbits, then there is a perfect set of elements in distinct orbits.

Although Morely's Theorem follow's immediately from Burgess' Theorem, it is instructive to see how to prove it using only Silver's Theorem. For each $\alpha < \omega_1$ let $\sim_\alpha$ be as in Definition 1.1.2. These are Borel equivalence relations, that is for each n the equivalence relation $(\mathcal{M}, \overline{a}) \sim (\mathcal{N}, \overline{b})$, where $\overline{a}, \overline{b}$ are of length n, is a Borel equivalence relation on $\mathbb{X}_\tau \times \omega^n$. This is an easy induction since

- $\sim_0$ is closed since $(\mathcal{M}, \overline{a}) \sim_0 (\mathcal{N}, \overline{b})$ if and only if $\mathcal{M} \models \phi(\overline{a}) \Leftrightarrow \mathcal{N} \models \phi(\overline{b})$ for all quantifier free ϕ;
- for α a limit, $\sim_\alpha = \bigcap_{\beta<\alpha} \sim_\beta$;
- $(\mathcal{M}, \overline{a}) \sim_{\alpha+1} (\mathcal{N}, \overline{c}) \Leftrightarrow \forall c \exists d\, (\mathcal{M}, \overline{a}, c) \sim_\alpha (\mathcal{N}, \overline{b}, d)$ and $\forall d \exists c\, (\mathcal{M}, \overline{a}, c) \sim_\alpha (\mathcal{N}, \overline{b}, d)$. Thus $\sim_\alpha$ is a countable intersection of countable unions of Borel sets and, hence, Borel.

For each $\mathcal{M} \models \phi$, if α is the Scott rank of $\mathcal{M}$, then

$$\mathcal{M} \sim_\alpha \mathcal{N} \Leftrightarrow \mathcal{M} \cong \mathcal{N}.$$

Suppose there are $\aleph_2$ non-isomorphic countable models of ϕ. Then for some $\alpha < \omega_1$ there are $\aleph_2$ models of Scott rank at most α. But then, by Silver's Theorem, there must be a perfect set of $\sim_\alpha$-inequivalent elements and hence a perfect set of non-isomorphic elements.

One consequence of Silver's Theorem is that if $\cong$ is a Borel equivalence relation on $\mathrm{Mod}(\phi)$, then Vaught's Conjecture holds for ϕ. When does this happen? The arguments above show that if there is $\alpha < \omega_1$ such that all models have Scott rank at most α, then the isomorphism relation is just $\sim_\alpha$, a Borel equivalence relation. We will argue in Theorem 11.3.6 that $\cong$ is Borel if and only if there is a countable bound on Scott rank.

3.3. Scattered sentences and Morley's proof

DEFINITION 3.3.1. We say that ϕ is *scattered* if $S_n(\mathbb{A}, \phi)$ is countable for all countable fragments $\mathbb{A}$.

Non-scattered theories have many non-isomorphic models.

LEMMA 3.3.2. *Let $\mathbb{A}$ be a countable fragment. If $S_n(\mathbb{A}, \phi)$ is uncountable, then $|S_n(\mathbb{A}, \phi)| = 2^{\aleph_0}$.*

PROOF. We know similar results for type spaces in first order logic. Those proofs though always use the compactness of first order logic to prove the compactness of the type spaces. Here we can not rely on compactness.

Let X be the set of $\mathbb{A}$-formulas with n-free variables. We identify $S_n(\mathbb{A}, \phi)$ as a subset of the Polish space 2^X. Consider the standard Borel space $\text{Mod}(\phi) \times \omega^n$ and the map F from $\text{Mod}(\phi) \times \omega^n$ onto $S_n(\mathbb{A}, \phi)$ given by $(\mathcal{M}, \overline{a}) \mapsto \text{tp}^{\mathcal{M}}(\overline{a}, \mathbb{A})$. By the arguments above this map is Borel. Thus $S_n(\mathbb{A}, \phi)$ is Σ^1_1 and, by the Perfect Set Theorem for Σ^1_1-sets (see [34] 14.13), must contain a non-empty perfect subset of 2^X. ⊣

COROLLARY 3.3.3. *If ϕ is not scattered there are $2^{\aleph_0}$ non-isomorphic countable models of ϕ. Indeed, there is a perfect set of non-isomorphic models.*

PROOF. There is a countable fragment $\mathbb{A}$ such that $|S_n(\mathbb{A}, \phi)| = 2^{\aleph_0}$. As each model only realizes countably many types, there must be $2^{\aleph_0}$ non-isomorphic countable models.

To find a perfect set of non-isomorphic models, consider the equivalence relation on $\text{Mod}(\phi)$ where $\mathcal{M} \approx \mathcal{N}$ if and only $\{\text{tp}^{\mathcal{M}}(\overline{a}, \mathbb{A}) : \overline{a} \in \mathcal{M}\} = \{\text{tp}^{\mathcal{N}}(\overline{a}, \mathbb{A}) : \overline{a} \in \mathcal{N}\}$, i.e., $\mathcal{M} \approx \mathcal{N}$ if and only if they realize the same $\mathbb{A}$-types. This is a Borel equivalence relation and if $\mathcal{M} \cong \mathcal{N}$, then $\mathcal{M} \approx \mathcal{N}$. But if $|S_N(\mathbb{A}, \phi)|$ is uncountable, $\approx$ has uncountably many equivalence class. Thus, by Silver's Theorem, there is a perfect set of $\approx$-inequivalent, and hence non-isomorphic models. ⊣

The last step of Morley's original proof is to show that if ϕ is scattered, then there are at most $\aleph_1$ non-isomorphic models. This proof uses a variant of the type of analysis used in the proof of Scott's Isomorphism Theorem.

We build a sequence of countable fragments $(\mathbb{A}_\alpha : \alpha < \omega_1)$ of $\mathcal{L}_{\omega_1,\omega}$. Let $\mathbb{A}_0$ be a countable fragment of $\mathcal{L}_{\omega_1,\omega}$-containing ϕ. If α is a limit ordinal, let $\mathbb{A}_\alpha = \bigcup_{\beta<\alpha} \mathbb{A}_\beta$. Given $\mathbb{A}_\alpha$, let $\mathbb{A}_{\alpha+1}$ be the smallest fragment of $\mathcal{L}_{\omega_1,\omega}$

containing all formulas

$$\Theta_p(\overline{v}) = \bigwedge_{\psi \in p} \psi(\overline{v})$$

for all $p \in S_n(\mathbb{A}, \phi)$ and $n \in \omega$. Since ϕ is scattered, $\mathbb{A}_{\alpha+1}$ is a countable fragment.

LEMMA 3.3.4. *For each $\mathcal{M} \models \phi$ and $\overline{a} \in \mathcal{M}$ there is $\alpha < \omega_1$ such that if $\mathrm{tp}^{\mathcal{M}}(\overline{a}, \mathbb{A}_\alpha) = \mathrm{tp}^{\mathcal{M}}(\overline{b}, \mathbb{A}_\alpha)$, then $\mathrm{tp}^{\mathcal{M}}(\overline{a}, \mathbb{A}_\beta) = \mathrm{tp}^{\mathcal{M}}(\overline{b}, \mathbb{A}_\beta)$ for all $\beta < \omega_1$.*

We call the least such α the Morley height of $\overline{a} \in M$.

PROOF. Let $\alpha = \sup\{\beta + 1 : \exists \overline{b} \in \mathcal{M}\ \mathrm{tp}^{\mathcal{M}}(\overline{a}, \mathbb{A}_\beta) = \mathrm{tp}^{\mathcal{M}}(\overline{b}, \mathbb{A}_\beta)$ but $\mathrm{tp}^{\mathcal{M}}(\overline{a}, \mathbb{A}_{\beta+1}) \neq \mathrm{tp}^{\mathcal{M}}(\overline{b}, A_{\beta+1})\}$. ⊣

The Morley height of $\mathcal{M}$ is the sup of all Morley height's of tuples in $\mathcal{M}$.

LEMMA 3.3.5. *Suppose $\mathcal{M}$ has Morley height α and $\mathcal{N} \equiv_{\mathbb{A}_{\alpha+1}} \mathcal{M}$. If $\overline{a}, \overline{b} \in \mathcal{N}$ and $\mathrm{tp}^{\mathcal{N}}(\overline{a}, \mathbb{A}_\alpha) = \mathrm{tp}^{\mathcal{N}}(\overline{b}, \mathbb{A}_\alpha)$, then $\mathrm{tp}^{\mathcal{N}}(\overline{a}, \mathbb{A}_{\alpha+1}) = \mathrm{tp}^{\mathcal{N}}(\overline{b}, \mathbb{A}_{\alpha+1})$. Thus $\mathcal{N}$ has Morley height α as well.*

PROOF. Let $p = \mathrm{tp}^{\mathcal{N}}(\overline{a}) = \mathrm{tp}^{\mathcal{N}}(\overline{b})$. Then, for any $\psi \in \mathbb{A}_{\alpha+1}$

$$\mathcal{M} \models \forall \overline{v} \forall \overline{w}\, [(\Theta_p(\overline{v}) \leftrightarrow \Theta_p(\overline{w})) \rightarrow (\psi(\overline{v}) \leftrightarrow \psi(\overline{w}))].$$

But this sentence is in $\mathbb{A}_{\alpha+1}$ and, hence, is also true in $\mathcal{N}$. ⊣

LEMMA 3.3.6. *If $\mathcal{M}$ and $\mathcal{N}$ are countable models of ϕ, $\mathcal{M}$ has Morley height α and $\mathcal{M} \equiv_{\mathbb{A}_{\alpha+1}} \mathcal{N}$, then $\mathcal{M} \cong \mathcal{N}$.*

PROOF. Via a back-and -orth argument, we build a sequence of finite functions $f_0 \subseteq f_1 \subseteq \cdots$ from $\mathcal{M}$ to $\mathcal{N}$ so that $f = \bigcup_{n \in \omega} f_n$ is the desired isomorphism. We will always have that if $\overline{a}$ is the domain of f_i, then $\mathrm{tp}^{\mathcal{M}}(\overline{a}, \mathbb{A}_\alpha) = \mathrm{tp}^{\mathcal{N}}(f_i(\overline{a}), \mathbb{A}_\alpha)$. Since $\mathcal{M} \equiv_{\mathbb{A}_{\alpha+1}} \mathcal{N}$, we can take $f_0 = \emptyset$. Suppose we have f_i with domain $\overline{a}$ and $f(\overline{a}) = \overline{b}$ and we want to add c to the domain of f_α. Let p be the $\mathbb{A}_\alpha$ type of $(\overline{a}, c)$. Then

$$\mathcal{M} \models \exists \overline{v} \exists w \Theta_p(\overline{v}, w).$$

Thus there is $(\overline{s}, t) \in \mathcal{N}$ such that $\mathrm{tp}^{\mathcal{N}}(\overline{s}, t, \mathbb{A}_\alpha) = \mathrm{tp}^{\mathcal{M}}(\overline{a}, c)$. In particular, $\mathrm{tp}^{\mathcal{N}}(\overline{s}, \mathbb{A}_\alpha) = \mathrm{tp}^{\mathcal{N}}(\overline{b}, \mathbb{A}_\alpha)$ and, by the previous lemma $\mathrm{tp}^{\mathcal{N}}(\overline{s}, \mathbb{A}_{\alpha+1}) = \mathrm{tp}^{\mathcal{N}}(\overline{b}, \mathbb{A}_{\alpha+1})$. But the formula $\exists w \Theta_p(\overline{v}, w) \in \mathrm{tp}^{\mathcal{N}}(\overline{s}, \mathbb{A}_{\alpha+1})$. Thus we can find d such that $\mathrm{tp}^{\mathcal{N}}(\overline{b}, d) = p$.

Similarly, we can add elements of $\mathcal{N}$ to the domain of any f_i. Thus by the usual back-and-forth bookkeeping, we can build $f_0 \subset f_1 \subset \cdots$ so that $\bigcup f_n$ is the desired isomorphism. ⊣

COROLLARY 3.3.7. *If ϕ is scattered, then ϕ has at most $\aleph_1$ countable models.*

PROOF. Suppose for contradiction that T has at least $\aleph_2$ models. For some $\alpha < \omega_1$ there are uncountably many models of Morley height at most α. But any model of Morley height α is determined up to isomorphism by its $\mathbb{A}_{\alpha+1}$-theory and there are only countably many choices for the theory, a contradiction. ⊣

This finishes Morley's proof. Either ϕ is scattered and there are at most $\aleph_1$ countable models or ϕ is not scattered and there are continuum many models.

We will examine several equivalents of being scattered.

PROPOSITION 3.3.8. *Let $\phi \in \mathcal{L}_{\omega_1,\omega}$ then ϕ is scattered if and only if $\sim_\alpha$ has countably many equivalence classes of models of ϕ for all $\alpha < \omega_1$;*

PROOF. ($\Leftarrow$) Suppose $\mathbb{A}$ is a countable fragment such that $S_n(\phi, \mathbb{A})$ is uncountable. Let α be a bound on the quantifier rank of formulas. Since

$$(\mathcal{M}, \overline{a}) \sim_\alpha (\mathcal{N}, \overline{b}) \Leftrightarrow (\mathcal{M}, \overline{a}) \equiv_\alpha (\mathcal{N}, \overline{b})$$

there are uncountably many $\sim_\alpha$-classes.

($\Rightarrow$) Suppose α is least such that for some n there are uncountably many $\sim_\alpha$ classes of $(\mathcal{M}, a_1, \ldots, a_n) \models \phi$.

If $\alpha = 0$ then there are uncountably many quantifier free types and ϕ is not scattered.

If $\alpha = \beta + 1$, let $\mathbb{A}$ be the smallest fragment containing ϕ and all Scott sentences $\Phi^{\mathcal{M}}_{\overline{m},\beta}$ for $(\mathcal{M}, \overline{m}) \models \phi$ for $\overline{m} \in \mathcal{M}^{<\omega}$. Since there are only countably many $\sim_\beta$-classes, $\mathbb{A}$ is countable. Suppose $(\mathcal{M}, \overline{a}) \not\sim_\alpha (\mathcal{N}, \overline{b})$. Then, without loss of generality, there is a $c \in \mathcal{M}$ such $(\mathcal{M}, \overline{a}, c) \not\sim_\beta (\mathcal{N}, \overline{b}, d)$ for any $d \in \mathcal{N}$. Thus the formula

$$\exists w\, \Phi^{\mathcal{M}}_{\overline{a},c,\beta}(\overline{v}, w) \in \mathrm{tp}^{\mathcal{M}}(\overline{a}, \mathbb{A}) \setminus \mathrm{tp}^{\mathcal{N}}(\overline{b}, \mathbb{A}).$$

Since the are uncountably many $\sim_\alpha$ classes there are uncountably many $\mathbb{A}$-types and ϕ is not scattered.

If α is a limit ordinal, let $\mathbb{A}$ be the countable fragment generated by all formulas $\Phi^{\mathcal{M}}_{\overline{a},\beta}$ for $\beta < \alpha$. If $(\mathcal{M}, \overline{a}) \not\sim_\alpha (\mathcal{N}, \overline{b})$, then $(\mathcal{M}, \overline{a}) \not\sim_\beta (\mathcal{N}, \overline{b})$ for some $\beta < \alpha$ and we proceed as above. ⊣

We showed that if ϕ is not scattered there is a perfect set of non-isomorphic models. We now show that the converse is also true.

PROPOSITION 3.3.9. *If there is a perfect set of non-isomorphic models, then ϕ is not scattered.*

PROOF. We will give a different proof in Proposition 11.3.12. We quickly sketch here an argument using forcing and absoluteness.

CLAIM. *"ϕ is scattered" and "there is a perfect set of non-isomorphic models of ϕ" are absolute.*

ϕ is scattered if and only if for all countable sets of codes A for $\mathcal{L}_{\omega_1,\omega}$ formulas $\mathbb{A}$ there is no perfect set of elements $(\mathcal{M}, \overline{a}) \models \phi$ where all $\mathrm{tp}^{\mathcal{M}}(\overline{a}, A)$ are distinct. This can be shown to be $\Pi^1_2(\phi)$ and, hence, absolute by Shoenfield's Absoluteness Theorem (see [30] 25.20). Note, one ingredient of a full proof would be noting that the set of codes for $\mathcal{L}_{\omega_1,\omega}$-formulas is a Π^1_1 condition (see [55] 8.2).

There is a perfect set of non-isomorphic models is easily seen to be by $\Sigma^1_2(\phi)$ and hence absolute.

Suppose ϕ is scattered and there is a perfect set of non-isomorphic models of ϕ. By Morley's argument, this is only possible if $2^{\aleph_0} = \aleph_1$. Consider a forcing extension of the universe $\mathbb{V}[G] \models 2^{\aleph_0} > \aleph_1$. By absolutness, ϕ is still scattered in $\mathbb{V}[G]$ and there is a perfect set of non-isomorphic elements but this contradicts Morley's results. $\dashv$

Putting this all together.

COROLLARY 3.3.10. *The following are equivalent*

(i) *ϕ is scattered;*
(ii) *$\sim_\alpha$ has countably many equivalence classes of models of ϕ for all $\alpha < \omega_1$;*
(iii) *there is no perfect set of non-isomorphic countable models of ϕ.*

When studying Vaught's Conjecture we often look instead at the *Strong Vaught Conjecture*: If $\phi \in \mathcal{L}_{\omega_1,\omega}$, then either there are countably many models of ϕ or there is a perfect set of non-isomorphic models of ϕ. Clearly the strong version implies the usual one. Moreover, if the original version of Vaught's Conjecture is provable for ϕ in ZFC, then so is strong form. Suppose not. If there is no perfect set of non-isomorphic models, then ϕ is scattered. Suppose we can find $(\mathcal{M}_\alpha : \alpha < \omega_1)$ a sequence of pairwise non-isomorphic countable models of ϕ. Pass to a forcing extension $\mathbb{V}[G]$ of the universe making $2^{\aleph_0} > \aleph_1$ without collapsing cardinals (for example, add $\aleph_2$ Cohen reals). By Shoenfield Absoluteness, $\mathbb{V}[G] \models \phi$ is scattered and there are uncountably many non-isomorphic models of ϕ. But then, by Morely's analysis there are exactly $\aleph_1$ non-isomorphic countable models, contradicting the fact that Vaught's Conjecture is true in all models of ZFC.

For this reason we tend to focus on the strong conjecture. This way it makes sense to study counterexamples without worrying about whether the Continuum Hypothesis is true. This leads to the following definition.

DEFINITION 3.3.11. A *Vaught counterexample* is a scattered ϕ with uncountably many non-isomorphic countable models. Equivalently, it is a sentence with uncountably many non-isomorphic countable models but no perfect set of non-isomorphic models.

We will examine uncountable models of Vaught counterexamples in Chapter 7.

Chapter 4

THE MODEL EXISTENCE THEOREM

4.1. Consistency properties and model existence

Although we don't have the Compactness Theorem or a useful Completeness Theorem, Henkin-style arguments can still be used in some contexts to build models. In this section we describe the general framework and give several applications.

Throughout this section we will assume that τ is a countable vocabulary and that C is an infinite set of constant symbols of τ (though perhaps not all of the constant symbols of τ). The following idea, due to Makkai, is the key. It tells us exactly what we need to do a Henkin argument.

Definition 4.1.1. A *consistency property* Σ is a collection of countable sets σ of $\mathcal{L}_{\omega_1,\omega}$-sentences with the following properties. For $\sigma \in \Sigma$:

(C0) if $\mu \subseteq \sigma$, then $\mu \in \Sigma$;
(C1) if $\phi \in \sigma$, then $\neg\phi \notin \sigma$;
(C2) if $\neg\phi \in \sigma$, then there is $\mu \in \Sigma$ such that $\sigma \cup \{{\sim}\phi\} \subseteq \mu$;
(C3) if $\bigwedge_{\phi \in X} \phi \in \sigma$, then for all $\phi \in X$ there is $\mu \in \Sigma$ such that $\sigma \cup \{\phi\} \subseteq \mu$;
(C4) if $\bigvee_{\phi \in X} \phi \in \sigma$, then there is $\mu \in \Sigma$ and $\phi \in X$ such that $\sigma \cup \{\phi\} \subseteq \mu$;
(C5) if $\forall v\, \phi(v) \in \sigma$, then for all $c \in C$ there is $\mu \in \Sigma$ such that $\sigma \cup \{\phi(c)\} \subseteq \mu$;
(C6) if $\exists v\, \phi(v) \in \sigma$, then there is $\mu \in \Sigma$ and $c \in C$ such that $\sigma \cup \{\phi(c)\} \subseteq \mu$;
(C7) let t be a term with no variables and let $c, d \in C$,
  (a) if $c = d \in \sigma$, then there is $\mu \in \Sigma$ such tht $\sigma \cup \{d = c\} \subseteq \mu$;
  (b) if $c = t, \phi(t) \in \sigma$, then there is $\mu \in \Sigma$ such that $\sigma \cup \{\phi(c)\} \subseteq \mu$;
  (c) there is $\mu \in \Sigma$ and $e \in C$ such that $\sigma \cup \{e = t\} \subseteq \mu$.

Lemma 4.1.2. *Let Σ be a consistency property with $\sigma \in \Sigma$, $c, d, e \in C$.*

(i) *There is $\mu \in \Sigma, \sigma \subseteq \mu$ with $c = c \in \mu$.*
(ii) *Suppose $c = d, d = e \in \sigma$. Then $\sigma \cup \{c = e\} \in \Sigma$.*
(iii) *If $\phi, \phi \rightarrow \psi \in \sigma$, then there is $\mu \in \Sigma$, $\sigma \subseteq \mu$ with $\psi \in \mu$.*

Proof. (i) By (C7c) there is $d \in C$ such that $\sigma \cup \{d = c\} \in \Sigma$. Then by (C7a) $\sigma \cup \{c = d, d = c\} \in \Sigma$. Applying (C7b) with the "$v = c$" as $\phi(v)$, $\sigma \cup \{c = d, d = c, c = c\} \in \Sigma$. ⊣

EXERCISE 4.1.3. Prove (ii) and (iii) of the Lemma.

The next Exercise shows we really only need to verify (C1)–(C7).

EXERCISE 4.1.4. Suppose Σ_0 satisfies (C1)–(C7). Let

$$\Sigma = \{\Delta : \exists \mu \in \Sigma_0\, \Delta \subseteq \mu\}.$$

Prove that Σ is a consistency property.

EXERCISE 4.1.5. Suppose Σ_0 is a consistency property.

(a) Show that $\Sigma_1 = \{\sigma \in \Sigma_0 : \sigma \text{ is finite}\}$ is a consistency property.
(b) Show that $\Sigma_2 = \{\sigma \in \Sigma_0 :$ only finitely many constants from C occur in $\sigma\}$ is a consistency property.

THEOREM 4.1.6 (Model Existence Theorem). *If Σ is a consistency property and $\sigma \in \Sigma$, there is a countable $\mathcal{M} \models \sigma$.*

PROOF. Let Λ be the smallest set of $\mathcal{L}_{\infty,\omega}$-sentences such that:

- $\sigma \subseteq \Lambda$;
- if ϕ is a subsentence of a sentence in Λ, then $\phi \in \Lambda$;
- if $\phi(\overline{v})$ is a subformula of a sentence in Λ and $\overline{c} \in C$, then $\phi(\overline{c}) \in \Lambda$;
- If $\neg\phi \in \Lambda$, then ${\sim}\phi \in \Lambda$;
- if $c, d \in C$, then $c = d \in \Lambda$.

Let $\phi_0, \phi_1, \dots$ list all sentences in Λ. We assume that each sentence is listed infinitely often. Let $t_0, t_1, \dots,$ list all τ-terms with no variables.

Using the fact that Σ is a consistency property, we build

$$\sigma = \sigma_0 \subseteq \sigma_1 \subseteq \cdots \subseteq \sigma_n \subseteq \cdots$$

such that each $\sigma_i \in \Sigma$ and

(A) if $\sigma_n \cup \{\phi_n\} \in \Sigma$, then $\phi_n \in \sigma_{n+1}$, in this case:
  (a) if ϕ_n is $\bigvee_{\phi\in X} \phi$, then $\phi \in \sigma_{n+1}$ for some $\phi \in X$;
  (b) if ϕ_n is $\exists v\, \phi(v)$, then $\phi(c) \in \sigma_{n+1}$ for some $c \in C$;

(B) $c = t_n \in \sigma_{n+1}$ for some $c \in C$.

Let $\Gamma = \bigcup_{n=1}^{\infty} \sigma_n$. We will build a model of Γ. For $c, d \in C$ we say $c \sim d$ if $c = d \in \Gamma$.

CLAIM 1. *$\sim$ is an equivalence relation.*

Let $c \in C$. The sentence $c = c$ is ϕ_n for some n. By Lemma 4.1.2, and (A) $c = c \in \sigma_{n+1} \subseteq \Gamma$. Thus $c \sim c$.

If $c = d \in \Gamma$, then we can find n such that $c = d \in \sigma_n$ and $d = c$ is ϕ_n. Then by (C7a) and condition (A) $d = c \in \Gamma$.

If $c = d, d = e \in \Gamma$, choose n such that $c = d, d = e \in \sigma_n$ and $c = e$ is ϕ_n. Then by Lemma 4.1.2 and (A) $c = e \in \Gamma$.

Let $[c]$ denote the $\sim$-class of c. Let $M = \{[c] : c \in C\}$.

If d is a constant symbol of τ (perhaps not in C), then by (B) there is $c \in C$ such that $c = d \in \Gamma$. We interpret $d^{\mathcal{M}}$ as $[c]$.

Let f be an n-ary function symbol of τ and let $c_1, \ldots, c_n \in C$. By (B) there is $d \in C$ such that $d = f(c_1, \ldots, c_n) \in \Gamma$. Suppose $d_1 \in C$ and $d_1 = f(c_1, \ldots, c_n)$ is also in Γ, using (C7b), Lemma 4.1.2 and (A) we see that $d = d_1 \in \Gamma$ and $d \sim d_1$. Also note that if $c_0 = f(c_1, \ldots, c_n) \in \Gamma$, and $d_i \sim c_i$ for $i = 0, \ldots, n$ then, a similar argument shows $d_0 = f(d_1, \ldots, d_n) \in \Gamma$. Thus we can define $f^{\mathcal{M}} : M^n \to M$ by

$$f([c_1], \ldots, [c_n]) = [d] \Leftrightarrow f(c_1, \ldots, c_n) = d \in \Gamma.$$

CLAIM 2. *Let $t(v_1, \ldots, v_n)$ be a term, $c_1, \ldots, c_n, d \in C$. If $d = t(c_1, \ldots, c_n) \in \Gamma$, then $t^{\mathcal{M}}([c_1], \ldots, [c_n]) = [d]$.*

We prove this by induction on complexity of terms. Suppose $t = f(t_1, \ldots, t_m)$. For $i \leq m$ there is $d_i \in C$ such that $d_i = t_i(\overline{c}) \in \Gamma$. By induction, $[d_i] = t_i^{\mathcal{M}}([\overline{c}])$. By (C7b) and (A) $d = f(d_1, \ldots, d_m) \in \Gamma$. Thus

$$t^{\mathcal{M}}([\overline{c}]) = f^{\mathcal{M}}([d_1], \ldots, [d_m]) = [d]$$

as desired.

For $c_1, \ldots, c_n$ and R an n-ary relation symbol of τ, we define

$$R^{\mathcal{M}}([c_1], \ldots, [c_n]) \Leftrightarrow R(c_1, \ldots, c_n) \in \Gamma.$$

As above, we can show this does not depend on the choice of representative from $[c_i]$.

CLAIM 3. *If $\phi \in \Gamma$, then $\mathcal{M} \models \phi$.*

We prove this by induction on complexity.[7]

- ϕ is $t_1 = t_2$. There are $c_1, c_2 \in C$ such that $c_1 = t_1, c_2 = t_2 \in \Gamma$. Then $c_1 = c_2 \in \Gamma$ and by claim 2, $t_1^{\mathcal{M}} = t_2^{\mathcal{M}}$, so $\mathcal{M} \models \phi$.
- ϕ is $R(t_1, \ldots, t_n)$.

EXERCISE 4.1.7. Prove $\mathcal{M} \models R(t_1^{\mathcal{M}}, \ldots, t_n^{\mathcal{M}})$.

- ϕ is $\bigwedge_{\psi \in X} \psi$. By (C3) and (A), $\psi \in \Gamma$ for each $\psi \in X$. By induction, $\mathcal{M} \models \psi$ for all $\psi \in X$. Thus $\mathcal{M} \models \phi$.
- ϕ is $\bigvee_{\psi \in X} \psi$. By (C4) and (A), there is $\psi \in X$ such that $\psi \in \Gamma$. By induction $\mathcal{M} \models \psi$. Thus $\mathcal{M} \models \phi$.
- ϕ is $\forall v\, \psi(v)$. By (C5) and (A), $\psi(c) \in \Gamma$ for all $c \in C$. By induction, $\mathcal{M} \models \psi(c)$ for all $c \in C$. Since every element of M is named by a constant in C, $\mathcal{M} \models \phi$.
- ϕ is $\exists v\, \psi(v)$. By (A) there is $c \in C$ such that $\psi(c) \in \Gamma$. By induction, $\mathcal{M} \models \psi(c)$. Thus $\mathcal{M} \models \phi$.
- ϕ is $\neg\psi$. By (C2) and (A), ${\sim}\psi \in \Gamma$. This now breaks into cases depending on ψ.

---

[7]We need to be slightly careful how we define "complexity". For example, $\exists v\, \psi(v)$ is more complex than any $\psi(c)$ and $\neg \bigvee_{\psi \in X} \psi$ is more complicated than $\neg\psi$ for any $\psi \in X$.

- ψ is $t_1 = t_2$. There are constants $c_1, c_2 \in C$ such that $c_i = t_i \in \Gamma$. Since $\sim\psi$ is $t_1 \neq t_2$, by (C7) and (A), $c_1 \neq c_2 \in \Gamma$. Suppose, for contradiction, that $\mathcal{M} \models \psi$. By claim 2 $[c_i] = t_i^{\mathcal{M}}$. Thus $\mathcal{M} \models c_1 = c_2$. Hence $c_1 \sim c_2$ and $c_1 = c_2 \in \Gamma$. This contradicts (C1).
- ψ is $R(t_1, \ldots, t_m)$. Then $\neg R(t_1, \ldots, t_m) \in \Gamma$. There are $c_1, \ldots, c_m \in C$ such that $c_i = t_i \in \Gamma$. Then $\neg R(c_1, \ldots, c_m) \in \Gamma$. By (C1), $R(c_1, \ldots, c_m) \notin \Gamma$. Thus
$$\mathcal{M} \models \neg R(c_1, \ldots, c_m).$$
By claim 2,
$$\mathcal{M} \models \neg R(t_1, \ldots, t_m).$$
Thus $\mathcal{M} \models \phi$.
- ψ is $\neg\theta$. Then $\theta \in \Gamma$. By induction $\mathcal{M} \models \theta$ and $\mathcal{M} \models \phi$.
- ψ is $\bigwedge_{\theta \in X} \theta$. Then $\bigvee_{\theta \in X} \sim\theta \in \Gamma$ and $\sim\theta \in \Gamma$ for some $\theta \in X$. By induction, $\mathcal{M} \models \neg\theta$. Thus $\mathcal{M} \models \phi$.
- ψ is $\bigvee_{\theta \in X} \theta$. Then $\bigwedge_{\theta \in X} \sim\theta \in \Gamma$ and $\sim\theta \in \Gamma$ for all $\theta \in X$. By induction $\mathcal{M} \models \neg\theta$ for all $\theta \in X$. Thus $\mathcal{M} \models \phi$.
- ψ is $\exists v\theta(v)$. Then $\forall v{\sim}\theta(v) \in \Gamma$ and $\sim\theta(c) \in \Gamma$ for all $c \in C$. By induction, $\mathcal{M} \models \theta(c)$ for all $c \in C$. Thus $\mathcal{M} \models \phi$.
- ψ is $\forall v\theta(v)$. Then $\exists v{\sim}\theta(v) \in \Gamma$ and $\sim\theta(c) \in \Gamma$ for some $c \in C$. But then $\mathcal{M} \models \neg\theta(c)$ and $\mathcal{M} \models \phi$.

This completes the induction. Thus $\mathcal{M} \models \Gamma$. ⊣

We can easily modify the Henkin argument above to prove the following useful variant.

THEOREM 4.1.8 (Extended Model Existence Theorem). *Let τ be as above. Let T be a countable set of $\mathcal{L}_{\omega_1,\omega}(\tau)$-sentences. Suppose Σ is a consistency property such that for all $\sigma \in \Sigma$ and $\phi \in T$, $\sigma \cup \{\phi\} \in \Sigma$. Then there is $\mathcal{M} \models T$.*

PROOF. Let $\theta_0, \theta_1, \ldots$ enumerate T. Modify the construction of $\sigma_0 \subseteq \sigma_1 \subseteq \cdots$ such that $\theta_n \in \sigma_{n+1}$ for all n. Then $T \subset \Gamma$ and $\mathcal{M} \models T$. ⊣

4.2. Omitting types and atomic models

As a first application of the Model Existence Theorem we will show how to prove a generalization of the Omitting Types Theorem of first order logic.

THEOREM 4.2.1 (Omitting Types Theorem). *Let $\mathbb{A}$ be a countable fragment of $\mathcal{L}_{\omega_1,\omega}(\tau)$, let $T \subset \mathbb{A}$ be a satisfiable theory and let $\Theta_n(v_1, \ldots, v_{m_n})$ be a set of $\mathbb{A}$-formulas for each $n \in \omega$. Suppose that for all $\mathbb{A}$-formulas $\psi(v_1, \ldots, v_{m_n})$ such that $T + \exists\overline{v}\,\psi(\overline{v})$ is satisfiable, there is $\theta \in \Theta_n$ such that $T + \exists\overline{v}\,(\psi(\overline{v}) \wedge \theta(\overline{v}))$ is satisfiable.*

Then

$$T + \bigwedge_{n\in\omega} \forall\overline{v} \bigvee_{\theta\in\Theta_n} \theta(\overline{v})$$

is satisfiable.

Proof. Let $\tau^* \supset \tau$ be obtained by adding a countable set of new constant symbols C and let $\mathbb{A}^*$ be the smallest fragment containing $\mathbb{A}$ such that if $\phi(\overline{v})$ is a formula in $\mathbb{A}$ and $\overline{c} \in C$, then $\phi(\overline{c}) \in \mathbb{A}^*$.

Let

$$\Delta = \left\{ \bigvee_{\theta\in\Theta_n} \theta(c_1, \ldots, c_{m_n}) : n \in \omega, \overline{c} \in C \right\}$$

and Σ be the set of all sets σ where

$$\sigma = \sigma_0 \cup T \cup \Delta$$

where σ_0 is a finite set of $\mathbb{A}^*$-sentences such that $T \cup \sigma_0$ is satisfiable. Note that σ_0 contains only finitely many constants from C.

We claim that Σ is a consistency property. Once we have shown this, the Model Existence Theorem will give us a model of $T \cup \Delta$ where every element is the interpretation of a constant in C. This is the desired model.

We will show that (C4) holds. The rest of the verification is routine and left as an exercise. Suppose σ is as above and $\chi = \bigvee_{\psi\in X} \psi$.

Case 1. $\chi \in \sigma_0 \cup T$.

In this case we know there is $\mathcal{M} \models \sigma_0 \cup T$ and $\mathcal{M} \models \psi$ for some $\psi \in X$. Then

$$\sigma_0 \cup \{\psi\} \cup T \cup \left\{ \bigvee_{\theta\in\Theta_n} \theta(c_1, \ldots, c_{m_n}) : n \in \omega, \overline{c} \in C \right\} \in \Sigma$$

as desired.

Case 2. $\chi = \bigvee_{\theta\in\Theta_n} \theta(\overline{c})$ for some n and $\overline{c} \in C$.

Let $\overline{d}$ be the constants from C occurring in σ_0 but not in $\overline{c}$. Let

$$\Phi(\overline{c}, \overline{d}) = \bigwedge_{\phi\in\sigma_0} \phi.$$

Then

$$T + \exists\overline{x}\exists\overline{y}\, \Phi(\overline{x}, \overline{y})$$

is satisfiable. By our assumptions, there is $\theta \in \Theta_n$ such that

$$T + \exists\overline{x}\ \left(\theta \wedge \exists\overline{y}\, \Phi(\overline{x}, \overline{y})\right)$$

is satisfiable. Thus $T + \Phi(\overline{c}, \overline{d}) \wedge \theta(\overline{c})$ is satisfiable and

$$\sigma_0 \cup \{\theta(\overline{c})\} \cup T \cup \Delta \in \Sigma.$$

Thus (C4) holds. ⊣

Exercise 4.2.2. Complete the verification that Σ is a consistency property.

EXERCISE 4.2.3. Show that Theorem 4.2.1 is a generalization of the classical Omitting Types Theorem for first order logic.

Atomic and prime models. Once we have the Omitting Types Theorem, many of Vaught's proofs about prime and atomic models for first order logic can easily be adapted in the infinitary setting.

Let $\mathbb{A}$ be a countable fragment of $\mathcal{L}_{\omega_1,\omega}$. Let $T \subset \mathbb{A}$ be a satisfiable theory that is $\mathbb{A}$-complete, i.e., $T \models \phi$ or $T \models \neg\phi$ for any $\mathbb{A}$-sentence ϕ.

DEFINITION 4.2.4. A satisfiable formula $\theta(\overline{v})$ is $\mathbb{A}$-*complete* if for any $\mathbb{A}$-formula $\phi(\overline{v})$ either

$$T \models \theta(\overline{v}) \rightarrow \phi(\overline{v}) \text{ or } T \models \theta(\overline{v}) \rightarrow \neg\phi(\overline{v}).$$

A formula $\phi(\overline{v})$ is $\mathbb{A}$-*completable* if there is a complete $\theta(\overline{v}) \in \mathbb{A}$ with $T \models \theta(\overline{v}) \rightarrow \phi(\overline{v})$. We say that ϕ is *incompletable* if it is satisfiable but non-completable.

DEFINITION 4.2.5. $\mathcal{M} \models T$ is $\mathbb{A}$-*atomic* if for all $\overline{a} \in \mathcal{M}$ there is an $\mathbb{A}$-complete formula $\theta(\overline{v})$ such that $\mathcal{M} \models \theta(\overline{a})$

DEFINITION 4.2.6. $\mathcal{M} \models T$ is $\mathbb{A}$-*prime* if for every $\mathcal{N} \models T$ there is an $\mathbb{A}$-elementary embedding of $\mathcal{M}$ into $\mathcal{N}$.

THEOREM 4.2.7. *$\mathcal{M} \models T$ is $\mathbb{A}$-prime if and only if it is countable and $\mathbb{A}$-atomic.*

PROOF. ($\Rightarrow$) Suppose $\mathcal{M}$ is an $\mathbb{A}$-prime model of T. Since T has a countable model, $\mathcal{M}$ must be countable. Suppose $\mathcal{M}$ is not atomic. Let $\overline{a} \in \mathcal{M}$ and suppose that $\overline{a}$ satisfies no $\mathbb{A}$-complete formula. Let $\Delta = \{\phi(\overline{v}) \in \mathbb{A} : \mathcal{M} \models \phi(\overline{a})\}$. We apply the Omitting Types Theorem to T and $\bigvee_{\phi\in\Delta} \neg\phi(\overline{v})$. Suppose $T + \exists v\, \psi(\overline{v})$ is satisfiable.

If $\mathcal{M} \models \neg\psi(\overline{a})$, then there is $\mathcal{N} \models T + \exists\overline{v}\,(\psi(\overline{v}) \wedge \neg\neg\psi(\overline{v}))$, so we satisfy the hypotheses of the Omitting Types Theorem.

If $\mathcal{M} \models \psi(\overline{a})$, then ψ is not $\mathbb{A}$-complete, so we can find ϕ such that $\mathcal{M} \models \phi(\overline{a})$ and

$$T \not\models \psi(\overline{v}) \rightarrow \phi(\overline{v}) \text{ and } T \not\models \psi(\overline{v}) \rightarrow \phi(\overline{v}).$$

Thus $T + \exists\overline{v}\,(\psi(\overline{v}) \wedge \neg\phi(\overline{v}))$ is satisfiable and we again satisfy the hypothesis of the Omitting Types Theorem.

Thus there is a

$$\mathcal{N} \models T + \forall\overline{v} \bigvee_{\mathcal{M}\models\phi(\overline{a})} \neg\phi(\overline{a})$$

and there is no $\mathbb{A}$-elementary embedding of $\mathcal{M}$ into $\mathcal{N}$.

($\Leftarrow$) Suppose $\mathcal{M} \models T$ is countable and $\mathbb{A}$-atomic and $\mathcal{N} \models T$. We must build an $\mathbb{A}$-elementary embedding of $\mathcal{M}$ into $\mathcal{N}$. Let $a_0, a_1, \ldots$ enumerate the elements of $\mathcal{M}$. We build a family of partial $\mathbb{A}$-elementary maps

$$f_0 \subset f_1 \subset \cdots$$

maps into $\mathcal{N}$ where $\mathrm{dom}(f_i) = \{a_0, \ldots, a_{i-1}\}$. Since T is complete, we can start with $f_0 = \emptyset$.

Given f_n, let $b_i = f_n(a_i)$ for $i < n$. There is an $\mathbb{A}$-complete formula $\theta(v_0, \ldots, v_{n-1})$ such that $\mathcal{M} \models \theta(a_0, \ldots, a_{n-1})$ and an $\mathbb{A}$-complete formula $\psi(v_0, \ldots, v_n)$ such that $\mathcal{M} \models \psi(a_0, \ldots, a_n)$. Since θ_0 is $\mathbb{A}$-complete,

$$\mathcal{M} \models \forall \overline{v}(\theta(v_0, \ldots, v_{n-1}) \rightarrow \exists v_n \, \psi(v_0, \ldots, v_n)).$$

Since f_n is $\mathbb{A}$-elementary, $\mathcal{N} \models \theta(b_0, \ldots, b_{n-1})$. Thus we can find $b_n \in \mathcal{N}$ such that $\mathcal{N} \models \psi(b_0, \ldots, b_n)$. Extend f_n to f_{n+1} by sending a_n to b_n. Since $\theta(\overline{v})$ is $\mathbb{A}$-complete, this is $\mathbb{A}$-elementary. The map $f = \bigcup f_n$ is the desired $\mathbb{A}$-elementary embedding of $\mathcal{M}$ into $\mathcal{N}$. ⊣

EXERCISE 4.2.8. Show that any two $\mathbb{A}$-prime models of T are isomorphic.

DEFINITION 4.2.9. T is $\mathbb{A}$-*atomic* if every satisfiable formula in $\mathbb{A}$ is completable.

THEOREM 4.2.10. *T has a countable $\mathbb{A}$-atomic model if and only if T is $\mathbb{A}$-atomic.*

PROOF. ($\Rightarrow$) Suppose $\mathcal{M}$ is a countable $\mathbb{A}$-atomic model. If $\psi(\overline{v})$ is satisfiable, then, since T is $\mathbb{A}$-complete, there is $\overline{a} \in \mathcal{M}$ such that $\mathcal{M} \models \psi(\overline{a})$. But $\overline{a}$ satisfies an $\mathbb{A}$-complete formula $\theta(\overline{v})$ and we must have $T \models \theta(\overline{v}) \rightarrow \phi(\overline{v})$. Thus ϕ is $\mathbb{A}$-completable.

($\Leftarrow$) For each n let Δ_n be the set of all complete formulas $\theta(v_1, \ldots, v_n)$. If $\psi(v_1, \ldots, v_n)$ is satisfiable, then there is $\theta(\overline{v}) \in \Delta_n(\overline{v})$ such that $T \models \theta(\overline{v}) \rightarrow \psi(\overline{v})$. In particular,

$$T + \exists \overline{v} \, (\theta(\overline{v}) \wedge \psi(\overline{v}))$$

is satisfiable and, by the Omitting Types Theorem, there is an $\mathbb{A}$-atomic $\mathcal{M} \models T$. ⊣

We next show that if there are only countably many $\mathbb{A}$-types then there is an $\mathbb{A}$-atomic model. One way to do this in the first order case is to assume there is an incompletable formula and then build a perfect tree of inconsistent incompletable formulas. By compactness, each branch gives rise to a different type. So we have that a non-atomic theory has $2^{\aleph_0}$ types. Here we use the Omitting Types Theorem directly to prove that an $\mathbb{A}$-theory with countably many types is atomic. We could combine this with Morley's argument from Lemma 3.3.2 to conclude that a non-$\mathbb{A}$-atomic theory has $2^{\aleph_0}$ $\mathbb{A}$-types.

THEOREM 4.2.11. *If for all $n \in \omega$ there are only countably many $\mathbb{A}$-types, then T is $\mathbb{A}$-atomic and has an $\mathbb{A}$-prime model.*

PROOF. Suppose $\phi(\overline{v})$ is satisfiable but not $\mathbb{A}$-completable. Let $\gamma_0(\overline{v})$, $\gamma_1(\overline{v}), \ldots$ list all $\mathbb{A}$-types containing ϕ. Since none of those types contains a complete formula for any satisfiable formula $\psi(\overline{v})$ and any $i \in \omega$ we can find

$\theta \in \gamma_i$ such that $T + \exists \overline{v}\,(\psi(\overline{v}) \wedge \neg\theta(\overline{v}))$ is satisfiable. By the Omitting Types Theorem we can find

$$\mathcal{M} \models T + \forall \bigwedge_{n\in\omega} \overline{v} \bigvee_{\theta\in\gamma_n} \neg\theta(\overline{v}).$$

Clearly $\mathcal{M}$ omits each γ_i. But these are all $\mathbb{A}$-types containing ϕ. But T is complete and ϕ is satisfiable, thus $\mathcal{M} \models \exists\overline{v}\phi(\overline{v})$, a contradiction. ⊣

LEMMA 4.2.12. *Suppose* $\mathbb{A}$ *is a fragment of* $\mathcal{L}_{\omega_1,\omega}$ *and* $\delta = \sup\{\text{qr}(\phi) : \phi \in \mathbb{A}\}$. *If* T *is an* $\mathbb{A}$*-atomic theory, then there is* $\mathcal{M} \models T$ *with* $\text{SR}(\mathcal{M}) \leq \alpha$.

PROOF. For $\overline{a} \in \mathcal{M}$ there is $\phi(\overline{v}) \in \mathbb{A}$ of quantifer rank $\alpha < \delta$ such that $\mathcal{M} \models \phi(\overline{a})$ and if $\mathcal{M} \models \phi(\overline{b})$ then $\overline{a}$ and $\overline{b}$ realize the same $\mathbb{A}$-type. Thus $\text{SR}(\mathcal{M}) < \delta$. ⊣

We give one application from Nadel [63].

THEOREM 4.2.13. *If* $\phi \in \mathcal{L}_{\omega_1,\omega}$ *is complete and* $\mathcal{M} \models \phi$, *then the Scott rank of* $\mathcal{M}$ *is at most* $\text{qr}(\phi) + \omega$, *indeed* $\text{SR}(\mathcal{M}) \leq \text{qr}(\phi) + \omega$.

PROOF. Let $\mathbb{A}$ be the smallest fragment containing ϕ. Note that every formula in $\mathbb{A}$ has quantifer rank less than $\text{qr}(\phi) + \omega$. There are only countably many $\mathbb{A}$-types. Thus if $\mathcal{M}$ is the unique countable model of ϕ, then $\mathcal{M}$ is $\mathbb{A}$-atomic and, by the Lemma, $\text{SR}(\mathcal{M}) \leq \text{qr}(\phi) + \omega$. ⊣

4.3. The Interpolation Theorem

The next application of the Model Existence Theorem is the $\mathcal{L}_{\omega_1,\omega}$ version of Craig's Interpolation Theorem from first order logic. This result was first proved by Lopez-Escobar by different means.

THEOREM 4.3.1. *Suppose* ϕ_1 *and* ϕ_2 *are* $\mathcal{L}_{\omega_1,\omega}$*-sentences with* $\phi_1 \models \phi_2$. *There is an* $\mathcal{L}_{\omega_1,\omega}$*-sentence* θ *such that* $\phi_1 \models \theta$, $\theta \models \phi_2$ *and every relation, function and constant symbol occurring in* θ *occurs in both* ϕ_1 *and* ϕ_2.

PROOF. Let C be a countably infinite collection of new constant symbols. Let τ_i be the smallest vocabulary containing C and all symbols in ϕ_i and let $\tau = \tau_1 \cap \tau_2$.

Let Σ be the collection of finite $\sigma = \sigma_1 \cup \sigma_2$ where σ_i is a set of τ_i-sentences and if ψ_1, ψ_2 are τ-sentences such that $\sigma_1 \models \psi_1$ and $\sigma_2 \models \psi_2$, then $\psi_1 \wedge \psi_2$ is satisfiable.

CLAIM 4.3.2. Σ *is a consistency property.*

We will verify properties (C3) and (C4) and leave the remainder as an exercise.

(C3) Suppose $\bigwedge_{\psi\in X} \psi \in \sigma \in \Sigma$, where $\sigma = \sigma_1 \cup \sigma_2$ as above. Suppose $\bigwedge_{\psi\in X} \psi \in \sigma_1$. Let $\sigma_1' = \sigma_1 \cup \{\psi\}$. We claim that $\sigma_1' \cup \sigma_2 \in \Sigma$. Suppose

$\sigma_1' \models \theta_1$ and $\sigma_2 \models \theta_2$. Then $\sigma_1 \models \theta_1$. By assumption, $\theta_1 \wedge \theta_2$ is satisfiable. Thus $\sigma_1' \cup \sigma_2 \in \Sigma$.

This is similar if $\bigwedge_{\psi \in X} \psi \in \sigma_2$.

(C4) Suppose $\bigvee_{\psi \in X} \psi \in \sigma_1$. Let $\sigma_{1,\psi} = \sigma_1 \cup \{\psi\}$. We claim that some $\sigma_{1,\psi} \cup \sigma_2 \in \Sigma$. Suppose not. Then for each ψ there are $\mathcal{L}_{\omega_1,\omega}(\tau)$-sentences $\theta_{1,\psi}, \theta_{2,\psi}$ such that $\sigma_{1,\psi} \models \theta_{1,\psi}$, $\sigma_2 \models \theta_{2,\psi}$ and $\theta_{1,\psi} \wedge \theta_{2,\psi}$ is unsatisfiable. Then $\theta_{1,\psi} \models \neg\theta_{2,\psi}$. Since

$$\sigma_1 \models \bigvee_{\psi \in X} \psi,$$

$$\sigma_1 \models \bigvee_{\psi \in X} \theta_{1,\psi}.$$

But

$$\sigma_2 \models \bigwedge_{\psi \in X} \theta_{2,\psi}$$

and

$$\bigvee_{\psi \in X} \theta_{1,\psi} \models \neg \bigwedge_{\psi \in X} \theta_{2,\psi},$$

contradicting that $\sigma \in \Sigma$.

EXERCISE 4.3.3. Finish the proof that Σ is a consistency property.

We now finish the proof of the Interpolation Theorem. Since $\phi_1 \models \phi_2$, by the Model Existence Theorem, $\{\phi_1, \neg\phi_2\} \notin \Sigma$. Thus there are $\mathcal{L}_{\omega_1,\omega}(\tau)$-sentences θ_1 and θ_2 such that $\phi_1 \models \theta_1$, $\neg\phi_2 \models \theta_2$ and $\theta_1 \wedge \theta_2$ is unsatisfiable.

Thus

$$\phi_1 \models \theta_1, \theta_1 \models \neg\theta_2, \text{ and } \neg\theta_2 \models \phi_2.$$

It follows that

$$\phi_1 \models \theta_1 \text{ and } \theta_1 \models \phi_2.$$

We would be done except θ_1 may contain constants from C. Let $\theta_1 = \psi(\overline{c})$, where $\psi(\overline{v})$ is a τ-formula with no constants from C. Then

$$\phi_1 \models \forall \overline{v}\, \psi(\overline{v}) \text{ and } \exists \overline{v}\, \psi(\overline{v}) \models \phi_2.$$

Since

$$\forall \overline{v}\, \phi(\overline{v}) \models \exists \overline{v}\, \phi(\overline{v})$$

we can take

$$\forall \overline{v}\, \psi(\overline{v})$$

as the interpolant. ⊣

The Interpolation Theorem gives rise to a separation theorem for $\mathrm{PC}_{\omega_1,\omega}$-classes, analogous to the Separation Theorem for Σ^1_1-sets in Descriptive Set Theory.

COROLLARY 4.3.4. *Suppose $\mathcal{K}_0$ and $\mathcal{K}_1$ are disjoint $\mathrm{PC}_{\omega_1,\omega}$-classes of τ-structures. There is $\phi \in \mathcal{L}_{\omega_1,\omega}$ such that $\mathcal{M} \models \phi$ for $\mathcal{M} \in K_0$ and $\mathcal{M} \models \neg\phi$ for $\mathcal{M} \in K_1$.*

PROOF. Let $\tau_0, \tau_1 \supseteq \tau$ and let $\phi_i \in \mathcal{L}_{\omega_1,\omega}(\tau_i)$ such that $\mathcal{K}_i$ is the class of τ-reducts of models of ϕ_i. We may assume that $\tau_1 \cap \tau_2 = \tau$. Since $\mathcal{K}_0$ and $\mathcal{K}_1$ are disjoint we have that $\phi_0 \models \neg\phi_1$, i.e., if there is any expansion of $\mathcal{M}$ that makes ϕ_0 true, then there is no expansion of $\mathcal{M}$ making ϕ_1 true. By the Interpolation Theorem there is $\phi \in \mathcal{L}_{\omega_1,\omega}(\tau)$ such that $\phi_0 \models \phi$ and $\phi \models \neg\phi_1$. Then every structure in $\mathcal{K}_0$ is a model of ϕ and no structure in $\mathcal{K}_1$ is a model of ϕ. ⊣

Indeed we can use these results to prove an invariant version of the Separation Theorem for Σ^1_1-sets. In Exercise 3.1.5 we noted that the collection of countable models in a $\mathrm{PC}_{\omega_1,\omega}$-class is an invariant Σ^1_1-sets. The converse is also true.

LEMMA 4.3.5. *Every invariant Σ^1_1 subset A of $\mathbb{X}_\tau$ is the set of countable models in a $\mathrm{PC}_{\omega_1,\omega}$-class.*

PROOF. For notational simplicity we assume $\tau = \{R\}$, where R is a binary relation symbol. Fix $p : \omega^2 \to \omega$ a bijective pairing function. For $R \subseteq \omega \times \omega$, let $f_R : \omega \to 2$ be defined by $f_R(p(i, j)) = 1$ if and only if $R(i, j)$.

Since A is analytic, there is a tree

$$T \subseteq \{(\eta, \nu) : \eta \in 2^{<\omega}, \nu \in \omega^{<\omega}, |\eta| = |\nu|\}$$

such that $(\omega, R) \in A$ if and only if there is g such that (f_R, g) is a path through T.

Let $\tau^* = \{s, c, f, g, S_n : n = 1, 2, \dots\}$ where s, f, g are unary functions, c is a constant symbol and S_n is a $2n$-ary relation symbol. Let Θ be a sentence asserting:

- s is one-to-one, every element except c has a preimage, there are no cycles and $\forall x \bigvee_{i=0}^{\infty} s^{(i)}(c) = x$;
- $\forall x\, f(x) = c \vee f(x) = s(c)$;
- $\bigwedge_{i,j}[R(s^{(i)}(c), s^{(j)}(c)) \leftrightarrow f(s^{(p(i,j))}(c)) = s(c)]$;
- $\bigwedge_{n=1}^{\infty} \big[\bigwedge_{(\eta,\nu)\in T} S_n(s^{\eta(0)}(c), \dots, s^{\eta(n-1)}(c), s^{\nu(0)}(c), \dots, s^{\nu(n-1)}(c)) \wedge \bigwedge_{\eta\in 2^n, \nu\in\omega^n, (\eta,\nu)\notin T} \neg S_n(s^{\eta(0)}(c), \dots, s^{\eta(n-1)}(c), s^{\nu(0)}(c), \dots, s^{\nu(n-1)}(c))\big]$;
- $\bigwedge_{n=1}^{\infty} S_n(f(c), f(s(c)), \dots, f(s^{n-1}(c)), g(c), g(s(c)), \dots, g(s^{n-1}(c)))$.

Suppose $(\omega, R) \in A$. Let f_R be as above. Interpret c as 0 and s as $x \mapsto x + 1$. Choose g such that (f_R, g) is a path through T. Interpret S_n by

$S_n(i_0, \ldots, i_{n-1}, j_0, \ldots, j_{n-1})$ if and only if $(\bar{i}, \bar{j}) \in T$. The resulting structure is a model of Θ.

Suppose $\mathcal{M}^* = (\omega, R, s, f, g, S_1, \ldots) \models \Theta$. Define $\eta_0 \subset \eta_1 \subset \cdots$ and $\nu_0 \subset \nu_1 \subset \cdots$ such that $f(s^i(c)) = s^{\eta(i)}(c)$ and $g(s^i(c)) = s^{\nu(i)}(c)$. Let $\widehat{f} = \bigcup \eta_i$ and $\widehat{g} = \bigcup \nu_i$. Then $(\widehat{f}, \widehat{g})$ is a path through T. Define $\widehat{R} \subseteq \omega \times \omega$ such that $f_{\widehat{R}} = \widehat{f}$ and let $\mathcal{N} = (\omega, R^*)$. Then $\mathcal{N} \in A$. But s gives an isomorphism between $\mathcal{N}$ and $\mathcal{M} = (\omega, R)$. Thus, since A is invariant, $\mathcal{M} \in A$. ⊣

COROLLARY 4.3.6. *If A and B are disjoint invariant Σ^1_1-sets there is an invariant Borel C with $A \subseteq C$ and $B \cap C = \emptyset$.*

We next prove the converse to Corollary 3.1.3

THEOREM 4.3.7. *Every invariant Borel subset of $\mathbb{X}_\tau$ is of the form* Mod(ϕ) *for some $\mathcal{L}_{\omega_1,\omega}(\tau)$-sentence ϕ.*

PROOF. If A is invariant Borel, then A and $\mathbb{X}_\tau \setminus A$ are disjoint invariant Σ^1_1 sets. We can find $\mathrm{PC}_{\omega_1,\omega}$-classes $\mathcal{K}_0$ and $\mathcal{K}_1$ such that $A = \mathbb{X}_\tau \cap \mathcal{K}_0$ and $\mathbb{X}_\tau \setminus A = \mathbb{X}_\tau \cap \mathcal{K}_1$. Since there are no countable structures in $\mathcal{K}_0 \cap \mathcal{K}_1$, by Löwenheim-Skolem, $\mathcal{K}_0$ and $\mathcal{K}_1$ are disjoint classes. Thus by Corollary 4.3.4 there is an $\mathcal{L}_{\omega_1,\omega}(\tau)$-sentence ϕ such that $A \subset \mathrm{Mod}(\phi)$ and $\mathrm{Mod}(\phi) \cap (\mathbb{X}_\tau \setminus A) = \emptyset$. Clearly $A = \mathrm{Mod}(\phi)$.[8] ⊣

4.4. The undefinability of well-ordering

Let $\tau = \{<, \ldots\}$.

THEOREM 4.4.1. *Suppose ϕ is an $\mathcal{L}_{\omega_1,\omega}$-sentence and for all $\alpha < \omega_1$ there is $\mathcal{M} \models \phi$ where $(\alpha, <)$ embeds into $<^{\mathcal{M}}$. Then there is $\mathcal{N} \models \phi$ where $(\mathbb{Q}, <)$ embeds into $<^{\mathcal{N}}$.*

COROLLARY 4.4.2. *If ϕ is an $\mathcal{L}_{\omega_1,\omega}$-sentence and $<^{\mathcal{M}}$ is well-ordered for all $\mathcal{M} \models \phi$, then there is $\alpha < \omega_1$ such that $<^{\mathcal{M}}$ has order type at most α for all $\mathcal{M} \models \phi$.*

PROOF. Let τ_0 be our original vocabulary. Form τ by adding a countable set of new constants C and distinct constants $D = \{d_q : q \in \mathbb{Q}\}$. Let Σ be all σ of the form

$$\sigma_0 \cup \{\phi\} \cup \{d_q < d_r : q < r\}$$

where σ_0 is a finite set of $\mathcal{L}_{\omega_1,\omega}$-sentences using only finitely many constants from $C \cup D$ such that if we let

$$\theta(\bar{c}, d_{i_1}, \ldots, d_{i_m}) = \bigwedge_{\psi \in \sigma_0} \psi$$

[8] Alternatively, using Exercise 1.0.14, we could just assume all models in $\mathcal{K}_0$ and $\mathcal{K}_1$ are countable to begin with.

where $i_1 < \cdots < i_m$ then for all $\alpha < \omega_1$ there is an $\mathcal{M}$ such that the following property $(*)$ holds:

$$\mathcal{M} \models \phi \wedge \exists \overline{x}\, \theta(\overline{x}, b_1 \ldots, b_m)$$

and $A \subset \mathcal{M}$ well ordered by $<^{\mathcal{M}}$, $\overline{b} \in A$ and

$$\alpha \leq b_1, b_1 + \alpha \leq b_2, \ldots, b_{m-1} + \alpha \leq b_m.$$

In particular, taking $\sigma_0 = \emptyset$, we see that, by assumption, $\{\phi\} \cup \{d_q < d_r : q < r\} \in \Sigma$. We claim that Σ is a consistency property. Once we have shown this we will know there is a model of ϕ containing a densely ordered set.

We do several of the non-routine claims and leave the rest of the verification that Σ is a consistency property as an exercise.

(C4) Suppose $\bigvee_{\psi \in X} \psi \in \sigma$ where $\sigma \in \Sigma$. Then for each α there is $\psi_\alpha \in X$ and $\mathcal{M}$ such that $(*)$ holds and $\mathcal{M} \models \psi_\alpha$. There is $\psi \in X$ such that $\psi = \psi_\alpha$ for uncountably many α. Note that if ψ works for α it works for all $\beta < \alpha$. Thus $\sigma \cup \{\psi\} \in \Sigma$.

(C7c) Let $\sigma \in \Sigma$. Suppose $t = d_r$ and σ_0 uses only $d_{i_1}, \ldots, d_{i_m}$ where $i_0 < \cdots < i_m$. Suppose $i_s < r < i_{s+1}$. Let c be element of C not yet used. We claim that $\sigma \cup \{c = d_r\} \in \Sigma$.

Let $\alpha < \omega_1$. Pick $\beta > \alpha + \alpha$. By $(*)$ there is

$$\mathcal{M} \models \phi \wedge \exists \overline{x}\, \theta(\overline{x}, \overline{b})$$

where

$$\beta \leq b_1, b_1 + \beta \leq b_2, \ldots, b_{m-1} + \beta \leq b_m.$$

Let $b = b_s + \alpha$. Then $b_s + \alpha \leq b$ and $b + \alpha \leq b_{s+1}$ as desired. ⊣

EXERCISE 4.4.3. Complete the proof that Σ is a consistency property.

We will give a sharper version of this result in Theorem 11.1.3 and a useful extension to uncountable models in Corollary 6.1.9.

Chapter 5

HANF NUMBERS AND INDISCERNIBLES

In Exercise 1.1.13 we showed that the Upward Löwenheim–Skolem Theorem fails in $\mathcal{L}_{\omega_1,\omega}$, by giving a sentence with models of size $2^{\aleph_0}$ but no larger models. In this chapter we will show that there is a cardinal κ such that if ϕ has models of cardinality κ, then ϕ has arbitrarily large models. We call the least such cardinal the *Hanf number* of $\mathcal{L}_{\omega_1,\omega}$. It is general nonsense that there is an Hanf number.

EXERCISE 5.0.4. Let I be a set. For each $i \in I$, let $\mathcal{K}_i$ be a class of structures. Let $\mathcal{K} = \{\mathcal{K}_i : i \in I\}$. Prove there is a cardinal κ such that for all i, if $\mathcal{K}_i$ has a structure of size κ, then $\mathcal{K}_i$ contains arbitrarily large structures. The least such κ is the *Hanf number* for $\mathcal{K}$.

Recall that for κ an infinite cardinal and α an ordinal, we inductively define $\beth_\alpha(\kappa)$ by $\beth_0(\kappa) = \kappa$ and

$$\beth_\alpha(\kappa) = \sup_{\beta<\alpha} 2^{\beth_\beta(\kappa)}.$$

In particular $\beth_1(\kappa) = 2^\kappa$. We let $\beth_\alpha = \beth_\alpha(\aleph_0)$. Under the Generalized Continuum Hypothesis $\beth_\alpha = \aleph_\alpha$.

The main theorem, due to Morley, is that the Hanf number for $\mathcal{L}_{\omega_1,\omega}$ is $\beth_{\omega_1}$.

THEOREM 5.0.5. *Let $\phi \in \mathcal{L}_{\omega_1,\omega}$. If for all $\alpha < \omega_1$ there is $\mathcal{M} \models \phi$ with $|M| \geq \beth_\alpha$, then ϕ has models of all infinite cardinalities.*

The next exercise generalizes Exercise 1.1.13 and shows that $\beth_{\omega_1}$ is optimal.

EXERCISE 5.0.6. Let $\alpha < \omega_1$. Let $\tau = \{U_\beta : \beta \leq \alpha + 1\} \cup \{E\} \cup \{c_0, c_1, \dots\}$ where U_β is a unary relation symbol, E is a binary relation symbol and $c_0, c_1, \dots$ are constants. Let ϕ assert:

(i) $U_0 = \{c_0, c_1, \dots\}$ and $\forall x\, U_{\alpha+1}(x)$;

(ii) $U_\gamma \subseteq U_\beta$ for $\gamma < \beta$;

(iii) $\forall x\, (U_\beta(x) \leftrightarrow \bigvee_{\alpha<\beta} U_\alpha(x))$ for β a limit ordinal;

(iv) $\forall x \forall y\, [(U_{\beta+1}(x) \wedge E(y, x)) \rightarrow U_\beta(y)]$;

(v) if $\{x : E(x, y)\} = \{x : E(x, z)\}$, then $y = z$.

(a) Show that there is $\mathcal{M} \models \phi$ with $|M| = \beth_{\alpha+1}$.
(b) Show that every model of ϕ has cardinality at most $\beth_{\alpha+1}$.

Here is the main idea of the proof of Theorem 5.0.5.

- By expanding the signature we may assume that we have $\phi \in T$ where T is a satisfiable theory in a countable fragment with built-in Skolem functions.
- Under the assumptions of Theorem 5.0.5 we can find a model of T with an infinite set of indiscernibles.
- Taking Skolem hulls we get models of all infinite cardinalities.

The first and third steps are exactly as in first order model theory. In first order model theory the second step is accomplished using Ramsey's Theorem and a compactness argument. We will need a stronger partition theorem and an application of the Model Existence Theorem.

5.1. The Erdös–Rado Partition Theorem

The material in this section is verbatim from chapter 5 of [56].

For X a set and κ, λ (possibly finite) cardinals, we let $[X]^\kappa$ be the collection of all subsets of X of size κ. We call $f : [X]^\kappa \to \lambda$ a *partition* of $[X]^\kappa$. We say that $Y \subseteq X$ is *homogeneous* for the partition f if there is $\alpha < \lambda$ such that $f(A) = \alpha$ for all $A \in [Y]^\kappa$ (i.e., f is constant on $[Y]^\kappa$). Finally, for cardinals κ, η, μ, and λ, we write $\kappa \to (\eta)^\mu_\lambda$ if whenever $|X| \geq \kappa$ and $f : [X]^\mu \to \lambda$, then there is $Y \subseteq X$ such that $|Y| \geq \eta$ and Y is homogeneous for f.

Stated in this notation Ramsey's Theorem becomes $\aleph_0 \to (\aleph_0)^n_m$ for all $n, m \in \omega$.

When we begin partitioning sets into infinitely many pieces it becomes harder to find homogeneous sets.

PROPOSITION 5.1.1. $2^{\aleph_0} \not\to (3)^2_{\aleph_0}$.

PROOF. We define $F : [2^\omega]^2 \to \omega$ by $F(\{f, g\})$ is the least n such that $f(n) \neq g(n)$. Clearly, we cannot find $\{f, g, h\}$ such that $f(n) \neq g(n)$, $g(n) \neq h(n)$, and $f(n) \neq h(n)$. ⊣

On the other hand, if $\kappa > 2^{\aleph_0}$, then $\kappa \to (\aleph_1)^2_{\aleph_0}$. This is a special case of an important generalization of Ramsey's Theorem.

THEOREM 5.1.2 (Erdös–Rado Theorem). $\beth_n(\kappa)^+ \to (\kappa^+)^{n+1}_\kappa$.

PROOF. We prove this by induction on n. For $n = 0$, $\kappa^+ \to (\kappa^+)^1_\kappa$ is just the Pigeonhole Principle.

Suppose that we have proved the theorem for $n - 1$. Let $\lambda = \beth_n(\kappa)^+$, and let $f : [\lambda]^{n+1} \to \kappa$. For $\alpha < \lambda$, let $f_\alpha : [\lambda \setminus \{\alpha\}]^n \to \kappa$ by $f_\alpha(A) = f(A \cup \{\alpha\})$.

We build $X_0 \subseteq X_1 \subseteq \cdots \subseteq X_\alpha \subseteq \cdots$ for $\alpha < \beth_{n-1}(\kappa)^+$ such that $X_\alpha \subseteq \beth_n(\kappa)^+$ and each X_α has cardinality $\beth_n(\kappa)$. Let $X_0 = \beth_n(\kappa)$. If α is a limit ordinal, then $X_\alpha = \bigcup_{\beta<\alpha} X_\beta$.

Suppose we have X_α with $|X_\alpha| = \beth_n(\kappa)$. Because

$$\beth_n(\kappa)^{\beth_{n-1}(\kappa)} = (2^{\beth_{n-1}(\kappa)})^{\beth_{n-1}(\kappa)} = 2^{\beth_{n-1}(\kappa)} = \beth_n(\kappa),$$

there are $\beth_n(\kappa)$ subsets of X_α of cardinality at most $\beth_{n-1}(\kappa)$. Also note that if $Y \subset X_\alpha$ and $|Y| \leq \beth_{n-1}(\kappa)$, then there are at most $\beth_n(\kappa)$ functions $g : [Y]^n \to \kappa$ because

$$\kappa^{\beth_{n-1}(\kappa)} = 2^{\beth_{n-1}(\kappa)} = \beth_n(\kappa).$$

Thus, we can find $X_{\alpha+1} \supseteq X_\alpha$ such that $|X_{\alpha+1}| = \beth_n(\kappa)$, and if $Y \subset X_\alpha$ with $|Y| = \beth_{n-1}(\kappa)$ and $\beta \in \lambda \setminus Y$, then there is $\gamma \in X_{\alpha+1} \setminus Y$ such that $f_\beta | [Y]^n = f_\gamma | [Y]^n$.

Let $X = \bigcup_{\alpha<\beth_{n-1}(\kappa)^+} X_\alpha$. If $Y \subset X$ with $|Y| \leq \beth_{n-1}(\kappa)$, then $Y \subset X_\alpha$ for some $\alpha < \beth_{n-1}(\kappa)^+$. If $\beta \in \lambda \setminus Y$, then there is $\gamma \in X \setminus Y$ such that $f_\beta | [Y]^n = f_\gamma | [Y]^n$.

Fix $\delta \in \lambda \setminus X$. Inductively construct $Y = \{y_\alpha : \alpha < \beth_{n-1}(\kappa)^+\} \subseteq X$. Let $y_0 \in X$. Suppose that we have constructed $Y_\alpha = \{y_\beta : \beta < \alpha\}$. Choose $y_\alpha \in X$ such that $f_{y_\alpha} | [Y_\alpha]^n = f_\delta | [Y_\alpha]^n$.

By the induction hypothesis, there is $Z \subseteq Y$ such that $|Z| \geq \kappa^+$ and Z is homogeneous for f_δ. Say $f_\delta(B) = \gamma$ for all $B \in [Z]^n$. We claim that Z is homogeneous for f. Let $A \in [Z]^{n+1}$. There are $\alpha_1 < \cdots < \alpha_{n+1}$ such that $A = \{y_{\alpha_1}, \ldots, y_{\alpha_{n+1}}\}$. Then

$$f(A) = f_{y_{\alpha_{n+1}}}(\{y_{\alpha_1}, \ldots, y_{\alpha_n}\}) = f_\delta(\{y_{\alpha_1}, \ldots, y_{\alpha_n}\}) = \gamma.$$

Thus, Z is homogeneous for f. ⊣

We will use the following corollary.

COROLLARY 5.1.3. $\beth_{\alpha+n}^+ \to (\beth_\alpha^+)^{n+1}_{\beth_\alpha}$.

PROOF. This follows from Erdös–Rado as $\beth_{\alpha+n} = \beth_n(\beth_\alpha)$. ⊣

5.2. The Hanf number of $\mathcal{L}_{\omega_1,\omega}$

We now prove Theorem 5.0.5. We assume that the reader is used to the construction and applications of indiscernibles in first order logic (see for example chapter 5 of [56]).

EXERCISE 5.2.1. Suppose $\mathbb{A}$ is a countable fragment of $\mathcal{L}_{\omega_1,\omega}$ and T is an $\mathbb{A}$-theory with built-in Skolem functions. Suppose $\mathcal{M} \models T$ contains an infinite set of indiscernibles for the fragment $\mathbb{A}$. Then T has arbitrarily large models.

We begin by expanding the signature we may assume that:

- there is a countable fragment $\mathbb{A}$ of $\mathcal{L}_{\omega_1,\omega}$ and $T \subset \mathbb{A}$ a theory with built in Skolem functions such that $\phi \in T$ and T has models of cardinality $\beth_\alpha$ for all $\alpha < \omega_1$;
- we have two disjoint countably infinite sets of constant symbols $C = \{c_0, c_1, \dots\}$ and $D = \{d_0, d_1, \dots\}$, but formulas in $\mathbb{A}$ use only finitely many constant symbols from $C \cup D$.

Let $\Gamma = \{d_i \neq d_j : i < j\} \cup \{\theta(d_{i_1}, \dots, d_{i_n}) \leftrightarrow \theta(d_{j_1}, \dots, d_{j_n}) : \theta(v_1, \dots, v_n) \in \mathbb{A}, i_1 < \dots < i_n, j_1 < \dots < j_n$ and no constants from $C \cup D$ occur in $\theta\}$.

If we can find $\mathcal{M} \models T \cup \Gamma$, then the interpretation of D in $\mathcal{M}$ gives an infinite set of indiscernibles. We can then stretch the indiscernibles to build arbitrarily large models.

If σ is a finite set of $\mathbb{A}$-sentences, we let Θ_σ be the $\mathbb{A}$-formula with no constants from $C \cup D$ such that $\Theta_\sigma(\overline{c}, \overline{d})$ is the conjunction of σ.

Let Σ be the set of all finite sets σ of $\mathbb{A}$-sentences such that for all $\alpha < \omega_1$ there is $\mathcal{M} \models T$, $A \subseteq \mathcal{M}$ with $|A| = \beth_\alpha$, $<$ a linear order of A and

$$\mathcal{M} \models \exists \overline{v}\, \Theta_\sigma(\overline{v}, a_1, \dots, a_n)$$

for all $a_1 < \dots < a_n$ in A.

We need to show that Σ is a consistency property and if $\sigma \in \Sigma$ and $\psi \in T \cup \Gamma$, then $\sigma \cup \{\psi\} \in \Sigma$. Once we have established these claims we can use the Extended Model Existence Theorem 4.1.8 to conclude that there is $\mathcal{M} \models T \cup \Gamma$.

CLAIM. *Σ is satisfiable.*

The only tricky case is (C4). Suppose $\bigvee_{\psi \in X} \psi \in \sigma \in \Sigma$. Let $\Theta(\overline{c}, d_1, \dots, d_n)$ be conjunction of σ. For all $\alpha < \omega_1$ we can find $\mathcal{M}_\alpha \models T$, $A_\alpha \subset \mathcal{M}_\alpha$ of cardinality at least $\beth_{\alpha+n}$ and $<$ a linear order of A such that

$$\mathcal{M}_\alpha \models \exists \overline{v}\, \Theta(\overline{v}, a_1, \dots, a_n)$$

for all $a_1 < \dots < a_n \in A$. Let $f_\alpha : [A_\alpha]^n \to X$ such that if $f_\alpha(\overline{a}) = \psi(\overline{c}, d_1, \dots, d_n)$, then

$$\mathcal{M}_\alpha \models \exists \overline{v}[\Theta(\overline{v}, \overline{a}) \wedge \psi(\overline{v}, \overline{a})].$$

Since $\beth_{\alpha+n} \geq \beth^+_{\alpha+n-1}$, by Erdös–Rado we can find $A'_\alpha \subseteq A_\alpha$ of cardinality at least $\beth_\alpha$ and $\psi \in X$ such that

$$\mathcal{M}_\alpha \models \exists \overline{v}\, [\Theta(\overline{v}, a_1, \dots, a_n) \wedge \psi(\overline{v}, a_1, \dots, a_n)]$$

for all $a_1 < \dots < a_n$ in A'_α. By the Pigeonhole Principle, we can find a ψ that works for cofinally many α. Thus $\sigma \cup \{\psi(\overline{c}, d_1, \dots, d_n)\} \in \Sigma$.

CLAIM. *If $\sigma \in \Sigma$ and $\psi \in T \cup \Gamma$, then $\sigma \cup \{\psi\} \in \Sigma$.*

Suppose $\sigma \in \Sigma$. It is clear that if $\chi \in T$, then $\sigma \cup \{\chi\} \in \Sigma$. Suppose χ is

$$\psi(d_{i_1}, \dots, d_{i_m}) \leftrightarrow \psi(d_{j_1}, \dots, d_{j_m})$$

where $i_1 < \cdots < i_m$ and $j_1 < \cdots < j_m$ and no constants from $C \cup D$ occur in ψ.

Let $\Theta(\overline{c}, d_1, \ldots, d_n)$ be the conjunction of σ. We may assume all $i_k, j_k \leq n$. For all $\alpha < \omega_1$ there is $\mathcal{M}_\alpha \models T$, $A_\alpha \subset M_\alpha$ of cardinality $\beth_{\alpha+m}$ such that

$$\mathcal{M}_\alpha \models \exists \overline{v}\, \Theta(\overline{v}, a_1, \ldots, a_m)$$

for $a_1 < \cdots < a_m$ in A_α.

Let $f_\alpha : [A_\alpha]^m \to \{0, 1\}$ with $f_\alpha(\overline{a}) = 1$ if $\mathcal{M}_\alpha \models \psi(\overline{a})$. By Erdös–Rado we can find $A'_\alpha \subset A_\alpha$ of cardinality $\beth_\alpha$ homogeneous for f_α. Then

$$\mathcal{M}_\alpha \models \exists v\, [\Theta(\overline{v}, a_1, \ldots, a_n) \wedge (\psi(a_{i_1}, \ldots, a_{i_m}) \leftrightarrow \psi(a_{j_1}, \ldots, a_{j_n})]$$

for increasing sequences from A'_α. Thus $\sigma \cup \{\chi\} \in \Sigma$.

This completes the Proof of Theorem 5.0.5. ⊣

The following Corollary is proven directly, by essentially the same argument, in Theorem 5.2.14 of [56].

COROLLARY 5.2.2. *Suppose T is a first order theory in a countable signature and $p_1, p_2, \ldots$ are partial types. If for all $\alpha < \omega_1$ there are models of T of cardinality $\beth_\alpha$ omitting all of the p_i, then there are arbitrarily large models of T omitting all of the p_i.*

EXERCISE 5.2.3. Show that Theorem 5.0.5 follows immediately from Corollary 5.2.2 and the characterization in Theorem 1.2.1 of $\mathcal{L}_{\omega_1,\omega}$-elementary classes as reducts of classes of models of a first order theory omitting a set of types.

EXERCISE 5.2.4. Suppose $\phi \in \mathcal{L}_{\omega_1,\omega}$ is κ-categorical for some $\kappa \geq \beth_{\omega_1}$. Prove that if $\mathcal{M}, \mathcal{N} \models \phi$ have cardinality at least $\beth_{\omega_1}$ then $\mathcal{M} \equiv_{\omega_1,\omega} \mathcal{N}$. (Compare this to Exercise 2.2.15.)

By carefully choosing the order type of the indiscernibles, we can build large models realizing few types.

THEOREM 5.2.5. *If ϕ has arbitrarily large models, then for all κ there is a model of ϕ of cardinality κ realizing only countably many $\mathcal{L}_{\omega_1,\omega}$-types*

PROOF. We follow the proof of Theorem 5.0.5. We can find a countable vocabulary τ, a countable fragment $\mathbb{A}$ and $T \subseteq \mathbb{A}$ a theory with built in $\mathbb{A}$-Skolem functions such that $\phi \in T$ and T has a countable model with an infinite set of order indiscernibles.

Suppose $(A, <)$ is a linear order of cardinality κ that is n-transitive for all $n \in \omega$, i.e., for all $a_1 < \cdots < a_n$ and $b_1 < \cdots < b_n$ there is an automorphism σ of $(A, <)$ such that $\sigma(\overline{a}) = \overline{b}$. We see below that such orders exist. Stretch the indiscernibles of $\mathcal{M}$ to build a model $\mathcal{N}$ which is the hull of indiscernibles of order type $(A, <)$. Then any order automorphism of $(A, <)$ extends to an automorphism of $\mathcal{N}$. In particular for any Skolem term $f(v_1, \ldots, v_n)$, $a_1 < \cdots < a_n$ and $b_1 < \cdots < b_n$, $f(\overline{a})$ and $f(\overline{b})$ have the same $\mathcal{L}_{\omega_1,\omega}$-type. Thus $\mathcal{N}$ realizes only countably many $\mathcal{L}_{\omega_1,\omega}$-types. ⊣

EXERCISE 5.2.6. Show that for all κ there is a linear order of cardinality κ that is n-transitive for all n. [Hint: Consider the order type of an ordered field of cardinality κ.]

COROLLARY 5.2.7. *If ϕ has arbitrarily large models and is κ-categorical, there is a complete sentence ψ such that $\psi \models \phi$ and ψ has a model of cardinality κ.*

PROOF. We saw in Exercise 2.2.20 that a model realizing countable many $\mathcal{L}_{\omega_1,\omega}$-types is $\mathcal{L}_{\infty,\omega}$-equivalent to a countable model. ⊣

We will return to the question of building uncountable models realizing few types in Section 6.3.

We can refine the Hanf number question by looking at complete sentences.

EXERCISE 5.2.8. Let $\tau = \{+, 0, G_1, G_2, \dots\}$ and let ϕ be the $\mathcal{L}_{\omega_1,\omega}$ sentence asserting:

(i) we have an Abelian group where every element has order 2;
(ii) G_1 is an index 2 subgroup and G_{n+1} is an index 2 subgroup of G_n for each n;
(iii) $\bigcap G_n = \{0\}$.

Prove that ϕ is complete and every model of ϕ has cardinality at most $2^{\aleph_0}$.

Baumgartner [10], building on work of Malitz [52], showed that Hanf number for complete $\mathcal{L}_{\omega_1,\omega}$-sentences is still $\beth_{\omega_1}$. Knight [40] showed there is a complete sentence with models of size $\aleph_1$ but no larger. Hjorth [28] extended this by showing that for all $\alpha < \omega_1$ there is a complete sentence with models of size $\aleph_\alpha$ but no larger. We will give Hjorth's proof for $\aleph_1$ and an application to Vaught's Conjecture in Section 5.4. Without assumptions on cardinal exponentiation Hjorth's result is as far as we can go as Shelah [72] has shown that it consistent with ZFC that if ϕ has a model of size $\aleph_{\omega_1}$, then ϕ has a model of cardinality $2^{\aleph_0}$. See [76] and [77] for further results in this direction.

5.3. Morley's Two Cardinal Theorem

We give one more application of using Erdös–Rado to build useful indiscernibles.

DEFINITION 5.3.1. Let $\tau = \{U, \dots\}$ where U is a unary predicate. We say that a τ-structure $\mathcal{M}$ is a (κ, λ)-model if $|\mathcal{M}| = \kappa$ and $|U(\mathcal{M})| = \lambda$.

THEOREM 5.3.2 (Morley's Two Cardinal Theorem). *Let ϕ be an $\mathcal{L}_{\omega_1,\omega}$-sentence. Suppose for arbitrarily large $\alpha < \omega_1$ there is an infinite κ and $\mathcal{M}_\alpha$ a $(\beth_\alpha(\kappa), \kappa)$-model of ϕ. Then for all κ there is a $(\kappa, \aleph_0)$-model of ϕ.*

PROOF. We can extend the signature to find a countable fragment $\mathbb{A}$, an $\mathbb{A}$-theory T with built-in Skolem functions with $\phi \in T$ such that for arbitrarily large $\alpha < \omega_1$ there is a $(\beth_\alpha(\kappa), \kappa)$-model of T for some infinite κ.

Suppose $\mathcal{M} \models T$. Suppose $I \subseteq M$ and $<$ linearly orders I. We say that I is *indiscernible over* U if

$$\mathcal{M} \models \psi(x_1, \ldots, x_n, \overline{a}) \leftrightarrow \psi(y_1, \ldots, y_n, \overline{a})$$

for all $\psi(v_1, \ldots, v_n, \overline{u}) \in \mathbb{A}$, $\overline{a} \in U(\mathcal{M})$, $x_1 < \cdots < x_n, y_1 < \cdots < y_n$ in I.

EXERCISE 5.3.3. Suppose there is $\mathcal{M} \models T$ countable with $I \subseteq \mathcal{M}$ an infinite set of indiscernibles over U. Show that T has $(\kappa, \aleph_0)$ models for all infinite κ. [Hint: Prove that if $x_1 < \cdots < x_n \in I$, $\overline{a} \in U(\mathcal{M})$ and $f(\overline{x}, \overline{a}) \in U(\mathcal{M})$, then $f(\overline{y}, \overline{a}) = f(\overline{x}, \overline{a})$ for all $y_1 < \cdots < y_n$ in I.]

Add two new countable infinite sets of constant symbols C and D. Let $\Gamma = \{d_i \neq d_j : i < j\} \cup \{\forall \overline{u}[U(\overline{u}) \rightarrow (\psi(d_{i_1}, \ldots, d_{i_m}, \overline{u}) \leftrightarrow \psi(d_{j_1}, \ldots, d_{j_m}, \overline{u}) :$ for $\phi \in \mathbb{A}$ (with no contants from $C \cup D$), $i_1 < \cdots < i_m$, $j_1 < \cdots < j_m\}$.

We define a consistency property Σ. If σ is a finite set of $\mathbb{A}$-sentences with finitely many constants from $C \cup D$, then $\sigma \in \Sigma$ if and only if for arbitrarily large $\alpha < \omega_1$ there is $\mathcal{M} \models T$ and $A \subseteq M$ with $|A| \geq \beth_\alpha(U(\mathcal{M}))$ and $<$ an ordering of A such that

$$\mathcal{M} \models \exists \overline{v} \Theta(\overline{v}, \overline{a})$$

for all $a_1 < \cdots < a_n$ in A, where $\Theta(\overline{c}, d_1, \ldots, d_n)$ is the conjunction of all formulas in σ.

EXERCISE 5.3.4. Prove that Σ is a consistency property.

EXERCISE 5.3.5. Prove that if $\sigma \in \Sigma$ and $\psi \in T \cup \Gamma$, then $\sigma \cup \{\psi\} \in \Sigma$.

Combining the exercises as in the proof of Theorem 5.0.5, we conclude that for all infinite κ, there is a $(\kappa, \aleph_0)$-model of T. ⊣

The next exercise shows that the assumptions are optimal.

EXERCISE 5.3.6. Modify Exercise 5.0.6 to show that for all $\alpha < \omega_1$ there is a sentence $\phi \in \mathcal{L}_{\omega_1,\omega}$ with a $(\beth_\alpha, \aleph_0)$-model that does not have $(\kappa, \aleph_0)$-models for all infinite κ.

By a similar argument Morley proved a sharper result for first order theories. For a proof see Theorem 4.2.15 of [56].

THEOREM 5.3.7. *Let T be a first order theory in a countable language. Suppose that T has a $(\beth_n, \aleph_0)$-model for all $n \in \omega$. Then for all $\kappa \geq \lambda$, T has a (κ, λ)-model.*

5.4. Completely characterizing $\aleph_1$

THEOREM 5.4.1 (Knight [40]). *There is a complete $\mathcal{L}_{\omega_1,\omega}$-sentence ϕ such that ϕ has models of cardinality $\aleph_1$ but no larger models, indeed no model of size $\aleph_1$ has a proper extension that is a model of ϕ.*

We will give Hjorth's proof of a refinement on Knight's theorem and his application to Vaught counterexamples. Hjorth uses a variation of the Fraïssé construction. We begin with a quick review of the method.

DEFINITION 5.4.2. Let $\mathcal{K}$ be a class of finite structures closed under isomorphism. The class $\mathcal{K}$ has the *joint embedding property* if whenever $A_0, A_1 \in \mathcal{K}$ there is $B \in \mathcal{K}$ and an embeddings $f_i : A_i \to B$.

The class $\mathcal{K}$ has the *amalgamation property* if whenever $A, B_0, B_1 \in \mathcal{K}$ and $f_i : A \to B_i$ are embeddings, there is $C \in \mathcal{K}$ and embeddings $g_i : B_i \to C$. We say that $\mathcal{K}$ has the *disjoint embedding property* if we can choose g_0, g_1, and C such that $g_0(B_0 \setminus f_0(A)) \cap g_1(B_1 \setminus f_1(A)) = \emptyset$.

DEFINITION 5.4.3. We say that $\mathcal{M}$ is *$\mathcal{K}$-generic* if

(i) for every finite $A \subset \mathcal{M}$ there is $B \in \mathcal{K}$ such that $A \subseteq B \subset \mathcal{M}$;
(ii) for every $A \in \mathcal{K}$, there is an embedding of A into $\mathcal{M}$;
(iii) if $A, B \in \mathcal{K}$, $A \subset B$ and $A \subset \mathcal{M}$ there is $A \subset B_0 \subset \mathcal{M}$ such that there is an isomorphism $f : B \to B_0$ such that $f|A$ is the identity.

EXERCISE 5.4.4. Suppose $(\mathcal{M}_\alpha : \alpha < \beta)$ is a chain $\mathcal{M}_0 \subset \cdots \subset \mathcal{M}_\alpha \subset \cdots$ of models of $\Phi_{\mathcal{K}}$. Show that $\bigcup_{\alpha<\beta} \mathcal{M}_\alpha \models \Phi_{\mathcal{K}}$.

PROPOSITION 5.4.5. *Suppose there are $\aleph_0$ isomorphism types of structures in $\mathcal{K}$ and $\mathcal{K}$ has the joint embedding property and the amalgamation property. Then there is a $\mathcal{K}$-generic model of cardinality $\aleph_0$ and any two such models are isomorphic.*

PROOF. We build a countable $\mathcal{K}$-generic model $\mathcal{M}$ as a union of $A_0 \subset A_1 \subset \cdots$ of elements of $\mathcal{K}$. We can take $A_0 \in \mathcal{K}$ arbitrary. At stage s of our construction we are trying to insure an instance of (ii) or (iii).

Suppose we are given $A_s \in \mathcal{K}$ and $B \in \mathcal{K}$. Since $\mathcal{K}$ has the joint embedding property we can find $A_{s+1} \in \mathcal{K}$ such that A_s and B both embed in A_{s+1}.

Suppose $A \subset A_s$, $A \subset B$ where A_s, A and $B \in \mathcal{K}$. Since $\mathcal{K}$ has the amalgamation property we can find $A_{s+1} \supset A_s$ and $B' \subset A_{s+1}$ such that B and B' are isomorphic over A.

Let $\mathcal{M} = \bigcup_{s\in\omega} A_s$. Clearly $\mathcal{M}$ satisfies (i) and by carefully organizing our construction we can ensure $\mathcal{M}$ satisfies (ii) and (iii).

Suppose $\mathcal{M}$ and $\mathcal{N}$ are $\mathcal{K}$-generic models of cardinality $\aleph_0$. We build a sequence of partial embeddings

$$f_0 \subseteq f_1 \subseteq \cdots \subseteq f_n \subseteq \cdots$$

such that $A_s = \mathrm{dom}(f_s) \in \mathcal{K}$.

Let $A_0 \subset \mathcal{M}$ be an arbitrary element of $\mathcal{K}$. By (ii) there is an embedding of A_0 into $\mathcal{N}$. Suppose we have $f_s : A_s \to \mathcal{N}$ where $A_s \in \mathcal{K}$ and $A_s \subset \mathcal{M}$. Let $a \in \mathcal{M} \setminus A_s$. By (i) there is $A_{s+1} \subset \mathcal{M}_s$ such that $A_s \cup \{b\} \subseteq A_{s+1}$ and $A_{s+1} \in \mathcal{K}$. By (iii) we can find $f_{s+1} \supset f_s$ such that $f_{s+1} : A_{s+1} \to \mathcal{N}$. Similarly, we can extend f_s to add any element of $\mathcal{N}$ to the image. Thus we can construct the sequence of partial embedding such that $f = \bigcup f_s$ is an isomorphism between $\mathcal{M}$ and $\mathcal{N}$. $\dashv$

In our examples we will have $\emptyset \in \mathcal{K}$. In this case the amalgamation property implies the joint embedding property, so we can ignore the latter.

EXERCISE 5.4.6. Show that conditions (i), (ii) and (iii) give a simple description of $\Phi_{\mathcal{K}}$ a complete sentence satisfied by the countable $\mathcal{K}$-generic model.

PROPOSITION 5.4.7. *Suppose in addition that* $\mathcal{K}$ *has the disjoint amalgamation property and* $\emptyset \in \mathcal{K}$. *If* $\mathcal{M}$ *is the countable* $\mathcal{K}$*-generic model there is an embedding* $f : \mathcal{M} \to \mathcal{M}$ *such that* $\operatorname{img}(f)$ *is a proper subset of* $\mathcal{M}$.

If $\Phi_{\mathcal{K}}$ *is the Scott sentence of* $\mathcal{M}$, *then* $\Phi_{\mathcal{K}}$ *has a model of size* $\aleph_1$.

PROOF. Let $\tau' = \tau \cup \{P\}$ where P is a unary relation symbol. Let $\mathcal{K}'$ be the class of all expansions of structures in $\mathcal{K}$ to τ'.

EXERCISE 5.4.8. Show that since $\mathcal{K}$ has the disjoint amalgamation property and $\emptyset \in \mathcal{K}$, $\mathcal{K}'$ has joint embedding property and the amalgamation property.

Let $\mathcal{M}'$ be the unique $\mathcal{K}'$-generic model and let $\mathcal{M}$ be it's reduct to τ. We claim that there is a τ-embedding $f : \mathcal{M} \to P(\mathcal{M}')$. Then $\mathcal{M}$ is a proper embedding of $\mathcal{M}$ into itself.

Suppose $A \in \mathcal{K}$, $A \subset \mathcal{M}$ and $f : A \to P(\mathcal{M}')$ is a τ-embedding. Let $a \in \mathcal{M} \setminus A$ and let $B_1 \in \mathcal{K}', B_1 \subset \mathcal{M}'$ such that $A \cup \{a\} \subseteq B_1$. Let B be the τ-reduct of B_1. Let B_2 be a τ'-structure such that the τ-reduct is isomorphic to B and $P(B_2) = B_2$. Since $\mathcal{M}'$ is $\mathcal{K}'$-generic, we can extend f to $\widehat{f} : B \to P(\mathcal{M}')$. Carefully iterating this construction we can build a τ-embedding $f : \mathcal{M} \to P(\mathcal{M})$.

If $\mathcal{M}$ is the countable $\mathcal{K}$-generic model, we can build a chain

$$\mathcal{M}_0 \subset \mathcal{M}_1 \subset \cdots \subset \mathcal{M}_\alpha \subset \cdots$$

for $\alpha < \omega_1$ of countable models of Φ_K. At successor stages we use the construction above to build $\mathcal{M}_\alpha \subset \mathcal{M}_{\alpha+1}$ and at limit stages we use Exercise 5.4.4. Then $\mathcal{N} = \bigcup_{\alpha<\omega_1} \mathcal{M}_\alpha$ is the desired model of size $\aleph_1$. $\dashv$

For Hjorth's example we let $\tau_0 = \{S_i, R_i : i \in \omega\}$ where S_i is a binary relation symbol and R_i is an $i+2$-ary relation symbol. We will consider the class $\mathcal{K}_0$ of finite τ-structures with the following properties:

(i) $\forall a, b \bigvee_{i\in\omega} S_i(a, b)$;
(ii) $\forall a, b \bigwedge_{i\neq j}(S_i(a, b) \to \neg S_j(a, b))$;
(iii) $\forall a_0, a_1, b_0, \ldots, b_{k-1} [R_k(a_0, a_1, b_0, \ldots, b_{k-1}) \to (a_0 \neq a_1 \wedge b_i \neq b_j)]$, for all $i < j < k$;
(iv) $\forall a_0, a_1, b_0, \ldots, b_{k-1} [R_k(a_0, a_1, b_0, \ldots, b_{k-1}) \leftrightarrow R_k(a_0, a_1, b_{\sigma(0)}, \ldots, b_{\sigma(k-1)})]$ for all k and σ a permutation of $\{0, \ldots, k-1\}$;

(v) $\forall a_0, a_1, b_0, \ldots, b_{k-1}\,[R_k(a_0, a_1, b_0, \ldots, b_{k-1}) \rightarrow (S_i(a_0, b_j) \leftrightarrow S_i(a_1, b_j))]$ for all i, k and $j < k$;

(vi) $\forall a_0, a_1, b_0, \ldots, b_{k-1}, c\,[(R_k(a_0, a_1, b_0, \ldots, b_{k-1}) \wedge \bigwedge_{i<k} c \neq b_i) \rightarrow \bigwedge_{i\in\omega}(S_i(a_0, c) \rightarrow \neg S_i(a_1, c))]$ for all k;

(vii) $\forall a_0, a_1[a_0 \neq a_1 \rightarrow \bigvee_{k\in\omega} \exists b_0, \ldots, b_{k-1}\, R_k(a_0, a_1, b_0, \ldots, b_{k-1})]$.

We can think of structures in $\mathcal{K}_0$ as being complete graphs where we color the directed edge (a, b) with the unique color i such that $S_i(a, b)$. For each a_0 and a_1 there is a maximal set $\{b_0, \ldots, b_{k-1}\}$ such that each edge (a_0, b_i) has the same color as the edge (a_1, b_i) for $i = 1 \ldots k$ and $R_k(a_0, a_1, b_0, \ldots, b_k)$.

EXERCISE 5.4.9. (a) Show that there are only countably many isomorphism types in $\mathcal{K}_0$.

(b) Let A be a set of cardinality n. Color the complete directed graph on A such that each edge gets a distinct color. Let R_0 hold of all pairs (a, b) where $a \neq b$ and no other R_i relation holds. Show that $A \in \mathcal{K}_0$.

(c) Conclude that $\mathcal{K}_0$ is a countably infinite set of isomorphism types and $\emptyset \in \mathcal{K}_0$.

LEMMA 5.4.10. *$\mathcal{K}_0$ has the disjoint amalgamation property. Thus there is a countable $\mathcal{K}_0$-generic model $\mathcal{M}$. If $\Phi_{\mathcal{K}_0}$ is the Scott sentence of $\mathcal{M}$, then $\Phi_{\mathcal{K}_0}$ has models of size $\aleph_1$.*

PROOF. Without loss of generality, we may suppose we have $A, B_0, B_1 \in \mathcal{K}_0$ with $A = B_0 \cap B_1$. We make $B_0 \cup B_1$ into a τ_0-structure as follows.

- If $a_0 \in B_l$ and $a_1 \in B_{1-l}$ let $b_0, \ldots, b_{k-1}$ be the elements of A such that $S_i(a_0, b_j)$ and $S_i(a_1, b_j)$ for some i. Make $R_k(a_0, a_1, b_0, \ldots, b_{k-1})$ hold.
- For each pair $a_0 \in B_l, a_1 \in B_{1-l}$, choose a new i and make $S_i(a_0, a_1)$ hold.

It is easy to check that this makes $B_0 \cup B_1$ into a structure in $\mathcal{K}_0$. ⊣

LEMMA 5.4.11. *If $\mathcal{M} \models \Phi_{\mathcal{K}_0}$ is uncountable, then there is no proper extension $\mathcal{M} \subset \mathcal{N}$ where $\mathcal{N} \models \Phi_{\mathcal{K}_0}$. In particular, $\Phi_{\mathcal{K}_0}$ has no models of size $\aleph_2$.*

PROOF. Suppose $\mathcal{M} \subset \mathcal{N}$ and $a_0, a_1 \in \mathcal{M}$. For some k there are $b_0, \ldots, b_{k-1} \in \mathcal{M}$ such that $\mathcal{M} \models R_k(a_0, a_1, b_0, \ldots, b_{k-1})$. If $b \in \mathcal{N} \setminus \mathcal{M}$, by (vi) if $S_i(a_0, b)$, then $\neg S_i(a_1, b)$. Thus the function that sends $a \in \mathcal{M}$ to the unique i such that $S_i(a, b)$ is injective and $\mathcal{M}$ must be countable. ⊣

Thus $\Phi_{\mathcal{K}_0}$ is the desired complete $\mathcal{L}_{\omega_1,\omega}$-sentence with models of size $\aleph_1$ but no models of size $\aleph_2$.

Vaught counterexamples with no models of size $\aleph_2$. Let τ_0 and $\mathcal{K}_0$ be as above. Let $\tau_1 = \tau_0 \cup \{P, Q, F\}$ where P and Q are unary relation symbols and F is a binary relation symbol. Let $\mathcal{K}_1$ be the class of finite τ_1-structures A such that:

- A is the disjoint union of $P(A)$ and $Q(A)$;
- F defines a function from $P(A)$ into $Q(A)$;

- the relation symbols from τ_0 only hold of elements from $P(A)$;
- $P(A) \in \mathcal{K}_0$.

EXERCISE 5.4.12. (a) Show that $\mathcal{K}_1$ has the amalgamation and joint embedding properties (this argument should only need the fact that $\mathcal{K}_0$ has the disjoint amalgamation property and $\emptyset \in \mathcal{K}_0$).
(b) Let $\mathcal{M}$ be the countable $\mathcal{K}_1$-generic model and let $\Phi_{\mathcal{K}_1}$ be it's Scott sentence. Show that $P(\mathcal{M}) \models \Phi_{K_0}$.
(c) Show that in $\mathcal{M}$ the map from $P(\mathcal{M})$ to $Q(\mathcal{M})$ is surjective.
(d) Show Φ_{K_1} has models of size $\aleph_1$, but no uncountable model has a proper extension that is a model of Φ_{K_1}.

DEFINITION 5.4.13. Let $\mathcal{M}$ be a model and let $X \subset \mathcal{M}$ be $\mathcal{L}_{\omega_1,\omega}$-definable. We say that X is a set of *absolute indiscernibles* in $\mathcal{M}$ if for every permutation σ of X there is an automorphism f of $\mathcal{M}$ such that $f \supset \sigma$.

LEMMA 5.4.14. *Let $\mathcal{M}$ be the countable $\mathcal{K}_1$-generic model. Then $Q(\mathcal{M})$ is a set of absolute indiscernibles.*

PROOF. Let σ be a permutation of $Q(\mathcal{M})$. Suppose we have $A \subset \mathcal{M}$ such that $A \in \mathcal{K}_1$ and $f : A \to \mathcal{M}$ such $f|Q(A) = \sigma|Q(A)$. Suppose $A \subset B \subset \mathcal{M}$ and $B \in \mathcal{K}_1$. We need to show that there is $g : A \to \mathcal{M}$ extending f such that $f|Q(B) = \sigma|Q(B)$.

Let $c_1, \dots, c_m$ be the elements of $Q(B) \setminus Q(A)$. Note that $A' = A \cup \{c_1, \dots, c_m\} \in \mathcal{K}_1$. We can extend f to $f' : A' \to \mathcal{M}$ by defining $f'(c_i) = \sigma(c_i)$. We now use the fact that $\mathcal{M}$ is $\mathcal{K}_1$-generic to extend f' to $g : B \to \mathcal{M}$. Since $B \setminus A' \subset P(\mathcal{M})$, $g|Q(B) = \sigma|Q(B)$ as desired.

Similarly, we can add elements of $\mathcal{M} \setminus f(A)$ to the image of f. This allows us to build an automorphism of $\mathcal{M}$ that extends σ. ⊣

COROLLARY 5.4.15 (Hjorth [29]). *If there is a counterexample $\psi \in \mathcal{L}_{\omega_1,\omega}$ to Vaught's Conjecture, then there is a counterexample with no model of size $\aleph_2$.*

PROOF. Suppose ψ is a Vaught counterexample in $\mathcal{L}_{\omega_1,\omega}(\tau)$ where we may assume $\tau \cap \tau_1 = \emptyset$ and τ is relational. Consider the $\tau \cup \tau_1$-sentence Ψ with models $\mathcal{N}$ such that: that

- $\mathcal{N} \models \Phi_{\mathcal{K}_1}$;
- all of the τ-structure is on $Q(\mathcal{N})$;
- $Q(\mathcal{N}) \models \psi$.

Suppose $\mathcal{N}_1$ and $\mathcal{N}_2$ are countable models of Ψ. We claim that if $Q(\mathcal{N}_1) \cong Q(\mathcal{N}_2)$, then $\mathcal{N}_1 \cong \mathcal{N}_2$. Suppose $f_0 : Q(\mathcal{N}_1) \to Q(\mathcal{N}_2)$ is a τ-isomorphism. Since the τ_1-reducts of $\mathcal{N}_1$ and $\mathcal{N}_2$ are $\mathcal{K}_1$-generic, there is $f : \mathcal{N}_1 \to \mathcal{N}_2$ a τ_1-isomorphism. Then $\sigma_0 = f^{-1} \circ f_0$ a permutation of $Q(\mathcal{N}_1)$. Since $Q(\mathcal{N}_1)$ is a set of absolute indiscernibles, there is $\sigma \supset \sigma_0$ an automorphism of the τ_1-reduct of $\mathcal{N}_1$. But then $f \circ \sigma$ is a τ_1-isomorphism between $\mathcal{N}_1$ and $\mathcal{N}_2$. But $f \circ \sigma \supset \sigma_0$, so this is the desired isomorphism between $\mathcal{N}_1$ and $\mathcal{N}_2$

For any countable $\mathcal{A} \models \psi$, there is a countable $\mathcal{N} \models \Psi$ with $Q(\mathcal{N}) \cong \mathcal{A}$. Since ψ has uncountably many non-isomorphic countable models but no perfect set of non-isomorphic models the same is true of Ψ. But if $\mathcal{M} \models \Psi$, $|Q(\mathcal{M})| \leq |P(\mathcal{M})| \leq \aleph_1$. Thus Ψ is a Vaught counterexample with no models of size $\aleph_2$. ⊣

Hjorth's original proof was somewhat different. He used a slight variant of $\mathcal{K}_1$ and showed that the automorphism group of the $\mathcal{K}_1$-generic countable model had a closed subgroup with a surjective homomorphism onto S_∞. He was then able to construct the counterexample with no model of size $\aleph_2$ using some basic facts about the dynamics of Polish group actions. In [4] the authors notice that Hjorth's construction can be easily adapted to get a model with a set of absolute indiscernibles and from this the main result quickly follows.

EXERCISE 5.4.16. Show that the example Ψ has models of size $\aleph_1$.

We will see in Chapter 7 that every Vaught counterexample has models of size $\aleph_1$.

Part 2

BUILDING UNCOUNTABLE MODELS

Chapter 6

ELEMENTARY CHAINS

6.1. Elementary end extensions

The failure of compactness leaves us with no general tool for building elementary extensions and elementary chains. In this section we will give a criteria which will allow us to use the Model Existence Theorem to build elementary extensions in some special circumstances. Iterating this construction will be a useful tool for building models of cardinality $\aleph_1$.

DEFINITION 6.1.1. Let $\tau = (<, \dots)$. Let $\mathcal{M}$ be linearly ordered by $<$. We say that $\mathcal{M} \subset \mathcal{N}$ is an *end extension* if $\mathcal{M} \neq \mathcal{N}$ and $\mathcal{N} \models a < b$ for all $a \in M$ and $b \in \mathcal{N} \setminus \mathcal{M}$.

The following theorem of Keisler gives necessary and sufficient conditions for constructing elementary end extensions. For the purposes of the next few sections we will need to assume our fragments have a slightly strong closure property, namely

(†) $$\text{If } \exists \overline{v} \bigvee_{\psi \in X} \psi(\overline{v}) \in \mathbb{A}, \text{ then } \bigvee_{\psi \in X} \exists \overline{v}\, \psi(\overline{v}) \in \mathbb{A}.$$

For the next sections we let $\exists^* x\, \phi$ abbreviate $\forall y \exists x > y\, \phi$ and let $\forall^* x \phi$ denote the dual quantifier, $\exists y \forall x > y\, \phi$.

THEOREM 6.1.2. *Let $\tau = \{<, \dots\}$ be a countable vocabulary, let $\mathcal{M}$ be a countable τ-structure such that $<$ is a linear order, and let $\mathbb{A}$ be a countable fragment of $\mathcal{L}_{\omega_1,\omega}(\tau)$. The following are equivalent*:

1. *$\mathcal{M}$ has a countable $\mathbb{A}$-elementary end extension*;
2. *Each of the following is true in $\mathcal{M}$.*
   (a) $\forall x \exists y\, x < y$.
   (b) $\forall \overline{w}\, [\exists^* x \bigvee_{n\in\omega} \phi_n(x, \overline{w}) \to \bigvee_{n\in\omega} \exists^* x\, \phi_n(x, \overline{w})]$, *for* $\bigvee_{n\in\omega} \phi_n \in \mathbb{A}$.
   (c) $\forall \overline{w}\, [\exists^* x \exists y\, \phi(x, y, \overline{w}) \to (\exists y \exists^* x\, \phi(x, y, \overline{w}) \vee \exists^* y \exists x\, \phi(x, y, \overline{w}))]$, *for* $\phi \in \mathbb{A}$.

PROOF. $(1 \Rightarrow 2)$. Suppose $\mathcal{M} \prec_{\mathbb{A}} \mathcal{N}$ and $\mathcal{N}$ is an end extension of $\mathcal{M}$. The main idea is the following *overspill principle*: if $b \in \mathcal{N} \setminus \mathcal{M}$, $\overline{a}, c \in \mathcal{M}$ and $\mathcal{N} \models \phi(b, \overline{a})$, then $\mathcal{M} \models \exists x > c\, \phi(x, \overline{a})$. Thus $\mathcal{M} \models \exists^* x\, \phi(x, \overline{a})$.

(a) If $a \in \mathcal{M}$ and $b \in \mathcal{N} \setminus \mathcal{M}$, $a < b$. Since $\mathcal{M} \prec \mathcal{N}$, there is $c \in \mathcal{M}$ such that $a < c$.
(b) If $\mathcal{M} \models \exists^* x \bigvee \phi_n(x, \overline{a})$, then the same is true in $\mathcal{N}$ and we can find $b \in \mathcal{N} \setminus \mathcal{M}$ and $n \in \omega$ such that $\mathcal{N} \models \phi_n(b, \overline{a})$. By overspill $\mathcal{M} \models \exists^* x \phi_n(x, \overline{a})$.
(c) If $\mathcal{M} \models \exists^* x \exists y\, \phi(x, y)$, then $\mathcal{N} \models \exists^* x \exists y\, \phi(x, y)$. Thus there is $b \in \mathcal{N} \setminus \mathcal{M}$ such that $\mathcal{N} \models \exists y\, \phi(b, y)$. Suppose $\mathcal{N} \models \phi(b, c)$. If $c \in \mathcal{M}$, then, by overspill, $\mathcal{M} \models \exists^* x\, \phi(x, c)$. If $c \notin \mathcal{M}$, then, by overspill, $\mathcal{M} \models \exists^* y \exists x\, \phi(x, y)$.

$(2 \Rightarrow 1)$. We expand the vocabulary τ to τ' by adding a constant symbol for each element of $\mathcal{M}$ and a new constant symbol d. Let $\mathbb{A}'$ be the smallest fragment containing $\mathbb{A}$ and the new constant symbols. Let

$$T = \{\chi(d, \overline{m}) : \mathcal{M} \models \forall^* x \chi(x, \overline{m}) \text{ for } \chi \in \mathbb{A}, \overline{m} \in \mathcal{M}\}.$$

Note that if $\mathcal{N} \models T$ then $\mathcal{M} \prec_{\mathbb{A}} \mathcal{N}$. For $a \in \mathcal{M}$ let Θ_a be the sentence

$$\forall y \bigvee_{b \in M} (y = b \vee a < y).$$

If $\mathcal{N} \models T + \bigwedge_{a \in \mathcal{M}} \Theta_a$, then $\mathcal{M} \prec_{\mathbb{A}} \mathcal{N}$ and $\mathcal{N}$ is an end extension of $\mathcal{M}$. We will build $\mathcal{N}$ using the Omitting Types Theorem 4.2.1. We must show that T is satisfiable and that if $\exists y\, \phi(y) \in \mathbb{A}'$, $T \cup \exists y\, \phi(y)$ is satisfiable and $a \in \mathcal{M}$, then there is $b \in M$ such that $T + \exists y\, [\phi(y) \wedge (y = b \vee a < y)]$ is satisfiable.

CLAIM. *T is satisfiable.*

Let τ'' be obtained by adding C a countable set of new constant symbols and let $\mathbb{A}''$ be the smallest fragment containing $\mathbb{A}'$ and the new constants. Let Σ be all finite sets σ of $\mathbb{A}''$-sentences such that if $\theta_\sigma(c_1, \ldots, c_m, d)$ is the conjunction of all sentences in σ, then $\mathcal{M} \models \exists^* x \exists y_1 \ldots \exists y_m \theta_\sigma(y_1, \ldots, y_m, x)$.

We claim that Σ is a consistency property. Note that, by condition (a), $\exists^* x\, x = x$. Thus Σ is nonempty. As usual we will only check (C4). Suppose $\bigvee_{\psi \in X} \psi(c_1, \ldots, c_m, d) \in \sigma \in \Sigma$. Then

$$\mathcal{M} \models \exists^* x \exists v_1 \ldots \exists v_m\, \theta_\sigma(v_1, \ldots, v_m, x).$$

But then

$$\mathcal{M} \models \exists^* x \bigvee_{\psi \in X} \exists v_1 \ldots \exists v_m\, (\psi(\overline{v}, x) \wedge \theta_\sigma(\overline{v}, x))$$

and, by condition (b), there is $\psi \in X$ such that

$$\mathcal{M} \models \exists^* x \exists v_1 \ldots \exists v_m\, (\psi(\overline{v}, x) \wedge \theta_\sigma(\overline{v}, x)).$$

Thus $\sigma \cup \{\psi(\overline{c}, d)\} \in \Sigma$.

To show that T is satisfiable, we will show that if $\psi \in T$ and $\sigma \in \Sigma$, then $\sigma \cup \{\psi\} \in \Sigma$. The Extended Model Existence Theorem 4.1.8 will then guarantee the existence of $\mathcal{M} \models T$.

Suppose $\mathcal{M} \models \exists y \forall x > y\, \chi(x)$ and $\sigma \in \Sigma$. Then

$$\mathcal{M} \models \exists^* x(\chi(x) \wedge \exists \overline{v}\, \theta_\sigma(\overline{v}, x)).$$

Hence $\sigma \cup \{\chi(d)\} \in \Sigma$ as desired. Thus T is satisfiable.

Claim. *For $\overline{m} \in \mathcal{M}$, $\phi(x, \overline{y}) \in \mathbb{A}$, $T + \phi(d, \overline{m})$ is satisfiable if and only if $\mathcal{M} \models \exists^* x\, \phi(x, \overline{m})$.*

($\Rightarrow$) Since $\neg\phi(d, \overline{m}) \notin T$,

$$\mathcal{M} \not\models \exists x \forall y > x \neg\phi(y, \overline{m}).$$

Thus $M \models \exists^* y\, \phi(y, \overline{m})$.

($\Leftarrow$) We use the consistency property Σ. If $\mathcal{M} \models \exists^* x\, \phi(x, \overline{m})$, then $\{\phi(d, \overline{m})\} \in \Sigma$. By the arguments above and the Extended Model Existence Theorem 4.1.8, $T + \psi(d, \overline{m})$ is satisfiable.

Claim. *If $\exists y\, \phi(y) \in \mathbb{A}'$, $T \cup \exists y\, \phi(y)$ is satisfiable and $a \in \mathcal{M}$, then there is $b \in M$ such that $T + \exists y\, [\phi(y) \wedge (y = b \vee a < y)]$ is satisfiable.*

Suppose $T + \exists y\, \psi(y, d, \overline{m})$ is satisfiable and $a \in \mathcal{M}$. By the claim above, $\mathcal{M} \models \exists^* x \exists y \psi(y, x, \overline{m})$. Thus

$$\mathcal{M} \models \exists^* x\, \left[(\exists y\, (\psi(y, x, \overline{m}) \wedge a < y)) \vee (\exists y\, (\psi(y, x, \overline{m}) \wedge y \leq a))\right].$$

By condition (b) we are in at least one of two cases.

Case 1. $\mathcal{M} \models \exists^* x \exists y(\psi(y, x, \overline{m}) \wedge a < y)$

By the previous claim $T + \exists y(\psi(y, d, \overline{m}) \wedge a < y)$ is satisfiable, so certainly

$$T + \exists y(\psi(y, d, \overline{m}) \wedge (y = b) \vee a < y)$$

is satisfiable for any $b \in \mathcal{M}$.

Case 2. $\mathcal{M} \models \exists^* x \exists y\, (\psi(y, x, \overline{m}) \wedge y \leq a)$

By now apply condition (c). Clearly

$$\mathcal{M} \not\models \exists^* y \exists x\, ((\psi(y, x, \overline{m}) \wedge y \leq a).$$

Thus by (c) there is $b \in \mathcal{M}$ such that

$$\mathcal{M} \models \exists^* x\, (\psi(b, x, \overline{m}) \wedge b \leq a).$$

But the $T + \exists y(\psi(y, d, \overline{m} \wedge y = b)$ is satisfiable, so

$$T + \exists y(\psi(y, d, \overline{m}) \wedge (y = b) \vee a < y)$$

is satisfiable for some $b \in \mathcal{M}$.

We have verified the conditions of the Omitting Types Theorem. Thus there is

$$\mathcal{N} \models T \cup \bigwedge_{a \in \mathcal{M}} \forall y \bigvee_{b \in \mathcal{M}} (y = b \vee a < y).$$

Then $\mathcal{N}$ is an $\mathbb{A}$-elementary end extension of $\mathcal{M}$. $\dashv$

With mild extra assumptions on $\mathbb{A}$ we can iterated this method to build elementary extensions of size $\aleph_1$. In general if $\exists^* x \bigvee_{\theta \in X} \in \mathbb{A}$, we don't know that $\bigvee_{\theta \in X} \exists^* x \theta(x) \in \mathbb{A}$. We suppose that $\mathbb{A}$ has this additional closure property.

COROLLARY 6.1.3. *Let $\mathcal{M}$ be a countable τ-structure and let $\mathbb{A}$ be a countable fragment of $\mathcal{L}_{\omega_1,\omega}(\tau)$ with the above additional closure property. If $\mathcal{M}$ has an $\mathbb{A}$-elementary end extension, then $\mathcal{M}$ has an $\mathbb{A}$-elementary end extension of cardinality $\aleph_1$.*

PROOF. We can build an $\mathbb{A}$-elementary chain of countable elementary end extensions $(\mathcal{M}_\alpha : \alpha < \omega_1)$. Because $\mathbb{A}$ satisfies condition (†), the conditions (a)–(c) of Theorem 6.1.2 are $\mathbb{A}$-elementary, so given $\mathcal{M} \prec_{\mathbb{A}} \mathcal{M}_\alpha$ we can build an $\mathbb{A}$-elementary end extension of $\mathcal{M}_\alpha$. $\mathcal{N} = \bigcup_{\alpha<\omega_1} \mathcal{M}_\alpha$ is the desired $\mathbb{A}$-elementary end extension of cardinality $\aleph_1$. ⊣

Keisler's Theorem works in a slightly more general setting than linear orders. Let $\tau = \{R, \dots\}$ be a countable vocabulary, where R is a binary relation symbol.

DEFINITION 6.1.4. We say that $\mathcal{N}$ is an end extension of $\mathcal{M}$ if whenever $a \in \mathcal{N}$, $b \in \mathcal{M}$ and $a\ R\ b$, then $a \in \mathcal{M}$.

We say that $\mathcal{N}$ is a *strong end extension* if $\mathcal{M} \neq \mathcal{N}$ and $\mathcal{N} \models a\ R\ b \wedge \neg b\ R\ a$ for all $a \in \mathcal{M}$ and $b \in \mathcal{N} \setminus \mathcal{M}$.

If R is a linear ordering, then any end extension is strong.

EXERCISE 6.1.5. Modify the proof of Theorem 6.1.2 to show that if $\mathcal{M}$ is a countable τ-structure such that R is transitive and irreflexive, then $\mathcal{M}$ has a strong $\mathbb{A}$-elementary end extension if and only if (a′), (b) and (c) hold, where (a′) is $\forall x \forall y \exists z\ (x\ R\ z \wedge y\ R\ z)$; (b) and (c) are as above but we interpret $\exists^* x\ \phi$ as $\forall y \exists x\ (y\ R\ x \wedge \phi)$.

EXERCISE 6.1.6. Suppose $\mathcal{M} \models$ ZFC. Recall that we have a definable rank function $\rho : \mathcal{M} \to On^{\mathcal{M}}$ where $\rho(x)$ is the least ordinal α of $\mathcal{M}$ such that $x \in \mathbb{V}_{\alpha+1} \setminus \mathbb{V}_\alpha$. Suppose $\mathcal{M} \subset \mathcal{N}$. We say that $\mathcal{N}$ is a *top extension* if $\rho(m) < \rho(n)$ for all $m \in \mathcal{M}$ and $n \in \mathcal{N} \setminus \mathcal{M}$. Prove that every countable $\mathcal{M} \models$ ZFC has a top extension of cardinality $\aleph_1$. [Hint: Let $R(a, b)$ be the relation $\rho(a) < \rho(b)$ and show the version of Keisler's Theorem in the Exercise above applies.]

Vaught's two cardinal property. Let $\tau = \{U, \dots\}$ be countable where U is a unary predicate.

DEFINITION 6.1.7. We say that $\mathcal{M}$ is a (κ, λ)-model if $|\mathcal{M}| = \kappa$ and $|U(\mathcal{M})| = \lambda$.

In first order logic, Vaught's Two Cardinal Theorem says that if a countable theory T has a (κ, λ)-model where $\kappa > \lambda \geq \aleph_0$. Then T has an $(\aleph_1, \aleph_0)$-model.

The usual proofs break down as they all require some remnant of compactness. Keisler found the following proof using Theorem 6.1.2.

THEOREM 6.1.8. *If $\phi \in \mathcal{L}_{\omega_1,\omega}$ has a (κ, λ)-model where $\kappa > \lambda \geq \aleph_0$, then ϕ has an $(\aleph_1, \aleph_0)$-model.*

PROOF. Suppose $\mathcal{M}_0$ is a (κ, λ)-model of ϕ. Let $\mathbb{A}_0$ be a countable fragment of $\mathcal{L}_{\omega_1,\omega}$ with $\phi \in \mathbb{A}$. By Löwenheim–Skolem we can find $\mathcal{M}_1 \prec_{\mathbb{A}_0} \mathcal{M}_0$ a (λ^+, λ)-model. Let $<$ be a well ordering of $\mathcal{M}_1$ of order type λ^+ and let $\tau^* = \tau \cup \{<\}$. Note that, by the reguarlity of λ^+ all of the following are true in $\mathcal{M}_1$:

(a) $\neg\exists^* x\, U(x)$;
(b) $\forall \overline{v}\, [\exists^* x \bigvee_{\psi \in X} \psi(x, \overline{v}) \to \bigvee_{\psi \in X} \exists^* x\, \psi(x, \overline{v})]$, for all $\bigvee_{\psi \in X} \psi \in \mathcal{L}_{\omega_1,\omega}$;
(c) $\forall \overline{v}\, [\exists^* x \exists y\, \psi(x, y, \overline{v}) \to (\exists^* y \exists x\, \psi(x, y, \overline{v}) \vee \exists y \exists^* x\, \psi(x, y, \overline{v}))]$, for all $\psi \in \mathcal{L}_{\omega_1,\omega}$.

Let $\mathbb{A}$ be a countable fragment of $\mathcal{L}_{\omega_1,\omega}(\tau^*)$ such that $\phi \in \mathbb{A}$ and if $\bigvee_{\psi \in X} \psi \in \mathbb{A}$, then $\bigvee_{\psi \in X} \exists^* x \psi \in \mathbb{A}$.

Let $\mathcal{M} \prec_{\mathbb{A}} \mathcal{M}_1$ be countable. Then the three conditions above are true in $\mathcal{M}$ for $\mathbb{A}$-formulas. In particular, there is $b \in \mathcal{M}$ such that $\mathcal{M} \models \forall x\, (U(x) \to x < b)$. By Corollary 6.1.3 there is $\mathcal{M} \prec_{\mathbb{A}} \mathcal{N}$ such that $\mathcal{N}$ is an end extension and $|\mathcal{N}| = \aleph_1$. Since $\mathcal{N} \models \forall x\, (U(x) \to x < b)$, $U(\mathcal{N}) = U(\mathcal{M})$ is countable. ⊣

Undefinability of well order revisited. We can use end extensions to prove an uncountable version of the undefinability of well order (Theorem 4.4.1).

COROLLARY 6.1.9. *Let τ be a countable vocabulary containing a binary relation symbol $<$, let $\phi \in \mathcal{L}_{\omega_1,\omega}(\tau)$, and suppose $\mathcal{M} = (M, <, \dots) \models \phi$ where $<$ has order type ω_1. Then there is $\mathcal{N} \models \phi$ of cardinality $\aleph_1$ and an order preserving embedding of $(\mathbb{Q}, <)$ into $\mathcal{N}$.*

PROOF. Let $\mathbb{A}$ be a countable fragment of $\mathcal{L}_{\omega_1,\omega}(\tau)$ containing ϕ with the closure property above. We expand the signature by adding a new unary predicate P. Let Ψ assert that $\mathcal{M}$ is an $\mathbb{A}$-elementary end extension of $P(\mathcal{M})$. For any $\alpha < \omega_1$ we can find P_α such that

$$(\mathcal{M}, P_\alpha) \models \phi \wedge \Psi$$

and the order type of P_α is at least α.

By the undefinability of well order (Theorem 4.4.1), there is a countable model $(\mathcal{N}, P(\mathcal{N})) \models \phi \wedge \Psi$ where $(\mathbb{Q}, <)$ can be embedded into $(\mathcal{N}, <)$. Since $P(\mathcal{N}) \prec_{\mathbb{A}} \mathcal{N}$ and $P(\mathcal{N})$ has an $\mathbb{A}$-elementary end extension, conditions (a)–(c) of Theorem 6.1.2 hold in $P(\mathcal{N})$ and hence in $\mathcal{N}$. Thus $\mathcal{N}$ has an $\mathbb{A}$-elementary end extension, which, by Corollary 6.1.3 we may assume has cardinality $\aleph_1$. ⊣

6.2. Omitting types in end extensions

We continue the study of end extensions begun in Section 6.1. Keisler also characterized when a countable model has an uncountable elementary end extension omitting a type.

Fix $\tau = \{<, \dots\}$ be a countable vocabulary and let $\mathbb{A}$ be a countable fragment of $\mathcal{L}_{\omega_1,\omega}(\tau)$. Suppose for $n \in \omega$ we have $\Theta_n(v_1, \dots, v_{m_n})$ be a set of $\mathbb{A}$-formulas. We will be looking for models of

$$\chi = \bigwedge_{n\in\omega} \forall x_1 \dots \forall x_{m_n} \bigvee_{\theta\in\Theta_n} \theta(\overline{x}).$$

Note that the formulas $\bigvee_{\theta\in\Theta_n} \theta$ and χ need not be in $\mathbb{A}$.

Recall that an ordering $<$ is *ω_1-like* if it is uncountable but $\{x : x < a\}$ is countable for all a.

THEOREM 6.2.1. *Suppose $\mathcal{M}$ is a countable τ-structure. The following are equivalent*

1. *$\mathcal{M}$ has an ω_1-like $\mathbb{A}$-elementary end extension $\mathcal{N}$ where χ holds.*
2. *$\mathcal{M}$ satisfies the conditions from Theorem* 6.1.2 *and, in addition, for every $n \in \omega$, $\overline{a} \in \mathcal{M}$ if*

$$\mathcal{M} \models S\overline{y}\exists x_1 \dots \exists x_{m_n} \psi(\overline{x}, \overline{y}, \overline{a})$$

then there is $\theta \in \Theta_n$ such that

$$\mathcal{M} \models S\overline{y}\exists x_1 \dots \exists x_{m_n} \psi(\overline{x}, \overline{y}, \overline{a}) \wedge \theta(\overline{x})$$

where $S\overline{y}$ is a string of quantifiers of the form $\exists y_i$ and $\exists^* y_j$.

PROOF. $(1 \Rightarrow 2)$. Since $\mathcal{N}$ is an $\mathbb{A}$-elementary end extension, the conditions from Theorem 6.1.2 are satisfied in $\mathcal{M}$. Fix n and suppose

$$\mathcal{M} \models S\overline{y}\exists x_1 \dots \exists x_{m_n} \psi(\overline{x}, \overline{y})$$

(we suppress the parameters from $\mathcal{M}$ as they play no role). Since $\mathcal{N}$ is $\mathbb{A}$-elementary and $\mathcal{N} \models \chi$,

$$\mathcal{N} \models S\overline{y}\exists x_1 \dots \exists x_{m_n} \psi(\overline{x}, \overline{x}y) \wedge \bigvee_{\theta\in\Theta_n} \theta(\overline{x}).$$

Note that for any countable disjunction $\bigvee_{i=1}^{\infty} \phi_i(\overline{v}, w)$

$$\text{(i)} \qquad \mathcal{N} \models \forall \overline{v} \left[\exists w \bigvee_{i=1}^{\infty} \phi_i(\overline{v}, w) \leftrightarrow \bigvee_{i=1}^{\infty} \exists w\, \phi_i(\overline{v}, w)\right]$$

and

$$\text{(ii)} \qquad \mathcal{N} \models \forall \overline{v} \left[\exists^* w \bigvee_{i=1}^{\infty} \phi_i(\overline{v}, w) \leftrightarrow \bigvee_{i=1}^{\infty} \exists^* w\, \phi_i(\overline{v}, w)\right].$$

(i) is true in any model. (ii) is true in any ω_1-like model because $\exists^* x\, \phi(x)$ is equivalent to "there are uncountably many x such that $\phi(x)$". It now follows by induction that

$$\mathcal{N} \models \bigvee_{\theta \in \Theta_n} S\overline{y}\exists\overline{x}\, \psi(\overline{x}, \overline{y}) \wedge \theta(\overline{x}).$$

Thus

$$\mathcal{N} \models S\overline{y}\exists\overline{x}\, \psi(\overline{y}, \overline{x}) \wedge \theta(\overline{x})$$

for some $\theta \in \Theta_n$ and, since $\mathcal{M} \prec_{\mathbb{A}} \mathcal{N}$ so does $\mathcal{M}$.

($2 \Rightarrow 1$). Our basic strategy will be to show that if $\mathcal{M}$ is a countable and satisfies Condition 2, then it has a countable end extension $\mathcal{M} \prec_{\mathbb{A}} \mathcal{N}$ that also satisfies Condition 2. This will allow us to start building an $\mathbb{A}$-elementary chain of end extensions. It is easy to see that the union of an elementary chain of models satisfying Condition 2 will also satisfy Condition 2. Thus we can interate this process to build an $\mathbb{A}$-elementary chain of length ω_1. The union of this chain will be an ω_1-like model satisfying Condition 2.

If $\mathcal{N}$ is any model satisfying Condition 2, $n \in \omega$ and $a_1, \ldots, a_{m_n} \in \mathcal{N}$, let $\psi(\overline{x})$ be

$$x_1 = a_1 \wedge \cdots \wedge x_{m_n} = a_{m_n}.$$

Then for some $\theta \in \Theta_n$,

$$\mathcal{N} \models \exists\overline{x}\, (\psi(\overline{x}) \wedge \theta(\overline{x})),$$

i.e., $\mathcal{N} \models \theta(\overline{a})$. Thus $\mathcal{N} \models \chi$.

Thus it suffices to show that a countable model satisfying Condition 2 has an $\mathbb{A}$-elementary end extension satisfying Condition 2. We will do this by weaving an additional omitting types argument into the proof of Theorem 6.1.2.

Recall that in the proof of Theorem 6.1.2 we expanded the language by constants for the elements of $\mathcal{M}$ and an additional constant d. We introduced the theory

$$T = \{\phi(\overline{a}, d) : \phi \in \mathbb{A}, \overline{a} \in \mathcal{M} \text{ and } \mathcal{M} \models \exists^* x\, \phi(\overline{a}, x)\},$$

showed T is satisfiable and used the Omitting Types Theorem to find a countable model of

$$T + \bigwedge_{a \in M} \forall y \bigvee_{b \in M} (y = b \vee a \leq y).$$

We will recast Condition 2 as an omitting types problem and show how to find a model of T solving it. Finding an end extension satisfying Condition 2, will be accomplished by solving these two omitting types problems simultaneously. But once we know we can solve each of them, the Omitting Types Theorem tells us we can solve them both at once.

Fix n. We let $\overline{x} = x_1, \ldots, x_{m_n}$. Let $\psi(\overline{x}, \overline{y}, \overline{v})$ be an $\mathbb{A}$-formula with parameters from $\mathcal{M}$. We will show how to use the Omitting Types Theorem to build a model of

$$T + \forall \overline{v} \left[S\overline{y}\exists\overline{x}\, \psi(\overline{x}, \overline{y}, \overline{v}) \rightarrow \bigvee_{\theta \in \Theta_n} S\overline{y}\exists\overline{x}\, (\psi(\overline{x}, \overline{y}, \overline{v}) \wedge \theta(\overline{x})) \right].$$

Doing this for all n and all ψ will give the desired model of Condition 2.

To apply the Omitting Type Theorem we need to know that if we have $\phi(\overline{v}, d)$ an $\mathbb{A}$-formula with parameters from $\mathcal{M}$ and

$$T + \exists\overline{v}\, \phi(\overline{v}, d)$$

is satisfiable, then

$$T + \exists\overline{v}\, \left[\phi(\overline{v}, d) \wedge \left(S\overline{y}\exists\overline{x}\, \psi(\overline{x}, \overline{y}, \overline{v}) \rightarrow S\overline{y}\exists\overline{x}\, (\psi(\overline{x}, \overline{y}, \overline{v}) \wedge \theta(\overline{x}))\right)\right]$$

is satisfiable for some $\theta \in \Theta_n$.

Suppose $T + \exists\overline{v}\, \phi(\overline{v}, d)$ is satisfiable. We will show that either

(a) $\quad T + \exists\overline{v}\, [\phi(\overline{v}, d) \wedge \neg S\overline{y}\exists\overline{x}\, \psi(\overline{x}, \overline{y}, \overline{v})]$ is satisfiable

or

(b) $\quad T + \exists\overline{v}\, [\phi(\overline{v}, d) \wedge S\overline{y}\exists\overline{x}\, (\psi(\overline{x}, \overline{y}, \overline{v}) \wedge \theta(\overline{x}))]$ is satisfiable

for some $\theta \in \Theta_n$.

Recall from the proof of Theorem 6.1.2, that

$$T + \pi(d) \text{ is satisfiable } \Leftrightarrow \mathcal{M} \models \exists^* x\, \pi(x),$$

for any $\mathbb{A}$-formula with parameters from $\mathcal{M}$.

Suppose $T + \exists\overline{v}\, \phi(\overline{v}, d)$ is satisfiable, but (a) fails. Then

$$\mathcal{M} \models \exists^* w \exists\overline{v}\, \phi(\overline{v}, w)$$

and

$$T \models \forall\overline{v}\, [\phi(\overline{v}, d) \rightarrow S\overline{y}\overline{x}\psi(\overline{x}, \overline{y}, \overline{v})].$$

So

$$\mathcal{M} \models \exists^* w \exists\overline{v} S\overline{y}\exists\overline{x}\, (\phi(\overline{v}, w) \wedge \psi(\overline{x}, \overline{y}, \overline{v}))$$

and, by Condition 2, there is $\theta \in \Theta_n$ such that

$$\mathcal{M} \models \exists^* w \exists\overline{v} S\overline{y}\exists\overline{x}\, (\phi(\overline{v}, w) \wedge \psi(\overline{x}, \overline{y}, \overline{v}) \wedge \theta(\overline{x})).$$

But then

$$T + \exists\overline{v} S\overline{y}\exists\overline{x}\, [\phi(\overline{v}, d) \wedge \psi(\overline{x}, \overline{y}, \overline{v}) \wedge \theta(\overline{x})]$$

is satisfiable and (b) holds.

The Omitting Types Theorem now allows us to construct $\mathcal{N}$ a countable $\mathbb{A}$-elementary end extension of $\mathcal{M}$ where Condition 2 holds. $\dashv$

6.3. Uncountable models realizing few types

In this section we will collect a number of results where build uncountable models realizing countably few types. Results of this type can be very useful when studying both categoricity phenomena and Vaught counterexamples. We have already seen one example of such a theorem in Theorem 5.2.5 where we showed that if ϕ has arbitrarily large models, then it has arbitrarily large models realizing only countably many $\mathcal{L}_{\omega_1,\omega}$-types.

DEFINITION 6.3.1. If $\mathbb{A}$ is a fragment of $\mathcal{L}_{\omega_1,\omega}$ we say that $\mathcal{M}$ is $\mathbb{A}$-*small* if $\mathcal{M}$ realizes only countably many $\mathbb{A}$-types and we say that $\mathcal{M}$ is *small* if it is $\mathcal{L}_{\omega_1,\omega}$-small.

Recall that Exercise 2.2.20 shows that $\mathcal{M}$ is small if and only if there is a countable $\mathcal{N}$ with $\mathcal{M} \equiv_{\infty,\omega} \mathcal{N}$.

THEOREM 6.3.2. *If ϕ is an $\mathcal{L}_{\omega_1,\omega}$-sentence which has an uncountable model that is $\mathbb{A}$-small for every countable fragment $\mathbb{A}$, then ψ has a small model of cardinality $\aleph_1$.*

PROOF. Let $\mathcal{M} \models \phi$ of cardinality $\aleph_1$ be $\mathbb{A}$-small for all countable fragments of $\mathcal{L}_{\omega_1,\omega}$. Add a binary relation symbol to the signature and let $<$ be a well ordering of $\mathcal{M}$ of order type ω_1. For each $n \geq 1$ we also add $2n+1$-ary relation symbols $E_n(x, \overline{y}, \overline{z})$ and $n+1$-ary functions $f_n(x, \overline{y})$ and a constant 0. We let Ψ be an $\mathcal{L}_{\omega_1,\omega}$-sentence asserting:

(i) 0 is the $<$ least element and $E_n(0, \overline{a}, \overline{b})$ if and only if $\overline{a}$ and $\overline{b}$ satisfy the same quantifier free formulas in the original signature;
(ii) if $\alpha < \beta$ and $E_n(\beta, \overline{a}, \overline{b})$, then $E_n(\alpha, \overline{a}, \overline{b})$;
(iii) every α has a $<$-successor β and $E_n(\beta, \overline{a}, \overline{b})$ if and only if for all c there is d such that $E_{n+1}(\alpha, \overline{a}, c, \overline{b}, d)$ and for all d there is c such that $E_{n+1}(\alpha, \overline{a}, c, \overline{b}, d)$;
(iv) if β is a limit, then $E(\beta, \overline{a}, \overline{b})$ if and only if $E(\alpha, \overline{a}, \overline{b})$ for all $\alpha < \beta$;
(v) $<$ has an initial segment of order type ω and each f_n maps M^{n+1} into this segment;
(vi) $E_n(\beta, \overline{a}, \overline{b})$ if and only if $f_n(\beta, \overline{a}) = f_n(\beta, \overline{b})$.

(i)–(iv) assert that E_n gives the usual Scott analysis in $\mathcal{M}$ and (v) and (vi) assert that each $E_n(\beta, \cdot, \cdot)$ has only countably many equivalence classes.

By Corollary 6.1.9 there is an uncountable $\mathcal{N} \models \phi \wedge \Psi$ embedding $(\mathbb{Q}, <)$. In particular there is an infinite descending chain $d_0 > d_1 > \cdots$. Define $\overline{a} \sim^* \overline{b}$ if and only if $E_n(d_i, \overline{a}, \overline{b})$ for some i. Suppose $\theta(\overline{v})$ is an $\mathcal{L}_{\infty,\omega}$-formula. We prove by induction on subformulas that if $\overline{a} \sim^* \overline{b}$, then $\mathcal{N} \models \theta(\overline{a}) \leftrightarrow \theta(\overline{b})$. For quantifier free formulas this is true by (i) and (ii). By induction this is clear for finitary and infintiary Boolean connectives. Suppose $\theta(\overline{v})$ is $\exists y\, \chi(\overline{v}, y)$. If $\mathcal{N} \models \chi(\overline{a}, c)$, then by (iii) there is d such that $\mathcal{N} \models \chi(\overline{b}, d)$. Thus any

$\sim^*$-equivalent elements realize the same type. By (iv) and (v) there are only countably many $E_n(d_0, \cdot, \cdot)$-classes and, hence, only countably many $\sim^*$-classes. Thus $\mathcal{N}$ is small. ⊣

The existence of a reasonably large uncountable model is enough to imply that there is an uncountable model realizing few $\mathbb{A}$-types for a countable fragment $\mathbb{A}$.

THEOREM 6.3.3. *Let $\mathbb{A}$ be a countable fragment of $\mathcal{L}_{\omega_1,\omega}$. If ϕ has a model of cardinality $(2^{\aleph_0})^+$, then $\mathbb{A}$ has a model of cardinality $\aleph_1$ realizing only countably many $\mathbb{A}$-types.*

PROOF. Let $\mathcal{M}$ be model of ϕ of cardinality $\kappa > 2^{\aleph_0}$. There is $U \subset \mathcal{M}$ such that $|U| = \lambda \leq 2^{\aleph_0}$ and for all $\overline{a} \in \mathcal{M}$ there is $\overline{u} \in U$ such that $\text{tp}^{\mathcal{M}}(\overline{a}, \mathbb{A}) = \text{tp}^{\mathcal{M}}(\overline{u}, \mathbb{A})$. Let Θ be the sentence

$$\bigwedge_{n\in\omega} \forall a_1 \ldots \forall a_n \exists u_1 \ldots \exists u_n \left[\bigwedge_{i=1}^{n} U(u_i) \wedge \bigwedge_{\psi\in\mathbb{A}} \psi(\overline{a}) \leftrightarrow \psi(\overline{u}) \right].$$

Then $(\mathcal{M}, U)$ is a (κ, λ) model of $\phi \wedge \psi$ where $\kappa > \lambda$. By Theorem 6.1.8, there is $(\mathcal{N}, U^{\mathcal{N}})$ an $(\aleph_1, \aleph_0)$-model of $\phi \wedge \psi$. Then $\mathcal{N} \models \phi$ has cardinality $\aleph_1$ and $\mathcal{N}$ realizes countably many $\mathbb{A}$-types. ⊣

The next result is the analog of the classical theorem that an ω-stable theory realizes few types over any model.

THEOREM 6.3.4. *Let $\mathbb{A}$ be a countable fragment. Suppose for all countable $\mathcal{M} \models \phi$ there are only countably many $\mathbb{A}$-types over $\mathcal{M}$. Then for every model $\mathcal{N}$ there are only $|\mathcal{N}|$ $\mathbb{A}$-types over $\mathcal{N}$.*

PROOF. Suppose not. Let $\mathcal{N} \models \phi$ have cardinality κ where there are at least κ^+ $\mathbb{A}$-types over $\mathcal{N}$. For simplicity assume that there are κ^+ 1-types over $\mathcal{N}$. We can find $(\mathcal{N}_\alpha : \alpha < \kappa^+)$ such that $\mathcal{N} \prec_{\mathbb{A}} \mathcal{N}_\alpha$ and there is $a_\alpha \in \mathcal{N}_\alpha$ such that a_α and a_β realize different types over $\mathcal{N}$ for $\alpha \neq \beta$. We may also assume that $\mathcal{N}_\alpha \cap \mathcal{N}_\beta = \mathcal{N}$ for $\alpha \neq \beta$. Let $X = \bigcup_{\alpha<\kappa^+} \mathcal{N}_\alpha$. Consider a structure where we have:

- two sorts, one for X and one for $(\kappa^+, <)$;
- a binary relation $R \subset \kappa^+ \times X$ such that

$$R(\alpha, x) \text{ if and only if } x \in \mathcal{N}_\alpha;$$

- a predicate for $\mathcal{N}$ and $\mathcal{N} \models \phi$;
- all of the τ-structure on each $\mathcal{N}_\alpha$ and $\mathcal{N} \prec_{\mathbb{A}} \mathcal{N}_\alpha$;
- an injection function $f : \kappa^+ \to X$ such that $f(\alpha) = a_\alpha$;
- a bijection $g : \kappa^+ \to X$.

We can write down a sentence Θ such that if $(\mathcal{M}, \mathcal{N}^{\mathcal{M}}) \models \Theta$, then $\mathcal{N}^{\mathcal{M}} \models \phi$ and there are at least $|\mathcal{M}|$-many $\mathbb{A}$-types over $\mathcal{N}^{\mathcal{M}}$. Since Θ has a (κ^+, κ)-model, by Theorem 6.1.8, Θ has an $(\aleph_1, \aleph_0)$-model. But then there is a countable model $\mathcal{N}$ with uncountably many $\mathbb{A}$-types over $\mathcal{N}$. ⊣

EXERCISE 6.3.5. Let $\mathbb{A}$ be a countable fragment. Suppose for all $\mathcal{M} \models \phi$ and $A \subseteq \mathcal{M}$ countable there are only countably many $\mathbb{A}$-types over A. Then for all $\mathcal{M} \models \phi$ and all infinite $B \subseteq \mathcal{M}$ there are only $|B|$ $\mathbb{A}$-types over B.

THEOREM 6.3.6. *Let $\mathbb{A}$ be a countable fragment and suppose $\phi \in \mathbb{A}$ has at least one uncountable model. Then the set of $\mathbb{A}$-types realized in all uncountable models of ϕ is countable. Indeed, if ϕ has ω_1-like models, then the set of $\mathbb{A}$-types realized in all ω_1-like models is countable.*

PROOF. Suppose $\mathcal{N}$ is a model of cardinality $\aleph_1$. Let $<$ be a linear order of $\mathcal{M}$ of order type ω_1 and let $\widehat{\mathbb{A}}$ be a countable fragment of $\mathcal{L}_{\omega_1,\omega}(\tau \cup \{<\})$ containing $\mathbb{A}$. Let $\mathcal{M}$ be a countable $\widehat{\mathbb{A}}$-elementary initial segment of $\mathcal{N}$.

Let X be the set of all $\mathbb{A}$-types realized in every uncountable model of ϕ. Then for $p \in X$ we can not find an ω_1-like end extension of $\mathcal{M}$ omitting p. By Theorem 6.2.1 there is a formula ψ_p with parameters from $\mathcal{M}$ such that

$$\mathcal{M} \models S\overline{y}\exists\overline{x}\, \psi_p(\overline{x}, \overline{y})$$

but for all $\gamma(\overline{x}) \in p$,

$$\mathcal{M} \models \neg S\overline{y}\exists\overline{x}\, (\psi_p(\overline{x}, \overline{y}) \wedge \gamma(\overline{x})).$$

We claim that the map $p \mapsto \psi_p$ is injective.

Suppose for contradiction that $p, q \in X$, $p \neq q$ and $\psi_p = \psi_q$. Let ψ denote $\psi_p = \psi_q$. Suppose $\gamma(\overline{x}) \in p$ and $\neg\gamma(\overline{x}) \in q$. Then

$$\mathcal{M} \models S\overline{y}\exists\overline{x}\, \psi(\overline{x}, \overline{y})$$

but

$$\mathcal{M} \models \neg S\overline{y}\exists\overline{x}\, (\psi(\overline{x}, \overline{y}) \wedge \neg\gamma(\overline{x}))$$

and

$$\mathcal{M} \models \neg S\overline{y}\exists\overline{x}\, (\psi(\overline{x}, \overline{y}) \wedge \gamma(\overline{x})).$$

Since

$$\mathcal{M} \models S\overline{y}\exists\overline{x}\, \psi(\overline{x}, \overline{y}),$$

$$\mathcal{M} \models S\overline{y}\exists\overline{x}\, [(\psi(\overline{x}, \overline{y}) \wedge \gamma(\overline{x})) \vee (\psi(\overline{y}, \overline{x}) \wedge \neg\gamma(\overline{x}))].$$

But $\exists$ and $\exists^*$ commute with finite disjunctions, Thus

$$\mathcal{M} \models S\overline{yx}\, (\psi(\overline{x}, \overline{y}) \wedge \gamma(\overline{x})) \text{ or } \mathcal{M} \models S\overline{yx}\, (\psi(\overline{x}, \overline{y}) \wedge \neg\gamma(\overline{x}))$$

a contradiction. ⊣

We will make use of a special case.

COROLLARY 6.3.7. *If $\mathcal{M}$ is a countable model with an uncountable $\mathbb{A}$-elementary end extension, then the set of $\mathbb{A}$-types realized in every uncountable ω_1-like $\mathbb{A}$-elementary end extensions is countable.*

PROOF. Add constants for the elements of $\mathcal{M}$ and consider the $\mathbb{A}$-theory of $\mathcal{M}$ in this language together with sentences

$$\forall v \left(v < a \rightarrow \bigvee_{b<a} b = v \right)$$

for all $a \in \mathcal{M}$. Now apply the theorem. $\dashv$

THEOREM 6.3.8. *If* $\mathbb{A}$ *is a countable fragment* $\mathcal{L}_{\omega_1,\omega}$ *and* $\mathcal{M} \models \phi$ *realizes uncountably many* $\mathbb{A}$*-types, then* ϕ *has* $2^{\aleph_1}$ *non-isomorphic models of cardinality* $\aleph_1$.

PROOF. For notational simplicity we will assume there are uncountably many 1-types. Fix $\mathcal{M} \models \phi$ of cardinality $\aleph_1$ such that ϕ realizes uncountably many types. Let $\widehat{\tau} = \tau \cup \{<, U, G\}$ and expand $\mathcal{M}$ to a $\widehat{\tau}$-structure where $<$ is a linear order of $\mathcal{M}$ of order type ω_1, U is a subset where any two elements realize distinct $\mathbb{A}$-types, $G : U \rightarrow \mathcal{M}$ is a bijection. Let $\widehat{\mathbb{A}}$ be a countable fragment of $\mathcal{L}_{\omega_1,\omega}(\widehat{\tau})$ where we can express that any two elements of U realize distinct $\mathbb{A}$-types.

We will build a tree of countable models $(\mathcal{M}_f : f \in 2^{<\omega_1}\}$ and $\mathbb{A}$-types $(p_f : f \in 2^{<\omega_1})$ such that:

(i) if $f \subset g$, then $\mathcal{M}_g$ is a $\widehat{\mathbb{A}}$-elementary end extension of $\mathcal{M}_f$;
(ii) if $f \subseteq g$, then p_f is realized in $\mathcal{M}_g$;
(iii) $\mathcal{M}_f$ satisfies Condition 2 of Theorem 6.2.1 for types in $F_f = \{p_{f|\alpha^\frown i} :$ where $\alpha \in \mathrm{dom}(f)$ and $f(\alpha) \neq i\}$, the *forbidden types* for f.

Choose $\mathcal{M}_\emptyset$ a countable model such that $\mathcal{M}$ is an $\widehat{\mathbb{A}}$-elementary end extension. Note that any uncountable $\widehat{\mathbb{A}}$-elementary extension of $\mathcal{M}_\emptyset$ will realize uncountably many $\mathbb{A}$-types.

Let $p_\emptyset$ be an $\mathbb{A}$-type realized in $\mathcal{M}$.

If $\mathrm{dom}(f) = \alpha$ a limit ordinal, let $\mathcal{M}_f = \bigcup_{\beta<\alpha} \mathcal{M}_\beta$. Since Condition 2 is preserved under unions of chains and $F_g \subseteq F_h$ for $g \subset h$, $\mathcal{M}_f$ satisfies (i)–(iii).

Since $\mathcal{M}_f$ satisfies Condition 2, by Theorem 6.2.1, we can find an ω_1-like $\widehat{\mathbb{A}}$-elementary end extension $\mathcal{N}_0$ of cardinality $\aleph_1$ omitting all of the types forbidden in F_f. Since $\mathcal{N}_0$ realizes uncountably many $\mathbb{A}$-types, by Corollary 6.3.7 it realizes an $\mathbb{A}$-type $p_{f^\frown 0}$ that is not realized in every ω_1-like $\widehat{\mathbb{A}}$-elementary end extension of $\mathcal{M}_f$. Let $\mathcal{M}_{f^\frown 0}$ be a $\widehat{\mathbb{A}}$-elementary initial segment of $\mathcal{N}_0$ containing $\mathcal{M}_f$ and realizing $p_{f^\frown 0}$.

Let $\mathcal{N}_1$ be an ω_1-like $\widehat{\mathbb{A}}$-elementary end extension of $\mathcal{M}_f$ of cardinality $\aleph_1$ omitting $p_{f^\frown 0}$. By Corollary 6.3.7, there is an $\mathbb{A}$-type $p_{f^\frown 1}$ realized in $\mathcal{N}_1$ that can be omitted in an ω_1-like $\widehat{\mathbb{A}}$-elementary end extension of $\mathcal{M}_{f^\frown 0}$. Let $\mathcal{M}_{f^\frown 1}$ be an $\widehat{\mathbb{A}}$-elementary initial segment of $\mathcal{N}_1$ containing $\mathcal{M}_f$ and a realization of $p_{f^\frown 1}$. By construction, (i)–(iii) hold.

For $h \in 2^{\omega_1}$, let $\mathcal{M}_h = \bigcup_{\alpha<\omega_1} \mathcal{M}_{h|\alpha}$. If $g \neq h \in 2^{\omega_1}$, then $\mathcal{M}_g$ and $\mathcal{M}_h$ realize different $\mathbb{A}$-types and hence are non-isomorphic as τ-structures. $\dashv$

COROLLARY 6.3.9. *If T is a first order theory in a countable language and there are uncountably many types over $\emptyset$, then T has $2^{\aleph_1}$ non-isomorphic models of cardinality $\aleph_1$.*

This is a precursor to Shelah's result [71] that the same is true for a non-ω-stable theory.

COROLLARY 6.3.10. *If ϕ has uncountable models and fewer than $2^{\aleph_1}$ models of cardinaility $\aleph_1$, then there is a complete $\mathcal{L}_{\omega_1,\omega}$-sentence ψ such that $\psi \models \phi$ and ψ has a model of cardinality $\aleph_1$. In particular, if ϕ is $\aleph_1$-categorical, there is a complete $\aleph_1$-categorical sentence $\psi \models \phi$.*

PROOF. If ϕ has uncountable models and $I(\phi, \aleph_1) < 2^{\aleph_1}$, then there are only countably many $\mathbb{A}$-types for each countable fragment $\mathbb{A}$. By Theorem 6.3.2 and Exercise 2.2.20 the unique model of cardinality $\aleph_1$ is $\mathcal{L}_{\infty,\omega}$-equivalent to a countable model. $\dashv$

6.4. Extending models of set theory

The main result of this section is due to Keisler and Morley [38]. The proof we give is due to Paul Larson (see [5] for expansions of this idea) and his proof that "ϕ has an uncountable model" is absolute.

Start with a countable ω-model $\mathcal{M}$ of ZFC. Our goal is to build a countable elementary extension where we add elements exactly to the sets that $\mathcal{M}$ thinks are uncountable. Originally this was proved by an omitting types argument. Here we use a variant of a Skolem ultrapower argument where the ultrafilter is built by forcing.

Let $\mathcal{P}(\omega_1)^{\mathcal{M}} = \{A \in \mathcal{M} : \mathcal{M} \models A \subseteq \omega_1\}$. Let $P = \{A \in M : \mathcal{M} \models A \subset \omega_1$ contains a stationary set$\}$ ordered by $A \leq B$ if and only if $A \subseteq B$. Suppose $G \subset P$ is an $\mathcal{M}$-generic filter.

LEMMA 6.4.1. *Let $B \in \mathcal{P}(\omega_1)^{\mathcal{M}}$. The set $D_B = \{A \in P : A \subseteq B$ or $A \subseteq \omega_1^{\mathcal{M}} \setminus B\}$ is dense. Thus G is a non-principle ultrafilter on $\omega_1^{\mathcal{M}}$.*

PROOF. Let $A_0 \in P$. Suppose $\mathcal{M} \models S \subseteq A_0$ is stationary. If neither $S \cap B$ nor $S \setminus B$ is stationary, then we can find closed unbounded sets $C_0, C_1 \in \mathcal{M}$ such that $C_0 \cap S \cap B = \emptyset$ and $(C_1 \cap S) \setminus B = \emptyset$. But there is $\alpha \in C_0 \cap C_1 \cap S$, a contradiction. $\dashv$

We use G to build a restricted ultrapower of $\mathcal{M}$. Let $\mathcal{F} = \{f \in \mathcal{M} : \mathrm{dom}(f) = \omega_1\}$. Define $\sim$ on $\mathcal{F}$ by

$$f \sim g \Leftrightarrow \{\alpha : f(\alpha) = g(\alpha)\} \in G.$$

It is easy to see that f is an equivalence relation. Let $[f]$ denote the equivalence class of f and let $\mathcal{N} = \{[f] : f \in \mathcal{F}\}$. Define $\in^{\mathcal{N}}$ by

$$[f] \in^{\mathcal{N}} [g] \Leftrightarrow \{\alpha : f(\alpha) \in g(\alpha)\} \in G.$$

EXERCISE 6.4.2. Show that if $f \sim f_1$ and $g \sim g_1$ then

$$\{\alpha : f(\alpha) \in g(\alpha)\} \in G \Leftrightarrow \{\alpha : f_1(\alpha) \in g_1(\alpha)\} \in G.$$

Thus $\in^{\mathcal{N}}$ is well definied.

Let $j : \mathcal{M} \to \mathcal{N}$ by $j(a) = [f_a]$ where $f_a(\alpha) = a$ for all α.

EXERCISE 6.4.3. Prove that j is an embedding.

EXERCISE 6.4.4. Adapt the proof of the Fundamental Theorem of Ultraproducts to show that j is an elementary embedding. [Hint: Dealing with existential quantifiers will require that $\mathcal{M}$ is a model of the axiom of choice.]

LEMMA 6.4.5. *If $\mathcal{M} \models a$ is countable and $\mathcal{N} \models [f] \in a$, then $\mathcal{N} \models [f] = j(b)$ for some $b \in \mathcal{M}$.*

PROOF. There is a $g : \omega \to a$ a bijection such that $g \in \mathcal{M}$. Since $[f] \in^{\mathcal{N}} a$. There is $A \in G$ such that $f(\alpha) \in a$ for $\alpha \in A$. We need only show that

$$\{S \in P : f \text{ is constant on } S\}$$

is dense below A. Let $B \leq A$. There is $h \in \mathcal{M}$ such that $h : B \to \omega$ and $f(\alpha) = g(h(\alpha))$ for $\alpha \in B$. But B contains a stationary set, and thus there is stationary $S \subseteq B$ such that h is constant on B. ⊣

EXERCISE 6.4.6. Let $i : \omega_1^{\mathcal{M}} \to \omega_1^{\mathcal{M}}$ be the identity map. Show that $[i] \in j(\omega_1^{\mathcal{M}})$ but $[i] \neq j(b)$ for any $b \in \mathcal{M}$.

LEMMA 6.4.7. *If $a \in \mathcal{M}$ and $\mathcal{M} \models a$ is uncountable, then there is $[f] \in \mathcal{N} \setminus \text{img}(j)$ with $f \in j(a)$.*

PROOF. Since $\mathcal{M} \models a$ is uncountable, there is $f \in \mathcal{M}$ such $f : \omega_1^{\mathcal{M}} \to a$ is injective. Clearly $[f] \in j(a)$, but $f \not\sim f_b$ for any $b \in \mathcal{M}$. ⊣

We have now proved the main theorem.

THEOREM 6.4.8 (Keisler-Morley). *Let $\mathcal{M}$ be a countable ω-model of* ZFC. *Then $\mathcal{M}$ has a countable elementary extension $\mathcal{N}$ such that for $a \in \mathcal{M}$ there is $b \in \mathcal{N} \setminus \mathcal{M}$ such that $\mathcal{N} \models b \in a$ if and only if*

$$\mathcal{M} \models a \text{ is uncountable.}$$

We can iterate the construction to build an elementary chain of length ω_1.

COROLLARY 6.4.9. *If $\mathcal{M}$ is countable there is $\mathcal{M} \prec \mathcal{N}$ of cardinality $\aleph_1$ such that*

$$\mathcal{M} \models a \text{ is countable} \Leftrightarrow \{b \in \mathcal{N} : \mathcal{N} \models b \in a\} \text{ is countable.}$$

The absoluteness of uncountable satisfiability. Fix $\phi \in \mathcal{L}_{\omega_1,\omega}$. The following result was proved by Larson and also by Friedman, Hyttinen and Koerwien [20].

COROLLARY 6.4.10. *It is absolute whether ϕ has an uncountable model.*

PROOF. (Sketch) Let ZFC* be a fragment of ZFC that is strong enough to prove all of the facts we needed in the above argument but weak enough that, say, $H(\aleph_2) \models$ ZFC*. We claim that ϕ has an uncountable model if and only if there is an ω-model $\mathcal{M}$ of ZFC* with $\phi \in \mathcal{M}$ and $\mathcal{M} \models$ there is an uncountable model of ϕ.

The ($\Rightarrow$) direction follows because we can take a suitable elementary submodel of $H(\aleph_2)$. For the ($\Leftarrow$) direction start with such an $\mathcal{M}$. By the above construction, we can build an uncountable elementary extension $\mathcal{M} \prec \mathcal{N}$ such that every set that $\mathcal{N}$ believes in uncountable is uncountable. In particular, $\mathcal{N}$ will contain $\mathbb{A}$ uncountable such that $\mathcal{N}$ believe $\mathcal{A} \models \phi$. The tree of subformulas of ϕ is a well founded tree which will be in the model $\mathcal{M}$ and hence in $\mathcal{N}$. We can now do an induction to show that indeed

$$\mathcal{A} \models \psi(\overline{a}) \Leftrightarrow \mathcal{N} \models \text{“}\mathcal{A} \models \psi(\overline{a})\text{”}$$

for all subformulas of ϕ. In particular, $\mathcal{A} \models \phi$.

Thus the existence of an uncountable model is equivalent to a $\Sigma^1_1(\phi)$-formula, and hence absolute. ⊣

The same argument can be used to show that it is absolute whether an $\mathcal{L}_{\omega_1,\omega}(Q)$-sentence has a model where Q is the quantifer “there exists uncountably many”.

Friedman, Hyttinen and Koerwien [20] go on to show it is not absolute whether a sentence has a model in $\aleph_2$ (even assuming GCH). The absoluteness of $\aleph_1$-categoricity is one of the many interesting open questions in the subject.

Baldwin and Larson [5] have expanded the contructions of this style to give a different proof of Theorem 6.3.8 and to generalize it to more expressive logics.

6.5. $\aleph_1$-categorical sentences have models in $\aleph_2$

In this section we will prove the following theorem of Shelah. Our treatment in this section closely follows [2].

THEOREM 6.5.1 (Shelah [69]). *If $\phi \in \mathcal{L}_{\omega_1,\omega}$ is $\aleph_1$-categorical, then ϕ has a model of cardinality $\aleph_2$.*

For the remainder of this section fix $\mathbb{A}$ a countable fragment of $\mathcal{L}_{\omega_1,\omega}$ and $\phi \in \mathbb{A}$.

DEFINITION 6.5.2. We say that $(\mathcal{M}, \mathcal{N},)$ is a *proper pair* witnessed by a if $\mathcal{M} \prec_{\mathbb{A}} \mathcal{N}$ and $a \in \mathcal{N} \setminus \mathcal{M}$.

LEMMA 6.5.3. *Suppose ϕ is κ-categorical, $\mathbb{A}$ is a fragment containing ϕ and $\mathcal{M} \prec_{\mathbb{A}} \mathcal{N}$ is a proper pair of models of cardinality κ, then ϕ has a model of cardinality κ^+.*

PROOF. Build a continuous elementary chain

$$\mathcal{M}_0 \prec_{\mathbb{A}} \mathcal{M}_1 \prec_{\mathbb{A}} \cdots \prec_{\mathbb{A}} \mathcal{M}_\alpha \prec_{\mathbb{A}} \cdots$$

for $\alpha < \kappa^+$, where each $\mathcal{M}_\alpha \models \phi$ has cardinality κ and $(\mathcal{M}_\alpha, \mathcal{M}_{\alpha+1}) \cong (\mathcal{M}, \mathcal{N})$. This is possible since ϕ is κ-categorical so all of the $\mathcal{M}_\alpha \cong \mathcal{M}$. The union is the desired model. ⊣

EXERCISE 6.5.4. Show that if ϕ is κ-categorical and ϕ has a model of cardinality greater than κ, then there is a proper pair $\mathcal{M} \prec_{\mathbb{A}} \mathcal{N}$ of cardinality κ.

DEFINITION 6.5.5. We say that a proper pair $(\mathcal{M}, \mathcal{N}, a)$ is a *maximal triple* if there is no proper pair $(\mathcal{M}_1, \mathcal{N}_1)$ witnessed by a with $\mathcal{M} \prec_{\mathbb{A}} \mathcal{M}_1$ and $\mathcal{N} \preceq_{\mathbb{A}} \mathcal{N}_1$

LEMMA 6.5.6. *If there is a proper pair of cardinality κ but no maximal triple of cardinality κ, then ϕ has a proper pair of cardinality κ^+.*

PROOF. We build continuous elementary chains

$$\mathcal{M}_0 \prec_{\mathbb{A}} \mathcal{M}_1 \prec_{\mathbb{A}} \cdots \prec_{\mathbb{A}} \mathcal{M}_\alpha \prec_{\mathbb{A}} \cdots \text{ and } \mathcal{N}_0 \preceq_{\mathbb{A}} \mathcal{N}_1 \preceq_{\mathbb{A}} \cdots \preceq_{\mathbb{A}} \mathcal{N}_\alpha \preceq_{\mathbb{A}} \cdots$$

of for $\alpha < \kappa^+$ where $\mathcal{M}_\alpha \prec_{\mathbb{A}} \mathcal{N}_\alpha$ is a proper pair of models of cardinality κ witnessed by a. At successor stages we use the fact that $(\mathcal{M}_\alpha, \mathcal{N}_\alpha, a)$ is not maximal. At limit stages we note that the union is still a proper pair witnessed by a. The union of the $\mathcal{M}_\alpha$ is the desired model of cardinality κ^+. ⊣

DEFINITION 6.5.7. We say that $\mathcal{M} \prec_{\mathbb{A}} \mathcal{N}$ is a *cut pair* if there is a descending sequence

$$\mathcal{M} \prec_{\mathbb{A}} \cdots \prec_{\mathbb{A}} \mathcal{N}_{i+1} \prec_{\mathbb{A}} \mathcal{N}_i \prec_{\mathbb{A}} \cdots \prec_{\mathbb{A}} \mathcal{N}_1 \prec_{\mathbb{A}} \mathcal{N}$$

such that $\bigcap_{i\in\omega} \mathcal{N}_i = \mathcal{M}$.

LEMMA 6.5.8. *If ϕ has a uncountable model, then there is a cut pair in $\aleph_0$.*

PROOF. Let $\mathcal{M} \models \phi$ have cardinality $\aleph_1$. We fix a well ordering of $\mathcal{M}$ of order type ω_1 and identify the universe of $\mathcal{M}$ with ω_1. We can find a continuous ω_1-chain of countable models

$$\mathcal{M}_0 \prec_{\mathbb{A}} \cdots \prec_{\mathbb{A}} \mathcal{M}_\alpha \prec_{\mathbb{A}} \cdots$$

such that $\mathcal{M} = \bigcup \mathcal{M}_\alpha$. Let R be a binary relation on $\mathcal{M}$ such that $R(a, \alpha)$ if and only if $a \in \mathcal{M}_\alpha$. Let $\tau^* = \tau \cup \{<, R\}$. We can write down an $\mathcal{L}_{\omega_1,\omega}(\tau)$-sentence asserting that

- $<$ is a linear order with least element;
- $\forall a \exists b\, (R(a, b) \wedge \forall c < b \neg R(a, c))$;
- $\{a : R(a, b)\}$ is a τ-structure that is a model of ϕ for all b and an $\mathbb{A}$-elementary submodel of $\mathcal{M}$;

- if $b < c$, then $\{a : R(a,b)\}$ is a proper $\mathbb{A}$-elementary submodel of $\{a : R(a,c)\}$.

By the Undefinability of Well Order Theorem 4.4.1, there is a countable $\mathcal{N} \models \Theta$ where $\mathbb{Q}$ embeds into $<$. Let $a_0 > a_1 > \cdots$ be an infinite descending sequence in $\mathcal{N}$. Let $\mathcal{N}_i = \{x \in \mathcal{N} : \mathcal{N} \models R(x, a_i)\}$, let $I = \{a \in \mathcal{N} : a < a_i \text{ for all } i\}$. and let $\mathcal{N}^* = \bigcup_{a \in I}\{x : R(x,a)\}$. Then

$$\mathcal{N}^* \prec_{\mathbb{A}} \cdots \prec_{\mathbb{A}} \mathcal{N}_n \prec_{\mathbb{A}} \cdots \prec_{\mathbb{A}} \mathcal{N}_1 \prec_{\mathbb{A}} \cdots \prec_{\mathbb{A}} \mathcal{N}$$

and $\bigcap_{i \in \omega} \mathcal{N}_i = \mathcal{N}^*$. Thus $(\mathcal{N}^*, \mathcal{N})$ is a cut pair. ⊣

We will show that if a sentence has a cut pair and a maximal triple then we can construct many models. We will need the following standard set theoretic fact that is often useful in proving many models theorems. See, for example, [56] 5.3.10 for a proof. Other examples of this method are given in Appendix A.

LEMMA 6.5.9. *Suppose κ is regular. There is a family $(X_\alpha : \alpha < \kappa)$ of disjoint stationary subsets of κ. For $A \subseteq \kappa$, let $X_A = \bigcup_{\alpha \in A} X_\alpha$. Then $(X_A : A \subset \kappa)$ is a family of 2^κ stationary subsets of κ such that $X_A \triangle X_B$ is stationary for $A \neq B$.*

LEMMA 6.5.10. *Suppose ϕ is κ-categorical and ϕ has both a cut pair and a maximal triple of cardinality κ, then there are 2^{κ^+} non-isomorphic models of ϕ of cardinality κ^+.*

PROOF. Let $(\mathcal{M}, \mathcal{N}, a)$ be a maximal triple and let $(\mathcal{M}', \mathcal{N}')$ be a cut pair. For any $X \subseteq \kappa^+$ we build $\mathcal{M}(X) \models \phi$ as follows: We build a continuous κ^+-chain

$$\mathcal{M}_0(X) \prec_{\mathbb{A}} \cdots \prec_{\mathbb{A}} \mathcal{M}_\alpha(X) \prec_{\mathbb{A}} \cdots$$

such that if $\alpha \in X$, then

$$(\mathcal{M}_\alpha(X), \mathcal{M}_{\alpha+1}(X)) \cong (\mathcal{M}, \mathcal{N}),$$

while if $\alpha \notin X$, then

$$(\mathcal{M}_\alpha(X), \mathcal{M}_{\alpha+1}(X)) \cong (\mathcal{M}', \mathcal{N}').$$

Let $\mathcal{M}(X) = \bigcup_{\alpha < \kappa^+} \mathcal{M}_\alpha(X)$.

Suppose X and Y are stationary subsets of κ^+ such that $X \setminus Y$ is stationary. We claim that $\mathcal{M}(X) \ncong \mathcal{M}(Y)$. For contradiction, suppose $f : \mathcal{M}(X) \to \mathcal{M}(Y)$ is an isomorphism. Then we can find $\alpha \in X \setminus Y$ such that $f|\mathcal{M}_\alpha(X)$ is an isomorphism between $\mathcal{M}_\alpha(X)$ and $\mathcal{M}_\alpha(Y)$. There is a such that $(\mathcal{M}_\alpha(X), \mathcal{M}_{\alpha+1}(X), a)$ is a maximal triple. Since $(\mathcal{M}_\alpha(Y), \mathcal{M}_{\alpha+1}(Y))$ is a cut pair, there is $\mathcal{M}^*$ such that $\mathcal{M}_\alpha(Y) \prec_{\mathbb{A}} \mathcal{M}^* \preceq_{\mathbb{A}} \mathcal{M}_{\alpha+1}(Y)$ such that $f(a) \notin \mathcal{M}^*$. Choose $\beta \geq \alpha + 1$ such that $f^{-1}(\mathcal{M}^*) \prec_{\mathbb{A}} \mathcal{M}_{\beta(X)}$. Since

- $\mathcal{M}_\alpha(X) \prec_{\mathbb{A}} f^{-1}(\mathcal{M}^*) \prec_{\mathbb{A}} \mathcal{M}_\beta(X)$
- $\mathcal{M}_{\alpha+1}(X) \preceq_{\mathbb{A}} \mathcal{M}_\beta(X)$ and
- $a \in \mathcal{M}_{\beta(X)} \setminus f^{-1}(\mathcal{M}^*)$

this contradicts the fact that $(\mathcal{M}_\alpha(X), \mathcal{M}_{\alpha+1}(X), a)$ is a maximal pair.

Using Lemma 6.5.9 we can construct 2^{κ^+} non-isomorphic models of cardinality κ^+. ⊣

PROOF OF THEOREM 6.5.1. Suppose ϕ is $\aleph_1$-categorical. By Corollary 6.3.10 we may assume that ϕ is complete. By Lemma 6.5.8, there is a cut pair in $\aleph_0$. Since there is a cut pair, by Lemma 6.5.10 there is no maximal triple of cardinality $\aleph_0$. But then, by Lemma 6.5.6. There is a proper pair of cardinality $\aleph_1$ and by Lemma 6.5.3 a model of cardinality $\aleph_2$. ⊣

In fact Shelah [69] proves that the same is true for $\mathcal{L}_{\omega_1,\omega}(Q)$ where Q is the quantifier "there exists uncountably many." See chapter 7 of [2] for a proof.

How many models will ϕ have in $\aleph_2$? Possibly, the maximal number.

THEOREM 6.5.11 (Shelah). *Suppose $2^\kappa < 2^{\kappa^+}$, ϕ is κ-categorical and amalgamation fails in κ. Then there are 2^{κ^+} non-isomorphic models of ϕ of cardinality κ^+.*

See chapter 17 of [2] for a proof.

Chapter 7

VAUGHT COUNTEREXAMPLES

Suppose $\phi \in \mathcal{L}_{\omega_1,\omega}$ is a Vaught counterexample, i.e., ϕ is scattered with uncountably many countable models. What can we say about uncountable models of ϕ? The first results due to Harnik and Makkai show that there are always models of ϕ of cardinality $\aleph_1$. This was refined in an unpublished result of Harrington who proved that a Vaught counterexample has models of arbitrarily large Scott rank below ω_2. While most model theorists believe Vaught's Conjecture fails, Harrington's result combined with Hjorth's Theorem from Section 5.4 raise the seductive possibility of proving Vaught's Conjecture by strengthening Harrington's result to build a model of size $\aleph_2$ contradicting Hjorth's Theorem. Of course, all of the constructions we know of models of size $\aleph_1$ are built up from countable approximations and we have few good techniques for building a model of size $\aleph_2$.

Harrington's result shows that if ϕ is a Vaught counterexample then $I(\phi, \aleph_1) \geq \aleph_2$. An interesting open problem is if this can be extended to show $I(\phi, \aleph_1) = 2^{\aleph_1}$. Using two fundamental results of Shelah, Baldwin [1] noticed that this is true for first order theories.

PROPOSITION 7.0.12. *Suppose T is a complete first order theory in a countable language which is a Vaught counterexample. Then $I(T, \aleph_1) = 2^{\aleph_1}$.*

PROOF. Shelah [73] proved Vaught's Conjecture for ω-stable theories. Thus T is not ω-stable. But Shelah [71] also proved that a non ω-stable theory has $2^{\aleph_1}$ non-isomorphic models of cardinality $\aleph_1$. ⊣

In the remainder of this chapter we will prove the results of Harnik, Makkai and Harrington.

7.1. Minimal counterexamples

We will prove the following theorem of Harnik and Makkai.

THEOREM 7.1.1. *Let ϕ be a Vaught counterexample. There is $\mathcal{M} \models \phi$ of cardinality $\aleph_1$ that is not $\mathcal{L}_{\infty,\omega}$-equivalent to a countable model.*

DEFINITION 7.1.2. Suppose $\phi \in \mathcal{L}_{\omega_1,\omega}$ is a Vaught counterexample. We say that ϕ is a *minimal counterexample* if for every sentence $\psi \in \mathcal{L}_{\omega_1,\omega}$ either $\phi \wedge \psi$ or $\phi \wedge \neg\psi$ has at most countably many countable models.

Minimal counterexamples were introduced by Harnik and Makkai for this proof, but they are used in a number of results around Vaught's Conjecture.

LEMMA 7.1.3. *If ϕ is a Vaught counterexample, there is a minimal counterexample θ such that $\theta \models \phi$.*

PROOF. For purposes of contradiction, suppose there is no minimal counterexample θ with $\theta \models \phi$. We will build a tree of counterexamples $(\phi_\eta : \eta \in 2^{<\omega})$ such that:

(i) $\phi_\emptyset = \phi$;
(ii) $\phi_\eta \models \phi_\nu$ for $\nu \subseteq \eta$;
(iii) $\phi_{\eta,0} \wedge \phi_{\eta,1}$ is unsatisfiable, for all η.

This is easy to do. Given ϕ_η, because ϕ_σ is not a minimal counterexample, there is ψ such that $\phi_\eta \wedge \psi$ and $\phi_\eta \wedge \neg\psi$ are Vaught counterexamples. Let $\phi_{\eta,0} = \phi_\sigma \wedge \psi$ and let $\phi_{\eta,1} = \phi_\eta \wedge \neg\psi$.

For $f \in 2^\omega$, let

$$T_f = \{\phi_\eta : \eta \subset f\}.$$

Suppose we additionally have

(iv) each T_f is satisfiable.

If each T_f is satisfiable, then $I(\phi, \aleph_0) = 2^{\aleph_0}$ so ϕ is not a Vaught counterexample. Indeed, if $\alpha < \omega_1$ is a bound on the quantifier ranks of the formulas $(\phi_\eta : \eta \in 2^{<\omega})$, then there are uncountably many models that are not $\equiv_\alpha$, and hence a perfect set of non-isomorphic models.

Unfortunately, condition (iv) is not automatic so we will need to build the tree with more care. Add C a countable set of new constant symbols to the language. Let $\Sigma = \{\sigma : \sigma$ is a finite set of $\mathcal{L}_{\omega_1,\omega}$-sentences using only finitely many constant symbols from C such that $\sigma \cup \{\phi\}$ has uncountably many countable models$\}$.

CLAIM. *Σ is a consistency property.*

We'll only check (C4). Suppose $\bigvee_{\psi \in X} \psi \in \sigma \in \Sigma$. Since there are uncountably many models of σ and X is countable, for some $\psi \in X$ there are uncountably many models of $\sigma \cup \{\phi\}$.

If $\sigma \in \Sigma$, and $\theta(\overline{v})$ is an $\mathcal{L}_{\omega_1,\omega}$-formula in the original vocabulary such that $\theta(\overline{c})$ is the conjunction of $\sigma \cup \{\phi\}$. Since there are uncountably many models of $\sigma \cup \{\phi\}$, there are uncountably many models of $\exists\overline{v}\,\theta(\overline{v})$. Thus $\exists\overline{v}\,\theta(\overline{v})$ is a Vaught counterexample.

We will build a sequence of countable fragments $\mathbb{A}_0 \subseteq \mathbb{A}_1 \subseteq \mathbb{A}_2 \subseteq \cdots$ and $\mathbb{A} = \bigcup \mathbb{A}_n$. We will let $\chi_0, \chi_1, \ldots$ list all $\mathbb{A}$-sentences and $t_0, t_1, \ldots$ list all $\mathbb{A}$-terms, both lists with infinite repetition. We will also build a tree $(\sigma_\eta : \eta \in 2^{<\omega})$ such that:

(i) $\sigma_\emptyset = \{\phi\}$;
(ii) $\sigma_\eta \subseteq \sigma_\nu$ if $\eta \subseteq \nu$;
(iii) if $|\eta| = n$ and $\sigma_\eta \cup \{\chi_n\} \in \Sigma$, then $\chi_n \in \sigma_{\eta,i}$ for $i = 0, 1$, in addition
  - if $\chi_n = \bigvee_{\psi \in X} \psi$, then there is $\psi \in X$ such that $\psi \in \sigma_{\eta,i}$ for $i = 0, 1$ and
  - if $\chi_n = \exists v\, \psi(v)$, then for some constant $c \in C$, $\psi(c) \in \sigma_{\eta,i}$ for $i = 0, 1$;

(iv) if $|\eta| = n$, then there is $c \in C$ such that $t_n = c \in \sigma_{\eta,i}$ for $i = 1, 2$;
(v) for each η there is an $\mathcal{L}_{\omega_1,\omega}$-sentence ψ with no constant symbols from C such that $\psi \in \sigma_{\eta,0}$ and $\neg\psi \in \sigma_{\eta,1}$.

Because Σ is a consistency property, given $\sigma_\eta \in \Sigma$ we can find $\sigma_\eta \subset \sigma'_\eta \in \Sigma$ satisfying (iii) and (iv). Let $\theta(\overline{v})$ be the formula in the original vocabulary such that $\theta(\overline{c})$ is the conjunction of σ'_η. We argued above that $\exists \overline{v}\, \theta(\overline{v})$ is a Vaught counterexample implying ϕ. Since we are assuming there are no minimal counterexamples, we can find a sentence ψ such that there are uncountably many countable models of both $\psi \wedge \exists \overline{v}\, \theta(\overline{v})$ and $\neg\psi \wedge \exists \overline{v}\, \theta(\overline{v})$. Let $\sigma_{\eta,0} = \sigma'_\eta \cup \{\psi\}$ and $\sigma_{\eta,1} = \sigma'_\eta \cup \{\neg\psi\}$.

The final bookkeeping detail is that we let $\mathbb{A}_n$ be the smallest fragment containing all formulas in all σ_η for all $|\eta| \leq n$. Although our fragment is expanding during the construction it is easy to build the necessary lists $\chi_0, \chi_1, \ldots$ and $t_0, t_1, \ldots$.

For $f \in 2^\omega$, let $T_f = \bigcup_{\eta \subset f} \sigma_\eta$. As in the proof of the Model Existence Theorem 4.1.6, (iii) and (iv) guarantee that there is a canonical countable model of T_f and condition (v) insures that these countable models are $\not\equiv_{\mathbb{A}}$ in the original vocabulary. Thus there are perfectly many non-isomorphic models of ϕ, a contradiction. ⊣

We will need a refinement of this Lemma.

DEFINITION 7.1.4. Let $\phi \in \mathcal{L}_{\omega_1,\omega}(\tau)$ be a Vaught counterexample. Suppose $\psi(\overline{v})$ is an $\mathcal{L}_{\omega_1,\omega}(\tau)$-formula. We say that ψ is *ϕ-large* if there are uncountably many models of $\phi \wedge \exists \overline{v}\, \psi(\overline{v})$. We say that a ϕ-large formula $\psi(\overline{v})$ is *minimal ϕ-large* if for all $\mathcal{L}_{\infty,\omega}$-formulas $\chi(\overline{v})$ exactly one of $\psi(\overline{v}) \wedge \chi(\overline{v})$ and $\psi(\overline{v}) \wedge \neg\chi(\overline{v})$ is ϕ-large.

COROLLARY 7.1.5. *If $\psi(\overline{v})$ is ϕ-large, then there is a minimal ϕ-large $\theta(\overline{v})$ such that $\theta(\overline{v}) \models \psi(\overline{v})$.*

PROOF. If ψ has free variables $v_1, \ldots, v_n$ add new constants $d_1, \ldots, d_n$ and apply Theorem 7.1.3 to $\phi \wedge \psi(\overline{d})$. ⊣

PROOF OF THEOREM 7.1.1. Let ϕ be a Vaught counterexample. By Theorem 7.1.3 we may, without loss of generality, assume that ϕ is a minimal counterexample.

Our proof will need fragments with extra closure properties. We say that a fragment $\mathbb{A}$ is *rich* if for any ϕ-large $\psi(\overline{v}) \in \mathbb{A}$, there is a minimal ϕ-large $\theta(\overline{v}) \in \mathbb{A}$ with $\theta \models \phi$. By Corollary 7.1.5, for any countable fragment $\mathbb{A}$ with $\phi \in \mathbb{A}$, we can find a countable $\mathbb{A}' \supseteq \mathbb{A}$ where $\mathbb{A}'$ is rich.

If $\mathbb{A}$ is a countable rich fragment of $\mathcal{L}_{\omega_1,\omega}$, let $T_{\mathbb{A}} = \{\psi \in \mathbb{A} : \psi$ is a sentence such that $\phi \wedge \psi$ has uncountably many countable models$\}$. Since ϕ is a minimal counterexample, for every $\psi \in \mathbb{A}$ exactly one of ψ and $\neg\psi$ is in $T_{\mathbb{A}}$. Moreover, if ψ is an $\mathbb{A}$-sentence not in $T_{\mathbb{A}}$, then ψ has only countably many countable models. Since $\mathbb{A}$ is countable, there are uncountably many countable models of $T_{\mathbb{A}}$.

If $\psi(\overline{v})$ is an $\mathbb{A}$-formula and $T_{\mathbb{A}} \models \exists v\, \psi(\overline{v})$, then $\psi(\overline{v})$ is ϕ-large. Thus all $\mathbb{A}$-formulas consistent with $T_{\mathbb{A}}$ are ϕ-large.

If $\theta(\overline{v})$ is a minimal ϕ-large formula, then for any $\chi(\overline{v}) \in \mathcal{L}_{\omega_1,\omega}$,

$$T_{\mathbb{A}} \models \theta(\overline{v}) \rightarrow \chi(\overline{v}) \text{ or } T_{\mathbb{A}} \models \theta(\overline{v}) \rightarrow \neg\chi(\overline{v}).$$

Thus the minimal ϕ-large formulas are complete and consistent $\mathbb{A}$-formula consistent with $T_{\mathbb{A}}$ is implied by a minimal ϕ-large formula in $\mathbb{A}$. Thus $T_{\mathbb{A}}$ is $\mathbb{A}$-atomic. By Theorem 4.2.11 there is an $\mathbb{A}$-prime model of $T_{\mathbb{A}}$.

We now describe the basic construction. For $\alpha < \omega_1$ we build countable rich fragments

$$\mathbb{A}_0 \subseteq \mathbb{A}_1 \subseteq \cdots \subseteq \mathbb{A}_\alpha \subseteq \cdots.$$

Let $\mathbb{A}_0$ be a countable rich fragment containing ϕ.

Given $\mathbb{A}_\alpha$ and $\mathcal{M}_\alpha$ the $\mathbb{A}_\alpha$-prime model of T_α. Let $\mathbb{A}_{\alpha+1}$ be a countable rich fragment containing Φ_α the Scott sentence of $\mathcal{M}_\alpha$. Let $\mathcal{M}_{\alpha+1}$ be the prime model of $T_{\mathbb{A}_{\alpha+1}}$. Since $T_{\mathbb{A}_\alpha} \subset T_{\mathbb{A}_{\alpha+1}}$ we may assume $\mathcal{M}_\alpha \prec_{\mathbb{A}_\alpha} \mathcal{M}_{\alpha+1}$. Of course $\Phi_\alpha \notin T_{\mathbb{A}_{\alpha+1}}$.

If β is a limit ordinal, let $\mathcal{M}_\beta = \bigcup_{\alpha<\beta} \mathcal{M}_\alpha$ and $\mathbb{A}_\beta = \bigcup_{\alpha<\beta} \mathbb{A}_\alpha$. For all $\overline{a} \in M_\beta$, there is a minimal ϕ-large formula $\theta(\overline{v}) \in A_\beta$ such that $\mathcal{M}_\beta \models \theta(\overline{a})$ and every ϕ-large formula in $\mathbb{A}_\beta$ is implied by a minimal ϕ-large formula in $\mathbb{A}_\beta$. Thus $\mathcal{M}_\beta$ is the prime model of $T_{\mathbb{A}_\beta}$.

Note that Φ_α the Scott sentence of $\mathcal{M}_\alpha$ is not it $T_{\mathbb{A}_{\alpha+1}}$. Thus $\neg\Phi_\alpha \in T_{\mathbb{A}_{\alpha+1}}$. Hence $\mathcal{M}_\alpha \not\cong \mathcal{M}_\beta$ if $\alpha \neq \beta$.

Let $\mathcal{M} = \bigcup_{\alpha<\omega_1} \mathcal{M}_\alpha$. Suppose Φ is the Scott sentence of a countable $\mathcal{N} \models \phi$. We claim that $\mathcal{M} \not\models \Phi$. Suppose for contradiction that $\mathcal{M} \models \Phi$. By Exercise 1.1.16, $\{\alpha < \omega_1 : \mathcal{M}_\alpha \models \Phi\}$ is closed unbounded. But if $\mathcal{M}_\alpha, \mathcal{M}_\beta \models \Phi$ then $\mathcal{M}_\alpha \cong \mathcal{M}_\beta$, a contradiction. Thus $\mathcal{M} \not\equiv_{\infty,\omega} \mathcal{N}$ for any countable model $\mathcal{N}$. ⊣

Remark 7.1.6. One might wonder if it would be possible to do a similar construction if we did not have a scattered ϕ. Namely, does every sentence

ϕ of $\mathcal{L}_{\omega_1,\omega}$ with models of arbitrarily large Scott rank have an uncountable model? Makkai [51] shows the answer is no. Let ϕ be the Scott sentence of the lexicographic order on $\mathbb{N} \times \mathbb{Z}$. This sentence has no uncountable models and Makkai shows that this structure has expansions of arbitrarily large countable Scott rank. Thus if let τ be universal countable vocabulary, the sentence ϕ, when viewed as an τ-sentence, has models of arbitrarily large Scott rank.

Once we know that Vaught counterexamples have uncountable models we can shows there is a small model of size $\aleph_1$.

COROLLARY 7.1.7 (Makkai). *Every Vaught counterexample has a small uncountable model.*

PROOF. Let ϕ be a counterexample. We saw in Theorem 7.1.1 that every counterexample has an uncountable model. Since ϕ is scattered, that model is $\mathbb{A}$-small for every countable fragment $\mathbb{A}$. Thus, by Theorem 6.3.2, ϕ has an uncountable small model. ⊣

Makkai's orginal proof in [49] used admissible model theory and $\Sigma_{\mathbb{A}}$-saturated models. The proof we give here is due to Baldwin [1].

EXERCISE 7.1.8. If ϕ is Vaught counterexample. There are $\aleph_1$-countable models, that are $\equiv_{\infty,\omega}$ to models of cardinality $\aleph_1$. [Hint: Suppose $\Phi_1, \Phi_2, \ldots$ are Scott sentences for countably many models of ϕ. Consider $\phi \wedge \bigwedge_{n=1}^{\infty} \neg\Phi_n$.]

7.2. Harrington's Theorem

Harrington showed that a counterexample to Vaught's Conjecture has models of arbitrarily large Scott rank below ω_2. His original proof used admissibility. This proof was never published. I learned of it from lecture notes from a course Victor Harnik gave at Dartmouth in 1974. Recently new proofs have been given in [47], [4] and [42] which avoid admissibility. The proof we will follow is due to Baldwin, Friedman, Koerwien and Laskowski [4].

The Morley tree. Fix $\phi \in \mathcal{L}_{\omega_1,\omega}$ scattered with uncountably many countable models, i.e., models of arbitrarily high countable Scott rank. We continue the analysis from Section 3.3. Recall that we constructed a series of countable fragments

$$\mathbb{A}_0 \subset \mathbb{A}_1 \subset \cdots \subset \mathbb{A}_\alpha \subset \cdots$$

for $\alpha < \omega_1$ such that $\phi \in \mathbb{A}_0$, $\mathbb{A}_{\alpha+1}$ is obtained by adding formulas $\bigwedge_{\psi\in p} \psi(\overline{v})$ for each $\mathbb{A}_\alpha$-type p and $\mathbb{A}_\alpha = \bigcup_{\beta<\alpha} \mathbb{A}_\beta$ for each limit ordinal.

We build a tree of theories $\mathcal{T}$. For $\alpha < \omega_1$, we let $\mathcal{T}_\alpha$ be the elements of $\mathcal{T}$ of height α. An $\mathbb{A}_\alpha$-theory T is in $\mathcal{T}_\alpha$ if and only if

(i) $\phi \in T$;
(ii) T is satisfiable;
(iii) T is $\mathbb{A}_\alpha$-complete, ie., $\psi \in T$ or $\neg\psi \in T$ for any $\psi \in \mathbb{A}_\alpha$;

(iv) for all $\beta < \alpha$, $T \cap \mathbb{A}_\beta$ is not complete, i.e., not $\aleph_0$-categorical.

Thus the terminal nodes on $\mathcal{T}$ are the $\aleph_0$-categorical theories. We let $\mathcal{T}_{<\alpha}$ denote $\bigcup_{\beta<\alpha} \mathcal{T}_\beta$. We let $\mathcal{T} = \bigcup_{\alpha<\omega_1} \mathcal{T}_\alpha$.

EXERCISE 7.2.1. Show that if $T \in \mathcal{T}_\alpha$ and $\beta < \alpha$, then $T \cap \mathbb{A}_\beta \in \mathcal{T}_\beta$.

Note that since ϕ is scattered, each $T \in \mathcal{T}_\alpha$ is $\mathbb{A}_\alpha$-atomic.

EXERCISE 7.2.2. (a) Suppose $\alpha < \beta$, $T_0 \in \mathcal{T}_\alpha$, $T_1 \in \mathcal{T}_\beta$ and $T_0 \subset T_1$. Let $\mathcal{M}_0$ be the $\mathbb{A}_\alpha$-prime model of T_0 and $\mathcal{M}_1$ be the $\mathbb{A}_\beta$-prime model of T_1, then there is an $\mathbb{A}_\alpha$-elementary embedding of $\mathcal{M}_0$ into $\mathcal{M}_1$.

(b) Suppose α is a limit ordinal and $(T_\beta : \beta < \alpha)$ is a chain of theories such that $T_\beta \in \mathcal{T}_\beta$ and $T_\gamma \subset T_\beta$ for all $\gamma < \beta$. Then $\bigcup_{\beta<\alpha} T_\beta \in \mathcal{T}_\alpha$. [Hint: the only difficulty should be proving that this theory is satisfiable, and (a) should be helpful in showing this.]

EXERCISE 7.2.3. If $\mathcal{M} \models \phi$ is countable, there is a terminal node T in the Morley tree such that $\mathcal{M} \models T$.

EXERCISE 7.2.4. Show $\mathcal{T}_\alpha \neq \emptyset$ for all $\alpha < \omega_1$.

EXERCISE 7.2.5. (a) Suppose α is a limit ordinal. Prove that there is $\beta < \alpha$ and $T \in \mathcal{T}_\beta$ with a unique extension in $\mathcal{T}_\alpha$. [Hint: if not build a perfect tree of theories in $\mathcal{T}_{<\alpha}$.]

(b) Show that there is $T \in \mathcal{T}$ and a closed unbounded set $C \subset \omega_1$ such that T has a unique extension in $\mathcal{T}_\alpha$ for each $\alpha \in C$. [Hint: Fodor's Theorem]

(c) Let $(T_\alpha : \alpha \in C)$ be the sequence of unique extensions of T to $\mathcal{T}_\alpha$. Show that $\bigcup_{\alpha\in C} T_\alpha$ is satisfiable. [Hint: Consider the chain of models $(\mathcal{M}_\alpha : \alpha \in C)$ where $\mathcal{M}_\alpha$ is $\mathbb{A}_\alpha$-prime.]

LEMMA 7.2.6. *If $\alpha < \omega_1$ is a limit ordinal and $\mathcal{M} \models T \in \mathcal{T}_\alpha$ then the Scott rank of $\mathcal{M}$ is at least α.*

PROOF. Let $\mathcal{M} \models \phi$. We prove that for all $\overline{a} \in \mathcal{M}$ and $\alpha < \omega_1$ there is $\Psi^{\mathcal{M}}_{\overline{a},\alpha}(\overline{v}) \in \mathbb{A}_{\alpha+1}$ such that $\mathcal{M} \models \Psi^{\mathcal{M}}_{\overline{a},\alpha}(\overline{a})$ and

$$\models \Psi^{\mathcal{M}}_{\overline{a},\alpha}(\overline{v}) \rightarrow \Phi^{\mathcal{M}}_{\overline{a},\alpha}(\overline{v}).$$

Clearly $\mathrm{tp}^{\mathcal{M}}(\overline{a}/\mathbb{A}_0)$ implies $\Phi^{\mathcal{M}}_{\overline{a},0}$ which is the conjunction of atomic and negated atomic formulas satisfied by $\overline{a}$. We let $\Psi^{\mathcal{M}}_{\overline{a},0}(\overline{v})$ be the conjunction of $\mathrm{tp}^{\mathcal{M}}(\overline{a}, \mathbb{A}_0)$.

Let X be the set of $\mathbb{A}_\alpha$-types $q(\overline{v}, w)$ with $\Psi^{\mathcal{M}}_{\overline{a},\alpha}(\overline{v}) \in q(\overline{v}, w)$ such that $q(\overline{a}, w)$ is not realized in $\mathcal{M}$. Since ϕ is scattered, X is countable. For $q \in X$ let $\theta_q(\overline{v}, w) \in \mathbb{A}_{\alpha+1}$ be the conjunction of q. Then $\mathrm{tp}^{\mathcal{M}}(\overline{a}, \mathbb{A}_{\alpha+1})$ implies

$$\forall w\left(\bigwedge_{b\in\mathcal{M}} \Psi_{\overline{a},b,\alpha}(\overline{v}, w) \wedge \bigwedge_{q\in X} \neg\theta_q(\overline{v}, w)\right).$$

Let $\Psi^{\mathcal{M}}_{\overline{a},\alpha+1}$ be the conjunction of $\text{tp}^{\mathcal{M}}(\overline{a}, \mathbb{A}_{\alpha+1})$. Note that

$$\models \Psi^{\mathcal{M}}_{\overline{a},\alpha+1}(\overline{v}) \rightarrow \forall w \bigvee_{b\in\mathcal{M}} \Psi_{\overline{a},b}(\overline{v}, \overline{w})$$

since we have insured that no other types are realized. Thus $\models \Psi^{\mathcal{M}}_{\overline{a},\alpha+1} \Rightarrow \Phi^{\mathcal{M}}_{\overline{a},\alpha+1}$.

If α is a limit ordinal then, by induction $\text{tp}^{\mathcal{M}}(\overline{a}, \mathbb{A}_\alpha)$ implies $\Phi^{\mathcal{M}}_{\overline{a},\alpha}$, so we can take $\Psi^{\mathcal{M}}_{\overline{a},\alpha}$ to be the conjunction of $\text{tp}^{\mathcal{M}}(\overline{a}, \mathbb{A}_\alpha)$.

Suppose now that $\alpha < \omega_1$ is a limit ordinal and $\mathcal{M} \models \phi$ has Scott rank $\beta < \alpha$. If $\overline{a} \in \mathcal{M}$, then

$$\models \Psi^{\mathcal{M}}_{\overline{a},\beta}(\overline{v}) \rightarrow \Phi^{\mathcal{M}}_{\overline{a},\beta}(\overline{v}),$$

but

$$\models \Phi^{\mathcal{M}}_{\overline{a},\beta}(\overline{v}) \rightarrow \theta(\overline{v})$$

for any formula $\theta \in \mathcal{L}_{\omega_1,\omega}$ such that $\mathcal{M} \models \theta(\overline{a})$. In particular, this is true if θ is $\Psi^{\mathcal{M}}_{\overline{a},\beta+1}$. Thus

$$\mathcal{M} \models \forall \overline{v}\, (\Psi^{\mathcal{M}}_{\overline{a},\beta}(\overline{v}) \rightarrow \Psi^{\mathcal{M}}_{\overline{a},\beta+1}(\overline{v}))$$

for any $\overline{a} \in \mathcal{M}$. We can do a back-and-forth argument to show that the $\mathbb{A}_{\beta+2}$-theory of $\mathcal{M}$ is $\aleph_0$-categorical. Thus $\mathcal{M}$ is not a model of any $T \in \mathcal{T}_\alpha$. ⊣

In Chapter 12 we will need the next result from [59].

COROLLARY 7.2.7. *Suppose ϕ is a minimal Vaught counterexample. There is a closed unbounded $C \subset \omega_1$ such that if $\mathcal{M}, \mathcal{N} \models \phi$ and $\mathcal{M}$ and $\mathcal{N}$ have Scott rank at least α, then $\mathcal{M} \equiv_\alpha \mathcal{N}$.*

PROOF. For each α let

$$T_\alpha = \{\psi \in \mathbb{A}_\alpha : \phi \wedge \psi \text{ has } \aleph_1 \text{ countable models}\}.$$

Since ϕ is a minimal counterexample, T_α is $\mathbb{A}_\alpha$-complete, satisfiable, and all but countably many countable models of ϕ are models of T_α. Thus there is γ such that if $\text{SR}(\mathcal{M}) \geq \gamma$ then $\mathcal{M} \models T_\alpha$.

Let $g : \omega_1 \rightarrow \omega_1$ such that if $\text{SR}(\mathcal{M}) \geq g(\alpha)$, then $\mathcal{M} \models T_\alpha$. Let

$$C = \{\alpha < \omega_1 : \alpha \text{ is a limit and } g(\beta) < \alpha \text{ for all } \beta < \alpha\}.$$

Suppose $\alpha \in C$, $\mathcal{M}, \mathcal{N} \models \phi$ and $\text{SR}(\mathcal{M}), \text{SR}(\mathcal{N}) \geq \alpha$. Then $\mathcal{M}$ and $\mathcal{N}$ are models of T_α. Thus $\mathcal{M} \equiv_{\mathbb{A}_\alpha} \mathcal{N}$. By the arguments above $\mathcal{M} \equiv_\alpha \mathcal{N}$. ⊣

Twisted direct limits. We make a short digression to discuss a construction we will use in the main proof.

Let μ be a limit ordinal.

- Let $(\tau_\alpha : \alpha < \mu)$ be sequence of vocabularies with

  $$\tau_0 \subseteq \cdots \subseteq \tau_\alpha \subset \cdots$$

  and $\tau_\alpha = \bigcup_{\beta<\alpha} \tau_\beta$ for α a limit ordinal.
- Suppose for each $\alpha < \mu$ we have $\mathbb{A}_\alpha$ a fragment of $\mathcal{L}_{\infty,\omega}(\tau_\alpha)$.

DEFINITION 7.2.8. If $\tau \subseteq \tau'$, $\mathbb{A}$ and $\mathbb{A}'$ are fragments of $\mathcal{L}_{\infty,\omega}(\tau)$ and $\mathcal{L}_{\infty,\omega}(\tau')$ respectively and $\pi : \mathbb{A} \to \mathbb{A}'$ is an injection, then we say that π is a *fragment embedding* if and only if

- π is the identity on atomic τ-formulas;
- the free variables of $\pi(\psi)$ are the same as the free variables of ψ for $\psi \in \mathbb{A}$;
- π commutes with first order Boolean connectives $\wedge, \vee, \neg$ and quantification $\exists v$ and $\forall v$;
- ψ is an infinite disjunction (conjunction) if and only if $\pi(\psi)$ is an infinite disjunction (conjunction); and if ψ is an infinite disjunction (conjunction) then θ is a disjunct (conjunct) of ψ if and only if $\pi(\theta)$ is a disjunct (conjunct) of $\pi(\psi)$, for $\psi, \theta \in \mathbb{A}$.

Suppose that we have fragment embeddings $\pi_{\alpha,\beta} : \mathbb{A}_\alpha \to \mathbb{A}_\beta$ for all $\alpha < \beta$. We say that $(\pi_{\alpha,\beta} : \alpha, \beta)$ is a *directed system of fragment embeddings* if and only if $\pi_{\alpha,\gamma} = \pi_{\beta,\gamma} \circ \pi_{\alpha,\beta}$ for $\alpha < \beta < \gamma < \mu$.

DEFINITION 7.2.9. Suppose $(\mathbb{A}_\alpha, \pi_{\alpha,\beta} : \alpha < \beta < \mu)$ is a directed system of fragment embeddings. Let $\tau^* = \bigcup_{\alpha<\mu} \tau_\alpha$. Suppose we also have a fragment $\mathbb{A}^*$ of $\mathcal{L}_{\infty,\omega}(\tau^*)$ and τ_α embeddings $\pi_{\alpha,*} : \mathbb{A}_\alpha \to \mathbb{A}^*$ such that $\pi_{\alpha,*} = \pi_{\beta,*} \circ \pi_{\alpha,\beta}$ for all $\alpha < \beta$ and every element of $\mathbb{A}^*$ is of the form $\pi_{\alpha,*}(\psi)$ for some $\alpha < \mu$ and $\psi \in \mathbb{A}_\alpha$.

We call $(\mathbb{A}^*, \pi_{\alpha,*})$ the *direct limit* of the system $(\pi_{\alpha,\beta} : \alpha, \beta)$.

- Fix a directed system of fragment embeddings $(\pi_{\alpha,\beta} : \alpha < \beta < \mu)$.
- Fix $\mathbb{A}^*$ and $(\pi_{\alpha,*} : \alpha < \mu)$ a direct limit.
- Suppose that $\mathcal{M}_\alpha$ is a τ_α-structure for all $\alpha < \mu$.
- Suppose $\sigma_{\alpha,\beta} : \mathcal{M}_\alpha \to \mathcal{M}_\beta$ is a commuting family of τ_α-embedding for $\alpha < \beta < \mu$.

DEFINITION 7.2.10. We say that $(\mathcal{M}_\alpha, \sigma_{\alpha,\beta} : \alpha < \beta < \mu)$ is a *twisted elementary system* if

$$\mathcal{M}_\alpha \models \psi(\overline{a}) \Leftrightarrow \mathcal{M}_\beta \models \pi_{\alpha,\beta}(\psi)(\sigma_{\alpha,\beta}(\overline{a}))$$

for all $\psi \in \mathbb{A}_\alpha$ and $\alpha < \beta < \mu$.

Note that since $\pi_{\alpha,\beta}$ is the identity on atomic τ_α-formulas this is consistent with being a τ_α-embedding.

The following lemma is a twisted version of the standard theorems on elementary chains.

LEMMA 7.2.11. *Suppose $(\mathcal{M}_\alpha, \sigma_{\alpha,\beta} : \alpha < \beta < \mu)$ is a twisted elementary system. There is a τ^*-structure $\mathcal{M}^*$ and τ_α-embeddings $\sigma_{\alpha,*} : \mathcal{M}_\alpha \to \mathcal{M}^*$ such that*

- *$\sigma_{\alpha,*} = \sigma_{\beta,*} \circ \sigma_{\alpha,\beta}$ for all $\alpha < \beta < \mu$;*
- *every element of $\mathcal{M}^*$ is in the image of $\sigma_{\alpha,*}$ for all sufficiently large $\alpha < \mu$;*

- *for* $\psi \in \mathbb{A}_\alpha$, $\overline{a} \in \mathcal{M}_\alpha$

$$\mathcal{M}_\alpha \models \psi(\overline{a}) \Leftrightarrow \mathcal{M}^* \models \pi_{\alpha,*}(\psi)(\sigma_{\alpha,*}(\overline{a})).$$

PROOF. Let $\mathcal{F}$ be all functions with domain $[\alpha, \mu)$ such that $f(\alpha) \in \mathcal{M}_\alpha$ and $f(\beta) = \sigma_{\alpha,\beta}(f(\alpha))$ for all $\alpha < \beta < \mu$.

EXERCISE 7.2.12. Show that if $f, g \in \mathcal{F}$ and $f(\gamma) = g(\gamma)$, then f and g agree on their common domain.

Thus we have an equivalence relation $f \sim g$ if and only if $\exists \gamma\, f(\gamma) = g(\gamma)$. Let $[f]$ denote the equivalence class of f and let the universe of $\mathcal{M}^*$ be the set of all equivalence classes. Define $\sigma_{\alpha,*}(a) = [f]$ where $f(\alpha) = a$ and $f(\beta) = \sigma_{\alpha,\beta}(a)$ for all $\alpha < \beta < \mu$.

We can define a τ^*-structure on $\mathcal{M}^*$ such that

$$\mathcal{M}_\alpha \models \theta(\overline{a}) \Leftrightarrow \mathcal{M}^* \models \theta(\sigma_{\alpha,*}(\overline{a}))$$

for all atomic formulas $\theta \in \mathbb{A}_\alpha$ and $\overline{a} \in \mathcal{M}_\alpha$.

EXERCISE 7.2.13. Argue that $\mathcal{M}^*$ is well-defined.

We now prove by induction on complexity of formulas that

$$\mathcal{M}_\alpha \models \psi(\overline{a}) \Leftrightarrow \mathcal{M}^* \models \pi_{\alpha,*}(\psi)(\sigma_{\alpha,*}(\overline{a}))$$

for $\psi \in \mathbb{A}_\alpha$, $\overline{a} \in \mathcal{M}_\alpha$.

Consider the case where $\psi \in \mathbb{A}_\alpha$ is an infinite disjunction. Then $\pi_{\alpha,*}(\psi)$ is also an infinite disjuction. Suppose $\mathcal{M}_\alpha \models \psi(\overline{a})$. Then there is a disjunct $\theta(\overline{v})$ such that $\mathcal{M}_\alpha \models \theta(\overline{a})$. Since we have a fragment embedding $\pi_{\alpha,*}(\theta)$ is a disjunct of $\pi_{\alpha,*}(\phi)$. By induction, $\mathcal{M}^* \models \pi_{\alpha,*}(\theta)(\sigma_{\alpha,*}(\overline{a}))$. Thus $\mathcal{M}^* \models \pi_{\alpha,*}(\phi)(\sigma_{\alpha,*}(\overline{a}))$.

Suppose $\mathcal{M}^* \models \pi_{\alpha,*}(\psi)(\sigma_{\alpha,*}(\overline{a}))$. Then $\mathcal{M}^* \models \chi(\sigma_{\alpha,*}(\overline{a}))$ for some disjunct χ of $\pi_{\alpha,*}(\psi)$. There is $\alpha \leq \beta < \mu$ and $\theta \in \mathbb{A}_\beta$ such that $\pi_{\beta,*}(\theta) = \chi$. Then by induction $\mathcal{M}_\beta \models \theta(\sigma_{\alpha,\beta}(\overline{a}))$ and $\mathcal{M}_\beta \models \pi_{\alpha,\beta}(\phi)(\sigma_{\alpha,\beta}(\overline{a}))$. Since we have a twisted elementary system, $\mathcal{M}_\alpha \models \phi(\overline{a})$.

EXERCISE 7.2.14. Complete the induction on complexity. ⊣

DEFINITION 7.2.15. We say that a twisted elementary system $(\mathcal{M}_\alpha, \sigma_{\alpha,\beta} : \alpha < \beta)$ is *atomic* if each $\mathcal{M}_\alpha$ is $\mathbb{A}_\alpha$ atomic and a formula $\theta(\overline{v}) \in \mathbb{A}_\alpha$ is $\mathbb{A}_\alpha$-complete if and only if $\pi_{\alpha,\beta}(\theta)$ is $\mathbb{A}_\beta$-complete for all $\alpha < \beta < \mu$.

LEMMA 7.2.16. *Suppose $(\mathcal{M}_\alpha, \sigma_{\alpha,\beta}, \alpha < \beta)$ is an atomic twisted elementary system. Then the direct limit $\mathcal{M}^*$ is atomic and $\theta(\overline{v}) \in \mathbb{A}_\alpha$ is $\mathbb{A}_\alpha$-complete if and only if $\pi_{\alpha,*}(\theta)$ is $\mathbb{A}^*$-complete.*

PROOF. The following exercise is the $(\Leftarrow)$ direction.

EXERCISE 7.2.17. Show that if $\theta(\overline{v}) \in \mathbb{A}_\alpha$ is not $\mathbb{A}_\alpha$-complete, then $\pi_{\alpha,*}(\theta)$ is not $\mathbb{A}^*$-complete.

For the converse, suppose $\theta(\overline{v}) \in \mathbb{A}_\alpha$ is not $\mathbb{A}_\alpha$-complete. Let $\chi(\overline{v}) \in \mathbb{A}^*$. Choose $\beta \geq \alpha$ and $\psi \in \mathbb{A}_\beta$ such that $\pi_{\beta,*}(\psi) = \chi$. Since $\pi_{\alpha,\beta}(\theta)$ is $\mathbb{A}_\beta$-complete,

$$\mathcal{M}_\beta \models \forall \overline{v}\, (\pi_{\alpha,\beta}(\theta)(\overline{v}) \rightarrow \psi(\overline{v})) \text{ or } \mathcal{M}_\beta \models \forall \overline{v}\, (\pi_{\alpha,\beta}(\theta)(\overline{v}) \rightarrow \neg\psi(\overline{v})).$$

Without loss of generality assume the former, then

$$\mathcal{M}^* \models \forall \overline{v}\, (\pi_{\alpha,*}(\theta)(\overline{v}) \rightarrow \chi(\overline{v}))$$

as desired.

If $\theta(\overline{v})$ isolates the type of $\overline{a}$ in $\mathcal{M}_\alpha$, then $\pi_{\alpha,*}(\theta)$ will isolate the type of $\pi_{\alpha,*}(\overline{a})$ in $\mathcal{M}^*$. ⊣

EXERCISE 7.2.18 (Laskowski). Suppose T is a complete atomic first order theory in a vocabulary τ of cardinality $\aleph_1$. We will prove that T has an atomic model, a result due independently to Knight [41], Kueker [45] and Shelah [70].[9]

(a) Show that we can find a filtration $(\tau_\alpha : \alpha < \omega_1)$ such that $\tau_\alpha \subseteq \tau_\beta$ for $\alpha < \beta$, $\tau_\alpha = \bigcup_{\beta<\alpha} \tau_\beta$ if α is a limit ordinal and if $\phi(\overline{v})$ is a τ_α-formula such that $T \models \exists \overline{v}\, \phi(\overline{v})$, then there is $\theta(\overline{v})$ a complete τ_α-formula such that $T \models \theta(\overline{v}) \rightarrow \phi(\overline{v})$.

(b) Let T_α be the set of τ_α-sentences in T. Show that each T_α is a countable atomic theory. Let $\mathcal{M}_\alpha$ be a prime model of T_α.

(c) Do a direct limit construction to prove that T has an atomic model.

The generic Morley tree. While we can build paths of height ω_1 through the Morley tree and these paths determine satisfiable theories, we might have uncountably many types in $\bigcup_{\alpha<\omega_1} \mathbb{A}_\alpha$ so we can not continue further. We will give a variant of the construction, that allows us to continue our construction up to ω_2 but at a significant cost. It will no longer be clear that we are building satisfiable theories. Indeed, we could do a similar construction beyond ω_2, but we know from Hjorth's Theorem 5.4.15, that ϕ may have no models beyond $\aleph_1$.

Let P be the set of finite functions with domain contained in ω and image contained in ω_1. Suppose G is a generic filter on P, then $\omega_1^{\mathbb{V}}$ is countable in $\mathbb{V}[G]$ and $\omega_1^{\mathbb{V}[G]} = \omega_2^{\mathbb{V}}$. For a, perhaps uncountable, fragment $\mathbb{A} \subset \mathcal{L}_{\omega_2,\omega}$ we will look in $\mathbb{V}[G]$ for satisfiable $\mathbb{A}$-complete theories T containing ϕ. The next lemma shows that these theories are already in $\mathbb{V}$, but, a priori, we can't conclude they have models in $\mathbb{V}$.

LEMMA 7.2.19. *Let $\mathbb{A} \in \mathbb{V}$ be a fragment of $\mathcal{L}_{\omega_2,\omega}$ of cardinality at most $\aleph_1$. Suppose $p \in P$, $\dot{T}$ is a P-name and $p \Vdash \dot{T}$ is a satisfiable $\mathbb{A}$-complete theory containing ϕ. Then there is $q \leq p$ such that $q \Vdash \psi \in \dot{T}$ or $q \Vdash \psi \notin \dot{T}$ for every $\psi \in \mathbb{A}$. If $G \subset P$ is a generic filter and $p \in G$, then $\dot{T}_G \in \mathbb{V}$, where $\dot{T}_G$ is the element of $\mathbb{V}[G]$ named by $\dot{T}$.*

[9] By contrast, Laskowski and Shelah [48] prove that for any $\kappa > \aleph_1$ there is a complete satisfiable atomic theory in a vocabulary of cardinality κ with no atomic model.

Similarly, if $\dot{t}$ is a P-name and $p \Vdash$ "$\dot{t}$ is an $\mathbb{A}$-type", then there is $q \leq p$ that forces $\psi(\overline{v}) \in \dot{t}$ or $\psi(\overline{v})) \in \dot{t}$ for all $\psi(\overline{v}) \in \mathbb{A}$ and $\dot{t}_G \in \mathbb{V}$ for any generic $G \subset P$.

PROOF. Let $\mathcal{A}$ be a transitive model of ZFC$^-$ of cardinality $\aleph_1$ such that $\phi, \mathbb{A}, P, p$ and $\dot{T}$ are in $\mathcal{A}$. Let $\mathcal{A}_0$ be a countable elementary submodel of $\mathcal{A}$ containing $\phi, \mathbb{A}, P, p$ and $\dot{T}$ and let $\mathcal{B}$ be the Mostowski collapse of $\mathcal{A}_0$. Thus we may assume we have a countable transitive model $\mathcal{B}$ of ZFC$^- + \phi$ is scattered + there is a fragment $\widehat{\mathbb{A}}$ of $\mathcal{L}_{\omega_2,\omega}$ of cardinality at most $\aleph_1 +$ $\exists p \in P\, p \Vdash \phi \in \dot{T} \subset \widehat{\mathbb{A}}$ is complete and satisfiable.

We claim that there is $q \in \mathcal{B}$ with $q \leq p$ such that $q \Vdash \phi \in \dot{T}$ or $q \Vdash \neg\phi \in \dot{T}$ for all $\phi \in \widehat{\mathbb{A}}$.

Suppose not. Let $\dot{M} \in \mathcal{B}$ be a P-name such that $p \Vdash \dot{M} \models \dot{T}$. Let $D_0, D_1, \ldots,$ list all dense subsets of P in $\mathcal{B}$. Working in $\mathbb{V}$ we build a tree $(p_\sigma : \sigma \in 2^{<\omega})$ where all $p_\sigma \in \mathcal{B}$ as follows:

- $p_\emptyset = p$;
- given p_σ with $|\sigma| = n$, choose $p'_\sigma \leq p_\sigma$ with $p'_\sigma \in D_n$;
- by assumption there is $\psi \in \widehat{\mathbb{A}}$ such that p'_σ does not force either $\psi \in \dot{T}$ or $\neg\psi \in \dot{T}$. Choose $p_{\sigma,0} \leq p'_\sigma$ such that $p_{\sigma,0} \Vdash \psi \in \dot{T}$ and choose $p_{\sigma,1} \leq p'_\sigma$ such that $p_{\sigma,1} \Vdash \neg\psi \in \dot{T}$.

For each $f \in 2^\omega$ we have a $\mathcal{B}$-generic filter $G_f = \{q : \exists n\, p_{f|n} \leq q\}$. Let $T_f = \dot{T}_{G_f}$ be the theory named by $\dot{T}$ in $\mathcal{B}[G_f]$ and let $\mathcal{M}_f \models T_f$ be the structure named by $\dot{M}$ in $\mathcal{B}[G_f]$.

Since $\mathcal{B}[G_f] \models \mathcal{M}_f \models T_f$ and this is absolute, $\mathcal{M}_f$ really is a model of T_f, so T_f is satisfiable. For each $\psi \in \widehat{\mathbb{A}}$,

$$\{q : q \Vdash \psi \in \dot{T} \text{ or } q \Vdash \neg\psi \in \dot{T}\}$$

is dense below p. Thus T_f is $\widehat{\mathbb{A}}$-complete. We also have $\phi \in T_f$ for all f and, by construction, $T_f \neq T_g$ for $f \neq g$. But this contradicts the fact that ϕ is scattered.

Thus there is $q \leq p$ such that in any generic extension

$$\dot{T}_G = \{\phi \in \mathbb{A} : q \Vdash \phi \in \dot{T}\}.$$

But, by the definability of forcing, $\dot{T}_G \in \mathbb{V}$.

The argument for types is identical. ⊣

We now build the *generic Morley tree*. It is worth noting, that even though we are using forcing, the entire construction takes place in $\mathbb{V}$. Let $\mathbb{A}_0^* = \mathbb{A}_0$. If α is a limit ordinal, then $\mathbb{A}_\alpha^* = \bigcup_{\beta<\alpha} \mathbb{A}_\beta^*$.

Suppose $\alpha < \omega_2$ and $\mathbb{A}_\alpha^*$ is a fragment of $\mathcal{L}_{\omega_2,\omega}$ of cardinality at most $\aleph_1$. Let $\mathcal{T}_\alpha^* = \{T \subset \mathbb{A}_\alpha^* : \phi \in T$ and $p \Vdash$ "T is a satisfiable $\mathbb{A}_\alpha^*$-complete theory and no $T \cap \mathbb{A}_\beta^*$ is complete for $\beta < \alpha$" for some $p \in P\}$. For each $p \in P$ there are at most $\aleph_1$ such T. Thus $|\mathcal{T}_\alpha^*| \leq \aleph_1$.

Let $S = \{\gamma(\overline{v}) \subset \mathbb{A}^*_\alpha : p \Vdash$ "γ is an $\mathbb{A}_\alpha$-type realized in some $\mathcal{M} \models \phi$" for some $p \in P\}$. As above $|S| \leq \aleph_1$. Let $\mathbb{A}^*_{\alpha+1}$ be the smallest fragment of $\mathcal{L}_{\omega_2,\omega}$ generated by $\mathbb{A}^*_\alpha$ and all formulas $\bigwedge_{\psi\in\gamma} \psi(\overline{v})$ for $\gamma \in S$. Note that $|\mathbb{A}^*_{\alpha+1}| \leq \aleph_1$.

EXERCISE 7.2.20. (a) Prove that if $p \Vdash$ "T is a satisfiable $\mathbb{A}^*_\alpha$-complete theory and no $T \cap \mathbb{A}^*_\beta$ is complete for $\beta < \alpha$", then this is also forced by $\emptyset$. [Hint: P is *almost-homogeneous*. See Exercises (E1) and (E2) of chapter VII of [46].]
(b) Prove that $\mathcal{T}^*$ is the Morley tree in $\mathbb{V}[G]$.
(c) Prove that $\mathbb{A}_\alpha = \mathbb{A}^*_\alpha$ and $\mathcal{T}_\alpha = \mathcal{T}^*_\alpha$ for all $\alpha < \omega_1$.

Suppose $T \in \mathcal{T}^*_\alpha$. In any generic extension $\mathbb{V}[G]$, T is an $\mathbb{A}^*_\alpha$-atomic theory. While, for the moment, we don't know that T is satisfiable in $\mathbb{V}$, it still in many ways behaves like an atomic theory. We say that $\theta(\overline{v}) \in \mathbb{A}^*_\alpha$ is a *generic atom* in T if

- $\exists \overline{v}\, \theta(\overline{v}) \in T$ and
- $\forall \overline{v}\, (\theta(\overline{v}) \to \psi(\overline{v})) \in T$ or $\forall \overline{v}\, (\theta(\overline{v}) \to \neg\psi(\overline{v})) \in T$ for all $\psi(\overline{v}) \in \mathbb{A}^*_\alpha$.

Note that if T is satisfiable the generic atoms are atoms.[10] In $\mathbb{V}[G]$ the generic atoms are exactly the atoms and the atoms are dense. Since the statement that the generic atoms are dense is Δ_0, the generic atoms are still dense in $\mathbb{V}$, i.e., for all $\psi(\overline{v}) \in \mathcal{T}$ if $\exists \overline{v}\, \psi(\overline{v}) \in T$, then there is a generic atom $\theta(\overline{v}) \in \mathbb{A}^*_\alpha$ such that

$$\forall \overline{v}\, (\theta(\overline{v}) \to \psi(\overline{v})) \in T.$$

This will also be true in any transitive substructure containing T and $\mathbb{A}^*_\alpha$.

LEMMA 7.2.21. *If $T \in \mathcal{T}^*_\alpha$, then T is satisfiable.*

PROOF. To simplify notation we let $\mathbb{A}$ denote $\mathbb{A}^*_\alpha$. The ordinal α plays no further role in the proof.

Let $\mathcal{B}$ be a transitive model of ZFC$^-$ of cardinality $\aleph_1$ such that $\mathbb{A}, T \in \mathcal{B}$. Let $(\mathcal{B}_\beta : \beta < \omega_1)$ be a continuous elementary chain of countable models such that $\mathbb{A}, T \in \mathcal{B}_0$, $\mathbb{A}_0 \subset \mathcal{B}_0$ and $\mathcal{B} = \bigcup_{\beta<\omega_1} \mathcal{B}_\beta$. For each $\beta < \omega_1$ consider B_β the transitive collapse of $\mathcal{B}_\beta$. Let $p_\beta : \mathcal{B}_\beta \to B_\beta$ be the Mostowski collapse isomorphism. If $\gamma < \beta$ we have an elementary embedding $\pi_{\gamma,\beta} : B_\gamma \to B_\beta$ given by $\pi_{\gamma,\beta} = p_\beta \circ p_\gamma^{-1}$.

For each $\beta < \omega_1$ let $T_\beta = p_\beta(T)$ and $A_\beta = p_\beta(\mathbb{A})$. Note that $\pi_{\gamma,\beta}$ maps A_γ into A_β for $\gamma < \beta$.

EXERCISE 7.2.22. (a) Let $\gamma < \beta$. Show that the map $\pi_{\gamma,\beta}$ is a fragment embedding of A_γ into A_β. [Hint: Use the fact that $\pi_{\gamma,\beta} : B_\gamma \to B_\beta$ is elementary.]
(b) Show that if $\beta < \omega_1$ is a limit ordinal, then A_β is the direct limit of $(A_\gamma, \pi_{\gamma,\delta} : \gamma < \delta < \beta)$ (where, here, A_{ω_1} denotes $\mathbb{A}$).

---

[10] Here we use *atom* interchangeably with "$\mathbb{A}_\alpha$-complete formula".

CLAIM 1. (i) *T_β is a satisfiable A_β-atomic theory for all $\beta < \omega_1$.*
(ii) *The atoms of T_β are exactly the formulas that B_β believes are generic atoms.*
(iii) *If ψ is an atom of T_γ and $\gamma < \beta$, then $\pi_{\gamma,\beta}(\psi)$ is an atom of T_β.*

$B_\beta \models$ "$\exists p \in P\ p \Vdash T_\beta$ is satisfiable". Since $\mathcal{B}_\beta$ is countable, in $\mathbb{V}$ we can build an $\mathcal{B}_\beta$-generic $G \subset P$ such that $\mathcal{B}_\beta[G] \models \exists \mathcal{N} \models T_\beta$. But then $\mathcal{N}$ really is a model of T_β in $\mathbb{V}$! Since ϕ is scattered and A_β is countable, T_β is A_β-atomic. As T_β is satisfiable the generic atoms are exactly the atoms. Moreover being a generic atom is a first order property, thus if $\psi \in A_\gamma$ is an atom, so is $\pi_{\gamma,\beta}(\psi)$.

Let $\mathcal{M}_\beta$ be the A_β-prime model of T_β for $\beta < \omega_1$.

CLAIM 2. *There is an atomic system of twisted elementary embedding $\sigma_{\gamma,\beta} : \mathcal{M}_\gamma \to \mathcal{M}_\beta$ for $\gamma < \beta < \omega_1$.*

We build the maps inductively. Suppose $\beta = \gamma + 1$. We will build $\sigma_{\gamma,\beta}$ and let $\sigma_{\delta,\beta} = \sigma_{\gamma,\beta} \circ \sigma_{\delta,\gamma}$ for $\delta < \gamma$.

We build a sequence of finite maps $f_0 \subset f_1 \subset \cdots$ from $\mathcal{M}_\gamma$ to $\mathcal{M}_\beta$ such that if $\overline{a} \in \operatorname{dom} f_n$, then

$$\mathcal{M}_\gamma \models \psi(\overline{a}) \Leftrightarrow \mathcal{M}_\beta \models \pi_{\gamma,\beta}(\psi)(f_n(\overline{a})).$$

If we can do this, we can let $\sigma_{\gamma,\beta} = \bigcup_{n\in\omega} f_n$.

Let $f_0 = \emptyset$. If ψ is a sentence in T_γ, then $\pi_{\gamma,\beta}(\psi) \in T_\beta$. Thus

$$\mathcal{M}_\gamma \models \psi \Leftrightarrow \mathcal{M}_\beta \models \pi_{\gamma,\beta}(\psi)$$

as desired.

Suppose $\overline{a}$ is the domain of f_n and $b \in \mathcal{M}$. There is $\theta(\overline{v}, w) \in A_\gamma$ isolating the A_γ-type of $(\overline{a}, b)$. By induction, for any formula $\psi(\overline{v}, w) \in A_\beta$

$$\mathcal{M}_\beta \models \forall \overline{v}, w\ (\pi_{\gamma,\beta}(\theta)(\overline{v}, w) \to \psi(\overline{v}, w))$$

or

$$\mathcal{M}_\beta \models \forall \overline{v}, w\ (\pi_{\gamma,\beta}(\theta)(\overline{v}, w) \to \neg\psi(\overline{v}, w))$$

and

$$\mathcal{M}_\beta \models \exists w\ \pi_{\gamma,\beta}(\theta)(f_n(\overline{a}), w).$$

Then we can extend f_n by sending b to c.

Suppose $\beta < \omega_1$ is a limit ordinal and we have constructed a commuting system of twisted elementary $\sigma_{\delta,\gamma} : \mathcal{M}_\delta \to \mathcal{M}_\gamma$ for all $\delta < \gamma < \beta$. By Lemma 7.2.16 we can build $\mathcal{M}^*$ an atomic twisted direct limit. But for $\gamma < \beta$ and $\psi \in A_\gamma$,

$$\mathcal{M}^* \models \pi_{\gamma,\beta}(\psi) \Leftrightarrow \mathcal{M}_\gamma \models \psi \Leftrightarrow \psi \in T_\gamma \Leftrightarrow \pi_{\gamma,\beta}(\psi) \in T_\beta.$$

Thus $\mathcal{M}^*$ is an A_β-atomic model of T_β, so we can take $\mathcal{M}_\beta$ to be $\mathcal{M}^*$. Let $\sigma_{\gamma,\beta} = \sigma_{\gamma,*}$ be the maps obtained in the twisted direct limit construction.

This completes the proof of construction of an atomic twisted elementary system $(\mathcal{M}_\gamma, \pi_{\gamma,\beta} : \gamma < \beta < \omega_1)$.

Let $\mathcal{M}^*$ be the twisted direct limit of the system. Letting $\pi_{\beta,*} : A_\beta \to \mathbb{A}$ be p_b^{-1} we see that $\mathbb{A}$ is the direct limit of the system $(A_\gamma, \pi_{\gamma,\beta} : \gamma < \beta < \omega_1)$. For $\psi \in \mathbb{A}$. Choose β such that $\psi \in \mathcal{B}_\beta$, then

$$\psi \in T \Leftrightarrow p_\beta(\psi) \in T_\beta \Leftrightarrow \mathcal{M}_\beta \models p_\beta(\psi) \Leftrightarrow \mathcal{M}^* \models \pi_{\beta,*}(p_\beta(\psi)) \Leftrightarrow \mathcal{M}^* \models \psi.$$

Thus $\mathcal{M}^*$ is the desired model of T. ⊣

EXERCISE 7.2.23. Adapt the argument from Lemma 7.2.6 to show that if $\alpha < \omega_2$ is a limit ordinal and $\mathcal{M} \models T \in \mathcal{T}_\alpha^*$, then $\mathcal{M}$ has Scott rank at least α.

COROLLARY 7.2.24 (Harrington). *If ϕ is a Vaught counterexample, then there are models of ϕ of arbitrarily large Scott rank below ω_2.*

COROLLARY 7.2.25. *If ϕ is a Vaught counterexample, then $I(\phi, \aleph_1) \geq \aleph_2$.*

This leads to the following question which we state as a conjecture, though the only real evidence is that it is true in the first order case and true if $2^{\aleph_1} = \aleph_2$.

CONJECTURE. *If ϕ is a Vaught counterexample, then $I(\phi, \aleph_1) = 2^{\aleph_1}$.*

Interestingly, the number of $\equiv_{\infty,\omega}$-classes of models of size $\aleph_1$ is bounded.

PROPOSITION 7.2.26. *If ϕ is a Vaught counterexample, then the number of $\equiv_{\infty,\omega}$-classes of models of ϕ of cardinality $\aleph_1$ is exactly $\aleph_2$.*

PROOF. By Harrington's Theorem there are at least $\aleph_2$ equivalence classes. Suppose there are more. Let $\kappa > \aleph_2$ and suppose that for $\alpha < \kappa$ we have $\mathcal{M}_\alpha \models \phi$ of cardinality $\aleph_1$ with $\mathcal{M}_\alpha \not\equiv_{\infty,\omega} \mathcal{M}_\beta$ for $\alpha \neq \beta$.

Let P be finite partial functions from ω to ω_1 and let $G \subset P$ be a generic filter. Each $\mathcal{M}_\alpha$ is countable in $\mathbb{V}[G]$ and $\mathbb{V}[G] \models \mathcal{M}_\alpha \not\equiv_{\infty,\omega} \mathcal{M}_\beta$ for $\alpha \neq \beta$. Since $\kappa > \aleph_1^{\mathbb{V}[G]}$, this contradicts the fact that $\mathbb{V}[G] \models$ "ϕ is scattered". ⊣

Chapter 8

QUASIMINIMAL EXCELLENCE

8.1. Categoricity

Part of the resurgence of interest in infinitary logic comes from Zilber's study of structures associated to complex exponentiation. Zilber began by studying of universal cover of the multiplicative group of complex numbers. We have two sorts, one for the additive group of complex numbers $(\mathbb{C}, +)$ and a second for the multiplicative group $(\mathbb{C}^\times, \cdot)$ with predicates for all constructible subsets of $(\mathbb{C}^\times)^n$. We also have the exponential map $\exp : \mathbb{C} \to \mathbb{C}^\times$. Zilber [80] shows that there is $\phi \in \mathcal{L}_{\omega_1,\omega}$ true in this structure and categorical in all uncountable cardinalities.

In even more striking work, Zilber [79] described a class $\mathcal{K}$ of exponential fields that share many of the known properties of the complex exponential field, but also have good model theoretic properties. In particular, $\mathcal{K}$ has a unique model of each uncountable cardinality and in every $K \in \mathcal{K}$ every definable subset of K is either countable or co-countable. An intriguing open question is whether the unique element of $\mathcal{K}$ of cardinality $2^{\aleph_0}$ is isomorphic to complex exponentiation. The class $\mathcal{K}$ is not $\mathcal{L}_{\omega_1,\omega}$-axiomatizable, but can be axiomatized a sentence in $\mathcal{L}_{\omega_1,\omega}(Q)$ where Q is the quantifier "exists uncountably many" (see [36] or [33] for more on model theory with this generalized quantifier).[11]

The model theoretic core of Zilber's proofs is a categoricity theorem for quasiminimal excellent classes, which we will prove in this section. The proof has an algebraic flavor and in no way uses the syntax or semantics of $\mathcal{L}_{\omega_1,\omega}$. Our presentation will closely follow Kirby [39].

Pregeometries. We begin by recalling some basics on pregeometries.

DEFINITION 8.1.1. We say that $\mathrm{cl}_A : \mathcal{P}(A) \to \mathcal{P}(A)$ is a *pregeometry* if

(i) if $X \subseteq A$, then $X \subseteq \mathrm{cl}_A(X)$ and $\mathrm{cl}_A(\mathrm{cl}_A(X)) = \mathrm{cl}_A(X)$;
(ii) (monotonicity) if $X \subseteq Y \subseteq A$, then $\mathrm{cl}_A(X) \subseteq \mathrm{cl}_A(Y)$;

[11]Zilber's original papers contained some serious errors. Bays and Zilber [15] give a corrected proof of Zilber's theorem and an extension to positive characteristic. Bays and Kirby [14] give corrections to the paper on exponential fields.

(iii) (exchange) if $X \subseteq A, a, b \in A$ and $a \in \text{cl}_A(X \cup \{b\}) \setminus \text{cl}_A(X)$, then $b \in \text{cl}(A \cup \{a\})$;
(iv) (finite character) if $X \subseteq A$ and $a \in \text{cl}_A(X)$, then there is $X_0 \subseteq X$ finite such that $a \in \text{cl}_A(X_0)$.

DEFINITION 8.1.2. We say $X \subseteq A$ is *closed* if $\text{cl}_A(X) = X$.
We say that $X \subseteq A$ is *independent* if $x \notin \text{cl}_A(X \setminus \{x\})$ for all $x \in X$. We say that $Y \subseteq X$ is a *basis* for X if Y is independent and $\text{cl}_A(Y) = \text{cl}_A(X)$.

EXERCISE 8.1.3. (a) Show that any maximal independent subset of X is a basis.
(b) Show that if $I \subseteq A$ is independent, then there is a basis $J \supseteq I$.
(c) Prove that any two bases for X have the same cardinality.

DEFINITION 8.1.4. The *dimension* of $X \subseteq A$ is the cardinality of any basis of X. We let $\dim_{\text{A}}(\text{X})$ denote the dimension of X.

The following Exercise will be useful in our main argument.

EXERCISE 8.1.5. Suppose that $X \subset A$ is closed, I is independent over X and $I_0, I_1 \subseteq I$. Show that

$$\text{cl}_A(X \cup I_0) \cap \text{cl}_A(X \cup I_1) = \text{cl}_A(X \cup (I_0 \cap I_1)).$$

Quasiminimal classes. Let τ be a vocabulary which, for simplicity, we assume is relational. Let $\mathcal{K}$ be a class of pairs (A, cl_A) where A is a τ-structure and $\text{cl}_A : \mathcal{P}(A) \to \mathcal{P}(A)$ is a pregeometry.[12]
We make the following basic assumptions on $\mathcal{K}$.

- (Closure under isomorphism) If $(A, \text{cl}_A) \in \mathcal{K}$, $f : A \to B$ is τ-isomorphism and $\text{cl}_B : \mathcal{P}(B) \to \mathcal{P}(B)$ is defined by
$$b \in \text{cl}_B(X) \Leftrightarrow f^{-1}(b) \in \text{cl}_A(f^{-1}(X)),$$
then $(B, \text{cl}_B) \in \mathcal{K}$.
- (Downward closure) If $(A, \text{cl}_A) \in \mathcal{K}, B \subseteq A$ is closed and $\text{cl}_B = \text{cl}_A | \mathcal{P}(B)$, then $(B, \text{cl}_B) \in \mathcal{K}$.
- (Preservation under τ-embeddings) If $(A, \text{cl}_A), (B, \text{cl}_B) \in \mathcal{K}$, $X \subseteq A$, $f : X \to B$ is a partial τ-embedding, $x \in X$ and $Y \subseteq X$, then
$$x \in \text{cl}_A(Y) \Leftrightarrow f(x) \in \text{cl}_B(f(Y)).$$

If $(A, \text{cl}_A), (B, \text{cl}_B) \in \mathcal{K}$ and $f : A \to B$ is a τ-isomorphism, then $\text{cl}_B(f(X)) = f(\text{cl}_A(X))$ for all $X \subseteq A$ and f is also an isomorphism of the pregeometries and the pregeometry is determined by the τ-structure. Thus, we can simplify notation by saying $A \in \mathcal{K}$ rather than $(A, \text{cl}_A) \in \mathcal{K}$.

[12] While usually we don't consider $\emptyset$ as a structure, here we allow the possibility that $\text{cl}(\emptyset) = \emptyset$ in which case we would want $\emptyset \in \mathcal{K}$.

DEFINITION 8.1.6. If $A, B \in \mathcal{K}$, $X \subseteq A$ and $f : X \to B$ is a τ-embedding, we say that f is *closed* if and only if $f(Y)$ is closed whenever $Y \subseteq X$ is closed.

EXERCISE 8.1.7. Let $\mathcal{K}$ be the class of all (k, cl_k) where k is a field and $\mathrm{cl}_k(X)$ is the algebraic closure of X in k. Show that not every τ-embedding is closed.

EXERCISE 8.1.8. If $A, B \in \mathcal{K}$ and $f : A \to B$ is a τ-isomorphism, then f and f^{-1} are closed.

DEFINITION 8.1.9. We say that $\mathcal{K}$ is a *quasiminimal class* if, in addition, it satisfies the following properties.

- (Countable Closure Property) If $A \in \mathcal{K}$ and $X \subseteq A$ is finite, then $\mathrm{cl}_A(X)$ is countable;
- (Union of Chains) Let $A_\alpha \in \mathcal{K}$ for $\alpha < \beta$ where A_α is a closed subset of A_γ for $\alpha < \gamma < \beta$. Let $A = \bigcup_{\alpha<\beta} A_\alpha$ and
$$\mathrm{cl}_A(X) = \bigcup_{\alpha<\beta} \mathrm{cl}_{A_\alpha}(X \cap A_\alpha).$$
Then $(A, \mathrm{cl}_A) \in \mathcal{K}$.
- (Nondegeneracy) There is an infinite dimensional $A \in \mathcal{K}$.
- ($\aleph_0$-homogeneity over countable closed sets) Let $A_0, A_1 \in K$, let $B_i \subseteq A_i$ be either empty or a countable closed set and let $f : B_0 \to B_1$ be a τ-isomorphism.
  1. If $a_i \in A_i$ is independent over B_i for $i = 0, 1$, then $f \cup \{(a_1, a_2)\}$ is a partial τ-embedding.
  2. If $\overline{a} \in A_0$ and $g : B_0 \cup \{\overline{a}\} \to A_1$ is a τ-embedding extending f and $c \in \mathrm{cl}_A(B_0 \cup \{\overline{a}\})$, then there is $d \in A_1$ such that $g \cup \{(c, d)\}$ is a τ-embedding.[13]

EXERCISE 8.1.10. If $\mathcal{M}, \mathcal{N} \in \mathcal{K}$, then there is a τ-isomorphism between $\mathrm{cl}_{\mathcal{M}}(\emptyset)$ and $\mathrm{cl}_{\mathcal{N}}(\emptyset)$. [Hint: Build the isomorphism by a back-and-forth construction using Condition 2 of $\aleph_0$-homogeneity.]

For quasiminimal classes we can prove the countable case of the main result in this section.

LEMMA 8.1.11. *Let $\mathcal{K}$ be a quasiminimal class. Suppose*

- *$\mathcal{M}, \mathcal{N} \in \mathcal{K}$ and $\mathcal{M}$ is countable;*
- *$A \subset \mathcal{M}$ is either empty or closed;*
- *$I \subseteq \mathcal{M}$ is a basis for $\mathcal{M}$ over A;*
- *$f_0 : A \to \mathcal{N}$ is a closed embedding;*
- *$g : I \to \mathcal{N}$ is injective and $g(I)$ is independent over $f_0(A)$.*

Then there is a closed embedding $\widehat{f} : \mathcal{M} \to \mathcal{N}$ with $\widehat{f} \supset f_0 \cup g$.

---

[13] In [39], Kirby does not immediately include closure under unions of chains or nondegeneracy. As we will see, these axioms are not used in the proof that isomorphism type is determined by dimension, but will be used when proving that there are models of all possible dimensions.

PROOF. In the case when $A = \emptyset$ we can replace $\emptyset$ by its closure and use Exercise 8.1.10 to build a closed embedding $f_0 : \mathrm{cl}_{\mathcal{M}}(\emptyset) \to \mathcal{N}$. Thus we may assume A is closed.

We prove the Lemma by an inductive construction. Suppose at some stage of the construction we have $I_0 \subset I$, $\widehat{A} = \mathrm{cl}_{\mathcal{M}}(A \cup I_0)$ and $f : \widehat{A} \to \mathcal{N}$ a closed embedding extending $f_0 \cup g|I_0$ and $x \in I \setminus I_0$. By Condition 1 of $\aleph_0$-homogeneity, we can extend f to a τ-embedding by sending x to $g(x)$. Now, using Condition 2 of $\aleph_0$-homogeneity and a back-and-forth construction, we can extend $f \cup \{(x, g(x))\,]$ to an isomorphism between $\mathrm{cl}_{\mathcal{M}}(\widehat{A} \cup \{x\})$ and $\mathrm{cl}_{\mathcal{N}}(\widehat{A} \cup \{g(x)\})$.

The Lemma now follows with a bit of bookkeeping. ⊣

COROLLARY 8.1.12. *If $\mathcal{K}$ is a quasiminimal class, $\mathcal{M}, \mathcal{N} \in \mathcal{K}$ are countable and* $\dim(\mathcal{M}) = \dim(\mathcal{N})$, *then* $\mathcal{M} \cong \mathcal{N}$.

PROOF. Let I be a basis for $\mathcal{M}$, J a basis for $\mathcal{N}$ and $g : I \to J$ a bijection. Then g extends to an isomorphism between $\mathcal{M}$ and $\mathcal{N}$. ⊣

With a bit of care in the induction, we make the first step in proving categoricity for uncountable structures.

EXERCISE 8.1.13. Prove that if $\mathcal{K}$ is a quasiminimal class, $\mathcal{M}, \mathcal{N} \in \mathcal{K}$ and $\dim(\mathcal{M}) = \dim(\mathcal{N}) = \aleph_1$ then $\mathcal{M}$ and $\mathcal{N}$ are isomorphic.

LEMMA 8.1.14. *Suppose $\mathcal{K}$ is a quasiminimal class, $\mathcal{M}, \mathcal{N} \in \mathcal{K}$, $\mathcal{M}_0 \subset \mathcal{M}$ is empty or closed and countable, $f : \mathcal{M}_0 \to \mathcal{N}$ is a τ-embedding map, $\mathcal{N}$ is infinite dimensional over $f(\mathcal{M}_0)$, $\overline{a} \in \mathcal{M}$, $\overline{b} \in \mathcal{N}$ and f extends to a τ-embedding with $\overline{a} \mapsto \overline{b}$. If $c \in \mathcal{M}$, then there is $d \in \mathcal{N}$ such that f extends to a τ-embedding with $\overline{a}, c \mapsto \overline{b}, d$.*

PROOF. We may assume $\mathcal{M}$ is countable. We can do a back-and-forth argument using Condition 2 of $\aleph_0$-homogeneity to extend f to an isomorphism $\widehat{f} : \mathrm{cl}_{\mathcal{M}}(\mathcal{M}_0 \cup \{\overline{a}\}) \to \mathrm{cl}_{\mathcal{N}}(f(\mathcal{M}_0) \cup \{\overline{b}\})$ with $\widehat{f}(\overline{a}) = \overline{b}$. If $c \in \mathrm{cl}_{\mathcal{M}}(\mathcal{M}_0 \cup \{\overline{a}\})$, then we can take $d = f(c)$. Otherwise, let $d \in \mathcal{N}$ be independent from $\overline{b}$. By Condition 1 of $\aleph_0$-homogeneity, $\widehat{f} \cup \{(c, d)\}$ is a τ-embedding. ⊣

COROLLARY 8.1.15. *Suppose $\mathcal{K}$ is a quasiminimal class, $\mathcal{M}, \mathcal{N} \in \mathcal{K}$ are infinite dimensional, $\overline{a} \in \mathcal{M}$, $\overline{b} \in \mathcal{N}$ and $\overline{a}$ and $\overline{b}$ have the same quantifier free τ-types. Then for any $c \in \mathcal{M}$ there is $d \in \mathcal{N}$ such that $\overline{a}, c$ and $\overline{b}, d$ have the same quantifier free τ-type.*

Thus τ-embeddings form a back-and-forth system and $(\mathcal{M}, \overline{a}) \equiv_{\infty,\omega} (\mathcal{N}, \overline{b})$.

Excellence. We want to show that if $\dim(\mathcal{M}) = \dim(\mathcal{N})$, then $\mathcal{M} \cong \mathcal{N}$. We will use one more assumption to prove this when $\dim(\mathcal{M}) > \aleph_1$.

DEFINITION 8.1.16. Let $\mathcal{M} \in \mathcal{K}$. We say that $C \subseteq \mathcal{M}$ is a *crown* if there is an independent $I \subseteq C$ and $I_1, \ldots, I_n \subseteq I$ such that $C = \bigcup_{i=1}^{n} \mathrm{cl}_{\mathcal{M}}(I_i)$.

DEFINITION 8.1.17. We say that a quasiminimal class is *quasiminimal excellent* if it also satisfies the following property:

(Excellence) Let $\mathcal{M}, \mathcal{N} \in \mathcal{K}$, $C \subset \mathcal{M}$ a countable crown and $f : C \to \mathcal{N}$ a closed τ-embedding. For any $\overline{a} \in \mathrm{cl}_{\mathcal{M}}(C)$, there is $\overline{c} \in C$ such that if $\overline{b} \in \mathcal{N}$ and

$$(\overline{a}, \overline{c}) \mapsto (\overline{b}, f(\overline{c}))$$

is a τ-embedding, then we can extend f to $C \cup \{\overline{a}\}$ by sending $\overline{a}$ to $\overline{b}$.

We say that the quantifier free type of $\overline{a}$ over C is *determined over* $\overline{c}$.

LEMMA 8.1.18. *Let $\mathcal{K}$ be a quasiminimal excellent class. Suppose $\mathcal{M}, \mathcal{N} \in \mathcal{K}$, $C \subseteq A$ is a countable crown and $f : C \to \mathcal{N}$ is a closed embedding. There is a closed embedding $g : \mathrm{cl}_{\mathcal{M}}(C) \to \mathcal{N}$ extending f.*

PROOF. We will build $g : \mathrm{cl}_{\mathcal{M}}(C) \to \mathrm{cl}_{\mathcal{N}}(f(C))$. Thus, without loss of generality, we may assume $\mathcal{M} = \mathrm{cl}_{\mathcal{M}}(C)$ and $\mathcal{N} = \mathrm{cl}_{\mathcal{N}}(f(C))$. We will build a sequence of partial embeddings $g_n : A_n \to \mathcal{M}$ such that

$$g_0 \subseteq g_1 \subseteq g_2 \subseteq \cdots,$$

A_n is finite and $g_n \cup f$ is a partial closed embedding. By a back-and-forth construction, we will ensure that $g = \bigcup_{n \in \omega} g_n$ is an isomorphism.

Let $g_0 = \emptyset$. Suppose we have $g_n : A_n \to B$ and $a \in A \setminus A_n$. By excellence, there is $C_0 \subseteq C$ finite such that the quantifier free type of $A_n \cup \{a\}$ is determined over C_0. We may choose C_0 such that $A_n \cup \{a\} \subseteq \mathrm{cl}_{\mathcal{M}}(\overline{c})$. Let $g'_n = g_n \cup f | C_0$. By assumption, g'_n is a τ-embedding. By Condition 2 of $\aleph_0$-homogeneity, we can find $b \in \mathcal{N}$ such that $g'_n \cup \{(x, b)\}$ is a τ-embedding. Let $A_{n+1} = A_n \cup C_0 \cup \{a\}$ and let $g_{n+1} = g'_n \cup \{(a, b)\}$.

An analogous argument allows us to add elements of $\mathcal{N}$ to the image of g_n. Thus, a back-and-forth argument will allow us to build the desired embedding. ⊣

Zilber's Categoricity Theorem.

THEOREM 8.1.19. *Suppose $\mathcal{K}$ is a quasiminimal excellent class, $\mathcal{M}, \mathcal{N} \in \mathcal{K}$ and $\dim(\mathcal{M}) = \dim(\mathcal{N})$, then $\mathcal{M} \cong \mathcal{N}$.*

PROOF. As we have already proved this result if $\dim(\mathcal{M})$ is countable, we will assume $\dim(\mathcal{M})$ is uncountable. Let $\mathcal{M}_0 \subseteq \mathcal{M}$ be closed of dimension $\aleph_0$. We can find $f : \mathcal{M}_0 \to \mathcal{N}$ a closed embedding. Let I be a basis for $\mathcal{M}$ over $\mathcal{M}_0$ and let $g : I \to \mathcal{N}$ be an injection such that $g(I)$ is a basis for $\mathcal{N}$ over $f(\mathcal{M}_0)$.

Our strategy is to build a system of τ-embeddings $f_J : \mathrm{cl}_{\mathcal{M}}(\mathcal{M}_0 \cup J) \to \mathcal{N}$ for $J \subset I$ finite such that

- $f_{J_0} \subset f_J$ for $J_0 \subset J$;
- $f_\emptyset = f$;
- f_J extends $f \cup g|J$.

If we do so, then

$$\bigcup_{J \subset I \text{ finite}} f_J : \mathcal{M} \to \mathcal{N}$$

is the desired isomorphism.

We build the system of embeddings by induction on $|J|$. If $J = \{x\}$, then we can do a back-and-forth construction using Condition 2 of $\aleph_0$-homogeneity to build an isomorphism

$$f_J : \mathrm{cl}_{\mathcal{M}}(\mathcal{M}_0 \cup \{x\}) \to \mathrm{cl}_{\mathcal{N}}(f(\mathcal{M}_0) \cup \{g(x)\})$$

with $f_J(x) = g(x)$.

Suppose $J = \{x_1, \ldots, x_n\}$ and suppose that we have constructed $(f_X : X \subset J)$ satisfying the properties above. For $i \leq n$, let $J_i = J \setminus \{x_i\}$ and let $\mathcal{M}_i = \mathrm{cl}_{\mathcal{M}}(\mathcal{M}_0 \cup J_i)$. Consider the crown $C = \bigcup_{i=1}^n \mathcal{M}_i$. Note that $\mathrm{cl}_{\mathcal{M}}(C) = \mathrm{cl}_{\mathcal{M}}(\mathcal{M}_0 \cup J)$. We will show below that $f_C = \bigcup_{i=1}^n f_{J_i}$ is a well defined closed embedding of C into $\mathcal{N}$. Then, by Lemma 8.1.18, we can extend f_C to $f_J : \mathrm{cl}_{\mathcal{M}}(C) \to \mathcal{N}$. This will be the desired embedding.

We first argue that f_C is well defined. Suppose $x \in \mathcal{M}_i \cap \mathcal{M}_j$. Let $J' = J_i \cap J_j$. Then, by Exercise 8.1.5, $x \in \mathrm{cl}_{\mathcal{M}}(\mathcal{M}_0 \cup J')$. Then

$$f_{J_i}(x) = f_{J'}(x) = f_{J_j}(x)$$

and f_C is well defined.

We next show that f_C is a τ-embedding. For $m \leq n$, let

$$C_i = \mathcal{M}_1 \cup \cdots \cup \mathcal{M}_m$$

and we let $f_{C_m} = f_{J_1} \cup \cdots \cup f_{J_m}$. We will prove, by induction, that each f_{C_m} is a τ-embedding. When $m = 1$, this is clear. Next, suppose f_{C_m} is a τ-embedding. Let $\overline{a} \in C_m$ and $\overline{b} \in \mathcal{M}_{m+1}$. Let $A \subset \mathcal{M}_0$ be closed and finite dimensional such that $\overline{a}, \overline{b} \in \mathrm{cl}_{\mathcal{M}}(A \cup J)$. Let $y \in \mathcal{M}_0$ be independent over A and let $B = \mathrm{cl}_{\mathcal{M}}(A \cup J \cup \{y\})$. Since x_{m+1} and y are independent over $A \cup J_{m+1}$, Lemma 8.1.11 allows us to find an automorphism σ of B that fixes $\mathrm{cl}_{\mathcal{M}}(A \cup J_{m+1})$ and interchanging x_{m+1} and y.

We claim there is a closed embedding $e : \mathrm{cl}_{\mathcal{M}}(\mathcal{M}_0 \cup J) \to \mathcal{N}$ extending $f_{C_m} \cup g|J$. If $m = 1$ we are just extending f_{C_1} by sending x_1 to $g(x_1)$ and then extending to the closure and we can do this in any quasiminimal class. For $m > 1$, $\mathrm{cl}_M(\mathcal{M}_0 \cup J) = \mathrm{cl}_{\mathcal{M}}(C_m)$ and we obtain e from Lemma 8.1.18. Let

$$\mu = e \circ \sigma^{-1} \circ e^{-1} \circ f_{J_{m+1}} \circ \sigma.$$

Wherever this composition is well defined it will be a τ-embedding.

We next show that $\overline{a} \in \mathrm{dom}\ \mu$. Let $J_{i,j} = J \setminus \{x_i, x_j\}$. We know that

$$\overline{a} \in \bigcup_{i=1}^{m} \mathrm{cl}_{\mathcal{M}}(A \cup J_{i,m+1} \cup \{x_{m+1}\}).$$

Thus,

$$\sigma(\overline{a}) \in \bigcup_{i=1}^{m} \mathrm{cl}_M(A \cup J_{i,m+1} \cup \{y\}) \subset \bigcup_{i=1}^{m} \mathrm{cl}_{\mathcal{M}}(\mathcal{M}_0 \cup J_{i,m+1}) \subseteq C_m.$$

Since $f_{J_{m+1}}$ and f_{C_m} agree with f_{J_i} on $\mathrm{cl}_{\mathcal{M}}(\mathcal{M}_0 \cup J_{i,m+1})$,

$$f_{J_{m+1}}(\sigma(\overline{a})) = f_{C_m}(\sigma(\overline{a})).$$

Since e and $f_{C_{m+1}}$ both extend f_{C_m},

$$\begin{aligned}\mu(\overline{a}) &= e(\sigma^{-1}(e^{-1}(f_{J_{m+1}}(\sigma(\overline{a}))))) \\ &= e(\sigma^{-1}(e^{-1}(f_{C_m}(\sigma(\overline{a}))))) \\ &= e(\sigma^{-1}(\sigma(\overline{a}))) \\ &= e(\overline{a}) \\ &= f_{C_m}(\overline{a}).\end{aligned}$$

Thus

$$\mu(\overline{a}) = f_{C_m}(\overline{a}) = f_{C_{m+1}}(\overline{a}).$$

Since $\overline{b} \in \mathrm{cl}_{\mathcal{M}}(A \cup J_{m+1})$, $\sigma(\overline{b}) = \overline{b}$ holds. By preservation of closure, $e^{-1}(f_{J_{m+1}}(\overline{b}))$ is also in $\mathrm{cl}_{\mathcal{M}}(A \cup J_{m+1})$. Thus it is fixed by σ^{-1}. Thus

$$\begin{aligned}\mu(\overline{b}) &= e(\sigma^{-1}(e^{-1}(f_{J_{m+1}}(\overline{b})))) \\ &= e(e^{-1}(f_{J_{m+1}}(\overline{b}))) \\ &= f_{J_{m+1}}(\overline{b})\end{aligned}$$

and $\mu(\overline{b}) = f_{J_{m+1}}(\overline{b}) = f_{C_{m+1}}(\overline{b})$.

Let $\phi(\overline{v}, \overline{w})$ be a quantifier free formula. Since μ is a τ-embedding,

$$\mathcal{M} \models \phi(\overline{a}, \overline{b}) \Leftrightarrow \mathcal{N} \models \phi(\mu(\overline{a}), \mu(\overline{b})) \Leftrightarrow \mathcal{N} \models \phi(f_{C_{m+1}}(\overline{a}), f_{C_{m+1}}(\overline{b})).$$

Thus $f_{C_{m+1}}$ is a τ-embedding. It follows by induction that f_C is a τ-embedding.

Finally, we must show that f_C is closed. Let $A \subseteq C$ be closed. Suppose $b \in \mathrm{cl}_{\mathcal{N}}(f(A))$. We must show that $b \in f_C(A)$. There is $\overline{a} \in A$ such that $b \in \mathrm{cl}_N(f(\overline{a}))$. Since f_C is a τ-embedding, $\overline{a}, J \mapsto f_C(\overline{a}), g(J)$ is a τ-embedding over $\mathcal{M}_0$. By Lemma 8.1.14, there is $c \in \mathcal{M}$ such that we can extend this τ-embedding by sending $c \mapsto b$. By preservation under τ-embeddings, $c \in \mathrm{cl}_{\mathcal{M}}(\overline{a})$. Thus, since A is closed, $c \in \mathrm{cl}_{\mathcal{M}}(\mathcal{M}_0 \cup J_i)$ for some i. But then, by preservation under τ-embeddings,

$$b \in \mathrm{cl}_{\mathcal{N}}(f(\mathcal{M}_0) \cup g(J_i)) = f_{J_i}(\mathcal{M}_i)$$

for some i. ⊣

EXERCISE 8.1.20. Suppose $\mathcal{K}$ is a quasiminimal excellent class, $\mathcal{M} \in \mathcal{K}$ and I is a basis for $\mathcal{M}$. Show that any permutation of I extends to an automorphism of $\mathcal{M}$ and conclude that if $\dim(\mathcal{M}) = \kappa \geq \aleph_0$, then $|\mathrm{Aut}(\mathcal{M})| = 2^{\kappa}$.

EXERCISE 8.1.21. Let $\mathcal{K}$ be a quasiminimal excellent class and let $\mathcal{M} \in \mathcal{K}$ be infinite dimensional, then there is $\mathcal{M}$ isomorphic to a proper closed subset of itself.

Up until this point we have not used nondegeneracy or closure under unions of chains. These assumptions will be needed in the following Exercise.

EXERCISE 8.1.22 (Existence Theorem). Let $\mathcal{K}$ be a quasiminimal excellent class. Show that for all κ there is $\mathcal{M} \in \mathcal{K}$ of dimension κ. [Hint: Use Exercise 8.1.21 and union of chains.]

We have now proved Zilber's Categoricity Theorem.

COROLLARY 8.1.23 (Zilber's Catgoricity Theorem). *If $\mathcal{K}$ is a quasiminimal excellent class, then, up to isomorphism, $\mathcal{K}$ contains a unique structure of cardinality κ for all $\kappa \geq \aleph_1$.*

EXERCISE 8.1.24. Let $\mathcal{K}$ be a quasiminimal excellent class and let $\mathcal{M} \in \mathcal{K}$ and let $A \subset \mathcal{M}$ be finite.

(a) Suppose b and c are independent over A. Show that there is an automorphism σ of $\mathcal{M}$ fixing A pointwise such that $\sigma(b) = c$.
(b) Show that if $X \subset \mathcal{M}$ is A-definable, then either $X \subseteq \mathrm{cl}_{\mathcal{M}}(A)$ or $\mathcal{M} \setminus X \subseteq \mathrm{cl}_{\mathcal{M}}(A)$. Conclude that every definable subset of $\mathcal{M}$ is either countable or co-countable. (The same is true for $\mathcal{L}_{\infty,\omega}$-definable sets over a countable set of parameters.)

The property of having every definable set countable or co-countable is called *quasiminimality*.

EXERCISE 8.1.25. Let $\mathcal{K}$ be a quasiminimal excellent class. Let $\mathcal{M}, \mathcal{N} \in \mathcal{K}$, $A \subseteq \mathcal{M}$ is closed or empty, $f : A \to \mathcal{N}$ is a closed embedding, I is a basis for $\mathcal{M}$ over A, $g : I \to \mathcal{N}$ is a injection and $g(I)$ is independent over $f(A)$. Show that there is a closed embedding $\widehat{f} : \mathcal{M} \to \mathcal{N}$ extending $f \cup g$. [Hint: If A is countable, this follows from the proof of Theorem 8.1.19. If A is uncountable choose $\mathcal{M}_0 \subset A$ closed of dimension $\aleph_0$. Let I_0 be a basis for A over $\mathcal{M}_0$. Follow the proof of Theorem 8.1.19 insisting that $f_J = f | \mathrm{cl}_{\mathcal{M}}(\mathcal{M}_0 \cup J)$ for finite $J \subset I_0$.]

We have assumed for simplicity that our language is relational, but the results extent to more general languages provided we satisfy one additional condition.

EXERCISE 8.1.26. Let τ be an aribitrary countable vocabulary. Suppose $\mathcal{K}$ is a quasiminimal excellent class and for all $M_0, M_1 \in \mathcal{K}$ if A_i is the substructure of M_i generated by $\emptyset$, there is a τ-isomorphism between M_0 and M_1. Show that $\mathcal{K}$ is κ-categorical for all uncountable κ.

The redundancy of excellence. Given the key role excellence played in the proof of Zilber's Categoricity Theorem, it is remarkable that Bays, Hart, Hyttinen, Kesälä and Kirby show in [12] that it is unnecessary.

THEOREM 8.1.27. *Let $\mathcal{K}$ be a quasiminimal class. Then, up to isomorphism, there is a unique structure in $\mathcal{K}$ of each possible dimension. In particular, if $\kappa \geq \aleph_1$, then there is a unique structure in $\mathcal{K}$ of cardinality κ.*

Moreover, $\mathcal{K}$ is the class of models of an $\mathcal{L}_{\omega_1,\omega}(Q)$-sentence.

In an argument inspired by Shelah's proof of the Main Gap, they show that in any quasiminimal class the excellence axiom holds for crowns C in the following setting:

- $\mathcal{M} \in \mathcal{K}$ and $\mathcal{M}_0 \subset \mathcal{M}$ closed of dimension $\aleph_0$:
- a finite $B = \{b_1, \dots, b_n\} \subset \mathcal{M}$ independent over $\mathcal{M}_0$;
- $\mathcal{M}_i = \mathrm{cl}_{\mathcal{M}}(\mathcal{M}_0 \cup B \setminus \{b_i\})$;
- $C = \mathcal{M}_1 \cup \cdots \cup \mathcal{M}_n$.

Reviewing the proof of categoricity, we see that this form of excellence is all we use.

8.2. Covers of $\mathbb{C}^\times$

In this section we study the covering map $\exp : \mathbb{C} \to \mathbb{C}^\times$. We first view this as a two-sorted structure in a language where we have a sort for the additive group $(\mathbb{C}, \oplus)$, a second sort for the field $(\mathbb{C}, +, \cdot)$ and a surjective homomorphism $\exp : (\mathbb{C}, \oplus) \to (\mathbb{C}^\times, \cdot)$. Consider the vocabulary $\tau = \{H, K, \eta, \oplus, +, \cdot, \exp\}$ and the $\mathcal{L}_{\omega_1,\omega}(\tau)$-sentence ϕ where models of ϕ are pairs $((H, +), (K, +, \cdot))$ where

- $(H, \oplus)$ is a torsion free divisible abelian group;
- $(K, +, \cdot)$ is an algebraically closed field of characteristic 0;
- $\exp : H \to K^\times$ is a surjective homomorphism;
- exp has kernel $\mathbb{Z}\eta$.

THEOREM 8.2.1 (Zilber [80]). *ϕ is κ-categorical for all uncountable κ.*

Zilber proved this by showing that we had an appropriate quasiminimal excellent class. In the light of Theorem 8.1.27, we don't need to worry about excellence.

While it is natural to think of the two-sorted structure, it is also convenient to think of a 1-sorted structure

$$(H, \oplus, \eta, E, \lambda_q, \widehat{V} : q \in \mathbb{Q}, V \subseteq K^n \text{ an algebraic variety defined over } \mathbb{Q}),$$

where E is the equivalence relation

$$x\ E\ y \Leftrightarrow \exp(x) = \exp(y),$$

$\lambda_q : H \to H$ is the function $x \mapsto qx$ and

$$x_1, \dots, x_n \in \widehat{V} \Leftrightarrow (\exp(x_1), \dots, \exp(x_n)) \in V.$$

Let τ_0 be the natural vocabulary.

EXERCISE 8.2.2. Show that we can recover the two-sorted structure from the one-sorted structure.

EXERCISE 8.2.3. Let $\overline{a}$ and $\overline{a}' \in H$ be linearly independent over $\mathbb{Q}\eta$. Show that $\overline{a}$ and $\overline{a}'$ realize the same quantifier free τ_0-type if and only if $\exp(\mathbb{Q}\overline{a})$ and $\exp(\mathbb{Q}\overline{b})$ realized quantifier free type in the field language.

We define the closure relation in this context, for $\overline{a} \in H$, let $\mathbb{Q}\overline{a}$ denote the $\mathbb{Q}$-span of $\overline{a}$. We let

$$\mathrm{cl}(\overline{a}) = \exp^{-1}(\mathbb{Q}(\exp(\overline{a})^{\mathrm{alg}})).$$

EXERCISE 8.2.4. Show that cl is a pregeometry with the countable closure property.

Let $\mathcal{K}$ be the class of closed subsets of additive groups of models of ϕ equiped with the closure relation. We will prove that $\mathcal{K}$ is a quasiminimal class.

EXERCISE 8.2.5. Argue that $\mathcal{K}$ is nondegenerate and satisfies the union of chains property.

If $(H, K) \models \phi$ and $\overline{b} \in K^{\times}$, we say that $(\overline{b}^{\frac{1}{n}})_n \in \mathcal{N}$ is a *division system below* $\overline{b}$ if

$$(\overline{b}^{\frac{1}{mn}})^m = \overline{b}^{\frac{1}{n}}$$

for all m and n. If $\overline{b} = \exp(\overline{a})$, then we obtain a division system by setting $\overline{b}^{\frac{1}{n}} = \exp(a/n)$. We call this division system $\exp(\mathbb{Q}\overline{a})$.

EXERCISE 8.2.6. Suppose K is an algebraically closed field of characteristic 0 and $\overline{b} \in K$. Prove that there is a division system $(\overline{b}^{\frac{1}{n}}) - n$ below $\overline{b}$.

Thumbtack Lemma. In this section we will assume some familiarity with algebraic geometry. In our construction we will frequently need to insure that $(\overline{b}^{\frac{1}{n}})_n$ and $(\overline{c}^{\frac{1}{n}})_n$, division systems over $\overline{b}$ and $\overline{c}$, respectively, realize the same type over a subfield $k \subseteq K$. This would be easy if this were equivalent to $\overline{b}$ and $\overline{c}$ realizing the same type over k. But this is not, in general the case. Consider the irreducible algebraic variety V given by

$$(X_1^2 - 1) = X_2.$$

Let

$$W = \{(u_1, u_2) : (u_1^2, u_2^2) \in V\}.$$

Then W has two irreducible components

$$W_1 : Y_1^2 - Y_2 = 1$$

and

$$W_2 : Y_1^2 + Y_2 = 1.$$

Let $(b_{i,1}, b_{i,2})$ be a generic point of W_i. Then $\overline{\beta}_i = (b_{i,1}^2, b_{i,2}^2)$ is a generic point of V and we can find a division system for $\overline{\beta}_i$ such that $\overline{\beta}_i^{\frac{1}{2}} = (b_{i,1}, b_{i,2})$. But then $\overline{\beta}_1$ and $\overline{\beta}_2$ realize the same quantifier free field theoretic type, but the corresponding division systems do not.

The algebraic key to Zilber's proof is the following result known as the Thumbtack Lemma.[14] The idea is that if $\overline{b}$ is a generic point of an irreducible variety V and $(\overline{b}^{1/n})_n$ is a division system for $\overline{b}$, then there is an m such that the quantifier free type of $\overline{b}^{\frac{1}{m}}$ determines the quantifier free type of the division system.[15]

THEOREM 8.2.7 (Thumbtack Lemma). *Let k be either an algebraically closed field of characteristic 0 or $\mathbb{Q}$ adjoin all root of unity. Let $(\overline{a}^{1/n})_n$ be a division system for $\overline{a}$. Suppose $(b^{1/n})_n$ is a division system such that b is multiplicatively independent over the group generated by $\overline{a}$ and $k^\times$. Then there is an m such that the field theoretic type of b_m over $k((\overline{a}^{1/n})_n)$ determines the field theoretic type of the division system $(b^{1/n})_n$ over k.*

The proof when k is obtained from $\mathbb{Q}$ be adjoining roots of unity relies on classical Kummer theory and can be found in [80]. We will give the proof when k is an algebraically closed field of characteristic 0. This proof is due to Bays, Gavrilovich and Hils [11]. Their proof is quite general and works when the multiplicative group is replaced by a semi-abelian variety.

Recall that an abelian variety is a complete algebraic group (or equivalently, a projective algebraic group) like an elliptic curve and semi-abelian variety is a commutative algebraic group S where there is an abelian variety A and an exact sequence

$$0 \to \mathbb{G}_m^l \to S \to A \to 1$$

where $\mathbb{G}_m^l$ is the lth-power of the multiplicative group. We will write semi-abelian varieties as additive groups $(S, +, 0)$.[16]

We will really need only one fact about semi-abelian varieties. We let $S[n]$ denote the subgroup of n-torsion points of S.

EXERCISE 8.2.8. A key fact about abelian varieties is that if A is an abelian variety then $A[n]$ is finite for all n. (See, for example, [26] §A.7.) Show that the same is true for semi-abelian varieties.

---

[14]The original Thumbtack Lemma is more complicated as Zilber needed to prove excellence. This version suffices to prove quasiminimality.

[15]The intuition is that the division system might have many different strands, but once we put a thumbtack in $\overline{b}^{\frac{1}{m}}$ we have straightened it out so that only one strand remains. This suggestive name was suggested by Holland.

[16]Readers not familiar with abelian varieties can just think of S as a product of multiplicative groups.

Let S be a semi-abelian variety and $[n] : S \to S$ be the map $x \mapsto nx$ for $n = 2, 3, \ldots$. We say that an irreducible algebraic variety $X \subseteq S$ is *Kummer generic* if $[n]^{-1}X$ is irreducible for all n.

We say that X is free if X is not contained in a coset of a proper algebraic subgroup of S.

THEOREM 8.2.9. *If S is a semi-abelian variety and $X \subseteq S$ is irreducible and free, then for some n, there is V a Kummer generic irreducible component of $[n]^{-1}(X)$.*

We first show that this is enough to prove the Thumbtack Lemma when k is an algebraically closed field.

EXERCISE 8.2.10. If $\overline{a} = c, \overline{d}$ and c is multiplicatively dependent over the group generated by $\overline{d}$ and $k^{\times}$. Then c is in the field generated by k and a division systems for $\overline{d}$.

Thus, without loss of generality, we may assume that $\overline{a}$ is multiplicatively independent over $k^{\times}$, and, hence that $\overline{a}, b$ is multiplicatively dependent over $k^{\times}$. Let X be the locus of $\overline{a}, b$ over k, i.e., X is the irreducible algebraic variety defined over k with generic point $\overline{a}, b$. By Theorem 8.2.9, there is an m such that each irreducible component of $[m]^{-1}X$ is Kummer generic. Let V be the irreducible component of containing $(\overline{a}^{1/m}, b^{1/m})$.

Suppose $(\overline{c}^{\frac{1}{n}}, d^{\frac{1}{n}})_n$ is a division sequence such that $(\overline{c}^{1/m}, d^{1/m}) \in V$. Then $(\overline{c}^{\frac{1}{n}}, d^{\frac{1}{n}})_n$ and $(\overline{a}^{\frac{1}{n}}, b^{\frac{1}{n}})_n$ realize the same type over k. In particular if we take and division sequence $(d^{\frac{1}{n}})_n$ such that $(\overline{a}^{1/m}, d^{1/m}) \in V$, then $(\overline{d}^{\frac{1}{n}})_n$ and $(\overline{b}^{\frac{1}{n}})_n$ realize the same type over $k((\overline{a}^{\frac{1}{n}})_n)$. Thus the type of $b^{1/m}$ over $k((\overline{a}^{\frac{1}{n}})_n)$ determines the type of the division sequence $(b^{\frac{1}{n}})_n$ over $k((\overline{a}^{\frac{1}{n}})_n)$.

To prove Theorem 8.2.9 suffices to show that there is a bound B on the number of irreducible components of $[n]^{-1}(X)$ independent of n. If $[n]^{-1}(X)$ has the maximal number of irreducible components, then any such component is Kummer generic.

Note that $[m] : S[mn] \to S[n]$. The *Tate module* of S is the inverse limit of this system of maps

$$T = \lim_{\leftarrow} S[n].$$

We can think of elements of T as $(t_n : n = 1, 2, \ldots)$ where $t_1 = 1$, $t_n \in S[n]$ and $mt_{mn} = t_n$.

Suppose S and X are defined over an algebraically closed field k. Let b be a generic point of X and let

$$\mathcal{G} = \mathrm{Gal}(k(b)^{\mathrm{acl}}/k(b)).$$

In this context, a division system for b is a sequence $(b_1, b_2, \ldots)$ such that $b_1 = b$ and $mb_{mn} = b_n$. Choose $(b_n)_n$ a division sequence for b. We define a homomorphism $\theta : \mathcal{G} \to T$ where

$$\theta(\sigma) = (\sigma(b_n) - b_n)_n.$$

Note that $n\sigma(b_n) = \sigma(nb_n) = 1$, thus $\sigma(b_n) - b_n \in S[n]$ and

$$m\sigma(b_{mn}) = \sigma(mb_{mn}) = \sigma(b_n).$$

Thus the image of θ is in the Tate module. Suppose $\sigma(b_n) = \xi_n + b_n$ where $\eta_n \in S[n]$. Then

$$\begin{aligned}\theta(\tau\sigma) &= (\tau(\sigma(b_n)) - b_n)_n \\ &= (\tau(b_n + \xi_n) - b_n)_n \\ &= (\tau(b_n) + \xi_n - b_n)_n \qquad \text{since } \xi_n \in k \\ &= (\tau(b_n) - b_n)_n + (\sigma(b_n) - b_n)_n \\ &= \theta(\tau) + \theta(\sigma).\end{aligned}$$

Thus θ is a homomorphism. Note that both $\mathcal{G}$ and T are profinite groups and θ is continuous in the usual topology.

Let Z be the image of θ. Then Z is a closed subgroup of T. We first argue that to prove the theorem it suffices to show that Z has finite index in T.

Let $V_1, \ldots, V_M$ be the irreducible components of $[n]^{-1}X$. Note that all of the V_i are defined over k. The map $[n] : S \to S$ is a closed map (as it has finite fibers) and, since X is irreducible, there must be some component such that $[n]V_i = X$. Choose V an irreducible component of maximum dimension such that $[n]V = X$. Suppose v_i is a generic point of V_i. Since $nv_i \in [n]V$, there is $\xi \in S[n]$ such that $\xi + v_i \in V$. But since v_i is a generic point, we must have $\xi + V_i \subseteq V$. Thus $V_i \subseteq V - \xi$ an irreducible subset of $[n]^{-1}X$ of maximal dimension. It follows that the irreducible components of $[n]^{-1}X$ are of the form $V + \xi$ for $\xi \in S[n]$. We have $V = V + \xi$ if and only if there is $\sigma \in \mathcal{G}$ with $\sigma(c) = c + \xi$ where c is a generic point of V. Consider the homomorphism $\theta_n : \mathcal{G} \to S[n]$ given by $\theta_n = \sigma(c) - c$. Then the number of irreducible components is the index of the $\theta(\mathcal{G})$ in $S[n]$. This is at most the index of Z in T.

We now begin the proof that Z has finite index in T. Let d be the dimension of S and let $\Sigma : X^d \to S$ be the map $(x_1, \ldots, x_d) \mapsto \sum x_i$.

We will show that the image of X^d is Zariski dense in S. Translating if necessary, we may assume $0 \in X$. For $1 \le k < d$ let W_k be the Zariski closure of $\Sigma(X^k)$. We claim that if $\dim \mathrm{W_k} < \mathrm{d}$, then $\dim \mathrm{W_k} < \dim \mathrm{W_{k+1}}$. It then follows that $\dim \mathrm{W_d}$ is d and $\Sigma(X^d)$ is Zariski dense in S.

Suppose, for contradiction, that $\dim \mathrm{W_k} < \mathrm{d}$ and $\dim \mathrm{W_k} = \dim \mathrm{W_{k+1}}$. Since W_{k+1} is irreducible, $W_k = W_{k+1}$. Suppose w is a generic point of W_{k+2}. Then there are independent generic points $x_1, \ldots, x_{k+2} \in X$ such that $w = \sum x_i$. But $x_1 + \cdots + x_{k+1}$ is a generic point of W_{k+1} and hence of W_k. So there are independent generic points $y_1, \ldots, y_k \in X$ such that $\sum y_i = x_1 + \cdots + x_{k+1}$. Hence $w \in W_{k+1} = W_k$. It follows that $W_l = W_k$ for all $l \ge k$. Thus if $x, y \in W_k$ are independent generics, then $x + y \in W_k$. It follows that W_k is closed under addition. If $w \in W_k$, then $W_k + w \subseteq W_k$. Since these are irreducible of the same dimension, $W_k + w = W_k$. Thus

$-w \in W_k$ and W_k is a group.[17] But then W_k is contained in a proper algebraic subgroup of S containing X violating the freeness of X, a contradiction.

Let a be a generic point of S. Let $\widetilde{\mathcal{G}} = \mathrm{Gal}(k(a)^{\mathrm{alg}}/k(a))$. The group $\widetilde{\mathcal{G}}$ acts transitively on the irreducible components of $\Sigma^{-1}(a)$. Let V be an irreducible component of $\Sigma^{-1}(a)$ and let $\mathrm{Stab}_{\widetilde{\mathcal{G}}(V)}$ be the elements of $\widetilde{\mathcal{G}}$ that fix V setwise. Then $\mathrm{Stab}_{\widetilde{\mathcal{G}}(V)}$ has finite index in $\widetilde{\mathcal{G}}$.

Let $\widehat{\theta} : \widetilde{\mathcal{G}} \to T$ be the analogous homomorphism to the Tate module. We will argue that $\widehat{\theta}$ is surjective and that the kernel of the induced homomorphism $\widetilde{\mathcal{G}} \to T/Z$ contains $\mathrm{Stab}_{\widetilde{\mathcal{G}}(V)}$. It will follow that the index of Z in T is at most the index of $\mathrm{Stab}_{\widetilde{\mathcal{G}}(V)}$ in $\widetilde{\mathcal{G}}$.

Let $\overline{\beta} = (\beta_1, \dots, \beta_d)$ be a generic point of V. Then $\overline{\beta}$ is also a generic point of X^d over K. Choose $(\beta_{i,n})_n$ such that $\beta_{i,1} = \beta_i$ and $m\beta_{i,mn} = \beta_{i,n}$. If we let $a_n = \sum_i \beta_{i,n}$, then $a_1 = a$ and $ma_{mn} = a_n$.

We first argue that $\widehat{\theta}$ is surjective. Let $(a_n)_n$ be a sequence such that $ma_{mn} = a_n$ for all n and $a_1 = a$. Let $(\xi_n)_n \in T$. For each n, a_n and $a_n + \xi_n$ are both generic points of S. Thus there is $\sigma \in \widetilde{\mathcal{G}}$ with $\sigma(a_n) = a_n + \xi_n$. Let

$$U_k = \{\sigma \in \widetilde{\mathcal{G}} : \sigma(a_{k!}) = a_{k!} + \xi_{k!}\}.$$

Then each U_k is clopen in the profinite topology of $\widetilde{\mathcal{G}}$ and

$$U_1 \supseteq U_2 \supseteq \cdots U_n \supseteq \cdots$$

and, since $\widetilde{\mathcal{G}}$ is compact, there is $\sigma \in \bigcap_n U_n$. Clearly, $\widehat{\theta}(\sigma) = (\xi_n)_n$. Thus $\widehat{\theta}$ is surjective.

We next show that if $\sigma \in \mathrm{Stab}_{\widetilde{\mathcal{G}}(V)}$ then $\widehat{\theta}(\sigma) \in Z$.

If $\sigma \in \mathrm{Stab}_{\widetilde{\mathcal{G}}(V)}$, then $\mathrm{tp}(\overline{\beta}/k(a)^{\mathrm{alg}})$ is stabilized by σ. Thus, there is $\tau \in \mathrm{Gal}(k(\overline{\beta})^{\mathrm{alg}}/k(\overline{\beta}))$ extending σ. Then

$$\begin{aligned}
\widehat{\theta}(\sigma) &= \Big(\sigma\Big(\sum_i \beta_{i,n}\Big) - \sum_i \beta_{i,n}\Big)_n \\
&= \Big(\tau\Big(\sum_i \beta_{i,n}\Big) - \sum_i \beta_{i,n}\Big)_n \\
&= \sum_i \Big(\tau(\beta_{i,n}) - \sum_i \beta_{i,n}\Big)_n \\
&= \sum_i \Big(\tau(\beta_{i,n}) - \beta_{i,n}\Big)_n.
\end{aligned}$$

But each $(\tau(\beta_{i,n}) - \beta_{i,n})_n \in Z$. Thus, $\widehat{\theta}(\sigma) \in Z$, as desired.

---

[17] Indeed, any stable semigroup is a group (see, for example, [56] 7.5.2).

The proof of quasiminimality. By Exercise 8.1.26 we first need to check that if $A_0, A_1 \in \mathcal{K}$ then $\mathbb{Q}\eta$ realizes the same quantifier free type in both structures. Let η_i denote the interpretation of η in H_i. We need to show that $\alpha = (q_1\eta_0, \dots, q_n\eta_0)$ and $\beta = (q_1\eta_1, \dots, q_n\eta_1)$ realize the same quantifier free type of any sequence of rationals $q_1, \dots, q_n \in \mathbb{Q}$. To do this, by Exercise 8.2.3, it suffices to show that $\exp(\alpha)$ and $\exp(\beta)$ realize the same field type over $\mathbb{Q}$. Note that $\exp(q_i\eta_j)$ is a q_i^{th}-root of unity. Choose M such that for each $j \leq n$ there is an integer N_j such that $q_j = \frac{N_j}{M}$. We can find primitive M^{th}-roots of unity ζ_i such that $\exp(q_j\eta_i) = \zeta_i^{N_j}$. But $\mathbb{Q}(\zeta_0) = \mathbb{Q}(\zeta_1)$ and $\zeta_0 \mapsto \zeta_1$ is an isomorphism. Thus $\exp(\alpha)$ and $\exp(\beta)$ realize the same field theoretic type,

We have shown in the exercises above that $\mathcal{K}$ satisfies non-degeneracy, union of chains and countable closure. So, we are left with verifying $\aleph_0$-homogeneity over countable sets.

Let $A_0, A_1 \in \mathcal{K}$, let $B_i \supseteq A_i$ be either empty or closed and countable and let $f : B_0 \to B_1$ be an τ_0-isomorphism.

EXERCISE 8.2.11. Show that if $a_0 \in A_0$, $a_1 \in A_1$ and a_i is independent over B_i, then we can extend f to a partial τ_0-embedding with $a_0 \mapsto a_1$

The main thing to show is that if $\overline{a} \in A_0$ and $g : B_0 \cup \{\overline{a}\} \to B_1$ is a partial τ_0-embedding extending f and $c \in \text{cl}(B_0, \overline{a})$, then there is $d \in \text{cl}(B_0, g(\overline{a}))$ such that we can extend g to a partial τ_0 embedding by sending $c \mapsto d$.

If B_i is empty, then, as above, we can extend f to an isomorphism between $\mathbb{Q}\eta_0$ and $\mathbb{Q}\eta_1$. We replace B_i by $\mathbb{Q}\eta_i$.

Let $k_i = \exp(B_i)$.

By the appropriate version of the Thumbtack Lemma we can find m such that the type of $\exp(c)^{\frac{1}{m}}$ over $l = k_1(\exp(a_1)^{\frac{1}{m}}, \dots, \exp(a_n)^{\frac{1}{m}}, m = 1, 2, \dots)$ determines the type of the division system $(\exp(c)^{\frac{1}{n}})_n$ Since $\exp(c) \in l^{\text{acl}}$, there is $\beta \in \exp(B_1)$ such that natural field isomorphism between

$$l \text{ and } k_2(\exp(g(a_1))^{\frac{1}{m}}, \dots, \exp(g(a_n))^{\frac{1}{m}}, m = 1, 2, \dots)$$

extends by sending $\exp(c)^{\frac{1}{m}} \mapsto \beta$.

Choose $d \in B_1$ such that $\exp(d/m) = \beta$. We can then extend g be sending $c \mapsto d$.

Part 3

EFFECTIVE CONSIDERATIONS

Chapter 9

EFFECTIVE DESCRIPTIVE SET THEORY

A number of insights about the model theory of $\mathcal{L}_{\omega_1,\omega}$ come from the perspective of effective descriptive set theory or generalized recursion theory. We will look at some of these in Chapter 11. In this chapter and the next, we give a brief introduction to effective descriptive set theory and related results about hyperarithmetic sets. We assume that the reader has some familiarity with the basics of descriptive set theory of Borel, analytic and coanalytic sets as presented in, say, Kechris [34]. Sacks [68] Moschovakis [61], Kechris' portion of [58] and Mansfield and Weitkamp [54] are excellent references for effective descriptive set theory.

9.1. Recursion theory review

We recall some of the basic ideas we will need from recursion theory. We will take an informal approach where we don't specify a machine model, but all of this could easily be made precise. We assume that the reader has some intuitive idea what a "computer program" is. This could be a simple model such as Turing machine programs or an informal notion like a C++ program.

DEFINITION 9.1.1. A partial function $f : \omega \to \omega$ is *partial recursive* if there is a computer program P such that P halts on input n if and only if $n \in \mathrm{dom}(f)$ and if P halts on input n, then the output is $f(n)$. We say that a set $A \subseteq \omega$ is *recursive* if and only if its characteristic function is recursive.

We can code computer programs by integers so that each integer codes a program. Let P_e be the machine coded by e. Let ϕ_e be the partial recursive function computed by P_e. We write $\phi_e(n)\downarrow_s$ if P_e halts on input n by stage s and $\phi_e(n)\downarrow$ if P_e halts on input n at some stage. Our enumeration has the following features:

Fact 9.1.2. (i) (Universal Function) The function $(e, n) \mapsto \phi_e(n)$ is partial recursive.
(ii) The set $\{(e, n, s) : \phi_e(n)\downarrow_s\}$ is recursive.
(iii) (Halting Problem) The set $\{(e, n) : \phi_e(n)\downarrow\}$ is not recursive.

(iv) (Parameterization Lemma) If $F : \omega^2 \to \omega$ is partial recursive, then there is a total recursive $d : \omega \to \omega$ such that

$$\phi_{d(x)}(y) = F(x, y)$$

for all x, y.

DEFINITION 9.1.3. We say that $A \subseteq \omega$ is *recursively enumerable* if there is a partial recursive function $f : \omega \to \omega$ such that A is the image of f.

Fact 9.1.4. The following are equivalent

(i) A is recursively enumerable.
(ii) A is the domain of a partial recursive function.
(iii) $A = \emptyset$ or A is the image of a total recursive function.
(iv) there is a recursive B such that $A = \{n : \exists m\, (n, m) \in B\}$.

Fact 9.1.5. (a) If A and B are recursively enumerable, then so are $A \cup B$ and $A \cap B$.
(b) If $A \subseteq \omega \times \omega$ is recursively enumerable so is $\{n : \exists m\, (n, m) \in A\}$.
(c) If A is recursively enumerable and $f : \omega \to \omega$ is total recursive, then $f^{-1}(A)$ is recursively enumerable.

EXERCISE 9.1.6. If you have not seen them before prove the statements in the Facts above.

A program with oracle $x \in \omega^\omega$ is a computer program which, in addition to the usual steps, is allowed at any stage to ask the value of $x(n)$.

We say that f is *partial recursive in* x if there is a program with oracle x computing f and say that $A \subseteq \omega$ is *recursive in* x if the characteristic function of A is recursive in x. The facts above relativize to oracle computations. We write $\phi_e^x(n)$ for the value of the partial recursive function in x with oracle program P_e on input n. One additional fact is useful.

Fact 9.1.7 (Use Principle). If $\phi_e^x(n)\downarrow$, then there is m such that if $x|m = y|m$, then $\phi_e^y(n) = \phi_e^x(n)$.

PROOF. The computation of P_e with oracle x on input n makes only finitely many queries about x. Choose m greater than all of the queries made. $\dashv$

We may also consider programs with finitely many oracles.

For $x, y \in \omega^\omega$ we say x is *Turing-reducible* to y and write $x \leq_T y$ if x is recursive in y. For $x, y \in \omega^\omega$ define $x \oplus y$ to be $(x(0), y(0), x(1), y(1), \dots)$. The $x \oplus y$ is a least upper bound for x and y under $\leq_T$. We call $x \oplus y$ the *join* of x and y.

There is another useful notion of reducibility which is the analog of Wadge reducibility in descriptive set theory.

DEFINITION 9.1.8. We say A is *many-one reducible* to B if there is a total recursive f such that $n \in A$ if and only if $f(n) \in B$ for all $n \in \omega$. We write $A \leq_m B$ if A is many-one reducible to B.

Clearly if $A \leq_m B$, then $A \leq_T B$.

There is one subtle fact that will eventually play a key role.

THEOREM 9.1.9 (Recursion Theorem). *If $f : \omega \to \omega$ is total recursive, there is an e such that $\phi_e = \phi_{f(e)}$.*

PROOF. Let

$$F(x, y) = \begin{cases} \phi_{\phi_x(x)}(y) & \text{if } \phi_x(x)\downarrow, \\ \uparrow & \text{if } \phi_x(x)\uparrow. \end{cases}$$

By the Parameterization Lemma, there is a total recursive d such that

$$\phi_{d(x)}(y) = F(x, y).$$

Let $\psi = f \circ d$. There is m such that $\psi = \phi_m$. Since ψ is total $\phi_{d(m)} = \phi_{\phi_m(m)}$. Let $e = d(m)$. Then

$$\phi_e = \phi_{d(m)} = \phi_{\phi_m(m)} = \phi_{\psi(m)} = \phi_{f(d(m))} = \phi_{f(e)}. \qquad \dashv$$

For those of you who have not seen this before here are two samples of the many applications of the Recursion Theorem.

Suppose $F : \omega^2 \to \omega$. By the Parameterization Lemma there is a total recursive $d : \omega \to \omega$ such that $\phi_{d(m)}(n) = F(m, n)$ for all m and n. By the Recursion Theorem there is an e such that

$$\phi_e(n) = \phi_{d(e)}(n) = F(e, n)$$

for all n.

For a second example, consider

$$g(e, n) = \begin{cases} 1 & \text{if } e = n, \\ 0 & \text{if } e \neq n. \end{cases}$$

By the Parameterization Lemma, there is a total recursive f such that $\phi_{f(e)}(n) = g(e, n)$. By the Recursion Theorem there is an e such that $\phi_e(n) = 1$ if $n = e$ and $\phi_e(n) = 0$ if $n \neq e$. So this function "recognizes" its own code. The Recursion Theorem will be very useful in Chapter 7.

9.2. Computable functions on ω^ω

There is also a notion of computable function $f : \omega^\omega \to \omega^\omega$.

DEFINITION 9.2.1. We say that $f : \omega^\omega \to \omega^\omega$ is *computable* if there is an oracle program P such that if $x \in \omega^\omega$ and P is run with oracle x on input n, then P halts and outputs $f(x)(n)$.

We say that $f : \omega^\omega \to \omega^\omega$ is computable from z if there is a two oracle program P such that if $x \in \omega^\omega$ and P is run with oracles z and x on input n, then M halts and outputs $f(x)(n)$.

Recall that for $\sigma \in \omega^{<\omega}$ we let $N_\sigma = \{f \in \omega^\omega : \sigma \subset f\}$. The sets $\mathcal{N}_\sigma$ are the basic open sets in the topology on ω^ω.

LEMMA 9.2.2. *$f : \omega^\omega \to \omega^\omega$ is continuous if and only if there is $z \in \omega^\omega$ such that f is computable from z.*

PROOF. ($\Leftarrow$) Let P be the oracle program computing f from z. Suppose $f(x) = y$. By the Use Principle, for any m there is an n such that if $x|n = w|n$, then $f(w)|m = y|m$. Thus $N_{x|n} \subseteq f^{-1}N_{y|m}$ and f is continuous.

($\Rightarrow$) Let $X = \{(\tau, \sigma) : f^{-1}(N_\sigma) \supseteq N_\tau\}$. Since f is continuous, for all $x \in \omega^\omega$ if $f(x) = y$, then for all n there is an m such that $(x|m, y|n) \in X$.

We claim that f is computable from X. Suppose we are given oracles X and x and input n. We start searching X until we find $(\tau, \sigma) \in X$ such that $\tau \subset x$ and $|\sigma| > n$. Then $f(x)(n) = \sigma(n)$. ⊣

9.3. The arithmetic hierarchy

For the next few sections we will restrict our attention to Polish spaces $X = \omega^k \times (\omega^\omega)^l \times (2^\omega)^m$ where $k, l.m \geq 0$. Any such space is homeomorphic to one of $\omega, 2^\omega$ or ω^ω. (In [61] this theory is worked out for "recursively presented Polish spaces".)

Let $X = \omega^k \times (\omega^\omega)^l$. Let $S_X = \{(m_1, \dots, m_k, \sigma_1, \dots, \sigma_l) : m_i, \dots, m_k \in \omega, \sigma_1, \dots, \sigma_l \in \omega^{<\omega}\}$. For $\sigma = (m_1, \dots, m_k, \sigma_1, \dots, \sigma_l) \in S_X$, let

$$N_\sigma = \{(n_1, \dots, n_k, f_1, \dots, f_l) \in X : n_i = m_i \text{ if } i \leq k \text{ and } f_i \supset \sigma_i \text{ if } \leq l\}.$$

Then $\{N_\sigma : \sigma \in S_X\}$ is a clopen basis for the topology on X. Of course S_X is a countable set and there is a recursive bijection $i \mapsto \sigma_i$ between ω and S_X, such that all natural operations on sequences are given by recursive functions. Thus we can identify S_X with ω and talk about things like recursive subsets of S_X and partial recursive functions $f : \omega \to S_X$.

DEFINITION 9.3.1. We say that $A \subseteq X$ is Σ^0_1 if there is a partial recursive $f : \omega \to S_X$ such that $A = \bigcup_n N_{f(n)}$.

Note that here we are using a "lightface" Σ^0_1 rather than the "boldface" $\boldsymbol{\Sigma}^0_1$ that denotes the open subsets of X. Of course every Σ^0_1 set is open, but there are only countably many partial recursive $f : \omega \to S_X$ thus there are only countably many Σ^0_1 sets. Thus $\Sigma^0_1 \subset \boldsymbol{\Sigma}^0_1$. Relativizing these notions we get all open sets.

DEFINITION 9.3.2. If $x \in \omega^\omega$ we say that $A \subseteq X$ is $\Sigma^0_1(x)$ if there is $f : \omega \to S_X$ partial recursive in x such that $A = \bigcup_n N_{f(n)}$.

LEMMA 9.3.3. $\boldsymbol{\Sigma}^0_1 = \bigcup_{x \in \omega^\omega} \Sigma^0_1(x)$.

We will tend to prove things only for Σ^0_1 sets. The relativization to $\Sigma^0_1(x)$ sets is usually straightforward.

For the two interesting examples $X = \omega$ and $X = \omega^\omega$ we get slightly more informative characterizations.

LEMMA 9.3.4. (i) *$A \subseteq X$ is Σ^0_1 if and only if there is a recursively enumerable $W \subseteq S_X$ such that $A = \bigcup_{\eta \in W} N_\eta$. In particular $A \subseteq \omega$ is Σ^0_1 if and only if A is recursively enumerable.*

(ii) *$A \subseteq \omega^\omega$ is Σ^0_1 if and only if there is a recursive $S \subseteq \omega^{<\omega}$ such that $A = \bigcup_{\sigma \in S} N_\sigma$.*

PROOF. (i) This is clear since the recursively enumerable sets are exactly the images of partial recursive functions.

(ii) In this case $S_X = \omega^{<\omega}$. Clearly if $S \subseteq \omega^{<\omega}$ is recursive, there is $f : \omega \to \omega^{<\omega}$ partial recursive with image S and $\bigcup_{\sigma \in \omega^{<\omega}} N_\sigma$ is Σ^0_1.

Suppose $A = \bigcup_n N_{f(n)}$ where f is partial recursive let $S = \{\sigma :$ there is $n \leq |\sigma|$, the computation of $f(n)$ halts by stage $|\sigma|$ and $f(n) \subseteq \sigma$. It is easy to see that S is recursive. If $\sigma \in S$, then there is an n such that $f(n) \subseteq \sigma$, then $N_\sigma \subseteq N_{f(n)} \subseteq A$. On the other hand if f halts on input n, there is $m \geq n, |f(n)|$ such that f halts by stage m. If $\tau \supset \sigma$ and $|\tau| \geq m$, then $\tau \in S$. Thus

$$\bigcup_{\sigma \in S} N_\sigma \supset \bigcup_{\tau \supset f(n), |\tau| = m} N_\tau = N_{f(n)}.$$

It follows that $A = \bigcup_{\sigma \in S} N_\sigma$. ⊣

We have natural analogs of the finite levels of the Borel hierarchy.

DEFINITION 9.3.5. Let $X = \omega^k \times (\omega^\omega)^l$. We say that $A \subseteq X$ is Π^0_n if and only if $X \setminus A$ is Σ^0_n.

We say that $A \subseteq X$ is Σ^0_{n+1} if and only if there is $B \subseteq \omega \times X$ in Π^0_n such that

$$x \in A \text{ if and only if } \exists n\, (n, x) \in B.$$

We say that A is Δ^0_n if it is both Σ^0_n and Π^0_n.

We say that $A \subseteq X$ is *arithmetic* if $A \in \Delta^0_n$ for some n.

Recall that $T \subseteq \omega^{<\omega}$ is a *tree* if whenever $\sigma \in T$ and $\tau \subseteq \sigma$, then $\tau \in T$. We let Tr denote the set of all trees.

LEMMA 9.3.6. *$A \subseteq \omega^\omega$ is Π^0_1 if and only if there is a recursive tree $T \subseteq \omega^{<\omega}$ such that $A = [T]$.*

PROOF. Clearly,

$$[T] = \{f : \forall n\, f|n \in T\}$$

is Π^0_1. On the other hand, if S is a recursive set such that $X \setminus A = \bigcup_{\sigma \in S} N_\sigma$, let

$$T = \{\sigma \in \omega^{<\omega} : \forall m \leq |\sigma|\, \sigma|m \notin S\}.$$

Then T is a recursive tree and $A = [T]$. ⊣

Recall that a tree T is *pruned* if for all $\sigma \in T$ there is $f \in [T]$ with $\sigma \subset f$. If A is closed, then we can find a pruned tree such that $A = [T]$, but the next exercises shows that it is not always possible to find a recursive pruned tree.

EXERCISE 9.3.7. (a) Show that if T is a recursive pruned tree, then the left-most path through T is recursive.

(b) Let $T = \{\sigma \in \omega^{<\omega} :$ if $e < |\sigma|$ and $\phi_e(e)$ halts by stage $|\sigma|$, then $\phi_e(e)$ halts by stage $\sigma(e)\}$. Show that T is a recursive tree. Suppose $f \in [T]$. Show that $\phi_e(e)$ halts if and only if it halts by stage $f(e)$. Conclude that there are no recursive paths through T and, using (a), that there is no nonempty recursive pruned subtree of T.

We show that Σ^0_n and Π^0_n have closure properties analogous to those for the corresponding levels of the Borel hierarchy. The definition of computable function made in 9.2.1 makes sense for maps $f : X \to Y$ where both X and Y are of the form $\omega^k \times (\omega^\omega)^l$

LEMMA 9.3.8. (i) *Σ^0_n is closed under finite unions, finite intersections, and computable inverse images.*

(ii) *If $A \subseteq \omega \times X \in \Sigma^0_n$, then $\{x \in X : \exists n\, (n, x) \in A\} \in \Sigma^0_n$.*

(iii) *If $f : X \to \omega$ is computable and $A \subseteq \omega \times X$ is Σ^0_n then $\{x \in X : \forall m < f(x)\, (m, x) \in A\} \in \Sigma^0_n$.*

(iv) *Similarly Π^0_n is closed under union, intersection, computable inverse images, $\forall n$ and $\exists n < f(x)$.*

(v) $\Sigma^0_n \subseteq \Delta^0_{n+1}$.

PROOF. We prove (i)–(iii) this for Σ^0_1 and leave (iv), (v) and the induction steps for (i)–(iii) as an exercise.

(i) Suppose W_0 and W_1 are recursively enumerable subsets of S_X and $A_i = \bigcup_{\eta \in W_i} N_\eta$. Replacing W_i by the recursively enumerable set $\{\nu : \exists \eta \in W_i\, \eta \subseteq \nu\}$ if necessary we may assume that if $\nu \in W_i$ and $\eta \supset \nu$, then $\eta \in W_i$. Then

$$A_0 \cup A_1 = \bigcup_{\eta \in W_0 \cup W_1} N_\eta$$

and

$$A_0 \cap A_1 = \bigcup_{\eta \in W_0 \cap W_1} N_\eta$$

and $W_0 \cup W_1$ and $W_0 \cap W_1$ are recursively enumerable. Thus $A_0 \cup A_1$ and $A_0 \cap A_1$ are Σ^0_1.

If $f : X \to Y$ is computable with program P_e, let

$$G = \{(\eta, \nu) \in S_X \times S_Y : |\eta| = |\nu|, x \in N_\eta \Rightarrow f(x) \in N_\nu\}.$$

Then $(\eta, \nu) \in G$ if and only if for all $m < |\nu|$ the program P_e using oracle η halts on input m and outputs $\nu(m)$.[18] Thus G is recursively enumerable. Suppose $A = \bigcup_{\nu \in W} N_\nu$ where W is recursively enumerable, let $V = \{\eta : \exists \nu\, (\nu \in W \wedge (\eta, \nu) \in G\}$. Then V is recursively enumerable and $f^{-1}(A) = \bigcup_{\eta \in V} N_\eta$.

(ii) Suppose $A \subseteq \omega \times X$ is Σ_1^0. There is a recursively enumerable $W \subseteq S_{\omega \times X}$ such that $A = \bigcup_{\eta \in W} N_\eta$. Let $V = \{\nu \in S_X : \exists n\, (n, \nu) \in W\}$. Then V is recursively enumerable and

$$\{x : \exists n\, (n, x) \in A\} = \bigcup_{\nu \in V} N_\nu.$$

(iii) Suppose A and W are as in (ii) and $f : X \to \omega$ is computable by program P_e. Let $V = \{\nu \in S_X : \exists k\, P_e$ with oracle ν halts outputting k and $(m, \nu) \in W$ for all $m \leq k\}$. Then V is recursively enumerable and

$$\left\{x : \forall m < f(x)\, (m, x) \in A\} = \bigcup_{\nu \in V} N_\nu \right\}.$$

⊣

EXERCISE 9.3.9. Complete the proof of 9.3.8

DEFINITION 9.3.10. Let Γ be a collection of subsets of X. We say that $U \subseteq Y \times X$ is Γ-*universal* if:

(i) $\{x \in X : (y, x) \in U\} \in \Gamma$ for all $y \in Y$;
(ii) if $A \in \Gamma$, then there is $y \in Y$ such that $A = \{x \in X : (y, x) \in U\}$.

PROPOSITION 9.3.11. (i) *There is $U_n \subseteq \omega^\omega \times X$ a Σ_n^0-set that is Σ_n^0-universal.*
(ii) *There is $V_n \subseteq \omega \times X$ a Σ_n^0-set that is Σ_n^0-universal.*

PROOF. (i) Fix $f : \omega \to S_X$ a recursive bijection. The set $U_1 = \{(x, y) : \exists n\, (x(n) = 1 \wedge y \in N_{f(n)})\}$ is Σ_1^0 and Σ_1^0-universal.

If $U_n^* \subseteq \omega^\omega \times \omega \times X$ is Σ_n^0 and Σ_n^0-universal for $\omega \times X$, then

$$U_{n+1} = \{(x, y) : \exists n\, (x, n, y) \notin U_n^*\}$$

is Σ_{n+1}^0 and Σ_{n+1}^0-universal.

(ii) Let $V_1 = \{(n, x) : \exists m\, (\phi_n(m)\downarrow \wedge x \in N_{\phi_n(m)})\}$. Let $g : \omega \times \omega \to S_{\omega \times X}$ be partial recursive such that $g(n, m) = (n, \phi_n(m))$, then

$$V_1 = \bigcup_{n,m} N_{g(n,m)}$$

is Σ_1^0 and Σ_1^0-universal.

An induction as in (i) extends this to all levels of the arithmetic hierarchy. ⊣

Analogous results hold for Π_n^0. For $X = \omega^\omega$ or 2^ω we know that sets U_n are not Π_n^0.

[18] We assume that if the computation makes any queries about numbers $i \geq |\eta|$, then the computation does not halt.

COROLLARY 9.3.12. *There is $A \subseteq \omega$ such that A is Σ^0_n but not Δ^0_n.*

PROOF. Suppose $U \subseteq \omega \times \omega$ is Σ^0_n and universal Σ^0_n. Let $A = \{m : (m, m) \notin U\}$. If $U \in \Delta^0_n$, then $A \in \Sigma^0_n$ and $A = \{m : (i, m) \in U\}$ for some i. Then

$$i \in A \Leftrightarrow (i, i) \notin U \Leftrightarrow i \notin A,$$

a contradiction. Thus $U \in \Sigma^0_n \setminus \Delta^0_n$. ⊣

Let Γ be Π^0_n, Σ^0_n or Δ^0_n.If $A, B \subseteq \omega$, $B \in \Gamma$ and $A \leq_m B$, then $A \in \Gamma$.

We say that $A \subseteq \omega$ is *Γ-complete* if $A \in \Gamma$ and $B \leq_m A$ for all $B \subseteq \omega$ in Γ. Here are some well known examples from recursion theory.

Fact 9.3.13. (i) $\{e : \mathrm{dom}(\phi_e) \neq \emptyset\}$ is Σ^0_1-complete.
(ii) $\{e : \phi_e \text{ is total}\}$ is Π^0_2-complete.
(iii) $\{e : \mathrm{dom}(\phi_e) \text{ is infinite}\}$ is Π^0_2-complete.
(iv) If $U \subseteq \omega \times \omega$ is Γ-universal, then U is Γ-complete.

EXERCISE 9.3.14. Prove the statements in the last fact.

9.4. The effective projective hierarchy

We also have effective analogs of the projective point classes. Let $X = \omega^k \times (\omega^\omega)^l$ for some $k, l \in \omega$.

DEFINITION 9.4.1. We say that $A \subseteq X$ is Σ^1_1 if there is a $B \subseteq \omega^\omega \times X$ such that $B \in \Pi^0_1$ and $A = \{x : \exists y\, (y, x) \in B\}$.

We say $A \subseteq X$ is Π^1_n if $X \setminus A$ is Σ^1_n and we say that $A \subseteq X$ is Σ^1_{n+1} if there is a $B \subseteq \omega^\omega \times X$ with $B \in \Pi^1_n$ such that $A = \{x : \exists y\, (y, x) \in B\}$.

We say A is Δ^1_n if it is both Σ^1_n and Π^1_n.

All of these notions relativize and a set is $\boldsymbol{\Sigma}^1_n$ if and only $\Sigma^1_1(x)$ for some $x \in \omega^\omega$.

The next theorem summarizes a number of important properties of the classes Σ^1_n and Π^1_n. The proofs are exactly as in classical Descriptive Set Theory, and we leave the proofs as exercises.

THEOREM 9.4.2. (i) *The classes Σ^1_n and Π^1_n are closed under union, intersection, $\exists n \in \omega$, $\forall n \in \omega$ and computable inverse images.*
(ii) *If $A \subseteq X \times \omega^\omega$ is arithmetic, then $\{x : \exists y (x, y) \in A\}$ is Σ^1_1.*
(iii) *There is $U \subseteq \omega^\omega \times X$ a Σ^1_n-set that is $\boldsymbol{\Sigma}^1_n$-universal.*
(iv) *There is $V \subseteq \omega \times X$ a Σ^1_n-set that is Σ^1_n-universal.*
(v) *$\Sigma^1_n \subset \Delta^1_{n+1}$, but $\Sigma^1_n \neq \Delta^1_n$.*
(vi) *The set* WO *of well orders and the set* WF *of well founded trees are Π^1_1.*

EXERCISE 9.4.3. Prove 9.4.2.

Suppose $A \subseteq \omega^\omega$ is Σ^1_1. Then there is a recursive tree $T \subset \omega^{<\omega} \times \omega^{<\omega}$ such that $x \in A$ if and only if there is $y \in \omega^\omega$ such that (x, y) is a path through T. Let

$$T(x) = \{\eta : (x|n, \eta) \in T \text{ where } n = |\eta|\}.$$

Then $x \mapsto T(x)$ is continuous and $x \in A$ if and only if $T(x)$ is not well founded. Note that the map $x \mapsto T(x)$ is computable. Many of the proofs of classical facts about Σ^1_1 and Π^1_1-sets work equally well of Σ^1_1 and Π^1_1-sets. Here are statements of the effective versions.

THEOREM 9.4.4. (i) *If $A \subseteq X$ is Π^1_1, there is a computable $f : X \to Tr$ such that $x \in A$ if and only if $f(x) \in$ WF for all $x \in X$.*

(ii) *Π^1_1 has the reduction property, i.e., if A, B are Π^1_1, there are Π^1_1 A_0 and B_0 such that $A \cup B = A_0 \cup B_0$ and $A_0 \cap B_0 = \emptyset$.*

(iii) *Any two disjoint Σ^1_1 sets can be separated by a Δ^1_1-set.*

(iv) (Uniformization) *Any Π^1_1-subset of $\omega^\omega \times \omega^\omega$ can be uniformized by a Π^1_1-set, i.e., if $A \subseteq \omega^\omega \times \omega^\omega$ is Π^1_1, there is a Π^1_1 set $B \subseteq A$ such that $\pi(B) = \pi(A)$ and B is the graph of a function.*

(v) (Kreisel Uniformization) *If $A \subseteq \omega^\omega \times \omega$ is Π^1_1 and $\pi(A) = \omega^\omega$, then A has a Δ^1_1-uniformization.*

PROOF. We give the proof of (v). Let B be a Π^1_1 uniformization of A. Then

$$(x, n) \notin B \Leftrightarrow \exists m\, m \neq n \wedge (x, m) \in B.$$

Thus B is Δ^1_1. ⊣

Further analysis of Π^1_1-sets will require looking at an effective version of "ordinals".

9.5. Recursive ordinals

The set WF is Π^1_1. If $A \subseteq \omega^\omega$ is Π^1_1, we know that $A \leq_w$ WF. We will show that the reduction f can be chosen computable. There is a recursive tree T, such that

$$x \in A \Leftrightarrow \forall y\, (x, y) \notin [T] \Leftrightarrow T(x) \in \mathrm{WF}.$$

The function $x \mapsto T(x)$ is computable.

A similar construction gives rise to a Π^1_1-complete (for $\leq_m$) subset of ω.

Let $O = \{e \in \omega : \phi_e$ is the characteristic function of a well-founded tree $T_e \subseteq \omega^{<\omega}\}$. Then $e \in O$ if and only if

(i) $\forall \sigma\, \phi_e(\sigma)\downarrow$ and $\phi_e(\sigma) = 0$ or 1;

(ii) $\forall \sigma \in \omega^{<\omega}\, \forall \tau \in \omega^{<\omega}\, ((\sigma \subseteq \tau \wedge \phi_e(\tau) = 1) \to \phi_e(\sigma) = 1)$;

(iii) $\forall f : \omega \to \omega^{<\omega} \exists n\, (\phi_e(f(n)) = 0 \vee \phi_e(f(n+1)) = 0 \vee f(n) \not\subset f(n+1))$.

Conditions (i) and (ii) are Π^0_2 while (iii) is Π^1_1. Thus O is Π^1_1. The set O is (a variant of) *Kleene's O*.

PROPOSITION 9.5.1. *O is Π^1_1-complete.*

PROOF. We will argue that $\omega \setminus O$ is Σ^1_1-complete. Suppose $A \in \Sigma^1_1$. There is $B \subseteq \omega \times \omega^\omega$ in Π^0_1 such that $n \in A$ if and only if $\exists x\,(n, x) \in B$. There is a recursive tree $T \subseteq \omega \times \omega^{<\omega}$ such that $(n, x) \in A$ if and only if $(n, x|m) \in T$ for all $m \in \omega$. There is a recursive $f : \omega \to \omega$ such that $\phi_{f(n)}$ is the characteristic function of $\{\sigma : (n, \sigma) \in T\}$. Then $\phi_{f(n)}$ is the characteristic function of a tree T_n and

$$n \in A \Leftrightarrow T_n \notin \mathrm{WF} \Leftrightarrow f(n) \notin O. \qquad \dashv$$

O will play a very important role in effective descriptive set theory. As a first example, we will show how once we know the complexity of a set, we can find relatively simple elements of the set.

LEMMA 9.5.2. *Suppose $T \subseteq \omega^{<\omega}$ is a recursive tree. If $[T] \neq \emptyset$, there is $x \in [T]$ with $x \leq_T O$.*

PROOF. We build the "left-most" path though T.

For $\sigma \in T$ let $T_\sigma = \{\eta : \sigma^\frown\eta \in T\}$. There is a recursive function f such that $\phi_{f(\sigma)}$ is the characteristic function of T_σ for all $\sigma \in \omega^{<\omega}$. We build $\emptyset = \sigma_0 \subset \sigma_1 \dots$ with $\sigma_i \in T$ such that $[T_{\sigma_i}] \neq \emptyset$. Given σ_i, let $n \in \omega$ be least such that $\sigma_i{}^\frown n \in T$ and $f(\sigma_i{}^\frown n) \notin O$. $\dashv$

COROLLARY 9.5.3 (Kleene Basis Theorem). *If $A \subseteq \omega^\omega$ is Σ^1_1 and nonempty, there is $x \in A$ with $x \leq_T O$.*

PROOF. There is a Π^0_1-set $B \subseteq \omega^\omega \times \omega^\omega$ such that $x \in A$ if and only if $\exists y\,(x, y) \in B$. By the previous lemma there is $(x, y) \in B$ with $(x, y) \leq_T O$. Clearly $x \leq_T O$. $\dashv$

Using the Uniformization Theorem, we can find definable elements of Π^1_1-sets.

PROPOSITION 9.5.4. *If $A \subseteq \omega^\omega$ is Π^1_1, there is $x \in A$ such that $x \in \Delta^1_2$.*

PROOF. Uniformizing $\{0\} \times A$, we find $x \in \omega^\omega$ such that $B = \{(0, x)\}$ is Π^1_1. Then

$$\begin{aligned} x(n) = m &\Leftrightarrow \exists y\,((0, y) \in B \wedge y(n) = m) \\ &\Leftrightarrow \forall y\,((0, y) \notin B \vee y(n) = m). \end{aligned}$$

The first definition is Σ^1_2, while the second is Π^1_2. $\dashv$

We next need to understand the possible heights of recursive trees.

DEFINITION 9.5.5. We say that an ordinal α is *recursive* if there is a recursive set $A \subseteq \omega$ and $\prec$ a recursive linear order of A such that $(A, \prec) \cong (\alpha, <)$.

LEMMA 9.5.6. (a) *If α is a recursive ordinal and $\beta < \alpha$, then β is a recursive ordinal.*

(b) *If α is a recursive ordinal, then $\alpha + 1$ is a recursive ordinal.*

(c) *Suppose* $f : \omega \to \omega, g : \omega \to \omega$ *are recursive functions such that* $P_{f(n)}$ *is a program to compute the characteristic function of* A_n, $P_{g(n)}$ *is a program that computes the characterisitic function of* $\prec_n$ *a well-order of* A_n *and* $(A_n, \prec_n)$ *has order-type* α_n. *Then* $\sup \alpha_n$ *is a recursive ordinal.*

PROOF. (a) and (b) are routine. For (c) let $A = \{(n, m) : \phi_{f(n)}(m) = 1\}$ and let $(n, m) \prec (n', m')$ if and only if $n < n'$ or $n = n'$ and $m \prec_n m'$. Then $(A, \prec)$ is a recursive well-order. Let α be the order type of A. Then $\alpha_n \leq \alpha$ for all n. Since $\sup \alpha_n \leq \alpha$, $\sup \alpha_n$ is a recursive ordinal. ⊣

There are only countably many recursive well-orders. Thus there are only countably many recursive ordinals.

DEFINITION 9.5.7. Let ω_1^{ck} be the least non-recursive ordinal. We call this ordinal the *Church–Kleene ordinal.*

More generally, for any x we let ω_1^x be the least ordinal not recursive in x.

DEFINITION 9.5.8. We say that a countable ordinal $\alpha > \omega$ is *admissible* if $\alpha = \omega_1^z$ for some z. More generally, we say that α is x-admissible if and only if $\alpha = \omega_1^{x,z}$ for some z.[19]

DEFINITION 9.5.9. For $\sigma, \tau \in \omega^{<\omega}$ we say $\sigma \lhd \tau$ if $\tau \subset \sigma$ or there is an n such that $\sigma(n) < \tau(n)$, and $\sigma(m) = \tau(m)$ for all $m < n$. We call $\lhd$ the *Kleene–Brower order.*

EXERCISE 9.5.10. (a) Show that $\lhd$ is a linear order of $\omega^{<\omega}$.

(b) Show that if $T \subseteq \omega^{<\omega}$ is a tree, then T is well-founded if and only if $(T, \lhd)$ is a well-order. [Hint: If $\sigma_0, \sigma_1, \ldots$ is an infinite descending sequence in $(T, \lhd)$, define x inductively by $x(n) =$ least m such that $(x(0), \ldots, x(n-1), m) \lhd \sigma_i$ for some i. Prove that $x \in [T]$.]

Recall that $\rho(T)$, the rank of a well founded tree T is defined inductively as follows:

- $\rho(\emptyset) = 0$;
- $\rho(T) = \sup\{\rho(T_n) + 1 : n \in \omega\}$ where $T_n = \{\eta : n^\frown\eta \in T\}$.

We say $\rho(T) = \infty$ if T is not well founded.

EXERCISE 9.5.11. Prove that $\omega_1^{\mathrm{ck}} = \sup\{\rho(T) : T \subseteq \omega^{<\omega}$ a recursive well founded tree$\}$.

THEOREM 9.5.12. (i) *The set* $\{(S, T) : \rho(S) \leq \rho(T)\}$ *is* Σ_1^1.

(ii) *There is* $R \in \Sigma_1^1$ *such that if* $T \in \mathrm{WF}$, *then* $\{S : (S, T) \in R\} = \{S : \rho(S) < \rho(T)\}$.

[19] This is an abuse. The real definition is given in Appendix B. Here we are using Sacks' Theorem that a countable ordinal $\alpha > \omega$ is admissible if and only if $\alpha = \omega_1^x$ for some $x \subseteq \omega$. This theorem is given a model theoretic proof in Theorem B.3.11.

PROOF. (i) $\rho(S) \le \rho(T)$ if and only if there is an order preserving $f : S \to T$.

(ii) If T is well founded, then $\rho(S) < \rho(T)$ if and only if there is n such that $\rho(S) \le \rho(T_n)$. ⊣

COROLLARY 9.5.13 (Effective Σ^1_1-Bounding). (i) *If $A \subseteq O$ is Σ^1_1, then there is $\alpha < \omega_1^{\mathrm{ck}}$ such that $\rho(T) < \alpha$ for all $T \in A$.*

(ii) *If $A \subseteq \mathrm{WF}$ and $A \in \Sigma^1_1$, then there is $\alpha < \omega_1^{\mathrm{ck}}$ such that $\rho(T) < \alpha$ for all $T \in A$.*

PROOF. If either (i) or (ii) fails, then $O = \{e : \phi_e$ is the characteristic function of a recursive tree $\exists T\, \forall \sigma \in \omega^{<\omega}((\sigma \in T \leftrightarrow \phi_e(\sigma) = 1)$ and $\exists S \in A\, \rho(T) \le \rho(S))\}$ is Σ^1_1, a contradiction. ⊣

EXERCISE 9.5.14. Prove that if $A \subseteq \omega^\omega$ is Δ^1_1, then A is Σ^0_α for some $\alpha < \omega_1^{\mathrm{ck}}$.

Chapter 10

HYPERARITHMETIC SETS

Our first goal is to try to characterize the Δ^1_1-sets. In particular we will try to formulate the "light-faced" version of

$$\Delta^1_1 = \text{Borel}.$$

We begin by studying a method of coding Borel sets.

10.1. Borel codes

Let $X = \omega^k \times (\omega^\omega)^k$. Let S_X be as in Chapter 9.

DEFINITION 10.1.1. A *Borel code* for a subset of X is a pair $\langle T, l \rangle$ where $T \subseteq \omega^{<\omega}$ is a well-founded tree and $l : T \to (\{0\} \times \{0,1\}) \cup (\{1\} \times S_X)$ such that:

(i) if $l(\sigma) = \langle 0, 0 \rangle$, then $\sigma\frown 0 \in T$ and $\sigma\frown n \notin T$ for all $n \geq 1$;
(ii) if $l(\sigma) = \langle 0, 1 \rangle$, then $\exists n \sigma\frown n \in T$ for all $n \in \omega$;
(iii) if $l(\emptyset) = \langle 1, \eta \rangle$, then $\sigma, n \notin T$ for all $n \in \omega$.

Let BC be the set of all Borel codes. It is easy to see that BC is Π^1_1.

If $x = \langle T, l \rangle$ is a Borel code, we can define $B(x)$ the Borel set coded by x. If $\sigma \in T$, recall that $T_\sigma = \{\tau : \sigma\frown\tau \in T\}$. We let $l_\sigma : T_\sigma \to \{0\} \times 2 \cup \{1\} \times S_X$ by $l_\sigma(\tau) = l(\sigma\frown\tau)$. It is easy to see that $\langle T_\sigma, l_\sigma \rangle$ is also a Borel code.

DEFINITION 10.1.2. We define $B(\langle T, l \rangle)$ inductively on the height of T.

(i) $B(\langle \emptyset, \emptyset \rangle) = \emptyset$.
(ii) If $l(\sigma) = \langle 1, \eta \rangle$, then $B(\langle T, l \rangle) = N_\eta$.
(iii) If $l(\sigma) = \langle 0, 0 \rangle$, then $B(\langle T, l \rangle) = X \setminus B(\langle T_{\langle 0 \rangle}, l_{\langle 0 \rangle} \rangle)$.
(iv) If $l(\sigma) = \langle 0, 1 \rangle$, then

$$B(\langle T, l \rangle) = \bigcup_{\langle n \rangle \in T} B(\langle T_{\langle n \rangle}, l_{\langle n \rangle} \rangle).$$

EXERCISE 10.1.3. (a) Show that if $x \in BC$, then $B(x)$ is a Borel set.
(b) Show that if $A \subseteq X$ is Borel, then there is $x \in BC$ with $B(x) = A$.

LEMMA 10.1.4. *There are* $R \in \Sigma^1_1$ *and* $S \in \Pi^1_1$ *such that if* $x \in BC$ *then*

$$y \in B(x) \Leftrightarrow (x, y) \in R \Leftrightarrow (x, y) \in S.$$

In particular $B(x) \in \Delta^1_1(x)$.

PROOF. We define a set A such that $(x, y, f) \in A$ if and only if x is a pair $\langle T, l\rangle$ where $T \subseteq \omega^{<\omega}$ is a tree, $l : T \to \{0\} \times 2 \cup \{1\} \times S_X$ and $f : T \to 2$ such that for all $\sigma \in T$:

(i) if $l(\emptyset) = \langle 1, \eta\rangle$, then $f(\sigma) = 1$ if and only if $y \in N_\eta$;
(ii) if $l(\emptyset) = \langle 0, 0\rangle$, then $f(\sigma) = 1$ if and only if $f(\sigma^\frown 0) = 0$;
(iii) if $l(\emptyset) = \langle 0, 1\rangle$, then $f(\sigma) = 1$ if and only if $f(\sigma^\frown n) = 1$ for some n.

An easy induction shows that if $x = \langle T, l\rangle$ is a Borel code then $(x, y, f) \in A$ if and only if f is the function

$$f(\sigma) = 1 \Leftrightarrow y \in B(\langle T_\sigma, l_\sigma\rangle).$$

It is easy to see that A is arithmetic and if $x \in BC$, then

$$\begin{aligned} y \in B(x) &\Leftrightarrow \exists f\ ((x, y, f) \in A \wedge f(\emptyset) = 1) \\ &\Leftrightarrow \forall f\ ((x, y, f) \notin A \vee f(\emptyset) = 1). \end{aligned}$$

⊣

COROLLARY 10.1.5. *If* $x \in BC$ *is recursive, then* $B(x)$ *is* Δ^1_1.

PROOF. Let R and S be as in the previous lemma. Let $\phi_e = x$. Then

$$\begin{aligned} y \in B(x) &\Leftrightarrow \exists z\ ((\forall n\ \phi_e(n)\downarrow = z(n)) \wedge R(z, y)) \\ &\Leftrightarrow \forall z\ ((\forall n\ \phi_e(n)\downarrow = z(n)) \to S(z, y)). \end{aligned}$$

The first condition is Σ^1_1 and the second is Π^1_1. ⊣

10.2. Recursively coded Borel sets

Our goal is to show that Δ^1_1 is exactly the collection of Borel sets with recursive codes. That will follow from the following two results and Σ^1_1-Bounding.

THEOREM 10.2.1. *If* $A \subseteq Y$ *is a recursively coded Borel set and* $f : X \to Y$ *is computable, then* $f^{-1}(A)$ *is a recursively coded Borel set.*

PROPOSITION 10.2.2. *If* $\alpha < \omega_1^{ck}$, *then* WF_α *is a recursively coded Borel set.*

COROLLARY 10.2.3. *Suppose* $A \subseteq X$. *The following are equivalent*:

(i) A *is* Δ^1_1;
(ii) A *is a recursively coded Borel set.*

PROOF. We have already shown that every recursively coded Borel set is Δ^1_1. Suppose A is Δ^1_1. Since A is Π^1_1, there is a computable $f : X \to Tr$ such that $x \in A$ if and only if $f(x) \in \mathrm{WF}$. The set

$$f(A) = \{y : \exists x\ x \in A \wedge f(x) = y\}$$

is a Σ^1_1-subset of WF. By Σ^1_1-Bounding, there is $\alpha < \omega_1^{\mathrm{ck}}$ such that $f(A) \subseteq \mathrm{WF}_\alpha$. By 10.2.2 WF_α is recusively coded, and by 10.2.1 $A = f^{-1}(\mathrm{WF}_\alpha)$ is recursively coded. ⊣

For notational simplicity we will assume $X = \omega^\omega$, but all our arguments generalize easily.

Let $BC_{\mathrm{rec}} = \{(e, x) : \phi_e^x \text{ is a total function and } \phi_e^x \in BC\}$. Then

$$(e, x) \in BC_{\mathrm{rec}} \Leftrightarrow \phi_e^x \text{ is total } \wedge \forall z\ (\forall n\ \phi_e^x(n) = z(n)) \rightarrow z \in BC).$$

Thus BC_{rec} is Π^1_1.

If $(e, x) \in BC_{\mathrm{rec}}$, then $B_{\mathrm{rec}}(e, x)$ is the Borel set coded by ϕ_e^x. A similar argument shows that there are $R_{\mathrm{rec}} \in \Sigma^1_1$ and $S_{\mathrm{rec}} \in \Pi^1_1$ such that if $(e, x) \in BC_{\mathrm{rec}}$ then

$$y \in B_{\mathrm{rec}}(e, x) \Leftrightarrow R_{\mathrm{rec}}(e, x, y) \Leftrightarrow S_{\mathrm{rec}}(e, x, y).$$

We say $e \in BC_{\mathrm{rec}}$ and $x \in B_{\mathrm{rec}}(e)$ if $(e, \emptyset) \in BC_{\mathrm{rec}}$ and $x \in B_{\mathrm{rec}}(e, \emptyset)$.

The proofs of both 10.2.1 and 10.2.2 will use the Recursion Theorem to do a transfinite induction.

We begin with the base case of the induction.

LEMMA 10.2.4. *There is a recursive function $F : \omega \times S_Y \rightarrow \omega$ such that if $f : X \rightarrow Y$ is computable and e is a code for the program computing f, then $B_{\mathrm{rec}}(F(e, i)) = f^{-1}(N_\eta)$.*

PROOF. For notational simplicity we assume $X = Y = \omega^\omega$, this is no loss of generality. Let

$$W = \{\nu \in \omega^{<\omega} : \forall m < |\eta|\ \exists s \leq |\nu|\ \phi_e^\nu(m)\downarrow_s = \eta(m)\}.$$

Then W is recursive and $f^{-1}(N_\eta) = \bigcup_{\nu \in W} N_\nu$. Let $\nu_0, \nu_1, \ldots$ be a recursive enumeration of $\omega^{<\omega}$. Let $T = \{\emptyset\} \cup \{\langle n \rangle : \sigma_n \in W\}$ and let $l(\emptyset) = \langle 0, 1 \rangle$, $l(\langle n \rangle) = \langle 1, \nu \rangle$. Then $x = \langle T, l \rangle$ is a recusive code. Given e and η we can easily compute $F(e, \eta) = i$ such that $\phi_i = x$. ⊣

LEMMA 10.2.5. (i) *There is a total recursive function $H_c : \omega \rightarrow \omega$ such that if $e \in BC$, then $B_{\mathrm{rec}}(H_c(e)) = \omega^\omega \setminus B_{\mathrm{rec}}(e)$.*

(ii) *There is a total recursive function $H_u : \omega \rightarrow \omega$ such that if $\phi_e(n) \in BC_{\mathrm{rec}}$ for all n, then $B_{\mathrm{rec}}(H_u(e)) = \bigcup_n B_{\mathrm{rec}}(\phi_e(n))$.*

PROOF. (i) ϕ_e is a code for a pair $\langle T, l \rangle$. Let

$$T' = \{\emptyset\} \cup \{0^\frown\eta : \eta \in T\}$$

and $l'(\emptyset) = \langle 0, 0 \rangle$, $l'(0^\frown\eta) = l(\eta)$. It is easy to find H_c such that $H_c(e)$ codes $\langle T', l' \rangle$ and that if $e \in BC_{\mathrm{rec}}$, then $H_c(e)$ is a code for the complement.

(ii) Suppose $\phi_e(n)$ code a pair $\langle T_n, l_n \rangle$. Let

$$T = \{\emptyset\} \cup \{n^\frown\sigma : \sigma \in T_n\}$$

and let $l(\emptyset) = \langle 0, 1\rangle$ and $l(n^\frown\sigma) = l_n(\sigma)$. It is easy to find H_u such that $H_u(e)$ codes $\langle T, l\rangle$. If each $\langle T_n, l_n\rangle$ is a Borel code, then $\langle T, l\rangle$ codes their union. ⊣

Theorem 10.2.1 follows from the next lemma.

LEMMA 10.2.6. *If $x = \langle T, l\rangle$ is a recursive Borel code, there is a recursive function $G : \omega \times T \to \omega$ such that if $f : \omega^\omega \to \omega^\omega$ is a computed by program P_e, then $G(e, \sigma) \in BC_{\text{rec}}$ is a Borel code for $f^{-1}(B(\langle T_\sigma, l_\sigma\rangle))$ for all $\sigma \in T$.*

PROOF. We define a recursive function $g : \omega \times \omega \times T \to \omega$ as follows:

(i) If $l(\sigma) = \langle 1, \eta\rangle$, then $g(i, e, \sigma) = F(e, \eta)$;
(ii) If $l(\sigma) = \langle 0, 0\rangle$, then $g(i, e, \sigma) = H_c(\phi_i(e, \sigma, 0))$;
(iii) Suppose $l(\sigma) = \langle 0, 1\rangle$. Choose j such that $\phi_j(n) = \phi_i(e, \sigma, n)$. Then $g(i, e, \sigma) = H_u(j)$.

By the Recursion Theorem, there is $\widehat{i}$ such that $\phi_{\widehat{i}}(e, \sigma) = g(\widehat{i}, e, \sigma)$ for all e, σ. Let $G(e, \sigma) = \phi_{\widehat{i}}(e, \sigma)$.

We prove by induction on T, that $G(e, \sigma)$ is a code for $f^{-1}(B(\langle T_\sigma, l_\sigma\rangle))$. By (i) this is clear if $l(\sigma) = \langle 1, \eta\rangle$. We assume the claim is true for all $\tau \supset \sigma$.

If $l(\sigma) = \langle 0, 0\rangle$, then

$$G(e, \sigma) = g(\widehat{i}, e, \sigma) = H_c(\phi_{\widehat{i}}(e, \sigma)) = H_c(G(e, \sigma, 0)).$$

By induction, $H_c(G(e, \sigma^\frown n)$ is a code for

$$f^{-1}(B(\langle T_\sigma, l_\sigma\rangle)) = X \setminus f^{-1}(B(\langle T_{\sigma^\frown 0}, l_{\sigma^\frown 0}\rangle)).$$

If $l(\sigma) = \langle 0, 1\rangle$, then $G(e, \sigma, n)$ is a Borel code for $A_n = f^{-1}(B(\langle T_{\sigma^\frown n}, l_{\sigma^\frown n}\rangle))$. We choose j such that $\phi_j(n)$ is a code for A_n and $G(e, \sigma) = H_u(j)$ is a code for $\bigcup A_n$. ⊣

Theorem 10.2.2 follows from the next lemma.

LEMMA 10.2.7. *If T is a recursive well founded tree, then there is a recursive function $G : T \to BC_{\text{rec}}$, such that $B_{\text{rec}}(G(\sigma)) = \{S \in Tr : \rho(S) \leq \rho(T_\sigma)\}$.*

PROOF. For $\sigma \in \omega^{<\omega}$ let $f_\sigma : Tr \to Tr$ be the computable function $S \mapsto S_\sigma$.

Note that $\rho(S) \leq \rho(T)$ if and only if for all $n \in \omega$ there is $m \in \omega$ such that $\rho(S_{\langle n\rangle}) \leq \rho(T_{\langle m\rangle})$. Thus

$$\{S \in Tr : \rho(S) \leq \rho(T_\sigma)\} = \bigcap_{n\in\omega}\bigcup_{m\in N} f^{-1}_{\langle n\rangle}(\{S \in Tr : \rho(S) \leq \rho(T_{\sigma^\frown m})\}).$$

Fix c such that $B_{\text{rec}}(c) = \emptyset$. We define a recursive function $g : \omega \times T \to \omega$ as follows.

(i) If $\sigma \notin T$, then $g(i, \sigma) = c$.
(ii) Otherwise $g(i, \sigma)$ is a Borel code for

$$\bigcap_n \bigcup_m f^{-1}_{\langle n\rangle}(B_{\text{rec}}(\phi_i(\sigma, m))).$$

We can do this using the functions F, H_u and H_c above. Of course for some i, this may well be undefined.

By the Recursion Theorem there is $\widehat{i}$ such that $\phi_{\widehat{i}}(\sigma) = g(\widehat{i}, \sigma)$ for all σ.

An easy induction shows that $G = \phi_{\widehat{i}}$ is the desired function. $\dashv$

10.3. Hyperarithmetic sets

DEFINITION 10.3.1. We say $x \in \omega^\omega$ is *hyperarithmetic* if $x \in \Delta^1_1$. We say that x is *hyperarithmetic in* y, and write $x \leq_{\text{hyp}} y$ if $x \in \Delta^1_1(y)$.

We sometimes let HYP denote the hyperarithmetic elements of ω^ω.

EXERCISE 10.3.2. (i) Show that if $x \leq_{\text{hyp}} y \leq_{\text{hyp}} z$, then $x \leq_{\text{hyp}} z$.
(ii) Show that if $x \leq_T y$, then $x \leq_{\text{hyp}} y$.

LEMMA 10.3.3. $\{(x, y) : x \leq_{\text{hyp}} y\}$ *is* Π^1_1. *In particular,* HYP *is* Π^1_1.

PROOF. $x \leq_{\text{hyp}} y$ if and only if $\exists e\, (BC_{\text{rec}}(e, y) \wedge \forall n \forall m\, (x(n) = m \leftrightarrow (n, m) \in B_{\text{rec}}(e, y))$.

This definition is Π^1_1. $\dashv$

THEOREM 10.3.4. *Suppose* $A \subseteq \omega^\omega \times \omega^\omega$ *is* Π^1_1. *Then* $B = \{x : \exists y \leq_{\text{hyp}} x\, (x, y) \in A\}$ *is* Π^1_1.

PROOF. $x \in B$ if and only if $\exists e \in \omega \forall z \in \omega^\omega (\phi_e = z \rightarrow (z \in BC \wedge (\forall n \forall m\, ((y(n) = m \rightarrow S((n, m), z)) \wedge (y(n) \neq m \rightarrow \neg R((n, m), z)))) \wedge (x, y) \in A)$.

This definition is Π^1_1. $\dashv$

We next give a refinement of Kleene's Basis Theorem. First note that we can a recursive ill-founded tree need not have a hyperarithmetic branch.

LEMMA 10.3.5. *There is a recursive ill founded tree with no infinite hyperarithmetic branch.*

PROOF. Let $A = \{e : e$ codes a recursive tree and $\forall f \in \text{HYP}\, \exists n\, f(n) \not\subset f(n+1)\}$. Then A is Σ^1_1 and $O \subseteq A$. Thus there is $e \in A \setminus O$. $\dashv$

LEMMA 10.3.6. *If* $\omega_1^{\text{ck}} < \omega_1^x$, *then* $O \leq_{\text{hyp}} x$.

PROOF. Clearly O is $\Pi^1_1(x)$. There is T recursive in x such that $T \in \text{WF}$ and $\rho(T) > \omega_1^{\text{ck}}$. Then

$$O = \{e : e \text{ codes a recursive tree } S \text{ and } \rho(S) < \rho(T)\}$$

is $\Sigma^1_1(x)$. Thus $O \leq_{\text{hyp}} x$. $\dashv$

THEOREM 10.3.7 (Gandy's Basis Theorem). *If* $A \subseteq \omega^\omega$ *is* Σ^1_1 *and nonempty, there is* $x \in A$ *such that* $x \leq_T O$, $x <_{\text{hyp}} O$ *and* $\omega_1^{\text{ck}} = \omega_1^x$.

PROOF. Let $B = \{(x, y) : x \in A \wedge y \not\leq_{\text{hyp}} x\}$. By 10.3.3 B is Σ^1_1. By Kleene's Basis Theorem 9.5.3 there is $(x, y) \in B$ with $(x, y) \leq_T O$. If $O \leq_{\text{hyp}} x$, then $y \leq_T O \leq_{\text{hyp}} x$, so $y \leq_{\text{hyp}} x$, a contradiction. By the previous lemma $\omega_1^{\text{ck}} = \omega_1^x$. ⊣

We give an extension of the Gandy Basis Theorem in Exercise 10.4.9

Remark 10.3.8. There are other characterization of the hyperaritmetic sets that show how natural this class is from many points of view. For example:.

(i) $\text{HYP} = \omega^\omega \cap \mathbb{L}_{\omega_1^{\text{ck}}}$;

(ii) $x \in \text{HYP}$ if and only if $x \leq_T 0^{(\alpha)}$ for some $\alpha < \omega_1^{\text{ck}}$, where $(0^{(\alpha)} : \alpha < \omega_1^{\text{ck}})$ is an iteration of the jump operator, i.e., $0^{(\alpha+1)}$ is the Halting Problem relative to $0^{(\alpha)}$ and, $0^{(\alpha)}$ is a canonically chosen uniform upper bound of $(0^{(\beta)} : \beta < \alpha)$ for limit ordinals α.

We will prove characterization (i) in the appendices in Corollary B.2.4. The exact definition of the sequence $(0^{(\alpha)} : \alpha < \omega_1^{\text{ck}})$ requires fixing ordinal notations and a more careful definition of O than we have given here. For a proof see [27] IV.4.21 or [68].

10.4. The Effective Perfect Set Theorem

The following theorem is very important.

THEOREM 10.4.1 (Harrison). *Let $A \subseteq \omega^\omega$ be Σ_1^1. If A is countable, then every element of A is hyperarithmetic. In particular, if A contains a nonhyperarithmetic element, then A contains a perfect set.*

There are many ways to prove this. The approach we will take is by first looking at the problem of finding strategies in closed games. We review the basics on determinacy in closed games.

Suppose $A \subseteq \omega^\omega$. We consider the game $G(A)$ where Players I and II alternate playing elements of ω.

Player I	Player II
x_0	
	x_1
x_2	
	x_3
$\vdots$	$\vdots$

Together they play $x = (x_0, x_1, \dots)$ and Player I wins if $x \in A$.

DEFINITION 10.4.2. A *strategy* for Player I is a function $\tau : \omega^{<\omega} \to \omega$.

Player I uses the strategy by opening with $\tau(\emptyset)$. If Player II responds with x_0, then Player I replies $\tau(x_0)$. If Player II next plays, x_1, then Player II replies $\tau(x_0, x_1) \dots$.

The full play looks like:

Player I	Player II
$\tau(\emptyset)$	
	x_0
$\tau(x_0)$	
	x_1
$\tau(x_0, x_1)$	
	x_2
$\tau(x_0, x_1, x_2)$	
$\vdots$	$\vdots$

DEFINITION 10.4.3. We say that τ is a *winning strategy* for Player I if Player I wins any game played using the strategy τ, i.e., for any $x_0, x_1, x_2, \ldots \in \omega$, the sequence

$$\tau(\emptyset), x_0, \tau(x_0), x_1, \tau(x_0, x_1), x_2, \tau(x_0, x_1, x_2), \ldots$$

is in A.

There are analogous definitions of strategies and winning strategies for Player II.

DEFINITION 10.4.4. We say that the game $G(A)$ is *determined* if either Player I or Player II has a winning strategy.

THEOREM 10.4.5 (Gale-Stewart Theorem). *If $A \subseteq \omega^\omega$ is closed, then $G(A)$ is determined.*

PROOF. Let T be a tree such that $A = [T]$. Suppose Player II has no winning strategy. We will show that Player I has a winning strategy. Suppose $\sigma \in \omega^{<\omega}$ and $|\sigma|$ is even. We consider the game $G_\sigma(A)$ where Players I and II alternate playing elements of ω to build $x \in \omega^\omega$ and Player I wins if $\sigma^\frown x \in A$.

Let $P = \{\sigma : |\sigma| \text{ is even and Player II has a winning strategy in } G_\sigma(A)\}$. If $\sigma \notin T$, then Player II has already won $G_\sigma(A)$. In particular, always playing 0 is a winning strategy for Player II. Thus $\omega^{<\omega} \setminus T \subseteq P$.

CLAIM. *Suppose that for all $n \in \omega$ there is $m \in \omega$ such that $\sigma^\frown n^\frown m \in P$. Then $\sigma \in P$.*

Player II has a winning strategy in $G_\sigma(A)$; namely if Player I plays n and Player II plays the least m such that Player II has a winning strategy in $G_{\sigma^\frown n^\frown m}$, and then uses the strategy in this game.

We describe a winning strategy for Player I. This strategy can be sumarized as "avoid losing postions".

Since Player II does not have a winning strategy $\emptyset \notin P$. Player I's strategy is to avoid P. If we are in position σ where $\sigma \notin P$ and $|\sigma|$ is even, then by the claim there is a least n such that $\sigma^\frown n^\frown m \notin P$ for all m. Player I plays n. No matter what m Player II now plays the new position is not in P. If Player I continues Playing playing this way they will play $x \in \omega^\omega$ such that $x|2n \notin P$

for all n. In particular $x|2n \in T$ for all n. Thus $x \in [T]$ and this is a winning strategy for Player I. ⊣

For our purposes it will be useful to understand the complexity of the winning strategies.

THEOREM 10.4.6 (Strategic Basis Theorem). *Let A be Π^0_1. If Player II has a winning strategy in $G(A)$, then Player II has a hyperarithmetic winning strategy.*

PROOF. Let T be a recursive tree such that $A = [T]$. Suppose Player II does not have a hyperarithmetic winning strategy. We will show that Player I has a winning strategy. Suppose $\sigma \in \omega^{<\omega}$ and $|\sigma|$ is even. We consider the game $G_\sigma(A)$ where Players I and II alternate playing elements of ω to build $x \in \omega^\omega$ and Player I wins if $\sigma^\frown x \in A$.

Let $P = \{\sigma : |\sigma|$ is even and Player II has a hyperarithmetic winning strategy in $G_\sigma(A)\}$. If $\sigma \notin T$, then Player II has already won. In particular, always playing 0 is a hyperarithmetic winning strategy for Player II. Thus $\omega^{<\omega} \setminus T \subseteq P$.

CLAIM. *Suppose that for all $n \in \omega$ there is $m \in \omega$ such that $\sigma, n, m \in P$. Then $\sigma \in P$.*

Let $B = \{(n, m, e) : e$ is a hyperarithmetic code for τ and $\forall y$ if we play $G_{\sigma,n,m}(A)$ where Player I plays y and Player II plays using τ, then the result is in $A\}$. The set B is Π^1_1. and $\forall n \exists m \exists e (n, m, e) \in B$. By selection there is a Δ^1_1-function $f : \omega \to \omega^2$ such that $(n, f(n)) \in B$ for all $n \in \omega$. Player II has a hyperarithmetic winning strategy in $G_\sigma(A)$; namely if Player I plays n and $f(n) = (m, e)$, then Player II plays m, and then uses the strategy coded by e.

We describe a winning strategy for Player I.

Since Player II does not have a hyperarithmetic winning strategy $\emptyset \notin P$. Player I's strategy is to avoid P. If we are in position σ where $\sigma \notin P$ and $|\sigma|$ is even, then by the claim there is a least n such that $\sigma, n, m \notin P$ for all m. Player I plays n. No matter what m Player II now plays the new postion is not in P. If Player I continues playing this way they will play $x \in \omega^\omega$ such that $x|2n \notin P$ for all n. In particular $x|2n \in T$ for all n. Thus $x \in [T]$ and this is a winning strategy for Player I. ⊣

EXERCISE 10.4.7. Suppose A is Π^0_1 and Player I has a winning strategy in $G(A)$. Then Player I has a winning strategy hyperaritmetic in O.

PROOF OF EFFECTIVE PERFECT SET THEOREM. Suppose A is Σ^1_1. Let $B \subseteq \omega^\omega \times \omega^\omega$ be Π^0_1 such that

$$A = \{x : \exists y\, (x, y) \in B\}.$$

Consider the following game. At stage s Player I plays $(\sigma_{s,0}, \sigma_{s,1}, y_s)$ where $\sigma_{s,0}, \sigma_{s,1} \in \omega^{<\omega}$ are incomparable and $y_s \in \omega$. Player II plays $i_s = 0, 1$. At the end of the game let

$$x = \sigma_{0,i_0}{}^\frown\sigma_{1,i_1}{}^\frown\sigma_{2,i_2} \dots \text{ and } y = y_0, y_1, y_2, \dots$$

Player I wins if $(x, y) \in B$. This is an open game.

CLAIM 1. *If Player I has a winning strategy, then A has a perfect subset.*

Consider the function $f : 2^\omega \to \omega^\omega \times \omega^\omega$ where $f(z) = (x(z), y(z))$ is the play of the game where Player II plays $z(i)$ at stage i and Player I uses her winning strategy. Then f is continuous and if $z_1 \neq z_2$ and s is least such that $z_1(s) \neq z_2(s)$, then we insure at stage s that $x(z_1) \neq x(z_2)$. Thus f is a continuous injection from 2^ω into A, so A has a perfect subset.

Suppose τ is a winning strategy for Player II. Consider a position p in the game where it is Player I's turn to move at stage $s+1$. Let $\mu = \sigma_{0,i_0}\frown\cdots\frown\sigma_{s,i_s}$ and then $\nu = (y_0, \ldots, y_s)$ be the initial segments of the final play determined at this position. Suppose $(x, y) \in B$, $x \supset \mu$ and $y \supset \nu$. We say that (x, y) is *rejected* at position p if for any possible move $(\sigma_{s+1,0}, \sigma_{s+1,1}, y_{s+1})$ by Player I, Player II using τ will make reply so that $x \not\subset \mu\frown\sigma_{s+1,i_{s+1}}$.

CLAIM 2. *If $(x, y) \in B$, there is a position p such that (x, y) is rejected at p.*

Suppose not. Consider a play of the game where Player II uses τ. Whenever it is Player I's turn to move we are at a position where (x, y) has not been rejected. Since (x, y) is not rejected at the empty position, there is some move by Player I where he plays $y(0)$ such that when II responds we will still have an initial segment of x. Since (x, y) is not rejected at this new position, there is some move that Player I can make, so that II's response will still be an initial segment of (x, y). Continuing this way the eventual play of the game will be (x, y) and Player I will win, a contradiction.

CLAIM 3. *For each position p there are at most countably many x such that some (x, y) is rejected at position p and each is arithmetic in τ.*

For each possible k, we will show that there is at most one $(x, y) \in B$ with $y \supset \nu\frown k$ such that y is rejected at p, where ν is as above.

Suppose (x, y) is rejected at position p at stage $s+1$. Let μ be as above and suppose that we have determined that $x \subset \mu\frown\eta$. There is a unique n such that for all m, if Player I plays $(\eta\frown n, \eta\frown m, k)$ at stage $s+1$, then Player II, using τ will play 1. Then we know that $x \subset \mu\frown\eta\frown 1$. This argument shows $x \leq_T \tau'$.

Since every element of A is rejected at one of the countably many possible positions, A must be countable and every element is arithmetic in τ

We can easily modify the proof of Theorem 10.4.6 to show that either Player I has a winning strategy or Player II has a hyperarithmetic winning strategy.

Thus if A is countable, $A \subseteq \text{HYP}$. ⊣

EXERCISE 10.4.8. Show that if A is Σ^1_1 and uncountable, then there is a continuous injection $f : 2^\omega \to A$ with f computable in x for some $x \leq_{\text{hyp}} O$.

EXERCISE 10.4.9. Show that A is an uncountable Σ^1_1-set then there is a perfect set $P \subseteq A$ such that $\omega_1^x = \omega_1^{\text{ck}}$ for all $x \in P$. [Hint: Show that $\{x \in A_1 : \omega_1^x = \omega_1^{\text{ck}}\}$ contains an non-hyperarithmetic element.]

EXERCISE 10.4.10. Modify the proof of the Effective Perfect Set Theorem, using the Banach–Mazur game, to prove that if $A \subseteq \omega^\omega$ is a nonmeager Δ^1_1-set, then there is a hyperarithmetic $x \in A$.

COROLLARY 10.4.11. *Suppose* $A \subseteq \omega^\omega \times \omega^\omega$ *is* Δ^1_1 *and* $\{y : (x, y) \in A\}$ *is countable for all* $x \in \omega^\omega$. *Then*

(i) *the projection* $\pi(A) = \{x : \exists y\, (x, y) \in A\}$ *is* Δ^1_1 *and*
(ii) A *has a* Δ^1_1*-uniformization.*

PROOF. (i) Clearly $\pi(A)$ is Σ^1_1, but by the Effective Perfect Set Theorem

$$\exists y\, (x, y) \in A \leftrightarrow \exists y \leq_{\text{hyp}} x\, (x, y) \in A.$$

The later condition is Π^1_1.

(ii) Let $A^* = \{(x, e) : e \in BC_{\text{rec}}(x) \wedge \forall y (y = B_{\text{rec}}(e, x) \rightarrow (x, y) \in A\}$. Then A^* is Π^1_1 and has a Π^1_1 uniformization B. But

$$(x, e) \notin B \Leftrightarrow x \notin \pi(A) \vee \exists i \neq e\, (x, i) \in B.$$

Thus B is Δ^1_1. Let

$$C = \{(x, y) : \exists e\, (x, e) \in B \wedge y = B_{\text{rec}}(e, x)\}.$$

Then C is a Δ^1_1-uniformization of A. ⊣

Relativizing these corollaries lead to interesting results about Borel sets.

COROLLARY 10.4.12. *Suppose* $A \subseteq \omega^\omega \times \omega^\omega$ *is a Borel set such that every section is countable. Then* X *the projection of* X *is Borel and* X *can be uniformized by a Borel set.*

COROLLARY 10.4.13. *Suppose* $f : \omega^\omega \rightarrow \omega^\omega$ *is continuous,* A *is Borel and* $f|A$ *is one-to-one. Then* $f(A)$ *is Borel.*

PROOF. $f(A)$ is the projection of $\{(x, y) : x \in A \wedge f(x) = y\}$ and sections are singletons. ⊣

If $x \in \omega^{\omega^2}$ we identify x with $(x_0, x_1, \ldots)$ in $(\omega^\omega)^\omega$ where $x_n(m) = x(n, m)$.

LEMMA 10.4.14. *Suppose* A *is a* Δ^1_1*-subset of* HYP. *There is a hyperarithmetic* $x \in \omega^{\omega^2}$ *such that* $A \subseteq \{x_0, x_1, \ldots\}$.

PROOF. Let $B = \{(x, i) \in \omega^\omega \times \omega : x \in A \wedge i \in BC_{\text{rec}} \wedge \forall n \forall m\, (x(n) = m \leftrightarrow (n, m) \in B_{\text{rec}}(i))\}$. Then B is Π^1_1 and $\pi(B) = A$. By selection, there is a Δ^1_1 function $s : A \rightarrow \omega$, uniformizing B.

Let $C = \{i : \exists x \in A\, s(x) = i\}$. Clearly C is Σ^1_1. Since

$$i \in C \leftrightarrow \exists x \in \text{HYP}\, (x \in A \wedge s(x) = i),$$

C is Δ^1_1. Let

$$x(i, n) = \begin{cases} 0 & \text{if } i \notin C, \\ m & \text{if } i \in D \wedge (n, m) \in B_{\text{rec}}(i). \end{cases}$$

Then $A \subseteq \{x_0, x_1, \ldots\}$. ⊣

EXERCISE 10.4.15. Show that the same is true if A is Σ^1_1. [Hint: First show that any Σ^1_1 subset of HYP is contained in an Δ^1_1 subset of HYP.]

Thin Π^1_1-sets. While the Perfect Set Theorem for Σ^1_1 sets is provable in ZFC, it is possible, say if $\mathbb{V} = \mathbb{L}$, that there are uncountable Π^1_1-sets with no perfect subsets. We say that a set is *thin* if contains no perfect subset. We will show that there is a largest thin Π^1_1-set.

Let $C_1 = \{x : \forall y\, (\omega_1^x \leq \omega_1^y \rightarrow x \leq_{\text{hyp}} y)\}$.

EXERCISE 10.4.16. (a) Show that $\{(x, y) : \omega_1^x \leq \omega_1^y\}$ is Σ^1_1.
(b) Show that C_1 is Π^1_1.

LEMMA 10.4.17. *C_1 is thin.*

PROOF. Suppose not. Then there is a perfect tree T such that $[T] \subseteq C_1$. Relativizing Exercise 10.4.9 we can find a perfect set $P \subseteq [T]$ such that $\omega_1^x \leq \omega_1^T$ for all $x \in P$. Since $P \subseteq C_1$, $x \leq_{\text{hyp}} T$ for all $x \in P$, a contradiction. ⊣

LEMMA 10.4.18. *If A is a thin Π^1_1 set, then $A \subseteq C_1$.*

PROOF. There is a recursive $f : \omega^\omega \rightarrow Tr$ such that $x \in A$ if and only $f(x) \in \text{WF}$. Suppose $x \in A \setminus C_1$. There is y such that $\omega_1^x \leq \omega_1^y$ and $x \not\leq_{\text{hyp}} y$. But then $\rho(f(x)) < \omega_1^x \leq \omega_1^y$. Thus there is $\alpha < \omega_1^y$ such that $x \in f^{-1}(\text{WF}_\alpha)$. But WF_α is $\Delta^1_1(y)$ and contains an element $x \notin \text{HYP}(y)$. Thus, by the relativized Effective Perfect Set Theorem, $f^{-1}(\text{WF}_\alpha)$ contains a perfect set. ⊣

EXERCISE 10.4.19. We will show that for all $\alpha < \aleph_1^{\mathbb{L}}$, there is $x \in C_1$ with $\omega_1^x \geq \alpha$.

(a) Show that there is a Π^1_1-set $A \subseteq \omega^\omega \times \omega^\omega$ such that the projection of A is $\omega^\omega \cap \mathbb{L}$ and the projection map is injective. [Hint: Recall that $\omega^\omega \cap \mathbb{L}$ is Σ^2_1 and apply uniformization.]
(b) Show that A is thin. [This is the classic example of a Π^1_1 set without a perfect subset.]
(c) Conclude that for all $\alpha < \aleph_1^{\mathbb{L}}$, there is $x \in C_1$ with $\omega_1^x \geq \alpha$.

There is a useful alternative characterization of C_1 which makes it clear that $C_1 \subset \mathbb{L}$. We say that x is *quickly constructible* if $x \in \mathbb{L}_{\omega_1^x}$ and let Q be the set of quickly constructible reals. We will argue in Appendix B that $C_1 = Q$. Once we know $C_1 \subseteq \mathbb{L}$, we can argue (see Exercise B.4.4) that any Σ^1_2 containing a non-constructible real has a perfect subset.

The Hard Core Theorem. The following result was proved by Gandy, Kreisel and Tait [21].

THEOREM 10.4.20 (Hard Core Theorem). *If X is a nonempty Σ^1_1 and $x \notin$ HYP, then there is $y \in X$ such that $x \not\leq_{\text{hyp}} y$.*

The main application of the Hard Core Theorem is to models of second order arithmetic.

COROLLARY 10.4.21. *If T is an ω-consistent Π^1_1-theory of second order arithmetic extending Δ^1_1-Comprehension, then*

$$\text{HYP} = \bigcap\{X : (\omega, X) \models T\}.$$

Our proof of the Hard Core Theorem is inspired by proofs of higher level analogs (under determinacy assumptions) in [35].

DEFINITION 10.4.22. We say that $X \subset \omega^\omega$ is Π^1_1-bounded if for any $\Pi^1_1(a)$-predicate $P(x, y)$ the set $\{x : \exists y \in X\, P(x, y)\}$ is $\Pi^1_1(a)$.

By Theorem 10.3.4, HYP is Π^1_1-bounded. Letting P be the predicate "$x = y$" we see that any Π^1_1-bounded set is Π^1_1.

DEFINITION 10.4.23. For $X \subseteq \omega^\omega$ be nonempty, let

$$\text{Hull}(X) = \{x : \forall y \in X\, x \leq_{\text{hyp}} y\}.$$

Note that if X is nonempty and Σ^1_1, then Hull(X) is Π^1_1.

LEMMA 10.4.24. *If X is a nonempty Σ^1_1-set, then* Hull(X) *is Π^1_1-bounded.*

PROOF. Let $P(x, y)$ be $\Pi^1_1(a)$. Then

$$\exists y \in \text{Hull}(X)\, P(x, y) \Leftrightarrow \forall z\, (z \in X \rightarrow \exists y \leq_{\text{hyp}} z\; y \in \text{Hull}(X) \wedge P(x, y))$$

and the right hand side is $\Pi^1_1(a)$. ⊣

The Hard Core Theorem we will follow from the next Lemma.

LEMMA 10.4.25. *Every Π^1_1-bounded set is contained in* HYP.

PROOF. Let B be Π^1_1-bounded, and let $B_1 = \{x : \exists y \in B\, x \leq_{\text{hyp}} y\}$. If $P(x, y)$ is $\Pi^1_1(a)$, then

$$\exists y \in B_1\, P(x, y) \Leftrightarrow \exists z \in B\, \exists y \leq_{\text{hyp}} z\, P(x, y).$$

Thus B_1 is also Π^1_1-bounded and we may, without loss of generality, assume B is downward closed under $\leq_{\text{hyp}}$.

We next claim that B has no perfect subsets. Suppose $f : 2^\omega \rightarrow B$ is continuous and 1-1. Let a be a code for f and let $P(x, y)$ be Π^1_1. Then

$$\exists y\, P(x, y) \Leftrightarrow \exists z \in B\, \forall y\, (f(y) = z \rightarrow P(x, y)).$$

The righthand side is $\Pi^1_1(a)$. But this shows every Σ^1_2 set is Π^1_1, a contradiction. Thus B is thin.

Since B is a thin Π^1_1-set, $B \subseteq C_1$. By Kleene's Basis Theorem 9.5.3, there is $x \notin B$ with $x <_T O$. Since B is downward closed $O \not\leq_{\text{hyp}} y$ for all $y \in B$. Thus $\omega_1^y = \omega_1^{\text{ck}}$ for all $y \in B$. But $B \subseteq C_1$, thus $B \subseteq$ HYP. ⊣

If X is a nonempty Σ^1_1, then Hull(X) is Π^1_1-bounded and hence contained in HYP. In other words, if $x \leq_{\text{hyp}} y$ for all $y \in x$, then $x \in$ HYP. Thus we have proved the Hard Core Theorem.

Chapter 11

EFFECTIVE ASPECTS OF $\mathcal{L}_{\omega_1,\omega}$

11.1. Coding $\mathcal{L}_{\omega_1,\omega}$-formulas

Before discussing effective aspects of the model theory of $\mathcal{L}_{\omega_1,\omega}$ we should take the time to discuss one method of coding $\mathcal{L}_{\omega_1,\omega}$-formulas. Throughout this section we will only consider formulas with finitely many free variables–though this requirement could easily be relaxed.

Fix τ a countable vocabulary. We will assume τ is recursive. All of the results in this section will relativize if it is not. We can define a coding of $\mathcal{L}_{\omega_1,\omega}(\tau)$-formulas that is analogous to the construction of Borel codes.

We assume that we have fixed a Gödel coding $\ulcorner\phi\urcorner$ of atomic τ-formulas.

A *labeled tree* is a non-empty tree $T \subseteq \omega^{<\omega}$ with functions l and v with domain T such that for any $\sigma \in T$ one of the following holds:

- σ is a terminal node and label $l(\sigma) = \ulcorner\psi\urcorner$ where ψ is an atomic τ-formula and $v(\sigma)$ is the set of free variables in ψ;
- $l(\sigma) = \neg$, $\sigma^\frown 0$ is the unique successor of σ in T and $v(\sigma) = v(\sigma^\frown 0)$;
- $l(\sigma) = \exists v_i$, $\sigma^\frown 0$ is the unique successor of σ in T and $v(\sigma) = v(\sigma^\frown 0)\setminus\{i\}$;
- $l(\sigma) = \bigwedge$, σ has at least one successor and $v(\sigma) = \bigcup_{\sigma^\frown i\in T} v(\sigma^\frown i)$ is finite.

A *formula code* ϕ is a well founded labeled tree (T, l, v). A *sentence code* is a formula where $v(\emptyset) = \emptyset$.

EXERCISE 11.1.1. The set of labeled trees is arithmetic. The set of formula codes is Π^1_1 as is the set of sentence codes.

PROPOSITION 11.1.2. *There is $R(x, y) \in \Pi^1_1$ and $S(x, y) \in \Sigma^1_1$. Such that if ϕ is a sentence and $\mathcal{M} \in \mathbb{X}_\tau$, then*

$$\mathcal{M} \models \phi \Leftrightarrow R(\mathcal{M}, \phi) \Leftrightarrow S(\mathcal{M}, \phi).$$

In particular, $\{(\mathcal{M}, \phi) : \phi$ is a sentence and $\mathcal{M} \models \phi\}$ is Π^1_1, but for any fixed ϕ, $\mathrm{Mod}(\phi) = \{\mathcal{M} \in \mathbb{X}_\tau : \mathcal{M} \models \phi\}$ *is $\Delta^1_1(\phi)$.*

PROOF. We define a predicate "f is a *truth definition* for the labeled tree (T, l, v) in $\mathcal{M}$" as follows.

- The domain of f is pairs (σ,η) where $\sigma \in T$ and $\eta : v(\sigma) \to \mathcal{M}$ is an assignment of the free variables at node σ and $f(\sigma,\eta) \in \{0,1\}$.
- If $l(\sigma) = \ulcorner\psi\urcorner$ an atomic τ-formula, then $f(\sigma,\eta) = 1$ if and only if ψ is true in $\mathcal{M}$ when we use η to assign the free variables.
- If $l(\sigma) = \neg$, then $f(\sigma,\eta) = 1$ if and only if $f(\sigma^\frown 0,\eta) = 0$.
- If $l(\sigma) = \exists v_i$ there are two cases. If $v_i \in v(\sigma^\frown 0)$, then $f(\sigma,\eta) = 1$ if and only if there is $a \in \mathcal{M}$ such that $f(\sigma^\frown 0,\eta^*) = 1$, where $\eta^* \supset \eta$ is the assignment where $\eta^*(v_i) = a$. Otherwise, $f(\sigma,\eta) = f(\sigma^\frown 0,\eta)$.
- If $l(\sigma) = \bigwedge$, then $f(\sigma,\eta) = 1$ if and only if $f(\sigma^\frown i, \eta|v(\sigma^\frown i)) = 1$ for all i such that $\sigma^\frown i \in T$.

This predicate is arithmetic. If ϕ is a sentence, there is a unique truth definition f for ϕ in $\mathcal{M}$. Let

$R(x,y) \Leftrightarrow x \in \mathbb{X}_\tau$ and y is a labeled tree, $v(\emptyset) = \emptyset$ and $f(\emptyset,\emptyset) = 1$ for all truth definitions f for y in x,

and

$S(x,y) \Leftrightarrow y$ is a labeled tree, $v(\emptyset) = \emptyset$ and there is a truth definition f for y in x such that $f(\emptyset,\emptyset) = 1$. ⊣

We will say that a sentence is recursive (hyperarithmetic) if it has a recursive (hyperarithmeic) code. We can sharpen some of our earlier theorems.

THEOREM 11.1.3. *Let $\tau = (<,\dots)$ be a recursive vocabulary and let ϕ be a hyperarithmetic $\mathcal{L}_{\omega_1,\omega}$-sentence such that every model of ϕ is well ordered by $<$. There is $\alpha < \omega_1^{\mathrm{ck}}$ such that every model of ϕ has order type at less that α.*

PROOF. Let $A = \{x : x$ codes a linear order and $\exists\mathcal{M} \in \mathrm{Mod}(\phi)\,\exists f\ f$ is an order embedding of x into $\mathcal{M}\}$. Then A is a Σ^1_1 set of well orders. Thus, by Σ^1_1-Bounding, there is a recursive bound on the order types in $\mathrm{Mod}(\phi)$. ⊣

EXERCISE 11.1.4. For each $\alpha < \omega_1^{\mathrm{ck}}$, there is a recursive ϕ such that any model of ϕ is isomorphic to α. Thus, the above bound is optimal.

Effective construction of models. If T is a first order theory in a recursive language, then we can find an arithmetic model of T. In fact, we can find a completion T^* of T with $T^* \leq_T T'$, the recursion theoretic jump of T, and $\mathcal{M} \models T^*$ with $\mathcal{M} \leq_T T^*$. For $\mathcal{L}_{\omega_1,\omega}$, this is more complicated.

Suppose $\phi \in \mathcal{L}_{\omega_1,\omega}$ is hyperarithmetic. Using Kleene's Basis Theorem 9.5.3 there is $\mathcal{M} \models \phi$ with $\mathcal{M} \leq_T O$.

DEFINITION 11.1.5. For any τ-structure $\mathcal{M}$ we let $\omega_1^{\mathcal{M}}$ be the infimum of ω_1^z where z codes an isomorphic copy of $\mathcal{M}$.

PROPOSITION 11.1.6. *There is a satisfiable recursive sentence with no hyperarithmetic models.*

PROOF. By Corollary 10.3.5 there is an ill-founded recursive tree $T \subseteq \omega^{<\omega}$ with no hyperarithmetic paths. Let τ be the vocabulary where we have a unary predicate $\mathcal{P}_\eta$ for all $\eta \in T$. Let ϕ be the conjunction of

$$\forall x \bigwedge_{\eta \in T} (P_\eta(x) \rightarrow \bigwedge_{\nu \subset \eta} P_\nu(x))$$

and

$$\forall x \bigwedge_{n=1}^{\infty} \bigvee_{\eta \in T \cap \omega^n} P_\eta(x).$$

If $\mathcal{M} \models T$, then there is a path through T recursive in $\mathcal{M}$. Thus ϕ has no hyperarithmetic models. ⊣

Nadel showed that it easier to find models of complete sentences.

THEOREM 11.1.7. *If ϕ is a complete sentence, then there is $\mathcal{M} \models \phi$ with $\mathcal{M} \leq_{\text{hyp}} \phi$.*

PROOF. Add countably many new constant symbols $c_0, c_1, \ldots$ and let $\mathbb{A}$ be the smallest fragment containing ϕ such that if $\theta(v_1, \ldots, v_n) \in \mathbb{A}$, then $\theta(c_{i_1}, \ldots, c_{i_n}) \in \mathbb{A}$ for all $i_0, \ldots, i_n$. The fragment $\mathbb{A}$ is arithmetic in ϕ. Let Σ be all finite sets σ of $\mathbb{A}$-sentences such that if $\overline{c}$ are the new constants occurring in σ then $\phi \models \exists \overline{v} \bigwedge_{\psi(\overline{c}) \in \sigma} \psi(\overline{v})$.

Since ϕ is complete,

$$\phi \models \theta \Leftrightarrow \forall \mathcal{M} \, (\mathcal{M} \models \phi \rightarrow \mathcal{M} \models \theta) \Leftrightarrow \exists \mathcal{M} (\mathcal{M} \models \phi \wedge \mathcal{M} \models \theta).$$

Thus Σ is Δ^1_1 in ϕ. It is also easy to check that Σ is a consistency property.

The proof of the Model Existence Theorem shows we can build $\mathcal{M} \models \phi$ with $\mathcal{M}$ recursive in Σ and $\mathbb{A}$. Thus we have $\mathcal{M} \models \phi$ with $\mathcal{M} \leq_{\text{hyp}} \phi$. ⊣

11.2. Kreisel–Barwise Compactness

Kreisel found an interesting version of the Compactness Theorem holds when we look at theories that are Π^1_1-sets of recursive sentences and think of the Δ^1_1-subsets as "finite".[20]

Throughout this section we assume that τ is a recursive vocabulary.

LEMMA 11.2.1. *Suppose I is a Π^1_1 set of hyperarithmetic Borel codes and*

$$\bigcap_{x \in I_0} B(x) \neq \emptyset$$

for any $I_0 \subseteq I$ with $I_0 \in \Delta^1_1$. Then

$$\bigcap_{n \in I} B(x) \neq \emptyset.$$

[20]This is a special case of Barwise Compactness Theorem for countable admissible fragments which we discuss in Appendix B. Interestingly, Barwise was not aware of Kreisel's earlier result.

PROOF. Suppose not. Then $\forall x \exists y \in \text{HYP}\ y \in I \wedge x \notin B(y)$. The set $A = \{(x,e) : \exists y \in \text{HYP}\ e \text{ codes } y \wedge y \in I \wedge x \notin B(y)\}$ is Π^1_1 By Kreisel's Uniformization Theorem 9.4.4 (v), it has a Δ^1_1 uniformization P. Thus $X = \{y : \exists x \exists n\ P(x,n) \wedge n \text{ codes } y\}$ is a Σ^1_1 subset of I. By the Σ^1_1-Separation Theorem, there is a Δ^1_1 set Y such that $X \subseteq Y \subseteq I$. But then there is $x \in \bigcap_{y \in Y} B(y)$, but there is $y \in X$ such that $x \notin B(y)$, a contradiction. ⊣

COROLLARY 11.2.2 (Kreisel–Barwise Compactness). *Suppose T is a Π^1_1 set of hyperarithmetic sentences in $\mathcal{L}_{\omega_1,\omega}$ and every Δ^1_1 subset of T has a model. Then T has a model.*

PROOF. We can easily pass from a hyperarithmetic sentence ϕ to a hyperarithmetic code for $\text{Mod}(\phi)$. ⊣

COROLLARY 11.2.3. *Let T be as above and suppose every Δ^1_1 subset of T has a recursive model. Then T has a recursive model.*

PROOF. For notational simplicity assume that $\tau = \{R\}$ a single binary relation and let $\tau^* = \tau \cup \{c \cup s\}$ where c is a constant and s is a unary function. If $x \in 2^{\omega \times \omega}$ is recursive let Ψ_x be the conjunction of

- s is injective with no cycles and every element except c is in the image of s;
- $\bigwedge_{x(i,j)=1} R(s^i(c), s^j(c)) \wedge \bigwedge_{x(i,j)=0} \neg R(s^i(c), s^j(c))$.

If $(\omega, R, c, s) \models \Psi_x$, then (ω, R) has an isomorphic recursive copy. Let Θ be the hypeartithmetic sentence

$$\bigvee_{x \text{ recursive}} \Psi_x$$

and let $T^* = T \cup \{\Theta\}$. Every Δ^1_1 subset of T^* has a model. Thus T^* has a model $\mathcal{M}$. Since $\mathcal{M} \models \Psi_x$ for some recursive x, T has a recursive model. ⊣

Suppose T is a Π^1_1 set of hyperarithmetic codes for $\mathcal{L}_{\omega_1,\omega}$-sentences and T has a model. Let $X = \{\mathcal{M} : \forall \phi\ (\phi \in T \rightarrow \mathcal{M} \models \phi)\}$. Then X is a Σ^1_1 set and by the Gandy Basis Theorem there is $\mathcal{M} \models T$ with $\omega_1^{\mathcal{M}} = \omega_1^{\text{ck}}$.

We will use Kreisel–Barwise Compactness to construct a recursive ordering of order type $\omega_1^{\text{ck}} + \mathbb{Q} \times \omega_1^{\text{ck}}$.

LEMMA 11.2.4. *Suppose $(\omega, \prec)$ is an ill-founded recursive linear ordering with no infinite hyperarithmetic descending chains. Then $(\omega, \prec)$ is isomorphic to $\omega_1^{\text{ck}} + \mathbb{Q} \times \omega_1^{\text{ck}} + \beta$ for some recursive ordinal β.*

PROOF. Let $I = \{n : \text{there is no infinite descending chain in } \{m : m \prec n\}$. Clearly I is Π^1_1.

CLAIM. *I is not Σ^1_1.*

Suppose I is Σ^1_1. Since $\prec$ is ill-founded there is $m \in \omega$ with $I \prec m$. Consider

$$A = \{(x,y) : y \prec x \prec m \wedge y \notin I\}.$$

Since I is Σ^1_1, A is Π^1_1 and we can find a Π^1_1-uniformization B. Pick x_0 such that $I \prec x_0 \prec m$. Let x_{n+1} be the unique element such that $(x_n, x_{n+1}) \in B$.

We claim that $x_0, x_1, x_2, \dots$ is a hyperarithmetic descending sequence.

Let $C = \{\sigma \in \omega^{<\omega} : \sigma(0) = x_0 \wedge \forall n < |\sigma| - 1\, (\sigma(n), \sigma(n+1) \in B\}$. C is clearly Π^1_1. But

$$\sigma \notin C \Leftrightarrow \sigma(0) \neq x_0 \vee \exists n < |\sigma| - 1 \exists m \neq \sigma(n+1)\, (\sigma(n), m) \in B.$$

Thus C is Δ^1_1 and

$$x_n = m \Leftrightarrow \exists \sigma \in C\, \sigma(n) = m.$$

Since there are no hyperarithmetic descending sequences this is a contradiction.

CLAIM. *I is isomorphic to ω_1^{ck}.*

Clearly the order type of I is at most ω_1^{ck}, for otherwise there would be a recursive initial segment of order type ω_1^{ck}. Suppose α is a recursive ordinal greater than the order type of I. Then $x \in I$ if and only if there is an order preserving map from $\{y : y \prec x\}$ into (ω, R) and I is Σ^1_1, a contradiction.

Thus I is isomorphic to ω_1^{ck}.

Define an equivalence relation $x \sim y$ if and only if the interval

$$[\min_{\prec}(x, y), \max_{\prec}(x, y)]$$

is well ordered. This is a Π^1_1-equivalence relation.

CLAIM. *Each $\sim$-class has a least element.*

Suppose the sim class of x has no least element. Let

$$A = \{(n, m) : n \sim x \wedge y \sim x \wedge y \prec x\}.$$

There is B a Π^1_1-uniformization of A and arguing as above we can build an infinite hyperarithmetic descending chain.

If n is the least element of it's $\sim$-class, then we can consider the restriction of $\prec$ to $\{m : n \preceq m\}$. By our first claim either this set is well ordered or it has an initial segment of order type ω_1^{ck}.

If there is a maximal $\sim$-class, that class must be well ordered, and hence has order type α for some $\alpha < \omega_1^{\mathrm{ck}}$. Every other $\sim$-class must have order type ω_1^{ck}.

If the $\sim$-class of x has order type ω_1^{ck}, then there must be $x \prec y$ with $x \sim y$. Otherwise we would get a recursive well order of order type ω_1^{ck}.

Suppose $x \prec y$ and y is the least element of it's $\sim$-class. Then there is an infinite descending chain $x \prec \cdots \prec x_n \prec x_1 \prec x_0 \prec y$. The $\sim$-class of any x_i is strictly between the $\sim$-classes of x and y.

We have shown that $(\omega, \prec)$ is order isomorphic to $\omega_1^{\mathrm{ck}} + (\mathbb{Q} \times \omega_1^{\mathrm{ck}}) + \alpha$ for some recursive α. ⊣

COROLLARY 11.2.5 (Harrison Order). *There is a recursive ordering of ω of order type $\omega_1^{\mathrm{ck}} + (\mathbb{Q} \times \omega_1^{\mathrm{ck}})$ with no hyperarithmetic infinite descending chains. Such an ordering is called a Harrison order.*

PROOF. CLAIM. *If α is a recursive ordinal, then there is a recursive formula $\varphi_\alpha(x)$ such that in any linear order $\mathcal{M}$*

$$\mathcal{M} \models \varphi_\alpha(a) \Leftrightarrow \{m : m < a\} \text{ has order type } \alpha.$$

We define φ_α be effective transfinite recursion.

$$\varphi_0(x) = \forall y\, x \leq y;$$

$$\varphi_\alpha(x) = \forall y < x \bigvee_{\beta<\alpha} \varphi_\beta(y) \wedge \bigvee_{\beta<\alpha} \exists y < x\, \varphi_\beta(y).$$

We can do this in such a way that there is a function f with Π^1_1-graph such that if e is the code for a recursive well ordering of order type α, then $f(e)$ is a code for φ_α.

Let τ be the vocabulary with a binary relation symbol $<$, constants $c_0, c_1, \ldots$. We will define a τ-theory T asserting:

- $<$ is a linear order;
- $\forall x \bigvee_{i\in\omega} x = c_i$;
- $\bigvee_{i\in\omega} c_{f(i+1)} \not< c_{f(i)}$ for $f \in \omega^\omega$ hyperarithmetic;
- $\exists x\, \varphi_\alpha(x)$ for $\alpha < \omega_1^{\mathrm{ck}}$.

EXERCISE 11.2.6. (a) There is a function f with Π^1_1-graph such that if e is the code for a recursive well ordering of order type α, then $f(e)$ is a code for φ_α.

(b) Show that the theory T is Π^1_1.

(c) Show that if $T' \subset T$ is Δ^1_1, then T' only contains $\exists x\, \varphi_\alpha(\overline{x})$ for $\alpha < \beta$ for some recursive β. Conclude that T' has a recursive model.

By Kreisel–Barwise Compactness, there is a computable model $\mathcal{H}$ of T. Clearly $\mathcal{H}$ has an initial segment of order type ω_1^{ck}. Since ω_1^{ck} is not a recursive ordinal, $\mathcal{H}$ must be ill-founded. The map $n \mapsto c_n^{\mathcal{H}}$ is recursive, so a hyperarithmetic descending chain in $\mathcal{H}$ will give rise to a hyperarithmetic descending chain $c_{f(0)}, c_{f(1)}, \ldots$, a contradiction. By the previous Lemma, $\mathcal{H} \cong \omega_1^{\mathrm{ck}} + \mathbb{Q} \times \omega_1^{\mathrm{ck}} + \alpha$ for some recursive α. By taking an initial segment, we may assume $\alpha = 0$. ⊣

Let $\prec$ be the Harrison order on ω. Let T be the tree set of finite sequences $(n_0, \ldots, n_m)$ such that $n_i \succ n_{i+1}$ for $i = 0, \ldots, m-1$. Then T is a recursive ill founded tree with no hyperarithmetic paths.

EXERCISE 11.2.7. Show that the existence of the Harrison ordering also follows directly from Lemmas 10.3.5 and 11.2.4.

11.3. Effective analysis of Scott rank

Let τ be a recursive vocabulary. Let $\mathbb{X}_\tau$ be the Polish space of countable τ-structures with universe ω.

LEMMA 11.3.1. *For any $\mathcal{M} \in \mathbb{X}_\tau$, $\overline{a} \in M$ and $\beta < \omega_1^{\mathcal{M}}$, the sentence $\Phi^{\mathcal{M}}_{\overline{a},\beta}$ is recursive in $\mathcal{M}$.*

PROOF. We will actually give a sentence $\widehat{\Phi}^{\mathcal{M}}_{\overline{a},\beta}$ that is equivalent to $\Phi^{\mathcal{M}}_{\overline{a},\beta}$ The inductive definition is a bit more cumbersome because we don't distinguish succesor and limit ordinals.[21]

Fix an ordering recursive in $\mathcal{M}$ of order type $\beta + 1$. We give a definition by effective transfinite recursion. We will build an $\mathcal{M}$-recursive map $(\overline{a}, \alpha) \mapsto$ a code for $\widehat{\Phi}^{\mathcal{M}}_{\overline{a},\alpha}$ for $\overline{a} \in \mathcal{M}$ and $\alpha \leq \beta$.

$$\widehat{\Phi}^{\mathcal{M}}_{\overline{a},0}(\overline{v}) = \Phi^{\mathcal{M}}_{\overline{a},0}(\overline{v}).$$

For $\alpha > 0$

$$\widehat{\Phi}^{\mathcal{M}}_{\overline{a},\alpha}(\overline{v}) = \bigwedge_{\gamma<\alpha} \left[\widehat{\Phi}_{\overline{a},\gamma}(\overline{v}) \wedge \bigwedge_{b\in\mathcal{M}} \exists w\, \widehat{\Phi}^{\mathcal{M}}_{\overline{a},b,\gamma}(\overline{v}, w) \wedge \forall \overline{w} \bigvee_{b\in\mathcal{M}} \widehat{\Phi}^{\mathcal{M}}_{\overline{a},b,\gamma}(\overline{v}, w)\right].$$

It is easy to see that $\widehat{\Phi}^{\mathcal{M}}_{\overline{a},\alpha}(\overline{v})$ is equivalent to $\Phi^{\mathcal{M}}_{\overline{a},\alpha}(\overline{v})$.

Here is a sketch of the details. Let R be a recursive well order of ω of order type $\beta + 1$. Suppose $f : \omega \times \omega^{<\omega} \to \omega$ is partial recursive we define a new partial recursive $\widehat{f} : \omega \times \omega^{<\omega} \to \omega$. We think of $\widehat{f}(n, \overline{a})$ and $f(n, \overline{a})$ as coding formulas let ψ_e denote the formula coded by e. Let

- $\widehat{f}(0, \overline{a}) = g(\overline{a})$ a code for the conjunction of all atomic and negated atomic sentence true about $\overline{a}$ in $\mathcal{M}$;
- $\widehat{f}(n, \overline{a})$ codes

$$\bigwedge_{nRm} \left[\psi_{f(n,\overline{a})}(\overline{v}) \wedge \bigwedge_{b\in\mathcal{M}} \exists w\, \psi_{f(m,\overline{a},b)}(\overline{v}, w) \wedge \forall w \left(\bigvee_{b\in\mathcal{M}} \psi_{f(m,\overline{a},b)}(\overline{v}, w)\right)\right].$$

Note, this is just a formal manipulation of potential codes and makes sense even f is not outputting actual codes. These manipulations are clearly computable. Thus there is a total recursive $F : \omega \to \omega$ such that if $f = \phi_e$, then $\widehat{f} = \phi_{F(e)}$. By the Recursion Theorem 9.1.9, there is $\widehat{e}$ such that $\phi_{\widehat{e}} = \phi_{F(\widehat{e})}$.

We can then show by induction that $\phi_e(n, \overline{a})$ codes $\widehat{\Phi}^{\mathcal{M}}_{\overline{a},\alpha}$ where $\{m : m\, R\, n\}$ has order type α. ⊣

LEMMA 11.3.2. *Suppose $\mathcal{M}$ is computable in z and $\mathbb{A}$ is the fragement of $\mathcal{L}_{\omega_1,\omega}$-formulas recursive in z. If $\mathcal{M} \equiv_{\mathbb{A}} \mathcal{N}$, then $\mathcal{M} \equiv_{\omega_1^{\mathcal{M}}} \mathcal{N}$.*

PROOF. For all $\beta < \alpha$, $\Phi^{\mathcal{M}}_\beta$ is recursive in z. Since $\mathcal{N} \models \Phi^{\mathcal{M}}_\beta$, $\mathcal{N} \equiv_\beta \mathcal{M}$. Since $\mathcal{N} \equiv_\beta \mathcal{M}$ for all $\beta < \omega_1^{\mathcal{M}}$, $\mathcal{N} \equiv_{\omega_1^{\mathcal{M}}} \mathcal{M}$. ⊣

[21] We could avoid this by beeing more careful about ordinal notations.

THEOREM 11.3.3 (Nadel [64]). *If* $\mathcal{M} \in \mathbb{X}_\tau$, *then* $\mathrm{SR}(\mathcal{M}) \leq \omega_1^{\mathcal{M}} + 1$.

Before proving Nadel's Theorem, we argue that it is best possible. Consider the Harrision ordering $\mathcal{H} = (\omega, \prec) \cong \omega_1^{\mathrm{ck}} + (\mathbb{Q} \times \omega_1^{\mathrm{ck}})$, where $\mathbb{Q} \times \omega_1^{\mathrm{ck}}$ is ordered lexicographically. Suppose $a, b \in \omega$ correspond to elements of the form $(q, 0)$ for $q \in \mathbb{Q}$. Then there is an automorphism of $\mathcal{H}$ taking a to b. Also note that any automorphism of $\mathcal{H}$ fixes the ω_1^{ck} initial segment of $\mathcal{H}$. Suppose $\beta < \omega_1^{\mathrm{ck}}$ and

$$\Phi^{\mathcal{H}}_{a,\beta}(v) \models \Phi^{\mathcal{H}}_{a,\alpha}(v)$$

for all α. The sentence $\Phi^{\mathcal{H}}_{a,\beta}$ is recursive. But using this formula we can define the well-founded initial segment of $\mathcal{H}$ by

$$X = \{n : \forall v\, \Phi^{\mathcal{H}}_{a,\beta}(v) \rightarrow n < v\}.$$

This set is Δ^1_1 and $e \in O$ if and only if e codes a linear order embedable into X. But this gives a Σ^1_1-definition of O, a contradiction. Thus $\mathrm{SR}(\mathcal{H}) \geq \omega_1^{\mathrm{ck}} + 1$.

We now do some preparation for the proof of Theorem 11.3.3. Let WO^* be the set of all linear orders R with domain ω such that:

(i) 0 is the R-least element of ω;

(ii) if x is not R-maximal, then there is y such that $x \mathrel{R} y$ and there is no z such that $x \mathrel{R} z$ and $z \mathrel{R} x$, we say y is the R-successor of x and write $y = s_R(x)$. If $n \neq 0$ is not an R-successor we say it is an R-limit.

Note that WO^*, $s_R(n) = m$ and "n is an R-limit" are arithmetic.

We say that z is an *R-analysis of $\mathcal{M}$ and $\mathcal{N}$* if

(i) $z \subseteq \omega \times \bigcup_{n\in\omega}(\omega^n \times \omega^n)$;

(ii) $(0, \overline{a}, \overline{b}) \in z$ if and only if $\mathcal{M} \models \phi(\overline{a}) \leftrightarrow \mathcal{N} \models \phi(\overline{b})$ for all quantifier free ϕ;

(iii) if $(n, \overline{a}, \overline{b})$ and $m \mathrel{R} n$, then $(m, \overline{a}, \overline{b})$;

(iv) $(s_R(n), \overline{a}, \overline{b}) \in z$ if and only if for all $c \in \omega$ there is $d \in \omega$ such that $(n, \overline{a}^\frown c, \overline{b}^\frown d) \in z$ and for all $d \in \omega$ there is $c \in \omega$ such that $(n, \overline{a}^\frown c, \overline{b}^\frown d) \in z$;

(v) if n is an R-limit, then $(n, \overline{a}, \overline{b}) \in z$ if and only if $(m, \overline{a}, \overline{b}) \in z$ for all $m \mathrel{R} n$.

Note:

- $\{(z, R, M, N) : z$ is an R-analysis of $\mathcal{M}$ and $\mathcal{N}\}$ is arithmetic.
- Suppose R is a well-order of order type α. Let $\beta(n) < \alpha$ be the order type of $\{m : m \mathrel{R} n\}$. If z is an R-analysis of $\mathcal{M}$ and $\mathcal{N}$, then

  $$(n, \overline{a}, \overline{b}) \in z \text{ if and only if } (\mathcal{M}, \overline{a}) \sim_{\beta(n)} (\mathcal{N}, \overline{b}).$$

- Suppose $R \in \mathrm{WO}^*$ is not a well-ordering and z is an R-analysis of $\mathcal{M}$ and $\mathcal{N}$. If $\ldots n_2 \mathrel{R} n_1 \mathrel{R} n_0$ is a descending chain and $(n_0, \overline{a}, \overline{b}) \in z$, then there is an isomorphism $\sigma : \mathcal{M} \rightarrow \mathcal{N}$ with $\sigma(\overline{a}) = \overline{b}$. [Do a back and forth building $\overline{a}_0 \subset \overline{a}_1 \subset \overline{a}_2 \subset \cdots$ and $\overline{b}_0 \subset \overline{b}_1 \subset \overline{b}_2 \subset \cdots$ such that

$(n_i, \overline{a}_i, \overline{b}_i) \in z$ and $\bigcup \overline{a}_i = \bigcup \overline{b}_i = \omega$. Then $\overline{a}_i \mapsto \overline{b}_i$ is the desired isomorphism.]

- We say that z is an R-analysis of $\mathcal{M}$ if it is an analysis of $\mathcal{M}$ with itself.

PROOF OF THEOREM 11.3.3. We need to show that if $\overline{a}, \overline{b} \in \omega$ and $(\mathcal{M}, \overline{a}) \not\sim (\mathcal{M}, \overline{b})$, then there is $\alpha < \omega_1^{\mathcal{M}}$ such that $(\mathcal{M}, \overline{a}) \not\sim_\alpha (\mathcal{M}, \overline{b})$. Fix $\overline{a}, \overline{b}$ with $(\mathcal{M}, \overline{a}) \not\sim (\mathcal{M}, \overline{b})$.

$S = \{R \in \mathrm{WO}^* : \exists z$ an R-analysis of $\mathcal{M}$ and $n \in \omega$ such that n is R-maximal and $(n, \overline{a}, \overline{b}) \in z\}$.

If $R \in S$ and R is not well-founded, then by the remarks above there is an automorphism σ of $\mathcal{M}$ with $\sigma(\overline{a}) = \overline{b}$ and $(\mathcal{M}, \overline{a}) \sim (\mathcal{M}_y, \overline{b})$, a contradiction. Thus $S \subseteq \mathrm{WO}$.

But S is $\Sigma_1^1(\mathcal{M})$. Thus, by Σ_1^1-Bounding, there is $\alpha < \omega_1^{\mathcal{M}}$ such that every element of R has order type $\leq \alpha$. But then $(\mathcal{M}, \overline{a}) \not\sim_\alpha (\mathcal{M}, \overline{b})$.

Hence $\mathrm{sr}(\mathcal{M}) \leq \omega_1^{\mathcal{M}}$ and $\mathrm{SR}(\mathcal{M}) \leq \omega_1^{\mathcal{M}} + 1$. ⊣

Remark 11.3.4. In particular, $\mathrm{SR}(\mathcal{H}) = \omega_1^{\mathrm{ck}} + 1$. Makkai [51] showed how to construct computable structures $\mathcal{M}$ with $\mathrm{SR}(\mathcal{M}) = \omega_1^{\mathrm{ck}}$. Some easier variants of his construction are given in [17].

THEOREM 11.3.5 (Nadel [64]). *Suppose $\mathcal{M}, \mathcal{N} \in \mathbb{X}_\tau$ are recursive in x and $\mathcal{M} \equiv_{\mathbb{A}} \mathcal{N}$ where $\mathbb{A}$ is the fragement of $\mathcal{L}_{\omega_1,\omega}$-formulas recursive in x. Then $\mathcal{M} \cong \mathcal{N}$.*

PROOF. Let $A = \{R \in \mathrm{WO}^* : \exists z\, z$ is an R-analysis of $\mathcal{M}$ and $\mathcal{N}$ and $(n, \emptyset, \emptyset) \in R$ for all $n \in \omega\}$. If some $R \in A$ is not a well ordering then $\mathcal{M} \cong \mathcal{N}$ as desired. So assume that $A \subseteq \mathrm{WO}$. Since A is Σ_1^1 in x, there is $\beta < \omega_1^x$ such that every $R \in A$ has order type less that β. But then $\mathcal{M} \not\sim_\beta \mathcal{N}$. But $\Phi_\beta^{\mathcal{M}} \in \mathbb{A}$. Thus $\mathcal{N} \models \Phi_\beta^{\mathcal{M}}$ and $\mathcal{M} \sim_\beta \mathcal{N}$, a contradiction. ⊣

THEOREM 11.3.6. *Suppose τ is a recursive vocabulary and ϕ is a hyperarithmetic $\mathcal{L}_{\omega_1,\omega}$-sentence. Then $\cong$ is Δ_1^1 on* $\mathrm{Mod}(\phi)$ *if and only if there is $\alpha < \omega_1^{\mathrm{ck}}$ such that every model of ϕ has Scott rank at most α.*

Relativizing, for any $\phi \in \mathcal{L}_{\omega_1,\omega}$, $\cong$ is a Borel equivalence relation on $\mathrm{Mod}(\phi)$ *if and only if there is a countable bound on Scott rank.*

PROOF. ($\Leftarrow$) If all models have Scott rank below α, then $\cong$ is $\sim_\alpha$ a Δ_1^1 equivalence relation.

($\Rightarrow$) Let $A = \{R \in \mathrm{WO}^* : \exists \mathcal{M}, \mathcal{N} \in \mathrm{Mod}(\phi)\, \mathcal{M} \ncong \mathcal{N} \wedge \exists z\, z$ is an R-analysis of $\mathcal{M}$ and $\mathcal{N}$ and $(n, \emptyset, \emptyset) \in \mathbb{Z}$ for all $n \in \omega\}$. Then A is Σ_1^1. As above if $R \in A$ is not a well-ordering and z is an R-analysis of $\mathcal{M}$ and $\mathcal{N}$ where $(n, \emptyset, \emptyset) \in z$ for all n, then $\mathcal{M} \cong \mathcal{N}$. Thus A is a Σ_1^1 set of well-orders and, by Σ_1^1-Bounding, there is $\alpha < \omega_1^{\mathrm{ck}}$ such that every element of A has order type less that α. It follows that we can bound Scott rank below ω_1^{ck}. ⊣

EXERCISE 11.3.7. Show that $\{\phi \in \mathcal{L}_{\omega_1,\omega} : \phi$ is complete$\}$ is Π_1^1. [Hint: Argue that ϕ is complete if and only if

- $\exists \mathcal{M} \leq_{\text{hyp}} \phi\, \mathcal{M} \models \phi$;
- for all $\mathcal{M}, \mathcal{N}\, \exists R \exists z \leq_{\text{hyp}} \phi\ R \in \text{WO}^*$ and z is an R-analysis of $\mathcal{M}$ and $\mathcal{N}$ and there is n such that if $\overline{a}, c \in \mathcal{M}$ and $\overline{b} \in \mathcal{N}$ with $(n, \overline{a}, \overline{b}) \in z$ then there is $d \in \mathcal{N}$ such that $(n, \overline{a}, c, \overline{b}, d) \in z$.]

THEOREM 11.3.8 (Sacks [67]). *Suppose* $\text{SR}(\mathcal{M}) \leq \omega_1^{\mathcal{M},\phi}$ *for all* $\mathcal{M} \in \text{Mod}(\phi)$, *then there is* $\alpha < \omega_1^{\phi}$ *such that* $\text{SR}(\mathcal{M}) < \alpha$ *for all* $\mathcal{M} \models \phi$. *In particular,* $\cong$ *is a Borel equivalence relation on* $\text{Mod}(\phi)$ *and Vaught's Conjecture holds for* ϕ.

PROOF. For notational simplicity we assum $\phi \in \text{HYP}$. Suppose $\text{SR}(\mathcal{M}) \leq \omega_1^{\mathcal{M}}$ for all $\mathcal{M} \in \text{Mod}(\phi)$. Then for all $\mathcal{M}$ and $\overline{a} \in \mathcal{M}$ there is $\alpha < \omega_1^{\mathcal{M}}$ such that if $\overline{b} \in \mathcal{M}$ and $(\mathcal{M}, \overline{a}) \sim_\alpha (\mathcal{M}, \overline{b})$, then $(\mathcal{M}, \overline{a}) \sim (\mathcal{M}, \overline{b})$. We let $r(\overline{a})$ denote the least such α.

Let $S_1 = \{(\mathcal{M}, \overline{a}, e) : e \in \omega$ and if $\mathcal{M} \models \phi$, then
$phi_e^{\mathcal{M}}$ is the characteristic function of a well-order R and for any R-analysis z of $\mathcal{M}$ there is n such that for all $\overline{b}$ if $(n, \overline{a}, \overline{b}) \in z$, then $(s_R(n), \overline{a}, \overline{b})\}$.

The set S_1 is Π^1_1. If $\mathcal{M} \not\models \phi$, then $(\mathcal{M}, \overline{a}, e) \in S_1$ for all $e \in \omega$. If $\mathcal{M} \models \phi$ then $(\mathcal{M}, \overline{a}, e) \in S_1$ if and only if $\{e\}^{\mathcal{M}}$ is the characteristic function of an ordinal $\alpha > r(\overline{a})$.

By Kreisel's uniformization Theorem v, there is $f : \mathbb{X}_\tau(\phi) \times \bigcup_n \omega^n \to \omega$ such that $(\mathcal{M}, \overline{a}, f(\mathcal{M}, \overline{a})) \in S_1$ for all $x, \overline{a}$. Since the range is ω, $f(\mathcal{M}, \overline{a}) \neq e$ if and only if $\exists m \neq n\ f(x, \overline{a}) = m$, and f is Δ^1_1.

Let $S_2 = \{R : \exists \mathcal{M} \exists \overline{a}\ \mathcal{M} \models \phi \wedge R$ is isomorphic to the ordering coded by $\phi^{\mathcal{M}}_{f(\mathcal{M},\overline{a})}\}$. Then S_2 is a Σ^1_1 set of well-orderings and, by Σ^1_1-Bounding, there is $\alpha < \omega_1^{\text{ck}}$ such every R in S_1 has order type at most α. Thus for all $\mathcal{M}_x \models \sigma$ we have $r(\overline{a}) < \alpha$ for all $\overline{a}$. Thus $\text{SR}(\mathcal{M}) \leq_\alpha$ for all $\mathcal{M} \models \sigma$. ⊣

Note: This argument would work just as well for $\text{PC}_{\omega_1,\omega}$-classes.

EXERCISE 11.3.9. In Exercise 2.2.7 we showed that for and $\alpha < \omega_1$ there is a formula $\sigma_\alpha \in \mathcal{L}_{\omega_1,\omega}(\tau)$ such

$$\mathcal{M} \models \sigma_\alpha \Leftrightarrow \text{SR}(\mathcal{M}) \geq \alpha$$

for any τ-structure $\mathcal{M}$. Show that if $\alpha < \omega_1^x$, then σ_α is computable in x and $\{\sigma_\alpha : \alpha < \omega_1^x\}$ is $\Pi^1_1(x)$.

We will need the following result in Chapter 12.

THEOREM 11.3.10. *Suppose* $\phi \in \text{HYP}$, α *is admissible and for all* $\beta < \alpha$, ϕ *has a model of Scott rank at least* β. *Then there is* $\mathcal{M} \models \phi$ *with* $\text{SR}(\mathcal{M}) \geq \omega_1^{\mathcal{M}} = \alpha$.

PROOF. Choose x such that $\omega_1^x = \alpha$.
Define $\Theta_0^n(\overline{u}, \overline{v})$ $|u| = |v| = n$ to be

$$\bigwedge_{\psi \text{ atomic}} \psi(\overline{u}) \leftrightarrow \psi(\overline{v})$$

and define $\Theta^n_\beta(\overline{u}, \overline{v})$ to be

$$\bigwedge_{\gamma<\beta} \left[\Theta^n_\gamma(\overline{u}, \overline{v}) \wedge \forall x \exists y\, \Theta^{n+1}_\gamma(\overline{u}, x, \overline{v}, y) \wedge \forall y \exists x\, \Theta^{n+1}_\gamma)\overline{u}, x, \overline{v}, y)\right].$$

Note that if $\overline{a}, \overline{b} \in \mathcal{M}$ and $|\overline{a}| = |\overline{b}| = n$, then $\overline{a} \sim_\beta \overline{b}$ if and only if $\mathcal{M} \models \Theta^n_\beta(\overline{a}, \overline{b})$. Let σ_β be the sentence

$$\bigvee_{n<\omega} \exists \overline{u} \exists \overline{v}\, \Theta^n_\beta(\overline{u}, \overline{v}) \wedge \neg\Theta^n_{\beta+1}(\overline{u}, \overline{v}).$$

Then $\mathcal{M} \models \sigma_\beta$ if and only if $\mathrm{SR}(\mathcal{M}) > \beta$.

As in the proof of Theorem 11.3.1, if $\beta < \alpha$, then we can find a codes for σ_β recursive in x and we can find a $\Pi^1_1(x)$ set of codes for $\{\sigma_\beta : \beta < \alpha\}$. By Kreisel–Barwise Compactness there are models of $\{\phi\} \cup \{\sigma_\beta : \beta < \alpha\}$ and, by Gandy's Basis Theorem 10.3.7 there is $\mathcal{M} \models \{\phi\} \cup \{\sigma_\beta : \beta < \alpha\}$ with $\omega_1^{\mathcal{M}} \leq \alpha$. Since $\mathrm{SR}(\mathcal{M}) \leq \omega_1^{\mathcal{M}} + 1$, we must have $\mathrm{SR}(\mathcal{M}) \geq \omega_1^{\mathcal{M}} = \alpha$. ⊣

EXERCISE 11.3.11. Show that for any Vaught counterexample ϕ and $z \geq_T \phi$ there is $\mathcal{M} \in \mathrm{Mod}(\phi)$ with $\mathrm{SR}(\mathcal{M}) \geq \omega_1^{\mathcal{M},z} = \omega_1^z$. [Hint: Relativize the previous proof.]

Finally, we give the proof promised in Proposition 3.3.9.

PROPOSITION 11.3.12. *Suppose there is a perfect set of non-isomorphic countable models of ϕ. Then for some $\alpha < \omega_1$ there are uncountably many $\sim_\alpha$-classes.*

PROOF. Let P be a perfect set of non-isomorphic models. Let $A = \{R \in \mathrm{WO}^* : \exists \mathcal{M}, \mathcal{N} \in P \text{ distinct} \wedge \exists z \text{ an } R\text{-analysis of } \mathcal{M} \text{ and } \mathcal{N} \text{ and } (n, \emptyset, \emptyset) \in z \text{ for all } n \in \omega\}$. As usual, if A contains an ill-founded order R and z is an R-analysis of $\mathcal{M}$ and $\mathcal{N}$ as above, then we can do a back-and-forth argument to show $\mathcal{M} \cong \mathcal{N}$, a contradiction. Thus $A \subseteq \mathrm{WO}$. But A is Σ^1_1. Thus by Σ^1_1-bounding, there is $\alpha < \omega_1$ such that every $R \in A$ has order type less than α. It follows that if $\mathcal{M}, \mathcal{N} \in P$ and $\mathcal{M} \neq \mathcal{N}$, then $\mathcal{M} \not\sim_\alpha \mathcal{N}$. ⊣

We can also give a stronger version of this.

PROPOSITION 11.3.13. *If there is a perfect set of non-isomorphic countable models of ϕ, then for some $\alpha < \omega_1$ there is a perfect set of models of Scott rank α.*

PROOF. Suppose T is a perfect tree such that $[T]$ the paths through T is a perfect set of non-isomorphic models of T. By Exercise 10.4.9 there is a perfect subset P of $[T]$ such that $\omega_1^x \leq \omega_1^T$ for all $x \in P$. But then, by Nadel's Theorem 11.3.3, each model coded in P has Scott rank at most ω_1^T. ⊣

EXERCISE 11.3.14. Give a second proof of Proposition 11.3.13 using Proposition 11.3.12.

Chapter 12

SPECTRA OF VAUGHT COUNTEREXAMPLES

Suppose $\mathcal{M}$ is a countable τ-structure.

DEFINITION 12.0.15. The *degree spectrum* of $\mathcal{M}$ is

$$\mathrm{Sp}(\mathcal{M}) = \{z : \exists \mathcal{N} \leq_T z\, \mathcal{M} \cong \mathcal{N}\},$$

i.e., the set of reals where there is a computable isomorphic copy of $\mathcal{M}$.

Suppose α is a countable ordinal. Then $z \in \mathrm{Sp}(\alpha)$ if and only if the ordinal α is recursive in z. Thus $z \in \mathrm{Sp}(\alpha)$ if and only if $\omega_1^z \geq \alpha$.

If we review the counterexamples we have seen to Vaught's Conjecture for $\mathrm{PC}_{\omega_1,\omega}$ classes in Exercises 3.2.4 and 3.2.5, we see, there too, that for any $\mathcal{M}$ there is an ordinal α such that $\mathrm{Sp}(\mathcal{M}) = \{z : \omega_1^z \geq \alpha\}$. This corresponds to our intuition that the models of Vaught counterexamples in someway behave like ordinals.

In this chapter we will prove a warm-up result to the main theorem of Montalbán [59] where, under the additional assumption PD (i.e., that all projective games are determined), he makes this intuition surprisingly precise.

THEOREM 12.0.16 (Montalbán [59]). (PD) *The following are equivalent*:

(i) *ϕ is a Vaught counterexample*;

(ii) *there is an oracle relative to which*

$$\{\mathrm{Sp}(\mathcal{M}) : \mathcal{M} \models \phi\} = \{\{z : \omega_1^z \geq \alpha\} : \alpha < \omega_1\}.$$

It is unknown if this result can be proved in ZFC alone. We will prove the (i) $\Rightarrow$ (ii) direction under the additional assumption that ϕ is a minimal counterexample.

12.1. Determinacy and Turing degrees

Montalbán's work relies on several results of Martin [57] using determinacy to understand the structure of the Turing degrees. We begin with a survey of these results. The results of this section are due to Martin, though some appear first in [75].

Pointed perfect sets.

DEFINITION 12.1.1. We say that a tree $T \subseteq 2^{<\omega}$ is a *perfect* if $T \neq \emptyset$ and for any $\sigma \in T$ there are incomparable $\sigma_0, \sigma_1 \in T$ with $\sigma_0, \sigma_1 \supset \sigma$.

A perfect tree T is *pointed* if $T \leq_T f$ for all $f \in [T]$, where we recall that $[T]$ is the set of infinite paths through T. We say that $P \subseteq 2^\omega$ is a *pointed perfect set* if $P = [T]$ for some pointed perfect tree.

EXERCISE 12.1.2. Suppose $f : 2^\omega \to 2^\omega$ is continuous, one-to-one. Let

$$T = \{\eta \in 2^{<\omega} : \exists x \in 2^\omega\, \eta \subset f(x)\}.$$

Show that T is a perfect tree.

LEMMA 12.1.3. *Suppose f is computable. Then the tree T in the exercise above is computable.*

PROOF. There is a program Q that using oracle x and input i computes $f(x)(i)$. For $\nu \in 2^{<\omega}$ we say $f(\nu)(i) = j$ if the computation of Q using oracle ν and input i halts in at most $|\nu|$ steps making no oracle queries outside the domain of ν and outputs j. Then

$$\eta \in T \Leftrightarrow \exists \nu \forall i < |\eta|\, f(\nu)(i) = \eta(i)$$

and

$$\eta \notin T \Leftrightarrow \exists n \forall \nu \in 2^n \exists i < |\eta|\, f(\nu)(i) \neq \eta(i).$$

It follows that T is computable. ⊣

COROLLARY 12.1.4. *Suppose $f : 2^\omega \to 2^\omega$ is continuous, one-to-one, computable in d and $d \leq_T f(x)$ for all $x \in 2^\omega$. Then the image of f is a pointed perfect set.*

LEMMA 12.1.5. *If T is a pointed perfect tree, then there is a T-computable, continuous, one-to-one $f : 2^\omega \to [T]$. Further, if $z \geq_T T$, then $f(z) \equiv_T z$.*

PROOF. Let $x \in 2^\omega$. We build a path $\emptyset = \eta_0 \subset \eta_1 \subset \cdots$ through T where $|\eta_i| = i$. Let $c_0 = 0$; c will count how many choices we have made so far.

If η_s has only one extension in T, let η_{s+1} be this unique extension and let $c_{s+1} = c_s$. If not, let $\eta_{s+1} = \eta_s{}^\frown x(c_s)$ and $c_{s+1} = c_s + 1$. Define $f(x) = \bigcup \eta_s$.

Clearly if $z \geq_T T$, then $f(z) \leq_T z$, since f is T-computable. On the other hand, since T is pointed, $f(x) \geq_T T$. Using $f(x)$ and T we can recover x. Start building $\eta_0 \subset \eta_1 \subset \cdots$ as above. By noticing what we did when we had choices, we can compute x. ⊣

Turing determinacy. Recall that if $A \subset 2^\omega$, then $G(A)$ is the game where players alternate playing 0 or 1

Player I	Player II
n_0	
	n_1
n_2	
	n_3
$\vdots$	$\vdots$

and Player II wins if $x \in A$ where $x(i) = n_i$ for all i.

If σ is a strategy for Player II, define $\sigma^* : 2^\omega \to 2^\omega$ by $\sigma^*(x)$ is the final play of the game when Player I plays $x(0), x(1), \dots$ and Player II uses σ.

Player I	Player II
$x(0)$	
	$\sigma(x(0))$
$x(1)$	
	$\sigma(x(0), x(1))$
$\vdots$	$\vdots$

If σ is a winning strategy for Player II, then $\sigma^* : 2^\omega \to A$. We can define analogous maps from strategies for Player I.

LEMMA 12.1.6. *If Player II has a winning strategy in $G(A)$, then A contains a pointed perfect set.*

PROOF. Let σ be a winning strategy for Player II. Suppose $\sigma \equiv_T b \in 2^\omega$. Let $g : 2^\omega \to A$ be the function

$$g(x) = \sigma^*(x(0), b(0), x(1), b(1), \dots).$$

The image of g is the set of plays in the game where Player II uses σ and Player I plays

$$x(0), b(0), x(1), b(1), \dots$$

Then g is continuous, injective and computable in σ. Moreover, $\sigma \equiv_T b \leq_T g(x)$ for all $x \in 2^\omega$. Thus A contains a pointed perfect set. ⊣

A similar argument shows that if Player I has a winning strategy, then $2^\omega \setminus A$ contains a pointed perfect set.

Note in the argument above that $g(x) \leq_T (x, \sigma)$, $x \leq_T g(x)$ and $\sigma \leq_T g(x)$.

DEFINITION 12.1.7. A subset C of 2^ω is a *cone* if there is $d \in 2^\omega$ such that $C = \{x \in 2^\omega : d \leq_T x\}$.

We say that $A \subseteq 2^\omega$ is *Turing invariant* if $y \in A$ whenever there is $x \in A$ with $x \equiv_T y$.

EXERCISE 12.1.8. Let $A \subseteq 2^\omega$ be Turing-invariant. At most one of A and $2^\omega \setminus A$ contains a cone.

COROLLARY 12.1.9 (Turing Determinacy). (PD) *If $A \subseteq 2^\omega$ is projective and Turing-invariant, then one of A and $2^\omega \setminus A$ contains a cone.*

PROOF. If σ is, say, a winning strategy for Player II in $G(A)$ and g is as above, then for any $x \geq_T \sigma$, $x \equiv_T g(x)$, so $x \in A$. Thus A contains the cone above σ. ⊣

EXERCISE 12.1.10. Assume AD the Axiom of Determinacy, i.e., that $G(A)$ is determined for all $A \subset 2^\omega$. Define $\mathcal{U} \subset \mathcal{P}(\omega_1)$ by

$$A \in \mathcal{U} \Leftrightarrow \{x : \omega_1^x \in A\} \text{ contains a cone.}$$

Show that $\mathcal{U}$ is a non-principle σ-complete ultrafilter on ω_1. This shows that AD $\Rightarrow \aleph_1$ is a measurable cardinal.

LEMMA 12.1.11. (PD) *Suppose $A \subseteq 2^\omega$ is projective and cofinal in the Turing degrees (i.e., for all x there is $x \leq_T y$ with $y \in A$). Then there is a pointed perfect set $P \subseteq A$.*

PROOF. Consider the game where Player I plays $e \in \omega$ and then plays $x \in 2^\omega$, and Player II plays $y \in 2^\omega$. Player I wins if $x \in A$ and $y = \phi_e^x$ (i.e., y is computable using Turing machine e and oracle x). We claim that Player I has a winning strategy. Suppose, for purposes of contradiction, that Player II has a winning strategy σ. Choose $x \in A$ such that $\sigma \leq_T x$. Consider the x-recursive function f where $f(e, n)$ is the nth move by Player II using σ if Player I first plays e and then plays $x(0), x(1), \ldots$, i.e, if Player II uses σ this is a play of the game

Player I	Player II
e	
	$f(e,0)$
$x(0)$	
	$f(e,1)$
$x(1)$	
	$f(e,2)$
$\vdots$	$\vdots$

By the Recursion Theorem 9.1.9, there is an e such that $\phi_e^x(n) = f(e, n)$ for all n. Let $y = \phi_e^x$. But since $x \in A$, Player I wins by playing e followed by x, contradicting the fact σ is a winning strategy of Player II.

Thus Player I has a winning strategy σ. Let e be the first move made by Player I using this strategy. Let $z \in 2^\omega$ be such that $z \equiv_T \sigma$. Let $g : 2^\omega \to 2^\omega$ be such that $e\frown g(y)$ is the element of 2^ω played by Player I if Player II plays

$$z(0), y(0), z(1), y(1), \ldots$$

and one uses σ. Since $y = \phi_e^{g(y)}$, g is one-to-one. It is not immediately clear that the image of g contains a perfect pointed set. But consider the map $h : 2^\omega \to 2^\omega$

where $h(x) = g(z(0), x(0), z(1), x(1), \dots)$. Then h is also continuous, one-to-one and computable from σ. Note that $(z(0), x(0), z(1), x(1), \dots) = \phi_e^{h(x)}$, thus σ is computable from $h(x)$, for all $x \in 2^\omega$. Thus the image of h is a pointed perfect set. ⊣

COROLLARY 12.1.12. (PD) *Suppose $f : 2^\omega \to \omega$ is projective. Then there is a pointed perfect set P such that f is constant on P.*

PROOF. If $f^{-1}(i)$ is cofinal in the Turing degrees, then it contains a pointed perfect set. Otherwise there is z_i such that if $z_i \leq_T x$, then $f(x) \neq i$. If no $f^{-1}(i)$ contains a pointed perfect set, then we can find $x_T \geq z_i$ for all i. But then $f(x) \neq i$ for any i, a contradiction. ⊣

THEOREM 12.1.13. (PD) *Suppose $f : 2^\omega \to \omega_1$ is Turing-invariant and there is a projective $g : 2^\omega \to$ WO such that $f(x)$ is the order type of the well ordering $g(x)$. If $f(x) < \omega_1^x$ for all x, then f is constant on a cone.*

PROOF. There is $h : 2^\omega \to \omega$ projective such that $\phi_{h(x)}^x \in$ WO has order type $f(x)$. By the last Corollary, there is a pointed perfect tree T and $e \in \omega$ such that $h(x) = e$ for $x \in [T]$. The map $x \mapsto \phi_e^x$ is continuous on $[T]$. Thus by Σ_1^1-bounding, there is $\alpha < \omega_1$ such that the order type of $\phi_e^x < \alpha$ for all $x \in [T]$.

For $\beta < \alpha$, let $A_\beta = \{x \geq_T T : f(x) = \beta\}$. This is a Turing-invariant, projective set. If it contains a cone we are done, if not there is a z_β such that $x \notin A_\beta$ if $x \geq_T z_\beta$. But then we can find $x \geq_T T$ such that $x \geq z_\beta$ for all $\beta < \alpha$. We can find $w \in [T]$ such that $w \equiv_T x$, but then $|f(w)| \neq \beta$ for all $\beta < \alpha$, a contradiction. ⊣

12.2. Montalbán's Theorem

DEFINITION 12.2.1. We say that an equivalence relation E on a set $X \subseteq 2^\omega$ is *ranked* if there is an E-invariant $r : X \to \omega_1$. We say that r is a *projective ranking* if there is a projective $g : X \to$ WO such that $r(x)$ is the order type of $g(x)$.

We say that (E, r) is *scattered* if $r^{-1}(\alpha)$ has only countably many E-classes for all $\alpha < \omega_1$.

Of course we are thinking about the case where $X = \mathrm{Mod}(\phi)$, E is isomorphism and $r(\mathcal{M})$ is the Scott rank of $\mathcal{M}$. Then ϕ is scattered if and only if (E, r) is scattered.

THEOREM 12.2.2. (PD) *Let (E, r) be a scattered projective ranked equivalence relation. Then there is a cone of z such that if $x \in X$, if $r(x) < \omega_1^z$, then there is $y \mathrel{E} x$ such that $y \leq_T z$.*

PROOF. Suppose not. Then for every cone C there is $z \in C$ and x such that $r(x) < \omega_1^z$ but there is no z-computable $y \mathrel{E} x$. By Turing Determinacy, there is a cone C_0 of such z. For $z \in C_0$ let α_z be least such that there is x with $r(x) = \alpha_z < \omega_1^z$ but no z-computable $y \mathrel{E} x$. By Theorem 12.1.13, there is a cone $C_1 \subseteq C$ such that $\alpha_z = \alpha$ for $z \in C_1$. But there are only $\aleph_0$ classes of rank at most α. Thus we can find $(y_i : i \in \omega)$ a set of representatives. But then we could find z in the cone greater than all of the y_i. But for any x, if $r(x) \leq \alpha$, then there is $y_i \leq_T z$ with $y_i \mathrel{E} x$, a contradiction. ⊣

We will use the following recursion theoretic lemma which we state without proof. This is Lemma 3.6 of [59] where Montalbán proves it using hyperarithmetic Cohen forcing and the Harrison order. Harington (Theorem 2.10 of [25]) proved a similar result by Steel forcing.

LEMMA 12.2.3. *If $\omega_1^x = \omega_1^y$, then there is z such that*

$$\omega_1^x = \omega_1^{x,z} = \omega_1^z = \omega_1^{y,z} = \omega_1^y.$$

LEMMA 12.2.4. *For any $z \in 2^\omega$ the set of z-admissible ordinals contains a closed unbounded subset.*

PROOF. The set of $\{\alpha : \mathbb{L}_\alpha[z] \prec \mathbb{L}_{\omega_1}[z]\}$ is a closed unbounded set of z-admissible ordinals. ⊣

For the following proof, recall that for any structure $\mathcal{M}$ we let $\omega_1^{\mathcal{M}}$ be the minimum of $\{\omega_1^z : z \text{ codes an isomorphic copy of } \mathcal{M}\}$.

THEOREM 12.2.5. (PD) *Suppose ϕ is a minimal Vaught counterexample. There is $z_0 \in 2^\omega$ such that*

$$z \in \mathrm{Sp}(\mathcal{M}) \Leftrightarrow \omega_1^z \geq \omega_1^{\mathcal{M},z_0}$$

for every $z_0 \leq_T z$ and every $\mathcal{M} \in \mathrm{Mod}(\phi)$.

PROOF. By Theorem 12.2.2, there is a cone C_4 with base z_4 such that if $z \in C_4$, $\mathcal{M} \in \mathrm{Mod}(\phi)$ and $\mathrm{SR}(\mathcal{M}) < \omega_1^z$, then $z \in \mathrm{Sp}(\mathcal{M})$. Without loss of generality, we may assume $\phi \leq_T z_4$.

By Exercise 11.3.11, for any $y \geq_T \phi$, there is $\mathcal{M} \in \mathrm{Mod}(\phi)$ such that $\mathrm{SR}(\mathcal{M}) \geq \omega_1^{\mathcal{M},y} = \omega_1^y$. Thus, taking $z = \mathcal{M} \oplus y$

$$\{z : \exists \mathcal{M} \in \mathrm{Mod}(\phi)\, \mathcal{M} \leq_T z \wedge \mathrm{SR}(\mathcal{M}) \geq \omega_1^{\mathcal{M},z} = \omega_1^z\}$$

is cofinal in the Turing degrees, projective and Turing invariant, and hence contains a cone C_3 with base z_3. Without loss of generality, we may assume $C_3 \subseteq C_4$.

Let

$$C_0 = \{z \geq_T \phi : \forall \mathcal{M} \in \mathrm{Mod}(\phi)\, [\mathrm{SR}(\mathcal{M}) \geq \omega_1^{\mathcal{M},z_3} \Rightarrow \exists \mathcal{N} \leq_T z\, \mathcal{M} \cong \mathcal{N}]\}.$$

CLAIM. *C_0 is cofinal in the Turing degrees.*

Let $z_2 \geq_T z_3$. We will find $z \in C_0$ with $z \geq_T z_2$. Recall from Corollary 7.2.7 that the set X of α such that if $\mathcal{M}, \mathcal{N} \in \mathrm{Mod}(\phi)$ and $\mathrm{SR}(\mathcal{M}), \mathrm{SR}(\mathcal{N}) \geq \alpha$, then $\mathcal{M} \equiv_\alpha \mathcal{N}$ is closed unbounded. Since the set of z_2-admissible ordinals is also closed unbounded, we can find $\alpha \in X$ where α is z_2-admissible.

Let $z_1 \geq_T z_2$ such that $\omega_1^{z_1} = \alpha$. We claim that $z_1 \in \mathcal{C}_0$. By Exercise 11.3.11, there is $\mathcal{M} \in \mathrm{Mod}(\phi)$ with $\mathrm{SR}(\mathcal{M}) \geq \omega_1^{\mathcal{M},z_1} = \omega_1^{z_1}$. Let $y = \mathcal{M} \oplus z_1$. Then $\mathcal{M} \leq_T y$ and $\omega_1^y = \omega_1^{z_1}$. Since $z_1 \in \mathcal{C}_3$, there is $\mathcal{N} \leq_T z_1$ with $\mathrm{SR}(\mathcal{N}) \geq \omega_1^{\mathcal{N},z_1} = \omega_1^{z_1}$. By Nadel's Theorem 11.3.3, $\mathrm{SR}(\mathcal{N}) \leq \omega_1^{\mathcal{N}} + 1$, thus we must have $\omega_1^{z_1} = \omega_1^{\mathcal{N}}$.

We will show that $\mathcal{M} \cong \mathcal{N}$. By Lemma 12.2.3, there is $g \geq z_3$ such that

$$\alpha = \omega_1^{z_1} = \omega_1^{z_1,g} = \omega_1^g = \omega_1^{y,g} = \omega_1^y.$$

Since $g \in \mathcal{C}_3$, there is $\mathcal{M}' \in \mathrm{Mod}(\phi)$ with $\mathrm{SR}(\mathcal{M}') \geq \omega_1^{\mathcal{M}'} = \omega_1^g$. Note that $\mathcal{M} \leq_T y$, $\mathcal{N} \leq_T z_1$ and $\mathcal{M}' \leq_T g$. All of these structures have Scott rank at least α, thus they are $\equiv_\alpha$-equivalent.

By Theorem 11.3.5, since $\mathcal{M}, \mathcal{M}' \leq_T y \oplus g$, $\alpha = \omega_1^{y,g}$ and $\mathcal{M} \equiv_\alpha \mathcal{M}'$, $\mathcal{M} \cong \mathcal{M}'$. By an analogous argument, $\mathcal{N} \cong \mathcal{M}'$. Thus $\mathcal{M} \cong \mathcal{N}$. This proves that $\mathcal{C}_0$ is unbounded in the Turing degees.

By Turing determinacy, $\mathcal{C}_0$ contains a cone with base z_0. Let $z \geq_T z_0$ and let $\mathcal{M} \in \mathrm{Mod}(\phi)$ such that $\omega_1^{\mathcal{M},z_0} \leq \omega_1^z$. If $\mathrm{SR}(\mathcal{M}) < \omega_1^z$, then since $z \in \mathcal{C}_4$, $z \in \mathrm{Sp}(\mathcal{M})$. If $\mathrm{SR}(\mathcal{M}) \geq \omega_1^z$, that

$$\omega_1^z + 1 \geq \omega_1^{\mathcal{M},z_0} + 1 \geq \omega_1^{\mathcal{M},z_3} + 1 \geq \mathrm{SR}(\mathcal{M}) \geq \omega_1^z.$$

Thus $\mathrm{SR}(\mathcal{M}) \geq \omega_1^{\mathcal{M},z_3} = \omega_1^z$. Since $z \in \mathcal{C}_0$, $z \in \mathrm{Sp}(\mathcal{M})$. $\dashv$

Appendix A

$\aleph_1$-FREE ABELIAN GROUPS

We will prove that there are $2^{\aleph_1}$ non-isomorphic $\aleph_1$-free abelian groups of cardinality $\aleph_1$. The proof works by coding stationary sets into the groups. This is a standard trick in "many model" arguments. We need the following set theoretic fact.

LEMMA A.0.1. *There is a family $(S_\alpha : \alpha < \omega_1)$ of disjoint stationary subsets of ω_1. For $S \subseteq \omega_1$, let $S_A = \bigcup_{\alpha \in A} S_\alpha$. Then $(S_A : A \subset \omega_1)$ is a family of $2^{\aleph_1}$ stationary subsets of ω_1 such that $S_A \triangle S_B$ is stationary for $A \neq B$.*

See, for example, [56] 5.3.10. The following exercise gives a simple example of how we can use stationary sets to build many non-isomorphic models.

EXERCISE A.0.2. Recall that a linear order $(X, <)$ is ω_1-*like* if and only if $|X| = \aleph_1$ and $|\{y : y < x\}| \leq \aleph_0$ for all $x \in X$. We will show that there are $2^{\aleph_1}$ non-isomorphic ω_1-like dense linear orders.

Let $(L, <)$ be a countable dense linear order with a least element but no greatest element. For $S \subseteq \omega_1$, let

$$A_\alpha(S) = \begin{cases} (L, <) & \text{if } \alpha \in S, \\ (\mathbb{Q}, <) & \text{if } \alpha \notin S \end{cases}$$

and let

$$X(S) = \sum_{\alpha < \omega_1} A_\alpha(S),$$

i.e., $X(S) = \bigcup_{\alpha < \omega_1} \{\alpha\} \times A_\alpha(S)$ ordered such that $(\alpha, a) < (\beta, b)$ if $\alpha < \beta$ or $\alpha = \beta$ and $a < b$ in $A_\alpha(S)$. We let $X_\alpha(S) = \sum_{\beta < \alpha} A_\beta(S)$.

(a) Show that $X(S)$ is ω_1-like.
(b) Suppose $f : X(S) \to X(S_1)$ is an isomorphism. Show that

$$\{\alpha : f|X_\alpha(S) \text{ is an isomorphism onto } X_\alpha(S_1)\}$$

is closed unbounded.
(c) Show that if $S \triangle S_1$ is stationary, that $X(S) \not\cong X(S_1)$.
(d) Conclude that there are $2^{\aleph_1}$ non-isomorphic ω_1-like dense linear orders.

Recall that an abelian group is is $\aleph_1$-free if every countable subgroup is free. We describe the construction of abelian groups of size $\aleph_1$ that are $\aleph_1$-free but not free. Throughout this appendix all groups are abelian. We constantly use the basic algebraic fact that a subgroup of a free abelian group is free.

We will need the following algebraic lemma.

LEMMA A.0.3. *There are free abelian groups* $K_1 \subset K_2 \subset \cdots \subset K_n \subset \cdots$, $K = \bigcup K_n$, *and* $F \supset K$ *such that*:

(i) F/K_m *is free for all* m, *but* F/K *is not*;
(ii) *each* K_m, F/K, K_{m+1}/K_m *is free abelian of rank* $\aleph_0$.

PROOF. Let G be the subgroup of $\mathbb{Q}$ generated by $\{\frac{1}{2^n} : n \in \omega\}$. Let $\widehat{F}$ be the free abelian group on generators $x_0, x_1, \ldots$, let $f : \widehat{F} \to G$ be the surjective homomorphism $x_i \mapsto \frac{1}{2^i}$ and let $\widehat{K}$ be the kernel of F. Then $\widehat{K}$ is the free abelian group on generators $\{x_0 - 2x_1, x_1 - 2x_2, \ldots\}$ and $\widehat{F}/\widehat{K} \cong G$ is not free.

Let $\widehat{K}_m$ be the free abelian group on $\{x_0 - 2x_1, \ldots, x_{m-1} - 2x_m\}$. Then $\widehat{F}/\widehat{K}_m$ is isomorphic to the free abelian group on $x_m, x_{m+1}, \ldots$.

The groups we have constructed satisfy (i), but not (ii). It is easy to make (ii) true by adding in large free factors. Here are the details. Let H_i be a free abelian group on $\aleph_0$ generators.

Let

$$K_m = \widehat{K}_m \oplus \oplus_{i=0}^m H_i,$$

let

$$K = \bigcup_{m=0}^{\infty} K_m = \widehat{K} \oplus \oplus_{i=0}^{\infty} H_i$$

and let $F = \widehat{F} \oplus \oplus_{i=0}^{\infty} H_i \oplus H$. These groups have the desired properties. $\dashv$

THEOREM A.0.4. *There are* $2^{\aleph_1}$ *nonisomorphic* $\aleph_1$*-free groups of cardinality* $\aleph_1$.

PROOF. Let $S \subset \omega_1$ be a set of limit ordinals. We build a group $G(S)$ as a union of a continuous chain of countable free abelian subgroups

$$G_0(S) \subset G_1(S) \subset \ldots G_\alpha(S) \subset \cdots$$

for $\alpha < \omega_1$. We will arrange the construction such that

$(*)$ it $\beta + 1 < \alpha$ then $G_\alpha(S)/G_{\beta+1}(S)$ is free abelian of infinite rank.

Let $G_0(S) = \mathbb{Z}$. If α is a limit ordinal, let $G_\alpha(S) = \bigcup_{\beta<\alpha} G_\beta(S)$. Choose $\gamma_0 < \gamma_1 < \cdots$ with $\sup \gamma_n = \alpha$ such that each γ_n is a successor ordinal. By $(*)$ $G_{\gamma_{n+1}}(S)$ is free over $G_{\gamma_n}(S)$. Thus $G_\alpha(S)$ is free abelian. It is easy to see that $(*)$ holds.

If $\alpha \notin S$, then $G_{\alpha+1}(S) = G_\alpha(S) \oplus H$, where H is free abelian on $\aleph_0$-generators. Clearly $(*)$ still holds.

Suppose $\alpha \in S$. Choose $\gamma_0 < \gamma_1 < \cdots$ successor ordinals with $\alpha = \sup \gamma_n$.

Let $F, K, K_0, K_1, \ldots$ be as in the Lemma. Since $G_{\gamma_0}(S)$ and K_0 are both free abelian on $\aleph_0$-generators, we can find an isomorphism $\phi_0 : G_{\gamma_0} \to K_0$. Since $G_{\gamma_{n+1}}(S)$ is free over $G_\gamma(S)$ of rank $\aleph_0$ and K_{n+1}/K_n is free of rank $\aleph_0$, we can extend ϕ_n to an isomorphism $\phi_{n+1} : G_{\gamma_n}(S) \to K_n$. Then $\phi = \bigcup \phi_n$ is an isomorphism from $G_\alpha(S)$ to K. We then define $G_{\alpha+1}(S) \supset G_\alpha(S)$ such that ϕ extends to an isomorphism from $G_{\alpha+1}(S) \to F$. Thus we have

$$(**) \qquad \text{if } \alpha \in S, \text{ then } G_{\alpha+1}(S)/G_\alpha(S) \text{ is not free.}$$

On the other hand if $\beta + 1 < \alpha + 1$, there is an n such that $\beta + 1 < \gamma_n$. Then $G_{\gamma_n}(S)$ is free over $G_{\beta+1}(S)$ and, by construction, $G_{\alpha+1}(S)$ is free over $G_{\gamma_n}(S)$. Thus $G_{\alpha+1}(S)$ is free over $G_{\beta+1}(S)$ and $(*)$ holds.

This concludes the construction. Let $G(S) = \bigcup \alpha < \omega_1 G_\alpha(S)$. Every countable subgroup $H \subset G(S)$ is contained in $G_\alpha(S)$ for some α and $G_\alpha(S)$ is free abelian. Thus $G(S)$ is $\aleph_1$-free.

Suppose $S \triangle S_1$ is stationary. We claim that $G(S) \not\cong G(S_1)$. Suppose, for contradiction, that f is an isomorphism and assume, without loss of generality, that $S \setminus S_1$ is stationary.

EXERCISE A.0.5. (a) Show there is $\alpha \in S \setminus S_1$ such the $f|G_\alpha(S)$ is an isomorphism onto $G_\alpha(S_1)$.

(b) Argue that $G_{\alpha+1}(S)/G_\alpha(S)$ is not free, while $H/G_\alpha(S_1)$ is free for all countable subgroups H with $G_\alpha(S_1) \subset H \subset G(S_1)$.

(c) Conclude that $G(S) \not\cong G(S_1)$ and that we can find $2^{\aleph_1}$ non-isomorphic $\aleph_1$-free abelian groups of cardinality $\aleph_1$.

(d) Show that we can also ensure all $G(S)$ are strongly $\aleph_1$-free. (See Exercise 2.4.9.) ⊣

Appendix B

ADMISSIBLILITY

Originally, I had not planned on including material on admissible sets and the model theory of admissible fragments. In the end, I felt that, as the material fits well both with the model theory and generalized recursion developed above, it would be worthwhile adding an appendix with a brief introduction. In particular, I wanted to give a model theoretic proof of Sacks' Theorem that a countable ordinal $\alpha > \omega$ is admissible if and only $\alpha = \omega_1^x$ for some $x \subseteq \omega$.

Throughout this chapter we will assume familiarity with basic set theory up through Gödel's constructible universe $\mathbb{L}$ and refer the reader to [46] and [30]. Many details will be left to the reader.

B.1. Kripke–Platek set theory

Recall that the collection of Δ_0-*formulas* is the smallest collections of formulas in the language of set theory containing the atomic formulas closed under Boolean connectives and bounded quantifiers $\exists x \in y$ and $\forall x \in y$. While $\phi(\overline{v})$ is a Σ_1-*formula* if it is of the form

$$\exists \overline{u}\, \psi(\overline{u}, \overline{v})$$

where $\psi(\overline{u}, \overline{v})$ is Δ_0. For T a theory we say that $\phi(\overline{v})$ is Δ_0^T if, in T, it is provably equivalent to a Δ_0-formula. We define Σ_1^T analogously. We say that ϕ is Δ_1^T if both ϕ and $\neg\phi$ are Σ_1^T. Similarly, we say that ϕ is Δ_1 on a structure $\mathcal{M}$ if both $\phi(\mathcal{M})$ and $\neg\phi(\mathcal{M})$ are definable by Σ_1-formulas.

Kripke–Platek set theory KP is the fragment of ZFC where we take the usual axioms of extensionality, pairing and union together with the following schema:

- (Foundation schema) $\forall \overline{a}\left[\exists x \phi(x, \overline{a}) \rightarrow \exists y(\phi(y, \overline{a}) \wedge \forall z \in y\, \neg\phi(z, \overline{a}))\right]$, for all formulas $\phi(v, \overline{w})$;
- (Δ_0-Separation) $\forall \overline{a} \forall x \exists y \forall z \left[z \in y \leftrightarrow (z \in x \wedge \phi(z, \overline{a}))\right]$, for all Δ_0-formulas $\phi(v, \overline{w})$;
- (Δ_0-Collection) $\forall \overline{a}\left[(\forall y \in x \exists z \phi(y, z, \overline{a})) \rightarrow \exists u \forall y \in x \exists z \in u \phi(y, z, \overline{a})\right]$, for all Δ_0-formulas $\phi(v, \overline{w})$.

If κ is an infinite cardinal, then $H(\kappa)$ the collection of sets of hereditary cardinality $< \kappa$ is a model of KP. For example, HF, the hereditarily finite sets, and HC, the hereditarily countable sets are models of KP. If $\kappa \geq \aleph_0$ is a cardinal, then $\mathbb{V}_\kappa$ is also a model of KP. We will also consider the theory KPI which is KP together with the axiom of infinity.

EXERCISE B.1.1. (i) Let $\phi(x, \overline{v}, \overline{w})$ be a Δ_0-formula show that

$$\text{KP} \vdash \forall x \in A \exists \overline{w}\, \phi(x, \overline{v}, \overline{w}) \leftrightarrow \exists y \forall x \in A \exists \overline{w} \in y\, \phi(x, \overline{v}, \overline{w}).$$

(ii) Prove that the set of Σ_1^{KP} formulas is closed under positive Boolean combinations, bounded quantification, and existential quantification.

EXERCISE B.1.2 (Δ_1-Separation). Suppose $\phi(v, \overline{u})$ and $\psi(v, \overline{w})$ are Σ_1-formulas, $\mathcal{M} \models \text{KP}$ and

$$\mathcal{M} \models \forall v \in a\, (\phi(v, \overline{b}) \leftrightarrow \neg\psi(v, \overline{c})),$$

then there is $X \in \mathcal{M}$ such that

$$\mathcal{M} \models \forall v \in a (v \in X \leftrightarrow \phi(v, \overline{b})).$$

EXERCISE B.1.3 (Σ_1-Collection). Show that

$$\text{KP} \models \forall \overline{a}\, \left[(\forall y \in x \exists z\, \phi(y, z, \overline{a})) \to \exists u \forall y \in x \exists z \in u\, \phi(y, z, \overline{a})\right]$$

for all Σ_1-formulas $\phi(u, v, \overline{w})$. [Hint: code y and the witness to the existential quantifier together.]

EXERCISE B.1.4. Show that we can prove the existence of transitive closures in KP. [Hint: For KPI, show that for any x there is a function $f : \omega \to \mathcal{M}$ such that $f(0) = x$ and $f(n+1) = f(n) \cup \bigcup_{y \in f(n)} y$ and $\text{TC}(x) = \bigcup_{n \in \omega} f(n)$. A simpler argument will work in KP $+ \neg$I.]

More generally, in KP we can do arbitrary Σ_1 recursive definitions.

EXERCISE B.1.5. If $\mathcal{M} \models \text{KP}$ and $G : \mathcal{M} \times \mathcal{M} \to \mathcal{M}$ is Σ_1, then there is a Δ_1 function $F : \mathcal{M} \to \mathcal{M}$ such that

$$F(x) = G(x, F|\text{TC}(x)).$$

Using recursive definitions we can prove many important closure properties.

EXERCISE B.1.6. Let $\mathcal{M} \models \text{KP}$ and let $On^{\mathcal{M}}$ denote the ordinals of $\mathcal{M}$. Prove that there is a Δ_1-function $r : \mathcal{M} \to On^{\mathcal{M}}$ such that

$$r(x) = \sup\{r(y) + 1 : y \in x\}.$$

EXERCISE B.1.7. (a) If $\mathcal{M} \models \text{KPI}$, then the map $\alpha \mapsto \mathbb{L}_\alpha^{\mathcal{M}}$ with domain $On^{\mathcal{M}}$ is a Δ_1-function on $\mathcal{M}$.

(b) Show that if $\mathcal{M} \models \text{KPI}$, then $\mathbb{L}^{\mathcal{M}} \models \text{KPI} + \mathbb{V} = \mathbb{L}$.

If $\mathcal{M} \models \mathrm{KP} + x$ is an ordinal, then, in nonstandard models, it is possible that x is not well ordered. There will be an initial segment of ordinals of $\mathcal{M}$ that are well ordered, we let $o(\mathcal{M})$ denote the order type of well founded ordinals of $\mathcal{M}$. If $\mathcal{M}$ is not an ω-model or the axiom of infinity fails in $\mathcal{M}$, then $o(\mathcal{M}) = \omega$.

More generally, for $\mathcal{M} \models \mathrm{KP}$ we can define a well-founded part of $\mathcal{M}$

$$\mathrm{wf}(\mathcal{M}) = \{x \in \mathcal{M} : r(x) < o(\mathcal{M})\}$$

i.e., $\mathrm{wf}(\mathcal{M})$ are the elements of $\mathcal{M}$ such that $\in$ is truly well founded on $\mathrm{TC}(x)$.

EXERCISE B.1.8. (i) If $a \in \mathrm{wf}(\mathcal{M}), b \in \mathcal{M}$ and $\mathcal{M} \models b \in a$, then $b \in \mathrm{wf}(\mathcal{M})$. Thus $\mathcal{M}$ is an end extension of $wf(\mathcal{M})$.

(ii) If $\phi(\overline{v})$ is a Δ_0 formula, $\overline{a} \in \mathrm{wf}(\mathcal{M})$ and

$$M \models \phi(\overline{a}) \Leftrightarrow \mathrm{wf}(\mathcal{M}) \models \phi(\overline{a}).$$

We will show that $\mathrm{wf}(\mathcal{M}) \models \mathrm{KP}$. The only interesting step is the next lemma.

LEMMA B.1.9. *If* $\mathcal{M} \models \mathrm{KP}$, *then* $\mathrm{wf}(\mathcal{M}) \models \Delta_0$*-collection.*

PROOF. If $\mathcal{M} = \mathrm{wf}(\mathcal{M})$, this is clear. Otherwise, $On^{\mathcal{M}}$ contains nonstandard ordinals. Suppose $\phi(v, w)$ is a Δ_0-formula (we suppress parameters) and

$$\mathrm{wf}(\mathcal{M}) \models \forall x \in a \exists y\, \phi(x, y).$$

Then

$$\mathcal{M} \models \forall x \in a \exists y\, \phi(x, y).$$

If α is any nonstandard ordinal then

$$\mathcal{M} \models \forall x \in a \exists y\, (\phi(x, y) \wedge r(y) < \alpha).$$

By Foundation, there is least ordinal $\alpha \in On^{\mathcal{M}}$ making the above true. Since there is no minimal nonstandard ordinal, $\alpha < o(\mathcal{M})$. Thus

$$\mathcal{M} \models \forall x \in a \exists y\, (\phi(x, y) \wedge r(y) < \alpha)$$

and by Σ_1-Collection there is $c \in \mathcal{M}$ such that

$$\mathcal{M} \models \forall x \in a \exists y \in c\, (\phi(x, y) \wedge r(y) < \alpha).$$

Let $b = \{y \in c : r(y) < \alpha\}$, $b \in \mathcal{M}$ by Δ_1-Separation and $b \in \mathrm{wf}(\mathcal{M})$ and

$$\mathrm{wf}(\mathcal{M}) \models \forall x \in a \exists y \in b\, \phi(x, y). \qquad \dashv$$

EXERCISE B.1.10. Show that if $\mathcal{M} \models \mathrm{KP}$, then so does $\mathrm{wf}(\mathcal{M})$ and if $\mathcal{M}$ is an ω-model of KPI, then so is $\mathrm{wf}(\mathcal{M})$.

DEFINITION B.1.11. An *admissible set* is a transitive set $\mathcal{A}$ such that $(\mathcal{A}, \varepsilon) \models \mathrm{KP}$.

EXERCISE B.1.12. If $\mathcal{M} \models \mathrm{KP}$, then $\mathrm{wf}(\mathcal{M})$ is isomorphic to an admissible set. [Hint: Argue that $\mathrm{wf}(\mathcal{M})$ is well founded, thus we can take the Mostowski collapse to get an isomorphic transitive model.]

B.2. Admissible sets and HYP

EXERCISE B.2.1. (a) Suppose $\mathcal{M}$ is an ω-model of KPI, $T \subset \omega^{<\omega}$ is a well founded tree and $T \in \mathcal{M}$. Show that $\rho(T) \in \mathcal{M}$. Conclude that $o(\mathcal{M}) \geq \omega_1^{\mathrm{ck}}$.
(b) Show that if $\mathcal{M}$ is an ω-model of KPI, then HYP $\subset \mathcal{M}$.

LEMMA B.2.2. *If $x \notin$ HYP, then there is an ω-model $\mathcal{M} \models$ KPI with $x \notin \mathcal{M}$.*

PROOF. We will use the Omitting Types Theorem.[22] Let $\mathbb{A}$ be the smallest fragment of $\mathcal{L}_{\omega_1,\omega}$ containing the axioms for KPI + the infinitary sentence asserting that we have an ω-model. Every formula in $\mathbb{A}$ is recursive. Let $\theta(v)$ be the formula

$$v \not\subseteq \omega \vee \bigvee_{n\in x} n \notin v \vee \bigvee_{n\in\omega\setminus x} n \in x.$$

Suppose $\psi(v) \in \mathbb{A}$ and there are ω-models of KPI + $\exists v\, \psi(v)$. To apply the Omitting Types Theorem, we must show that there are ω-models of KPI$+\exists v(\psi(v) \wedge \theta(v))$. Suppose not. But then

$$\begin{aligned} n \in x &\Leftrightarrow \exists \mathcal{M} \exists c \in \mathcal{M}\, (\mathcal{M} \text{ is an } \omega\text{-model of KPI} + \psi(c) \wedge n \in c) \\ &\Leftrightarrow \forall \mathcal{M} \forall c \in \mathcal{M}\, (\mathcal{M} \text{ is an } \omega\text{-model of KPI} + \psi(c) \rightarrow n \in c). \end{aligned}$$

"$\mathcal{M}$ is an ω-model of KPI$+\psi(c)$" is Δ^1_1. Thus the top condition is Σ^1_1 and the bottom condition is Π^1_1. Hence x is hyperarithmetic, a contradiction.

By the Omitting Types Theorem there is an ω-model $\mathcal{M} \models$ KPI$+\forall v\, \theta(v)$. Clearly $x \notin \mathcal{M}$. ⊣

COROLLARY B.2.3. *There is an ω-model $\mathcal{M} \models$ KPI such that $o(\mathcal{M}) = \omega_1^{\mathrm{ck}}$.*

PROOF. Let $\mathcal{M}$ be an ω-model of KPI such that $O \notin \mathcal{M}$. By Exercise B.2.1, $\omega_1^{\mathrm{ck}} \leq o(\mathcal{M})$, so we assume $\omega_1^{\mathrm{ck}} < o(\mathcal{M})$ and $\omega_1^{\mathrm{ck}} \in \mathcal{M}$. But then, by Exercise B.1.7, $\mathbb{L}_{\omega_1^{\mathrm{ck}}} \in \mathcal{M}$. If $T \subseteq \omega^{<\omega}$ is a recursive tree, then

$$\begin{aligned} &T \in \mathrm{WF} \Leftrightarrow \rho(T) < \omega_1^{\mathrm{ck}} \Leftrightarrow \\ &\qquad \exists f \in \mathbb{L}_{\omega_1^{\mathrm{ck}}}\, f : T \rightarrow \omega_1^{\mathrm{ck}} \wedge \forall \sigma \subset \tau \in T\, f(\sigma) > f(\tau). \end{aligned}$$

But this give rise to a Δ_1-definition of O. Thus $O \in \mathcal{M}$, a contradiction. ⊣

COROLLARY B.2.4. (i) *$\mathbb{L}_{\omega_1^{\mathrm{ck}}}$ is admissible.*
(ii) *If $\mathcal{A} \models$ KPI is admissible, then $\mathbb{L}_{\omega_1^{\mathrm{ck}}} \subseteq \mathcal{A}$.*
(iii) HYP $= \omega^\omega \cap \mathbb{L}_{\omega_1^{\mathrm{ck}}}$.

PROOF. (i) We can find an ω-model $\mathcal{M}$ of KPI with $o(\mathcal{M}) = \omega_1^{\mathrm{ck}}$. Let $\mathcal{A}$ be the transitive collapse of wf$(\mathcal{M})$. Then $\mathbb{L}^{\mathcal{A}}$ is admissible and, since $o(\mathcal{A}) = \omega_1^{\mathrm{ck}}$, $\mathbb{L}^{\mathcal{A}} = \mathbb{L}_{\omega_1^{\mathrm{ck}}}$.

[22] Here we could use the first order Omitting Types Theorem by also omitting the type of a nonstandard integer.

(ii) If $\mathcal{A} \models \mathrm{KPI}$ is admissible, then $\mathbb{L}_{\omega_1^{\mathrm{ck}}} \subseteq \mathbb{L}_{o(\mathcal{A})} \subseteq \mathcal{A}$.
(iii) By Exercise B.2.1, HYP$\subseteq \mathbb{L}_{\omega_1^{\mathrm{ck}}}$. Suppose $x \notin$ HYP. By Lemma B.2.2, there is an ω-model $\mathcal{M}$ of KPI with $x \notin \mathcal{M}$. Then $x \notin \mathbb{L}_{o(\mathcal{M})}$ and $o(\mathcal{M}) \geq \omega_1^{\mathrm{ck}}$. Thus $x \notin \mathbb{L}_{\omega_1^{\mathrm{ck}}}$. $\dashv$

EXERCISE B.2.5. Let $T \subseteq \omega^{<\omega}$ be a recursive ill-founded tree with no hyperarithmetic paths (see Lemma 10.3.5). Argue that $\mathbb{L}_{\omega_1^{\mathrm{ck}}} \models T$ is well founded. Conclude that, in contrast to Exercise B.2.1, that KP $\not\models$ Every well founded tree has an ordinal rank.

EXERCISE B.2.6. Relativize earlier arguments to show that $\mathbb{L}_{\omega_1^x}[x] \models \mathrm{KP}$ for any $x \subseteq \omega$ and that $y \leq_{\mathrm{hyp}} x$ if and only if $y \in \omega^\omega \cap \mathbb{L}_{\omega_1^x}$.

DEFINITION B.2.7. We say that α is admissible if there is an admissible set $\mathcal{A}$ with $\alpha = o(\mathcal{A})$.

We have shown that ω_1^x is *admissible* for all $x \subseteq \omega$. In Theorem B.3.11 we will show that every countable admissible $\alpha > \omega$ is of this form.

EXERCISE B.2.8. Let $X \subseteq$ HYP.

(a) Suppose X is Π^1_1. Show that X is Σ_1 on $\mathbb{L}_{\omega_1^{\mathrm{ck}}}$. [Hint: Show that if $x \in$ HYP, a tree T recursive in x is well-founded if and only in there is $\alpha \in On$ and $f : T \to \alpha$ order preserving with $f \in \mathbb{L}$.]
(b) Prove the converse. [Hint: Let $\phi(v)$ be a Σ_1-formula. Show $x \in X$ if and only if for any countable ω-model $\mathcal{M}$ of KPI with $x \in \mathcal{M}$, $\mathcal{M} \models \phi(x)$ and show that this condition is Π^1_1.]

B.3. Model theory of admissible fragments

We fix a natural coding of $\mathcal{L}_{\infty,\omega}$-formulas as set theoretic objects. For example, we might code $\bigvee_{\theta \in X} \theta$ as the ordered pair$\langle 3, X\rangle$.[23]

If $\mathcal{A}$ is an admissible set, then we let $\mathcal{L}_\mathcal{A}$ be the $\mathcal{L}_{\infty,\omega}$-formulas coded in $\mathcal{A}$. In the special case when $\mathcal{A}$ is HC, the hereditarily countable sets, $\mathcal{L}_\mathcal{A}$ is just $\mathcal{L}_{\omega_1,\omega}$.

Much like proofs of compactness from the completeness in first order logic, compactness in admissible fragments relies on a notion of provability. We had carefully avoided provability in the main text, but begin by giving a quick sketch of provability in $\mathcal{L}_{\omega_1,\omega}$.

[23]One key point is that we do not necessarily assume we have a listing $(\theta_\alpha : \alpha < \beta)$ of the elements of X.

Provability in $\mathcal{L}_{\omega_1,\omega}$. We inductively define $\vdash \phi$ for ϕ an $\mathcal{L}_{\omega_1,\omega}$-formula. In our proof system we have the following *axioms*:

1. $\vdash \phi$, where ϕ is an instance of a propositional tautology;
2. $\vdash \neg\phi \leftrightarrow \sim\phi$;
3. $\vdash \bigwedge_{\phi\in X} \phi \to \phi$, for each $\phi \in X$;
4. $\vdash \forall x\phi(x) \to \phi(t)$, where t is a term and $\phi(t)$ is obtained by replacing all free occurrences of x in ϕ by t;
5. $\vdash \phi(x) \wedge x = t \to \phi(t)$, where t and $\phi(t)$ are as above;
6. $\vdash x = x$;
7. $\vdash x = y \to y = x$.

In addition to the axioms we have three *inference rules*.

1. if $\vdash \phi$ and $\vdash \phi \to \psi$, then $\vdash \psi$;
2. if $\vdash \phi \to \psi$ and v is not free in ϕ, then $\vdash \phi \to \forall x\psi$;
3. if $\vdash \phi \to \psi$ for all $\psi \in X$, then $\vdash \phi \to \bigwedge_{\psi\in X} \psi$.

As in first order logic, the notion of provability completely captures the notion of validity.

THEOREM B.3.1 (Karp's Completeness Theorem). *Let ϕ be an $\mathcal{L}_{\omega_1,\omega}(\tau)$-sentence. Then $\vdash \phi$ if and only if ϕ is true in all τ-structures.*

PROOF. (Sketch) First show soundness by showing by induction on proofs that if $\vdash \phi$, then $\models \phi$. For the other direction, add a countable set of new constants C to our vocabulary. Let Σ be the collection of finite sets σ of $\mathcal{L}_{\omega_1,\omega}$-sentences such that only finitely many constants from C occur in σ and $\nvdash \neg \bigwedge_{\phi\in\sigma}$.

We claim that Σ is a consistency property. Once we have done that the Model Existence Theorem will allow us to conclude that if $\nvdash \phi$, then there is $\mathcal{M} \models \neg\phi$.

We check (C4). Suppose $\sigma \in \Sigma$ and $\bigvee_{\theta\in X} \theta \in \sigma$. Let $\Phi = \bigwedge_{\phi\in\sigma} \phi$. Suppose for contradiction that $\sigma \cup \{\theta\} \notin \Sigma$ for any $\theta \in X$. Then $\vdash \Phi \to \neg\theta$, for all $\theta \in X$. But then, using inference rule 3)

$$\vdash \Phi \to \bigwedge_{\theta\in X} \neg\theta.$$

But $\vdash \Phi \to \bigvee_{\theta\in X} \theta$ by axiom 3. But then, using axiom 1 and 2 and inference rule 1, we can conclude $\vdash \neg\Phi$, a contradiction since $\sigma \in \Sigma$. ⊣

EXERCISE B.3.2. Finish the proof that this a consistency property. [You should be a little careful as some of the other conditions require some effort. The full proof is in chapter 4 of [37].]

In first order logic, the Completeness Theorem tells us that the infinitary notion of "validity" is equivalent to the finitary notion of "proof". Important consequences include the compactness theorem and the fact that collection of vaild sentences is recursively enumerable. In $\mathcal{L}_{\omega_1,\omega}$ proofs are infinitary.[24] We know compactness fails and it is not clear what sort of effectiveness we might expect. Barwise showed that both of these features can be partially recovered when we look at admissible fragments.

Barwise Completeness.

THEOREM B.3.3 (Barwise Completeness). *Let $\mathcal{A}$ be an admissible set such that $\mathcal{A} \subseteq$ HC. The set of valid $\mathcal{L}_{\mathcal{A}}$ formulas in Σ_1 on $\mathcal{A}$.*

PROOF. (sketch) We will show that there is a notion $\vdash_{\mathcal{L}_{\mathcal{A}}}$ which is Σ_1 on $\mathcal{A}$ such that an $\mathcal{L}_{\mathcal{A}}$-sentence ϕ is valid if and only if $\vdash_{\mathcal{L}_{\mathcal{A}}} \phi$. We use Σ_1-recursion to construct a Δ_1-definable $P : On^{\mathcal{A}} \times \mathcal{A} \to \mathcal{A}$ where, intuitively, $\phi \in P(\alpha, A)$ if and only if there is a proof of height at most α of ϕ using formulas in A.

We inductively define $P(\alpha, A)$ to be the set of all $\phi \in x \cap \mathcal{L}_{\mathcal{A}}$ such that

- ϕ is an axiom for $\mathcal{L}_{\omega_1,\omega}$;
- $\exists\beta < \alpha\exists\psi \in A\,[\theta, \theta \to \phi \in P(\beta, A)]$;
- $\exists\beta < \alpha\exists\psi, \theta \in A\,[\phi$ is $\psi \to \forall v\theta \wedge v$ is not free in ψ and $\psi \to \theta \in P(\beta, A)]$;
- $\exists Y \in X\exists\psi \in A[\phi$ is $\psi \to \bigwedge_{\theta\in Y}\theta \wedge \forall\theta \in Y\exists\beta < \alpha\,\psi \to \theta \in P(\beta, A)]$.

Note that if $P(\alpha, A)$ and $A \subseteq B$, then $P(\alpha, B)$.

We then define $\vdash_{\mathcal{L}_{\mathcal{A}}} \phi$ to be

$$\exists\alpha\exists A\,(\alpha \in On \wedge \phi \in P(\alpha, A)).$$

Since P is Δ_1 on $\mathcal{A}$, $\vdash_{\mathcal{L}_{\mathcal{A}}}$ is Σ_1-definable. Clearly if $\vdash_{\mathcal{L}_{\mathcal{A}}} \phi$, then $\vdash \phi$ and $\models \phi$ so this proof system is sound. It suffices to show that if $\nvdash_{\mathcal{L}_{\mathcal{A}}} \neg\phi$, then there is $\mathcal{M} \models \phi$. We will use the Model Existence Theorem. Let C be a countable set of new constant symbols that we add to our vocabulary. Let Γ be the collection of all finite sets of $\mathcal{A}$-sentences using only finitely many constants from C such that

$$\nvdash_{\mathcal{L}_{\mathcal{A}}} \neg \bigwedge_{\phi\in\sigma} \phi.$$

It suffices to show that Γ is a consistency property.

We verify (C4). Suppose $\bigvee_{\theta\in X}\theta \in \sigma$ and $\sigma \in \Gamma$. Suppose, for contradiction, that $\sigma \cup \{\theta\} \notin \Gamma$ for all $\theta \in X$. Let Φ be $\bigwedge_{\phi\in\sigma}\phi$. Then $\vdash_{\mathcal{L}_{\mathcal{A}}} \Phi \to \neg\theta$, for all $\theta \in X$. Thus

$$\forall\theta \in X\exists\alpha\exists A\,\Phi \to \theta \in P(\alpha, A).$$

By Σ_1-Collection, we can find $\delta \in On^{\mathcal{A}}$ and $B \in \mathcal{A}$ such that

$$\forall\theta \in X\exists\alpha < \delta\exists A \in B\,\text{“}\Phi \to \theta\text{”} \in P(\alpha, A).$$

Let $\widehat{A} = \bigcup_{b\in B} b$. Then $\Phi \to \bigwedge_{\theta\in Y}\neg\theta \in P(\delta, C)$. Thus

$$\vdash_{\mathcal{L}_{\mathcal{A}}} \Phi \to \bigwedge_{\theta\in Y}\neg\theta.$$

---

[24] It would be natural to code them with well-founded trees.

Arguing as in the proof of the Completeness Theorem, we can then conclude that $\vdash_{\mathcal{L}_{\mathcal{A}}} \neg\Phi$, a contradiction since $\sigma \in \Gamma$. $\dashv$

EXERCISE B.3.4. Finish the proof that Γ is a consistency property.

COROLLARY B.3.5. *The set of valid $\mathcal{L}_{\omega_1,\omega}$-sentences is Δ_1 on* HC.

PROOF. ϕ is not valid if and only if there is a countable model of $\neg\phi$. Thus nonvalidity is also Σ_1 on HC. $\dashv$

The following more general version of the Completeness Theorem is also true.

EXERCISE B.3.6. Let $\mathcal{A}$ be a countable admissible set. Let $T \subset \mathcal{L}_{\mathcal{A}}$ be a set of $\mathcal{L}_{\mathcal{A}}$-sentences Σ_1 on $\mathcal{A}$. Show that

$$\{\phi : \phi \text{ an } \mathcal{L}_{\mathcal{A}}\text{-sentence and } T \models \phi\}$$

is Σ_1 on $\mathcal{A}$.

Barwise Compactness.

THEOREM B.3.7 (Barwise Compactness). *Suppose $\mathcal{A}$ is a countable admissible set, $T \subset \mathcal{L}_{\mathcal{A}}$ is Σ_1 on $\mathcal{A}$ and every $T_0 \subseteq T$ with $T_0 \in \mathcal{A}$ is satisfiable, then T is satisfiable.*

PROOF. Add a countable new set of constants C to the vocabulary. Let Γ be the collection of all sets of sentences σ that are Σ_1 on $\mathcal{A}$, use only finitely many constants from C and if $\gamma \subseteq \sigma$ and $\gamma \in \mathcal{A}$, then γ is satisfiable.

We will show that Γ is a consistency property. Barwise Compactness will then follow from the Model Existence Theorem.[25]. As usual, the only difficult aspect is the proof of (C4). Suppose $\sigma \in \Gamma$ and $\bigvee_{\theta \in X} \theta \in \sigma$. Suppose, for contradiction, that $\sigma \cup \{\theta\} \notin \Gamma$ for all $\theta \in X$. Thus for all $\theta \in X$ there is $X_\theta \in \mathcal{A}$ such that

$$\bigwedge_{\psi \in X_\theta} \models \neg\theta,$$

$$\mathcal{A} \models \forall\theta \in X \exists y \left[y \subseteq \sigma \wedge \vdash_{\mathcal{L}_{\mathcal{A}}} \bigwedge_{\phi \in y} \phi \rightarrow \theta \right].$$

Let $\psi(v, w)$ be a Δ_0-formula, possibly with suppressed parameters, such that $y \subset \sigma$ if and only if $\mathcal{A} \models \exists w\, \psi(y, w)$. By Σ_1-collection there is A such that for all $\theta \in X$ we can find $y, w \in A$ such that

$$\psi(y, w) \wedge \vdash_{\mathcal{L}_{\mathcal{A}}} \bigwedge_{\phi \in y} \phi \rightarrow \theta.$$

Let $B = \{y \in A : \exists w \in A\, \psi(y, w)\}$. Then $B \in \mathcal{A}$ by Δ_0-collection. Let $D = \bigcup_{y \in B} y$. Then $D \in \mathcal{A}$, $D \subset \sigma$ and $\theta \in X$ there is $y \subseteq D$ such that $\bigwedge_{\phi \in y} \phi \models \neg\theta$. Hence $D \models \neg \bigvee_{\theta \in X} \theta$ and $\sigma \notin \Gamma$, a contradiction. $\dashv$

[25]The countability of $\mathcal{A}$ is needed to apply the Model Existence Theorem.

EXERCISE B.3.8. Show that the Kreisel–Barwise Compactness Theorem (Theorem 11.2.2) is a special case of Barwise Compactness.

DEFINITION B.3.9. Let $\mathcal{A}$ be an admissible set. We say that $\mathcal{M}$ is $\Sigma_{\mathcal{A}}$-*compact* if whenever $p(v, \overline{w})$ is set of $\mathcal{L}_{\mathcal{A}}$-formulas that is Σ_1 on $\mathcal{A}$, $\overline{a} \in \mathcal{M}$ and $\mathcal{M}$ realizes $p_0(v, \overline{a})$ for all $p_0 \subseteq p$ with $p_0 \in \mathcal{A}$, then $\mathcal{M}$ realizes $p(v, \overline{a})$.

EXERCISE B.3.10. Let $\mathcal{A}$, T, C and Γ be as the statement and proof of the Barwise Compactness Theorem.

(a) Show that if $\sigma \in \Gamma$ and $p(v, \overline{w})$ is Σ_1 on $\mathcal{A}$ and $\overline{c} \in C$, then either there is $p_0 \subseteq p$ such that $p_0 \in \mathcal{A}$ and
$$\sigma \cup \left\{ \neg \exists v \bigwedge_{\phi \in p_0} \phi(v, \overline{c}) \right\} \in \Gamma$$
or
$$\sigma \cup p(d, \overline{c}) \in \Gamma$$
for some $d \in C$.

(b) Show that there is a $\Sigma_{\mathcal{A}}$-compact model $\mathcal{M} \models T$. [Hint: Modify the proof of the Model Existence Theorem using (a).]

Conclude that any satisfiable theory that is Σ_1 on a countable admissible set $\mathcal{A}$ has a $\Sigma_{\mathcal{A}}$-compact model.

$\Sigma_{\mathcal{A}}$-compact models can be used to give an easy proof of this key characterization of countable admissible ordinals.

THEOREM B.3.11 (Sacks). *If $\alpha > \omega$ is a countable admissible ordinal, then there is $x \subseteq \omega$ with $\omega_1^x = \alpha$.*

PROOF. Let $\mathcal{A}$ be a countable admissible set with $o(\mathcal{A}) = \alpha$. Let $\tau = \{\in, d, c_\beta : \beta < \alpha\}$ and let T be the $\mathcal{L}_{\mathcal{A}}$-theory asserting:

(i) KPI;
(ii) c_β is an ordinal, for $\beta < \alpha$;
(iii) $\forall x \left(x < c_\beta \leftrightarrow \bigvee_{\gamma < \beta} x = c_\gamma\right)$ for $\beta < \alpha$;
(iv) $d \subseteq \omega$;
(v) there is an ordering of ω recursive in d isomorphic to $(c_\beta, <)$ for all $\beta < \alpha$.

The theory T is Σ_1 on $\mathcal{A}$. If $x \subseteq \omega$ and $\omega_1^x > \alpha$, then $\mathbb{L}_{\omega_1^x}[x] \models T$. Thus T is satisfiable.

Let $\mathcal{M} \models T$ be $\Sigma_{\mathcal{A}}$-compact. Suppose $b \in On^{\mathcal{M}}$ and $c_\beta < b$ for all $\beta < \alpha$. Consider the type $p(v, b)$ asserting
$$c_\beta < v < b$$
for all $\beta < \alpha$. Then p is Σ_1 on $\mathcal{A}$. Suppose $p_0 \subseteq p$ and $p \in \mathcal{A}$. Then $p_0 \subseteq \{v < b\} \cup \{c_\gamma < v : \gamma < \delta\}$ for some $\delta < \alpha$ and p_0 is realized in $\mathcal{M}$ by c_δ. Thus p is realized in $\mathcal{M}$. It follows that every element of $On^{\mathcal{M}} \setminus \alpha$ is

ill-founded. Hence $o(\mathcal{M}) = \alpha$. Let $\mathcal{B}$ be the transitive collapse of $\mathrm{wf}(\mathcal{M})$. Then $o(\mathcal{B}) = \alpha$. Let $x \in \mathcal{B}$ be the image of $d^{\mathcal{M}}$ under the transitive collapse. Every $\beta < \alpha$ is recursive in x. Thus $\omega_1^x = \alpha$. ⊣

Sacks' original proof [66] was by forcing using perfect trees. Friedman and Jensen [19] gave a model theoretic proof. A proof using the Omitting Types Theorem is given in [37].

$\Sigma_{\mathcal{A}}$-compact models can be thought of as a generalization of recursively saturated models from first order model theory. It might seem that "$\Sigma_{\mathcal{A}}$-compact" models should be better called "$\Sigma_{\mathcal{A}}$-saturated", but that name is reserved for a stronger property introduced by Ressayre [65].

DEFINITION B.3.12. We say that $\mathcal{M}$ is $\Sigma_{\mathcal{A}}$-*saturated.* If it is $\Sigma_{\mathcal{A}}$-compact and, in addition, if $\overline{a} \in \mathcal{M}$, $I \in \mathcal{A}$ and $p_i(\overline{x}, \overline{a})$ is a set of $\mathcal{L}_{\mathcal{A}}$ formulas for $i \in I$ such that

$$X = \{(i, \phi) : \phi \in p_i\}$$

is Σ_1 on $\mathcal{A}$ and for all $Y \subseteq X$ with $Y \in \mathcal{A}$

$$\mathcal{M} \models \exists \overline{x} \bigvee_{i \in I} p_i(\overline{x}, \overline{a}) \cap Y,$$

then

$$\mathcal{M} \models \exists \overline{x} \bigvee_{i \in I} p_i(\overline{x}, \overline{a}).$$

EXERCISE B.3.13. Modify the proof from Exercise B.3.10 to prove that any satisfiable theory that is Σ_1 on $\mathcal{A}$ has a $\Sigma_{\mathcal{A}}$-saturated model.

$\Sigma_{\mathcal{A}}$-saturated models are the right generalization of recursively saturated models. Though to make this precise we need to redo the basic theory working over structures with urelements (see [8]). The key result about $\Sigma_{\mathcal{A}}$-saturated models is the following theorem of Ressarye generalizing the fact that countable recursively saturated models are resplendent.

THEOREM B.3.14. *Let $\mathcal{M}$ be a countable $\Sigma_{\mathcal{A}}$-saturated τ-structure. Let $\tau^* \supset \tau$ and let T be an $\mathcal{L}_{\mathcal{A}}(\tau^*)$-theory that is Σ_1 on $\mathcal{A}$ and $\mathcal{M} \models \phi$ for all $\phi \in \mathcal{L}_{\mathcal{A}}(\tau)$ such that $T \models \phi$. Then we can expand $\mathcal{M}$ be interpreting the new symbols in τ^* and make $\mathcal{M}$ a model of T.*

We refer the interested reader to [65], [50] and [62] for proofs and more on $\Sigma_{\mathcal{A}}$-saturated models and their applications.

It is possible to refine some of our earlier results for $\mathcal{L}_{\omega_1,\omega}$ to obtain sharper results for countable admissible fragments. For example, the interpolation theorem holds for $\mathcal{L}_{\mathcal{A}}$ and the Hanf number for $\mathcal{L}_{\mathcal{A}}$ is $\beth_{o(\mathcal{A})}$. The reader should consult [8] or [37] for proofs. Barwise [8] also discusses what can be done in $\mathcal{L}_{\mathcal{A}}$ when $\mathcal{A}$ is not countable.

B.4. $C_1 = Q$

In Chapter 10 we introduced the sets

$$C_1 = \{x : \forall y\,(\omega_1^x \le \omega_1^y \to x \le_{\text{hyp}} y)\}$$

and

$$Q = \{x : x \in \mathbb{L}_{\omega_1^x}\}.$$

We also showed that C_1 is the largest thin Π^1_1-set. As an application of admissible model theory, we will now prove Mansfield's result [53] that $Q = C_1$.

EXERCISE B.4.1. Show that $x \in Q$ if and only if

$$\forall \mathcal{M}\,[(\mathcal{M} \text{ an } \omega\text{-model of KPI} \wedge x \in \mathcal{M}) \to \mathcal{M} \models x \in \mathbb{L}].$$

Conclude that Q is Π^1_1.

EXERCISE B.4.2. Prove that Q is thin. [Hint: Suppose not and T is a perfect tree such that $[T] \subseteq Q$. Collapse $\aleph_1^{\mathbb{L}}$ and note that Shoenfield absoluteness implies that

$$\forall x\,(x \in [T] \to x \in Q)$$

is absolute.]

Conclude that $Q \subseteq C_1$.

THEOREM B.4.3. $C_1 = Q$

PROOF. It suffices to show that $C_1 \subseteq Q$. Let $x \in C_1$ and let $\alpha = \omega_1^x$. Consider the language $\mathcal{L} = \{\varepsilon, c\}$ where c is a new constant symbol.

Let T be the $\mathcal{L}_{\mathcal{A}}$-theory asserting we are in an ω-model of

$$\text{KPI} + c \subseteq \omega + \{\phi_\beta : \beta < \alpha\}$$

where ϕ_β be a sentence asserting that for some $e \in \omega$, e is the index for a linear order recursive in c of order type β. The theory T is Σ_1 on $\mathbb{L}_\alpha$.

Suppose we could build a model $\mathcal{M}$ of T such that

$$\forall v \left[v \not\subseteq \omega \vee \bigvee_{n \in x} n \notin v \vee \bigvee_{n \notin x} n \in v \right].$$

If we can do this we let y be the interpretation of $c \in \mathcal{M}$. Then $\omega_1^x \le \omega_1^y$. Since $x \in C_1$, we must have $x \le_{\text{hyp}} y$ and $x \in \mathcal{M}$ a contradiction. Thus we can not build such a model.

By the Omitting Types Theorem, there must be a formula $\theta(v) \in \mathbb{A}$ such that

$$T + \theta(v) \models v \subset \omega \wedge \bigwedge_{n \in x} n \in v \wedge \bigwedge_{n \notin x} n \notin v.$$

Thus

$$n \in x \Leftrightarrow T + \theta(v) \models n \in v$$

and

$$n \notin x \Leftrightarrow T + \theta(v) \models n \notin v.$$

From Exercise B.3.6, it follows that $n \in x$ and $n \notin x$ are both Σ_1 on $\mathcal{L}_\alpha$. Since x is Δ_1 on $\mathbb{L}_\alpha$, by Δ^1_1-Separation $x \in \mathbb{L}_\alpha$. Thus $x \in Q$ as desired. ⊣

EXERCISE B.4.4. Show that any Π^1_1 set containing a non-constructible real contains a perfect set. Using uniformization, show the same is true for Σ^1_2-sets.

REFERENCES

[1] J. BALDWIN, *The Vaught Conjecture: Do uncountable models count?*, ***Notre Dame Journal of Formal Logic***, vol. 48 (2007), pp. 79–92.

[2] ———, ***Categoricity***, AMS University Lecture Series, AMS, 2009.

[3] ———, *Amalgamation, absoluteness, and categoricity*, ***Proceedings of the 11th Asian Logic Conference***, World Scientific Publishing, Hackensack, NJ, 2012, pp. 22–50.

[4] J. BALDWIN, S. FRIEDMAN, M. KOERWIEN, and M. LASKOWSKI, *Three red herrings around Vaught's Conjecture*, ***Transactions of the American Mathematical Society***, vol. 268 (2016), pp. 3673–3694.

[5] J. BALDWIN and P. LARSON, *Iterated elementary embeddings and the model theory of infinitary languages*, ***Annals of Pure and Applied Logic***, vol. 167 (2016), pp. 309–334.

[6] J. BARWISE, *The Syntax and Semantics of Infinitary Languages*, ***Lecture Notes in Mathematics***, vol. 72, Springer-Verlag, Berlin and New York, 1968, pp. 166–181.

[7] ———, *Back and forth through infinitary logics*, ***Studies in Model Theory*** (M. Morley, editor), MAA, 1973.

[8] ———, ***Admissible Sets and Structures***, Springer-Verlag, Berlin, 1975.

[9] J. BARWISE and S. FEFERMAN, ***Handbook of Model Theoretic Logics***, Springer-Verlag, Berlin, 1985.

[10] J. BAUMGARTNER, *The Hanf number for complete $\mathcal{L}_{\omega_1,\omega}$ sentences (without GCH)*, ***The Journal of Symbolic Logic***, vol. 39 (1974), pp. 575–578.

[11] M. BAYS, M. GAVRILOVICH, and M. HILS, *Some definability results in abstract Kummer theory*, ***International Mathematics Research Notices***, vol. 14 (2014), pp. 3975–4000.

[12] M. BAYS, B. HART, T. HYTTINEN, M. KESÄLÄ, and J. KIRBY, *Quasiminimal structures and excellence*, ***Bulletin of the London Mathematical Society***, vol. 46 (2014), pp. 155–163.

[13] M. BAYS, B. HART, and A. PILLAY, *Universal covers of commutative finite Morley rank groups*, preprint.

[14] M. BAYS and J. KIRBY, *Excellence and uncountable categoricity of Zilber's exponential fields*, preprint.

[15] M. Bays and B. Zilber, *Covers of multiplicative groups of algebraically closed fields of arbitrary characteristic*, ***Bulletin of the London Mathematical Society***, vol. 43 (2011), pp. 689–702.

[16] J. Burgess, *Equivalences generated by families of Borel sets*, ***Proceedings of the American Mathematical Society***, vol. 69 (1978), pp. 323–326.

[17] W. Calvert, J. Knight, and J. Millar, *Computable trees of Scott rank* ω_1^{ck} *and computable approximation*, ***The Journal of Symbolic Logic***, vol. 71 (2006), pp. 283–298.

[18] M. Dickmann, *Larger infinitary languages*, in [9].

[19] H. Friedman and R. Jensen, *A note on admissible ordinals*, in [6].

[20] S. Friedman, T. Hyttinen, and M. Koerwien, *The nonabsoluteness of model existence in uncountable cardinals for* $\mathcal{L}_{\omega_1,\omega}$, ***Notre Dame Journal of Formal Logic***, vol. 54 (2013), pp. 137–151.

[21] R. Gandy, G. Kreisel, and W. W. Tait, *Set existence*, ***Bulletin de l'Académie Polonaise des Sciences. Série des Sciences Mathématiques, Astronomiques et Physiques***, vol. 8 (1960), pp. 577–582.

[22] S. Gao, *On automorphism groups of countable structures*, ***The Journal of Symbolic Logic***, vol. 63 (1998), pp. 891–896.

[23] ———, ***Invariant Descriptive Set Theory***, Pure and Applied Mathematics, vol. 293, CRC Press, Boca Raton, 2009.

[24] V. Harnik and M. Makkai, *A tree argument in infinitary model theory*, ***Proceedings of the American Mathematical Society***, vol. 67, 1977, pp. 309–314.

[25] L. Harrington, *Analytic determinacy and* $0^\sharp$, ***The Journal of Symbolic Logic***, vol. 43 (1978), pp. 685–693.

[26] M. Hindry and J. Silverman, ***Diophantine Geometry. An Introduction***, Graduate Texts in Mathematics, vol. 201, Springer-Verlag, New York, 2000.

[27] P. Hinman, ***Recursion Theoretic Hierarchies***, Springer-Verlag, Berlin, New York, 1978.

[28] G. Hjorth, *Knight's model, its automorphism group, and characterizing the uncountable cardinals*, ***Journal of Mathematical Logic***, vol. 2 (2002), pp. 113–144.

[29] ———, *A note on counterexamples to the Vaught conjecture*, ***Notre Dame Journal of Formal Logic***, vol. 48 (2007), pp. 49–51.

[30] T. Jech, ***Set Theory***, Springer Monographs in Mathematics, Springer-Verlag, Berlin, 2003.

[31] A. Kanamori, ***The Higher Infinite***, Perspectives in Mathematical Logic, Springer-Verlag, Berlin, 1994.

[32] I. Kaplinsky, ***Infinite Abelian Groups***, University of Michigan Press, Ann Arbor, 1954.

[33] M. Kaufmann, *The quantifier "there exist uncountably many" and some of its relatives*, in [9].

[34] A. S. Kechris, ***Classical Descriptive Set Theory***, Springer-Verlag, New York, 1995.

[35] A. S. KECHRIS, D. A. MARTIN, and R. SOLOVAY, *Introduction to Q-theory*, ***Cabal Seminar 79–81***, Lecture Notes in Mathematics, vol. 1019, Springer, Berlin, 1983, pp. 199–282.

[36] H. J. KEISLER, *Logic with the quantifier "there exist uncountably many"*, ***Annals of Mathematical Logic***, vol. 1 (1970), pp. 1–93.

[37] ———, ***Model Theory of Infinitary Languages***, North-Holland, Amsterdam, 1971.

[38] H. J. KEISLER and M. MORLEY, *Elementary extensions of models of set theory*, ***Israel Journal of Mathematics***, vol. 6 (1968), pp. 49–65.

[39] J. KIRBY, *On quasiminimal excellent classes*, ***The Journal of Symbolic Logic***, vol. 75 (2010), pp. 551–564.

[40] J. KNIGHT, *A complete $\mathcal{L}_{\omega_1,\omega}$-sentence characterizing $\aleph_1$*, ***The Journal of Symbolic Logic***, vol. 42 (1977), pp. 59–62.

[41] ———, *Prime and atomic models*, ***The Journal of Symbolic Logic***, vol. 43 (1978), pp. 385–393.

[42] J. KNIGHT, A. MONTALBÁN, and N. SCHWEBER, *Computable structures in generic extensions*, preprint.

[43] D. KUEKER, *Back-and-forth arguments in infinitary languages*, ***Infinitary Logic: In Memoriam Carol Karp*** (D. Kueker, editor), Lecture Notes in Mathematics, vol. 72, Springer-Verlag, Berlin, 1975.

[44] ———, *Countable approximations and Löwenheim–Skolem theorems*, ***Annals of Mathematical Logic***, vol. 11 (1977), pp. 57–103.

[45] ———, *Uniform theorems in infinitary logic*, ***Logic Colloquium '77***, Studies in Logic and the Foundations of Mathematics, vol. 96, North-Holland, Amsterdam, New York, 1978.

[46] K. KUNEN, ***Set Theory: An Introduction to Independence Proofs***, Studies in Logic and the Foundations of Mathematics, vol. 102, North-Holland, Amsterdam, 1983.

[47] P. LARSON, *Axiomatizing Scott processes*, preprint.

[48] M. C. LASKOWSKI and S. SHELAH, *On the existence of atomic models*, ***The Journal of Symbolic Logic***, vol. 58 (1993), pp. 1189–1194.

[49] M. MAKKAI, *An "admissible" generalization of a theorem on countable Σ^1_1 sets of reals with applications*, ***Annals of Mathematical Logic***, vol. 11 (1977), pp. 1–30.

[50] ———, *Infinitary Logic*, ***Handbook of Mathematical Logic*** (J. Barwise, editor), Studies in Logic and the Foundations of Mathematics, vol. 90, North-Holland, Amsterdam, New York, Oxford, 1977.

[51] ———, *An example concerning Scott heights*, ***The Journal of Symbolic Logic***, vol. 46 (1981), pp. 301–318.

[52] J. MALITZ, *The Hanf number for complete $\mathcal{L}_{\omega_1,\omega}$-sentences*, in [6].

[53] R. MANSFIELD, *Omitting types: application to descriptive set theory*, ***Proceedings of the American Mathematical Society***, vol. 47 (1975), pp. 198–200.

[54] R. MANSFIELD and G. WEITKAMP, ***Recursive Aspects of Descriptive Set Theory***, Oxford University Press, Oxford, 1985.

[55] D. MARKER, *Basic definability notions for* $\mathcal{L}_{\omega_1,\omega}$, appendix to [3].

[56] ———, ***Model Theory: An Introduction***, Springer Graduate Texts in Mathematics, Springer, Berlin, 2002.

[57] D. A. MARTIN, *The axiom of determinateness and reduction principles in the analytical hierarchy*, ***Bulletin of the American Mathematical Society***, vol. 74 (1968), pp. 687–689.

[58] D. A. MARTIN and A. S. KECHRIS, *Infinite games and effective descriptive set theory*, ***Analytic Sets*** (C. A. Rogers et al., editors), Academic Press, London, 1980.

[59] A. MONTALBÁN, *A computability theoretic equivalent to Vaught's conjecture*, ***Advances in Mathematics***, vol. 235 (2013), pp. 56–73.

[60] M. MORLEY, *The number of countable models*, ***The Journal of Symbolic Logic***, vol. 35 (1970), pp. 14–18.

[61] Y. MOSCHOVAKIS, ***Descriptive Set Theory***, North Holland, Amsterdam, 1980.

[62] M. NADEL, $\mathcal{L}_{\omega_1,\omega}$ *and admissible fragments*, in [9].

[63] ———, *More Löwenheim–Skolem results for admissible sets*, ***Israel Journal of Mathematics***, vol. 18 (1974), pp. 53–64.

[64] ———, *Scott sentences and admissible sets*, ***Annals of Mathematical Logic***, vol. 7 (1974), pp. 267–294.

[65] J.-P. RESSARYE, *Models with compactness properties relative to an admissible language*, ***Annals of Mathematical Logic***, vol. 11 (1977), pp. 31–55.

[66] G. SACKS, *Metarecursively enumerable sets and admissible ordinals*, ***Bulletin of the American Mathematical Society***, vol. 72 (1966), pp. 59–64.

[67] ———, *On the number of countable models*, ***Southeast Asian Conference on Logic (Singapore, 1981)***, Studies in Logic and the Foundations of Mathematics, vol. 111, North-Holland, Amsterdam, 1983, pp. 185–195.

[68] ———, ***Higher Recursion Theory***, Springer-Verlag, Berlin, 1990.

[69] S. SHELAH, *Categoricity in* $\aleph_1$ *of sentences in* $\mathcal{L}_{\omega_1,\omega}(Q)$, ***Israel Journal of Mathematics***, vol. 20 (1975), pp. 127–148.

[70] ———, *Classification theory and the number of nonisomorphic models*, vol. 92 (1978).

[71] ———, *The number of pairwise non-elementarily-embeddable models*, ***The Journal of Symbolic Logic***, vol. 54 (1989), pp. 1431–1455.

[72] ———, *Borel sets with large squares*, ***Fundamenta Mathematicae***, vol. 159 (1999), pp. 1–50.

[73] S. SHELAH, L. HARRINGTON, and M. MAKKAI, *A proof of Vaught's conjecture for* ω*-stable theories*, ***Israel Journal of Mathematics***, vol. 49 (1984), pp. 259–280.

[74] J. Silver, *Counting the number of equivalence classes of Borel and coanalytic equivalence relations*, ***Annals of Mathematical Logic***, vol. 18 (1980), pp. 1–28.

[75] T. Slaman and J. Steel, *Definable functions on degrees*, ***Cabal Seminar 81–85***, Lecture Notes in Mathematics, vol. 1333, Springer, Berlin, 1988, pp. 37–55.

[76] I. Souldatos, *Characterizing the powerset by a complete (Scott) sentence*, ***Fundamenta Mathematicae***, vol. 222 (2013), pp. 131–154.

[77] ———, *Notes on cardinals that are characterizable by a complete (Scott) sentence*, ***Notre Dame Journal of Formal Logic***, vol. 55 (2014), pp. 533–551.

[78] J. Steel, *On Vaught's conjecture*, ***Cabal Seminar 76–77***, Lecture Notes in Mathematics, vol. 689, Springer, Berlin, 1978, pp. 193–208.

[79] B. Zilber, *Pseudo-exponentiation on algebraically closed fields of characteristic zero*, ***Annals of Pure and Applied Logic***, vol. 132 (2005), pp. 67–95.

[80] ———, *Covers of the multiplicative group of an algebraically closed field of characteristic zero*, ***Journal of the London Mathematical Society***, vol. 74 (2006), pp. 41–58.

INDEX

Health Policy in Asia

The book assesses the policy actions of select Asian governments (China, India, Hong Kong, South Korea, Singapore and Thailand) to address critical health system functions from a policy design perspective. The findings show that all governments in the region have made tremendous strides in focussing their attention on the core issues and, especially, the interactions among them. However, there is still insufficient appreciation of the usefulness of public hospitals and their efficient management. Similarly, some governments have not made sufficient efforts to establish an effective regulatory framework which is especially vital in systems with a large share of private providers and payers. A well-run public hospital system and an effective framework for regulating private providers are essential tools to support the governance, financing, and payment reforms underway in the six health systems studied in this book.

M RAMESH is UNESCO Chair on Social Policy Design in Asia at the National University of Singapore. Previously, he was the Founding Head of the Department of Asian and Policy Studies at the Hong Kong Institute of Education and Chair of the Department of Government and International Relations at the University of Sydney. Specialising in public policy and governance in Asia with a particular focus on social policy, Ramesh has authored and edited many books. His widely used textbook Studying Public Policy has been translated into several languages. He is the Editor of the journals Policy and Society and Policy Design and Practice.

AZAD S. BALI in Senior Lecturer in Public Policy and holds a joint appointment with the School of Politics & International Relations and the Crawford School of Public Policy at the Australian National University (ANU). His research interests straddle comparative public policy and health policy in Asia. Some of this research research has been published in leading journals including Policy Sciences, Policy & Society, Public Policy and Administration, Social Policy & Administration among others. Prior to the ANU, Bali held a Melbourne Early Career Fellowship at the University of Melbourne and was Research Fellow at Murdoch University.

CAMBRIDGE STUDIES IN COMPARATIVE PUBLIC POLICY

The **Cambridge Studies in Comparative Public Policy** series was established to promote and disseminate comparative research in public policy. The objective of the series is to advance the understanding of public policies through the publication of the results of comparative research into the nature, dynamics and contexts of major policy challenges and responses to them. Works in the series will draw critical insights that enhance policy learning and are generalizable beyond specific policy contexts, sectors and time periods. Such works will also compare the development and application of public policy theory across institutional and cultural settings and examine how policy ideas, institutions and practices shape policies and their outcomes. Manuscripts comparing public policies in two or more cases as well as theoretically informed critical case studies which test more general theories are encouraged. Studies comparing policy development over time are also welcomed.

In Defense of Pluralism: Policy Disagreement and its Media Coverage
By Éric Montpetit

Asia after the Developmental State: Disembedding Autonomy
Edited by Toby Carroll and Darryl S. L. Jarvis

The Politics of Shale Gas in Eastern Europe: Energy Security, Contested Technologies and the Social Licence to Frack
By Andreas Goldthau

Healthy or Sick? Coevolution of Health Care and Public Health in a Comparative Perspective 2018
By Philipp Trein

Policy Accumulation and the Democratic Responsiveness Trap
By Christian Adam, Steffen Hurka, Christoph Knill and Yves Steinebach

Hyper-Active Governance: How Governments Manage the Politics of Expertise
By Matthew Wood

Policy Consultancy in Comparative Perspective: Patterns, Nuances and Implications of the Contractor State
By Caspar van den Berg, Michael Howlett, Andrea Migone, Michael Howard, Frida Pemer and Helen M. Gunter

A Matter of Style? Organizational Agency in Global Public Policy
By Louisa Bayerlein, Christoph Knill and Yves Steinebach

Policy Controversies and Political Blame Games
By Markus Hinterleitner

Government Statistical Agencies and the Politics of Credibility
By Cosmo Howard

Convergence and Diversity in the Governance of Higher Education
By Giliberto Capano and Darryl S.L. Jarvis

Public Governance as Co-creation: A Strategy for Revitalizing the Public Sector and Rejuvenating Democracy
By Christopher Ansell and Jacob Torfing

CAMBRIDGE
UNIVERSITY PRESS

University Printing House, Cambridge CB2 8BS, United Kingdom

One Liberty Plaza, 20th Floor, New York, NY 10006, USA

477 Williamstown Road, Port Melbourne, VIC 3207, Australia

314–321, 3rd Floor, Plot 3, Splendor Forum, Jasola District Centre, New Delhi – 110025, India

103 Penang Road, #05–06/07, Visioncrest Commercial, Singapore 238467

Cambridge University Press is part of the University of Cambridge.

It furthers the University's mission by disseminating knowledge in the pursuit of education, learning, and research at the highest international levels of excellence.

www.cambridge.org
Information on this title: www.cambridge.org/9781108483537
DOI: 10.1017/9781108692656

First published 2021

A catalogue record for this publication is available from the British Library.

Library of Congress Cataloging-in-Publication Data
Names: Ramesh, M., 1960– author. | Bali, Azad, author.
Title: Health policy in Asia : a policy design approach / M. Ramesh, National University of Singapore, Azad S. Bali, The Australian National University.
Description: New York : Cambridge University Press, 2021. | Series: Cambridge studies in comparative public policy | Includes bibliographical references and index.
Identifiers: LCCN 2021037804 (print) | LCCN 2021037805 (ebook) | ISBN 9781108483537 (hardback) | ISBN 9781108728775 (paperback) | ISBN 9781108692656 (epub)
Subjects: LCSH: Medical policy–Asia | BISAC: POLITICAL SCIENCE / General
Classification: LCC RA395.A52 R36 2021 (print) | LCC RA395.A52 (ebook) | DDC 362.1095–dc23
LC record available at https://lccn.loc.gov/2021037804
LC ebook record available at https://lccn.loc.gov/2021037805

ISBN 978-1-108-48353-7 Hardback

Health Policy in Asia

A Policy Design Approach

M RAMESH
National University of Singapore

AZAD S. BALI
The Australian National University

CAMBRIDGE
UNIVERSITY PRESS

Contents

Figures

Tables

Acknowledgements

Many people have contributed to this book. We are grateful to government officials and colleagues who took the time to speak with us across the six countries studied in this book.

We would like to especially thank Ake Blomqvist, Alex He, Altaf Virani, Elizabeth Quah, Assa Doron, Daya Maurya, Donald Low, Indrani Gupta, John Langenbrunner, John Murphy, Kwan Chang Yee, Mohnish Kedia, Mukul Asher, Phua Kai Hong, Qian Jiwei, Sarah Bales, Tim Tenbensel, and Wu Xun for thoughtful conversations on health policy over the years.

Chapters from this book were presented at conferences and workshops and we thank the following discussants and panellists for their comments and suggestions: Al Stark, Brian Head, Claire Dunlop, Eva Heidbreder, Giliberto Capano, Guy Peters, Isabelle Engeli, Ishani Mukherjee, Jenny Lewis, Jeremey Rayner, Mark Considine, Matthew Kerby, Olivia Jensen, and Paul Fawcett.

We have benefitted tremendously from recent collaborations with Michael Howlett, Giliberto Capano, Guy Peters, Jenny Lewis, Adam Hannah, Daniel Beland, Ben Cashore, and Ishani Mukherjee. Thank you.

We would be remiss to not acknowledge colleagues who have helped in facilitating the completion of this book in different ways: Kenneth Tan, Kanti Bajpai, Alfred Wu, Namrata Chindarkar, Janine O'Flynn, and Darren Halpin.

We are also grateful to three anonymous reviewers for their constructive and critical comments. We thank Helen Sullivan, Helen Dickinson, Anne Mills, Guy Peters, and Peter Berman for taking the time to read through the manuscript and offering generous endorsement.

We dedicate this book to Keya, Nik, and Sandy who got to test the health care systems in their own ways in 2020.

Abbreviations

ACE	Agency for Care Effectiveness
AIC	Agency for Integrated Care
ALOS	average length of stay
CDMP	Chronic Disease Management Programme
CGHS	Central Government Health Scheme
CHAS	Community Health Assist Scheme
CHCs	community health centres
Covid-19	coronavirus-19
CMS	Cooperative Medical System
C-section	caesarean delivery
CSMBS	Civil Servants Medical Benefit Scheme
CSSA	Comprehensive Social Security Allowance
DBP	disease-based payment
DH	Department of Health (Hong Kong)
DHS	district health system
DREC	Drug Reimbursement Evaluation Committee
DRG	diagnosis-related group
EIU	Economist Intelligence Unit
ESIC	Employees' State Insurance Corporation
ESIS	Employees' State Insurance Scheme
FDA	Food and Drug Administration
FDI	foreign direct investment
FHB	Food and Health Bureau
FFS	fee for service
GBD	Global Burden of Disease study
GDP	gross domestic product
GIS	Government Insurance Scheme
GP	general practitioner / general practice
GPO	Government Pharmaceutical Organization
H1N1	influenza A virus subtype H1N1

HA	Hospital Authority
HAI	Healthcare Accreditation Institute
HALE	health adjusted life expectancy
HCSA	Healthcare Services Act (Singapore)
HIPDC	Health Insurance Policy Deliberation Committee
HIRA	Health Insurance Review Agency
HIRA-VIP	HIRA-Value Incentive Program
HITAP	Health Intervention and Technology Assessment Program
HK$	Hong Kong dollar
HLEG	High Level Expert Group
HMC	Hospital Management Council
HPA	Health Protection Account
HPS	Health Protection Scheme
HSP	Hospital Security Plan
HSRI	Health Systems Research Institute
HTA	Health Technology Assessment
ICU	intensive care unit
IHiS	Integrated Health Information Systems
IHPP	International Health Policy Program
IMR	infant mortaility rate
INR	Indian rupee
IP	Integrated Shield Plan
ISO	International Organization for Standardization
KFDA	Korea Food and Drug Administration
KRW	Korean won
LIC	Low Income Card
LIS	Labour Insurance System
LOS	length of stay
MAF	Medication Assistance Fund
MDEC	Medical Device Expert Committee
MDGs	millennium development goals
Medicaid	Medical Aid program
MERS	Middle East Respiratory Syndrome
MFAS	Medical Financial Assistance Scheme
MFDS	Ministry of Food and Drug Safety (Korea)
ML	Medishield Life

MoH	Minstry of Health (India)
MOHH	Ministry of Health Holdings Private Limited
MOHW	Ministry of Health and Welfare (Korea)
MOPH	Ministry of Public Health (Thailand)
MRD	Medical Registration Division
MSA	Medical Savings Account
MSF	Ministry of Strategy and Finance (Korea)
NCMS	New Cooperative Medical Scheme
NECA	National Evidence-based Collaborating Agency
NEHR	National Electronic Health Record
NHA	National Health Assembly
NHC	National Health Commission
NHCO	National Health Commission Office
NHFPC	National Health and Family Planning Commission
NHG	National Healthcare Group
NHI	National Health Insurance
NHIC	National Health Insurance Corporation
NHIS	National Health Insurance Service
NHM	National Health Mission
NHP	National Health Plan
NHS	National Health Service
NHSA	National Healthcare Security Administration
NHSB	National Health Security Board
NHSO	National Health Security Office
NLEM	National List of Essential Medicines (Thailand)
NPM	New Public Management
NRCMS	New Rural Cooperative Medical Scheme
NRHM	National Rural Health Mission
NSHC	National Standards for Health Care
NUHS	National University Health System
NUS	National University of Singapore
OECD	Organisation for Economic Co-operation and Development
OOP	Out of Pocket
P4P	Payment for Performance
PAP	People's Action Party

PCC	Primary Care Cluster
PCN	Primary Care Network
PHC	Primary Health Care
PHCs	primary health centres
PHMC	Private Hospitals and Medical Clinics
PHMCA	Private Hospitals and Medical Clinics Act (Singapore)
PM-JAY	Pradhan Mantri Jan Arogya Yojana
PPP	Purchasing Power Parity
RBRVS	Resource-Based Relative Value Score
RHS	Regional Health Systems
RMB	renminbi
RSBY	Rashtriya Swasthya Bima Yojana
SDGs	Sustainable Development Goals
SDL	Standard Drug List
SHI	Social Health Insurance
SingHealth	Singapore Health Services
SLA	Service-Level Agreement
SOEs	State-Owned Enterprises
SOP	Standard Operating Procedure
SSO	Social Security Office
SSS	Social Security Scheme
TRTP	Thai Rak Thai Party
THE	Total Health Expenditure
THPF	Thai Health Promotion Foundation
UCS	Universal Coverage Scheme
UEBMI	Urban Employee Basic Medical Insurance
UHC	Universal Health Coverage
URBM	Urban workers
URBMI	Urban Resident Basic Medical Insurance
USD	$US
VHCS	Voluntary Health Card Scheme
VHIS	Voluntary Health Insurance Scheme
WCS	Workman's Compensation Scheme
WHO	World Health Organization
YLL	Years of Life Lost

1 Introduction
Health Policy Design

"Nobody knew healthcare could be so complicated," quipped President Donald Trump in November 2017. His comments echo the frustrations of governments across the world struggling to come to grips with the intractable challenges of health policy. The fundamental question that all governments face is how to organize, finance, and pay for health care so that the entire population has access to necessary care at an affordable cost. Answering this question is hard, as governments have learnt all too well after decades of policy reforms and trillions of dollars in spending.

The need to find ways to expand access to health care at affordable cost became urgent following the adoption of the Sustainable Development Goals (SDGs) by all members of the United Nations in 2015. SDG Target 3.8 explicitly commits governments to achieve Universal Health Coverage (UHC) – defined as ensuring that "all people obtain the health services they need without suffering financial hardship when paying for them" (WHO, 2020) – by 2030. This is an ambitious goal considering that nearly half of the world's population lacks access to essential health services, and each year about 100 million people are pushed into 'extreme poverty' due to out-of-pocket (OOP) payments for health care. Based on current trends, it is projected that more than one-third of the world population will continue to be deprived of essential care in 2030, the deadline for UHC (New, 2019). While the most severe deprivations are concentrated in low-income countries, significant pockets of hardship persist in middle- and high-income countries. Even in countries that have succeeded in largely achieving the UHC goal, governments are faced with organizational and financial challenges of maintaining services. Sustaining UHC is, thus, as much of a challenge as achieving it requires sound health policies to succeed in the long run.

We define the goal of health policy as ensuring that the entire population has access to necessary services at costs affordable to both households and the society as a whole. While households are rightly the primary focus of health policy, fiscal considerations cannot be ignored if the programmes are to be sustainable in the long run. By focusing on necessary care and affordability, we attribute primary importance to health and financial protection, and secondary importance to other health system goals such as efficiency and consumer choice. What is 'necessary' and 'affordable' of course varies across societies, shaped by a range of factors, including national income, household characteristics, and social values. It is also possible for societies to seek more than the minimum suggested here, but policymakers need to be mindful of the additional difficulties in the form of extra spending as well as policy incoherence and weaker accountability associated with expanded goals. Americans, for example, place strong emphasis on provider autonomy and consumer choice, which involves additional financial and administrative costs that must be accounted for in affordability calculations.

The objective of this book is to describe the health policies of six governments in Asia – China, Hong Kong, India, (South) Korea, Singapore, and Thailand – and assess their performance with respect to achieving, maintaining, and improving UHC. It focuses particularly on the policy tools that these governments use to address critical health system functions. As we will see in the subsequent chapters, all governments studied here show an increasing comprehension of the intricacies of the health policy problems they face and are developing a design strategy to deal with them. Overall, while Hong Kong and Singapore enjoy the best health policy record, it is Thailand and, to a lesser extent, China that show the most improvements in progress towards UHC. The dominance of private providers in India and Korea makes it difficult for their governments to employ the policy tools necessary for sustaining UHC, reflected in their continued high incidence of OOP payments.

The analysis presented in this book shows that while there has been notable progress in strengthening governance, financing, and payment arrangements, some critical functions remain overlooked. Particularly, many governments have paid insufficient attention to deficiencies in public hospitals and their management. Similarly, governments have made inadequate efforts to establish and deploy a regulatory framework for the health sector which is especially vital in systems with a large share of private providers and payers. A well-run public hospital system and an effective framework for regulating private providers are essential tools to support the governance, financing, and payment reforms undertaken in the six health systems studied in this book. This serves as a timely reminder to governments and international organizations who obsess over improvements in financing and payment issues and overlook ownership and regulatory concerns.

1.1 Perspectives on Health Policy

One of the steepest hurdles in health policy is that neither scholars nor policymakers fully appreciate the enormity and complexity of the sector. Without fully appreciating the difficulties, they propose solutions – such as tighter regulations, increased competition, or enhanced expenditure – which often aggravate rather than ameliorate problems. The sources of the healthcare problem are not only complex – rooted as they are in myriad biological, economic, ethical, political, and social issues – but also often intertwined in unfathomable ways. It is for good reasons that reforming health care is often likened to shaping a balloon: squeeze in one place and it pops up in another.

The complexity of the healthcare sector stems from several interlinked characteristics which are somewhat unique. First, the pervasive information asymmetry and moral hazards that characterize healthcare goods and services give rise to a series of interrelated market and government failures that are difficult to manage (Blomqvist, 1991, 2011; Zweifel and Manning, 2000). Second, the key agents in a health system – the patients, the healthcare providers, and the insurers – create multiple principal–agent relationships that aggravate these failures (Brinkerhoff et al., 2014; Bali and Ramesh, 2015a; McGuire, 2000). Third, these agents have fundamentally differing interests and incentives. For example, healthcare providers prefer unfettered autonomy in both medical and financial matters, while insurers or third-party payers face incentives to insure only healthy patients or pass on costs across the risk pool. Users, for their part, want access to the most sophisticated care available, regardless of need, and seek some third party to pay for it. Fourth, the material interests of the key stakeholders – especially healthcare providers, who largely determine the distribution of resources in the sector – are difficult to reconcile with those of the government or the society at large, especially around issues of supply and affordability (Pauly, 2009; Fuchs, 2011; Ramesh et al.,

2013). Fifth, health systems and their constituent institutions have evolved gradually in a path-dependent manner that locks in existing practices and stymies reform (Roberts et al., 2003; Hsiao and Shaw, 2007; Pierson, 2000; Haeder, 2012; Beland, 2016). Sixth, these policy challenges in health care are intertwined and cannot be addressed in isolation: they need to be targeted individually but in a coordinated manner (Bali and Ramesh, 2017). Moreover, these challenges are ubiquitous across all modes of governance – hierarchical or otherwise – as they stem from the innate characteristics of the health system and are not an extension of a particular governance arrangement (Ramesh et al., 2015; Wu and Ramesh, 2014).

Attitudes towards health policy challenges and how to deal with them have changed over the years. Initially, health policy problems were seen essentially as a supply-side issue rooted in insufficient medical facilities, personnel, and financial resources. With this understanding, policy efforts understandably concentrated on increasing supply of health services and mobilizing financial resources to pay for them. It was subsequently realized that households' potential demand for care was almost endless, as was the providers' enthusiasm to meet it, so long as some third party was paying for it. Healthcare expenditures rose rapidly and were projected to grow yet faster with ageing populations and rising expectations. It was also realized that many were unable to access services despite their availability for a variety of reasons which raised critical issues of access and equity. Furthermore, the expansion of the middle class and the corresponding rise in demand for comfortable and prompt services drew attention to the requirement for more responsiveness to users' needs.

Thus began the search for reforms that not only provide needed health services to all but do so in an economical, equitable, fiscally sustainable, and user-friendly manner. But the way they have gone about it has differed according to their training and professional experiences, as we will see in the following section.

Economists have engaged with these questions in health care largely conceptualizing the problems afflicting the sector as a series of market and government failures related to moral hazard, information asymmetry, and adverse selection (for example, see Pauly et al., 2012; Culyer et al., 2000). As early as the 1990s, there was a consensus among mainstream economists that the optimal response to the failures was market-based tools such as competition and the use of financial incentives. Policy reforms inspired by this line of thinking include managed care in the United States and GP Fundholding in the United Kingdom launched during the 1990s (Blomqvist, 2011; Glied, 2000). In Asia, many governments similarly liberalized the health sector and introduced market relationships between providers and patients. Contemporary economic analyses of health care focus more on providers' behaviour, particularly how they respond to different payment tools such as diagnosis-related groups (DRGs) and capitation. This line of scholarship, however, does not adequately recognize the policy context in which these tools exist and work (Fox and Reich, 2017). For instance, measures to correct malfeasance produce different results in different countries due to the different social and ethical contexts (Ramesh, 2013; Nguyen et al., 2017).

Political scientists, on the other hand, have engaged with these questions largely through the lens of historical institutionalism. This line of scholarship conceptualizes policy problems in terms of the actors, ideas, and institutions in a given sector (Esping-Andersen, 1996; Moran, 2000; Harris and Milkis, 1989; Jochim and May, 2010). Political competition and its impact on health policy has also been a focus of studies by political scientists (for example, see Haggard and Kaufman, 2008; Ramesh and Asher, 2000; Wong, 2006). Studies from this perspective offer keen insights into the evolution of health systems but stop short of offering recommendations for improvements, except for reiterating the importance of institutional and political factors in constraining policy choices.

In a similar vein, scholars in social policy and development studies explain policy developments and outcomes with reference to the state and non-state actors involved in the sector (Wendt, 2009). They tend to focus heavily on the amount of fiscal resources spent on a policy problem and its equity implications, ignoring the efficiency and sustainability dimensions of spending. While easy to measure, levels of spending on health is a poor indicator of performance, as it is easy to spend a lot on health without a corresponding improvement in outcomes, as the case of the USA reminds us. Similarly, many countries in Southeast Asia, including Singapore and Thailand, enjoy admirable health outcomes at relatively modest levels of spending (Ramesh and Wu, 2009; Wu and Ramesh, 2009). Health spending is a political decision only secondarily related to improving the population's health (Ramesh and Asher, 2000; Wong, 2006). Spending as a policy tool is useful only if it is accompanied by other complementary tools (Ramesh et al., 2013).

Over the past fifteen years, there has been a proliferation of works by international organizations such as the World Bank and the World Health Organization (WHO) focusing on 'health system', defined as "organizations, people and actions whose primary intent is to promote, restore or maintain health" (WHO, 2010). With long experience on the frontlines of healthcare reforms, these organizations are acutely aware of the multifaceted nature of health policy and the range of considerations that go into its design. They have accordingly developed concepts such as 'building blocks' and 'control knobs', which are now widely used in health policy discussions. The WHO's 'building blocks' of health systems comprise six components: service delivery, health workforce, health information systems, access to essential medicines, financing, and leadership/governance (WHO, 2010). In a similar effort, the World Bank has conceptualized health policy in terms of five 'knobs' that it argues policymakers can use to effect change in the health system: financing, payment, organization, regulation, and persuasion (Roberts et al., 2003). While the two frameworks represent a worthy effort to systematize health policy thinking, they are essentially descriptive labels for different functions in the sector that offer little insight into the relationships among them or their relative importance (Mounier-Jack, 2014; Witter et al., 2019).

Another body of works produced by international organizations centres on the concept of 'governance'. However, the word has been defined in so many fundamentally different ways that it is hard to generalize about its meaning or usefulness. The WHO Glossary entry for 'governance', for example, offers three distinct definitions (WHO, undated). Broadly speaking, one body of works concentrates on institutional relationships in health policy, while another espouses normative principles – 'good governance' – that they would like to see embodied in health systems and policy. Discussions on health governance are thick with references to collaboration, participation, transparency, and accountability. While these ideas are logically cogent and intuitively appealing, there is little empirical evidence showing that they make health care accessible on a sustainable basis.

There are also various analysts associated with the WHO and the World Bank who have produced sophisticated 'how-to' manuals on design of financing and payment systems for health care

(Langenbrunner and Somanathan, 2011; Langenbrunner et al., 2009; Langenbrunner and Wiley, 2002). While this body of work is directly relevant to strengthening health systems and building UHC, it lacks the problem-centred approach necessary for understanding and addressing the underlying causes of the problems afflicting health systems. Dealing with health policy problems requires an understanding of not only its technical dimensions but also the political and social contexts within which they exist and are addressed (Bali and Ramesh, 2019; Chindarkar et al., 2017). Only Roberts and his colleagues (Roberts et al., 2003) come close to taking the broad yet detailed approach necessary to address contemporary health policy challenges.

International organizations and researchers affiliated with them conceptualize the challenges of UHC too narrowly to be helpful. The World Health Report 2010, for example, posits three 'fundamental problems' that hinder the achievement of UHC.

> The first is the availability of resources .... The second barrier to universal coverage is an overreliance on direct payments at the time people need care .... The third impediment to a more rapid movement towards universal coverage is the inefficient and inequitable use of resources. (WHO, 2010)

For solutions, the Report proposes, "Countries must raise sufficient funds, reduce the reliance on direct payments to finance services, and improve efficiency and equity." The conception of both problem and solution is narrow to the point of being unhelpful. Removing barriers to access is critical, but issues of financial sustainability cannot be ignored if the programme is to be viable in the long run. Securing financial viability goes beyond 'promoting efficiency and eliminating waste' through accountability and transparency. It also requires corollary structural measures, related to provision, financing, and payment and the different imperatives faced by public and private health systems. More importantly, the issue of financing cannot be divorced from production and delivery of health services because it is on the supply side that many of the root causes of financial unsustainability lie. Inefficiencies in hospitals and insurance funds, for example, are a major problem and must be dealt with before additional public funds are allocated to them.

To achieve UHC on a sustainable basis, policymakers need to adopt a problem-solving approach to the entire system. They need a comprehensive understanding of the barriers to UHC and, more importantly, their root causes with the purpose of uprooting them. In other words, they need to engage in collective problem-solving to remove the conditions that stymie access to health care and threaten viability. Since the lack of universal access to health care is the result of a range of related but distinct underlying conditions, it is imperative that they are dealt with individually but in a concerted manner. And this is what we propose to do in our study of the six Asian countries' experience with achieving UHC.

1.2 The Policy Design Approach

The theoretical framing of this book draws on the policy design literature, particularly the 'new' design orientation within the policy sciences. Policy design describes the activities related to addressing the causes of a public problem with the purpose of eliminating them or at least reducing their adverse effects. Reiterating one of the fundamental tenets of problem-solving, it emphasizes the means (i.e. policy tools or instruments) most able to potentially address the targeted problem. The proliferation of market- and network-centred governance reforms in the 1990s and early 2000s caused a decline in the popularity of the design approach as it limited the range of choices available to governments (Howlett and Lejano, 2013). This in turn encouraged analysts to focus on broad institutional arrangements to organize economic and social activities rather than specific tools to address the key problems. Recent years have seen a resurgence of interest in policy design and policy tools as the shortcomings of the undiscerning efforts to promote deregulation, marketization, and privatization became increasingly apparent (Peters, 2018a).

The 'new' design orientation re-emphasizes the role of a policy tool or a collection of tools in addressing policy problems (Howlett et al., 2015). Policy tools or instruments – defined as the means by which governments implement policies and achieve goals (Howlett et al., 2009) – have been a central concern in the policy sciences since the 1950s (Dahl and Lindblom, 1953; Hood, 2007). While early works dwelled on individual policy tools and substitutability among them, recent works have focused on how they work together as 'policy mixes' or 'portfolios' in attaining specific policy goals (Howlett and Rayner, 2013; Schaffrin et al., 2014; Howlett, 2019). The changed emphasis of recent works reflects the increasingly complex and multifaceted problems that policymakers are called upon to solve on one hand and the limited 'degrees of freedom' they enjoy in solving them on the other (Chindarkar et al., 2017; Howlett et al., 2015). This approach highlights issues of coordination, coherence, and consistency within mixes of policy tools used to address collective problems (Peters, 2015; Howlett and del Rio, 2015; Bali and Ramesh, 2018).

The design approach to public policy offers insights as well as a framework for understanding the challenges within the health sector and how they can be addressed. We argue that, in health care, the root of the problem lies in the inherent characteristics of the service which give rise to a series of interrelated market and government failures that result in rising costs and restricted access. To reduce the incidence of such failures and mitigate their adverse effects, policymakers need to purposefully search for tools matching the problem and apply them diligently. What is the specific problematic condition and its root cause that needs to be tackled? What are the choice of tools available to deal with it? In what combination and to what extent are the different tools to be used to address the problem? These questions point to not only the need for assessing the intrinsic usefulness of each tool but how it complements and contradicts the other tools in use. The mix of tools, and the synergies and contradictions among them, is as important as their individual characteristics and uses. No less important is the policy context including policy legacies, and the entrenched interests of stakeholders who can be expected to support or resist changes to tools used depending on their self-interest.

1.3 Health Policy Design

Health policy is about eliminating or mitigating the causes of problematic conditions that impede the achievement of universal health care, here defined as availability of health care to the entire population at cost affordable to households as well as the society. Health policy design, it follows, is about selecting and deploying policy tools to remove or, more realistically, ameliorate the adverse effects of the critical problems that prevent the achievement of universal health care. Health policy tools are the specific means by which health care is organized, financed, and delivered to the population.

Table 1.1 ***Health system functions, policy tools, and principal resources***

Health system function	Policy tool	Principal resource base of the tool
Governance	Stewardship and coordination	Organization and nodality
Provision	Public and private ownership	Organization
Financing	Risk pooling	Treasure and authority
Payment	Retrospective and prospective payment	Treasure and authority
Setting standards	Regulation	Authority

Sources: Adapted from Hood, 1983; Roberts et al., 2003; Howlett, 2019.

From health systems perspective, there are five critical functions that policymakers need to attend to in order to achieve universal health care: *governance*, *provision*, *financing*, *payment*, *and setting standards* (Table 1.1). *Governance* is an overarching function comprising of providing direction to the sector and coordinating the disparate public and private activities that affect health care. The vast financial, personnel, information, and other resources that go into health care require a robust governance framework in which the roles and responsibilities are allocated thoughtfully and enforced diligently (Brinkerhoff and Bossert, 2008). *Organizing* the delivery of health care through public or private providers – or a combination of the two, as is more commonly the case – is another critical function that requires attention. *Financing* the healthcare system in an equitable yet sustainable manner through pooling of financial resources is the third critical function in health care. The fourth critical function requires policymakers to pay attention to how healthcare providers are paid. Without a *payment* system that sets out right incentives, users will be either poorly served or over-served at excessive costs to the society. Finally, to function effectively, health systems need to ensure the *safety and quality* of medications, treatments, and services delivered to patients. Health systems with significant private provision and financing also require governments to set the terms of market exchange with the goal of protecting patients. While these five functions are distinct and involve separate sets of actors, they overlap and are closely related and, as result, need to be approached holistically (de Savigny and Adam, 2009; Roberts et al., 2003).

Each of the five critical health policy functions requires a corresponding set of appropriate policy tools. Thus, the challenge of *governance* is addressed through active stewardship, and vertical and horizontal coordination. Similarly, the *provision* functions are accomplished by organizing public and private providers in ways that lead them to serve the public rather than their own interest. For private providers, this means curbing opportunities for profit maximization at the expense of users, whereas for public providers, it usually involves motivating personnel to remain responsive to patients' needs and pay more attention to management issues. The third critical task involves establishing and managing *risk pools* either through government budget or insurance to ensure that households are protected from financial hardships caused by OOP payments without undermining the programmes' financial sustainability. Fourth, prospective and retrospective *payment* tools are necessary to ensure that the mechanisms for paying providers do not create perverse incentives for providers or users. For fee-for-service (FFS) payments, for example, this means dampening the motivation to over-service patients, while the opposite is the case for capped prospective payment mechanisms, such as capitation payments. Finally, defining the minimum *standards* of conduct for providers and insurers requires governments to establish a regulatory framework to promote appropriate competition among them while protecting the physical and financial interests of users. These health design tasks are examined in detail in the discussion below.

As shown in Table 1.1, each of these tools relies on one or more 'governing resources' for effectiveness (Hood, 2007; Howlett, 2019; Peters, 2018b). Thus, for example, stewardship and coordination require organizational resources to create agencies tasked with coordinating stakeholders as well as the ability to collect and disseminate information. Similarly, prospective and retrospective payments involve the use of governments' financial (or treasure) resources as well as regulations (or authority) to underpin payment arrangements. The larger point here is that no policy tool is capable of achieving a goal by itself and requires the use of other supporting tools (Howlett, 2019).

The choice of tools is shaped by historical choices, policy styles of governments, and ideas and interests of the key actors in the health sector and the relationships among them (Campos and Reich, 2019; Sparkes et al., 2019; Fox and Reich, 2013; Howlett and Tosun, 2018;

Capano and Lippi, 2017; Bali and Hannah, 2021). In health care, the interests and incentives of the three key stakeholders – users (patients), providers, and third-party payers (insurers) – are inherently conflictual and impose an enduring constraint on health policy design. Patients want immediate low- or no-cost access to the most advanced medical care available, regardless of need, and resist efforts to make them pay. Similarly, providers are eager to meet their patients' wishes for expensive care and resist any effort to constrain their autonomy. Correspondingly, third-party payers – private and public insurance schemes – face incentives to insure only those with a lower likelihood of falling ill and/or to pass healthcare costs on to patients or the government. Left to their own device, the three parties would either not cooperate or collude to advance their own interests while eroding the interest of the society in the form of higher expenditures than necessary.

Restraining the self-serving behaviour of the key stakeholders is a vital health policy function that only an outside party, such as the government, can perform. This is not to suggest that governments dominate provision or financing but rather to play the role of a steward that works with societal partners in aggregating and resolving conflicting demands (Ramesh et al., 2013; Brinkerhoff and Bossert, 2014). While private players – providers, third-party payers, users, pharmaceutical firms, professional association and unions – may play an important role in designing and executing health policies, there is a need for a steward and arbiter that sets out a vision, develops strategies, and brings the key stakeholders together (Balabanova et al., 2009; Mor and Yip, 2015). Governments are uniquely suited to play this coordinating role (Capano, 2011).

Governments do not of course enjoy a free hand in choosing policy goals and tools, as they are constrained by a swathe of institutional, political, social, and financial constraints (Balabanova et al., 2009; Fox and Reich, 2013). In health care, they are particularly constrained by historical legacy and existing institutions (Greener, 2002; McKee et al., 2013). As Wilsford (1994, 251) notes, "actors are hemmed in by existing institutions and structures that channel them along established policy paths." These policy paths are shaped by a confluence of competing interests, ideas and are an outcome of extensive bargaining and political battles in the past (Hannah, 2020). As a result, governments have limited elbow room or 'degrees of freedom' in designing health policies (Capano and Howlett, 2020; Capano and Lippi, 2017). It is for this reason that health policy processes and reforms are characterized by *calibrations* (i.e. introducing marginal changes to existing policy settings such as contribution rates or changing benefit levels), *patching* (correcting flaws in existing programmes), and *layering* (where new elements are added to existing policies) rather than adoption of major changes that would require significant change in the ideas, actors, and institutions in the sector.

We now turn to the specific tools that governments have at their disposal in pursuing the five critical functions in health care.

1.3.1 Governance: Stewardship and Coordination

Health policy and the cognate fields of public policy and public administration maintain a bipolar attitude towards governance. On the one hand, they seek autonomy, voice, and participation for local actors, often packaged and presented as an integral part of decentralized and democratic governance (WHO Erurope, 2008). On the other hand, key works in the field call for coordination, coherence, integration, and so on in government policies (Howlett and Rayner, 2013; Peters, 2015). This bifurcated line of thinking is damaging to effective policy design, as sustainable UHC requires a combination of both local autonomy and central coordination, not a choice of one over the other.

Recognition that health care is a local matter is as old as health policy, evident in the fact that local governments everywhere, in both federal and non-federal systems, bear the primary responsibility for delivering health services. Indeed, it is inconceivable how it could be otherwise given that populations' healthcare needs vary vastly across regions which would be difficult to meet by distant governments at the centre. The personalized nature of health services necessitates health policymakers and managers to reside locally so that they can understand and empathize with the users' needs and be accountable to them.

But delivering modern health care is more than heeding the voice of local users, who themselves would not fully understand their own or their community's health needs. Making modern health care available to all who need is an inordinately complex task, requiring immense financial, technical, and administrative resources. While some local governments may have the necessary resources to deliver the health services, many do not and must rely on financial and other support from central authorities (Rao, 2019).

The central authorities, for their part, cannot simply serve as cash registers transferring funds to local governments and providers because they have their own priorities and responsibilities as well as fiduciary and other obligations. This leads them to impose conditions which take local needs and preferences into account but allows them to establish common national standards. They also need to ensure that the national policy goals are adhered to by local authorities, necessitating machinery for setting policy directions as well ensuring their implementation. In other words, the national governments and their provincial counterparts must set direction and steer the country towards it. In health policy and other literatures, this is referred to as *stewardship* (Roberts et al., 2003).

Reduced to its essence, stewardship refers to setting a common goal and organizing collective efforts to achieve it. In health care, this manifests in stating a vision and ensuring that the priorities and actions of all levels of government are aligned and geared towards achieving it. As Barbazza and Tello (2014) elaborate,

> The task of [health] system stewardship can then be said to include defining an overarching vision for the health system and establishing the boundaries in which the system's actors operate. Characterizing stewardship according to its unique quality of direction, the governance function then distinctly describes the specific efforts and processes that are applied to steer the system toward this given priority.

Identifying relevant policy tools, assessing their appropriateness, and adapting to local needs – otherwise known as policy design – are the responsibilities of the policy steward. The tools include mechanisms for implementing the policy in questionas well as engaging with the society and collaborating with key stakeholders. As WHO (2014:10) put it, stewardship is about "ensuring that a strategic policy framework exists and is combined with effective oversight, coalition building, regulation, attention to system-design and accountability." In a similar vein, Alvarez-Rosete et al. (2013) note, "The concept of health stewardship ... implies a broader overarching responsibility over the functioning of the health system as a whole and, ultimately, over the health of the population." This interpretation of stewardship is separate from and narrower than the large literature that conflates it with the concept of governance, especially 'good governance' (for example, see WHO Europe, 2008). The noble sentiments – transparency, participation, and efficiency – embodied in definitions of good governance while laudable are too expansive to be useful (Barbazza and Tello, 2014).

The reassertion of the role of the government in health and other policy areas – variously described as recentring, rebalancing, de-privatization, re-regulation, and so on (see Peters, 2004; Ramesh and Araral, 2010) – is not an effort to revert to centralized government of the past. It rather represents an intention to find practical ways to move forward towards developing practical solutions that strike a balance between centralization and decentralization, autonomy and control, government and society (Mor and Yip, 2015; Wu and Ramesh, 2014; Ramesh, 2008). The idea is to stem centrifugal tendencies without stifling local voice and discretion, best described by the phrase *centralized decentralization* (Chua et al., 2019).

Under this conception of stewardship, it is the national government that is ultimately responsible for ensuring that all residents within its territory have access to necessary health care. It is also its responsibility to ensure that critical functions are appropriately allocated so that the right level of government is responsible for the functions for which it is most suited, and there exist mechanisms for holding them accountable.

In parallel with the rise of vertical stewardship has been a shift towards greater emphasis on horizontal coordination of policies, arising from disappointment with earlier efforts to promote functional specialization in governments (Peters, 2015). Inspired by New Public Management ideas, governments around the world had promoted specialization among agencies with individual managers held responsible for the performance of their organization. A by-product of these reforms was fragmentation of governance and public services, highlighting the need for countermeasures. The increasing recognition of interconnectedness among problems, especially 'wicked' problems, also pointed to the need for increased integration (Verhoest, Van Thiel, Bouckaert, and Lægreid, 2012; Hood and Dickson, 2015). As Peters (2015: 5) put it, specialization artificially segmented problems when what was needed was a more integrated approach. The underlying idea behind policy coordination – often used synonymously with policy integration, policy coherence, holistic government, joined-up government, and whole of government – is to promote greater coherence in policies through concerted efforts.

Policy coordination is a process by which actors with different responsibilities but overlapping purpose come together to make joint decisions (Cejudo and Michel, 2017). The objective of the coordination efforts is to develop coherence across programmes such that they not only reduce contradictions but are, in a positive sense, complementary and mutually reinforcing (Peters, 2018b). While policy coordination can be achieved through both top-down and bottom-up measures, what matters is that the national steward owns responsibility for it.

There are various mechanisms for achieving policy coordination, including network governance, collaboration and co-production, and hierarchy (Peters, 2018b). While much discussion a decade ago centred on network and collaborative governance, it is increasingly realized that they complement rather than substituted hierarchical coordination due to the unavoidable centrality of the state in government (Capano et al., 2015; Kekez et al., 2018; Peters, 2018b). As Peters (2018b) summarizes, "The centre of government actually has a very large repertoire of instruments to use when confronting coordination problems. Some of these instruments have been in place for decades, while others are relatively recent innovations." One particularly widely used method for promoting horizontal coordination is the establishment of central agencies and super-ministries with the explicit mandate and authority to promote coordination among related policy functions.

In sum, achieving the goals of universal coverage requires stewardship in establishing mechanisms that promote both vertical and horizontal coordination without stifling local autonomy and specialization. As we will see in the following chapters, the specific arrangements and actions for meeting these difficult challenges and the results they produce vary considerably across nations in Asia and elsewhere.

1.3.2 Provision: Ownership and Management

In health care, the organizational tools a government uses to produce and deliver services have a major impact on the sector's performance and outcomes. Governments usually face two choices: allow private hospitals to emerge and function in response to market conditions or establish public entities to produce and deliver health services. Both are viable alternatives and governments end up relying on a combination of the two. However, they offer different sets of advantages and disadvantages, as we will see in the discussion below.

Healthcare markets in many respects function like typical markets where services are exchanged to the mutual advantage of both the buyer and the seller. Healthcare facilities produce and sell their services for profit while users pay for the services either out of pocket or through third-party payers. Multiple providers emerge to meet demand, giving users the opportunity to purchase services from their preferred supplier. The ensuing competition among providers leads to, from a theoretical perspective, not only higher user satisfaction but also lower prices and better quality. Private providers are especially adept at service features valued by users: expediency, responsiveness, and hospitality (Willis and Lewis, 2016). In practice, private arrangements work largely as expected, as patients tend to display a distinct preference for private providers when they have a choice (Rana et al., 2020). Public hospitals, in contrast, are widely perceived as slow, inefficient, and uncaring.

However, private providers pose serious problems for both households and the society as a whole. First, while private providers are efficient in serving those with the means to pay, they are of no or little use to those unable to pay, which is a large share of the population given the high cost of health services. Second, healthcare markets left alone encourage all stakeholders (patients, providers, and insurers) to engage in economic malfeasance. The deep information asymmetries prevalent in health care especially allow providers to pursue their self-interest at the expense of other stakeholders. Third, modern health care is a costly business requiring large investments in technology and personnel which raise entry barriers for investors. Without government intervention, one would expect insufficient supply of health services that are unprofitable or in regions with low-income or sparse populations.

Public hospitals and clinics are uniquely suited to address many of the deficiencies of private provision. First, public ownership allows governments to use public facilities to plug gaps in the private market. Unencumbered by the goal of generating profit or even recovering costs, they can effectively serve households and communities without the ability to pay for health services. Second, public hospitals offer opportunities for reducing capital and operating costs. The stronger position of governments to borrow cheaply and assume financial risks allows them to invest in health services and facilities in areas and regions avoided by private providers. Moreover, public hospitals tend to be large with potential for significant gains from economy of scale. Governments can reap further efficiency gains by coordinating services across facilities in a given geographic area. For example, expensive medical equipment may be shared among a network of government facilities in a region. Governments may also use their bulk buying power to reduce costs (Tangcharoensathien et al., 2015). These options are also available to private providers who can organize themselves in networks but is difficult to realize due to competition among them.

Third, public hospitals allow governments to establish standards of clinical care and practice and benchmark costs. Health care is an information-intensive industry whose needs are compromised by proprietary information held by private firms. While it is possible for governments to mandate providers to share information on key performance indicators, this is difficult in practice due to resistance from private providers. It may be easier for governments to establish and operate their own medical facilities to acquire the necessary information rather than depend on private providers for it (Howlett et al., 2020). Moreover, public hospitals can enhance competition in the sector if they are viewed by users as an attractive alternative to private providers.

Fourth, public hospitals allow governments to simplify the management of the sector. Rather than instituting and enforcing additional regulations to achieve desired goals, it may be easier for governments to establish their own facilities and direct the managers to behave in the desired manner. Government ownership allows policymakers' mechanisms to intervene directly through internal decisions and instructions, instead of negotiating with external parties and regulations, as is the case with private ownership. For example, in response to the Covid-19 pandemic, governments in India introduced price and quality regulations for private hospitals, something that was accomplished through routine directives to public hospitals (Thiagarajan, 2020).

In reality, however, it is common for public hospitals and clinics to fall far short of their potential for a variety of reasons. Public organizations face massive principal–agent problems that diminish their potential for efficient operation. At the root of the problem lie the difficulties in incentivizing managers to work in the public interest and holding them accountable for performance (Preker and Harding, 2003). Public hospitals are often staffed and run by career civil servants with ironclad job security and remuneration unrelated to performance. Furthermore, they are often managed by doctors with limited training in management, which leads them to focus on medical rather than operational and strategic issues. The traditional accountability mechanisms, with emphasis on process rather than outcomes, also discourage innovation and indeed motivation. In developing countries especially, hospital managers are encumbered by ritualistic processes that frustrate them and undermine performance. The problems are aggravated by lack of market discipline and hard budget constraints which protect hospitals from financial losses. In some egregious cases, outdated budgetary practices prevent hospital managers from spending money already allocated to them (Bali and Ramesh, 2015a; Rao, 2019).

Ultimately, governments have to work with both public and private hospitals in order to realize the goal of UHC. To achieve the goal, they need to deploy tools that harness the relative strengths of public and private organizations while minimizing their respective weaknesses. For public health facilities, this primarily means making managers responsive to patients' needs and encouraging them to pay more attention to costs.

Three sets of tools are particularly relevant to improving management of hospitals. The first is some form of *competition* – though not necessarily market competition that requires many preconditions absent from healthcare markets (see Blomqvist, 2011) – to incentivize public providers to focus on users' needs and preferences. For example, governments can require hospitals to publish information on waiting times, treatment costs, and standardized clinical outcomes so that users can use the information for determining where to avail services. Comparison to national and international best practices would also help. Encouraging hospitals to secure national and international accreditation is another measure that can be expected to improve public hospitals' performance.

The second is to more explicitly link public funding to hospitals' *performance*. Paying hospitals based on the number of beds rather than their utilization, for example, removes all links to performance. However, very close links between revenues and performance in health care, without mechanisms to monitor the performance, often leave patients or their insurers with expensive bills. As such, capped prospective payments (to be discussed shortly) should be the norm with a modicum of performance-based payment tacked to them. Thus, performance criteria may be included as a part of global budget and DRG payment conditions. Similarly, fiscal transfers may include more specific goals and performance standards for hospitals to follow.

The third is to balance administrative controls with operational *autonomy* for managers. Hospitals are complex organizations constantly faced with contingent circumstances that cannot be anticipated in advance. Limited managers' discretion in utilizing funds, hiring personnel, and entering into routine contracts undermines their morale and limits their effectiveness. Allowing hospital managers' broad discretion while holding them accountable through performance standards gives managers the flexibility to manage their facilities. User surveys are a useful tool for improving service quality and strengthening accountability in systems without any monetary payment by users. Despite their many flaws, user surveys are a valuable instrument for providers to receive feedback and for payers and governments to monitor performance (Boyer et al., 2006; Perneger et al., 2020).

Ultimately what matters in health systems is not who owns hospitals, but what is done with that ownership. In their efforts to achieve universal coverage, governments are understandably focused on deploying tools that harness the respective strengths of public and private hospitals, while minimizing their limitations. As the discussion in this section has showed, health systems with a greater proportion of public hospitals offer additional leverage to governments in their efforts to achieve universal coverage, provided they are backed by efforts to ensure that they remain responsive to patients' needs and moderate healthcare costs.

1.3.3 Financing: Risk Pooling

One of the key goals of universal healthcare reforms is to reduce the financial hardships that households and communities face in accessing health services. The financial hardships stem from several reasons specific to health care. First, while the cost of routine health services has become relatively cheaper over time, the cost of acute and specialized care has risen sharply, far in excess of consumer price inflation.

This is the result of better diagnoses of diseases and invention of new technologies and therapies to treat them, which while frequently more effective are also usually more expensive (Goyen and Debatin, 2009). The cost of treating a large number of common ailments – such as accidents, cancer, and cardiovascular and cerebrovascular diseases – using modern technology and medicines is too expensive for all but the richest households to afford.

Second, despite advances in our understanding of epidemiology and modelling of diseases, there are vast uncertainties about onset of illness at the individual level, making it difficult for households to plan for it, as they would for pensions, for example. This is particularly the case for hospital care, the single episode of which often costs more than a quarter of the average annual income. Even middle-class households that can afford to pay for primary care find themselves struggling if they require hospitalization for major illness. In the United States, as much as 92 per cent of personal bankruptcies are estimated to be triggered by medical events (Himmelstein, 2009). Building personal savings to cover all hospitalization costs is not possible for most households, necessitating pooling of resources in society.

The conventional approach to addressing the problem of households' vulnerability to health expenditures is for societies to pool the risks of falling ill through: social insurance, private insurance, medical savings accounts, and government expenditures. These arrangements allow members of a population group to share the costs of medical treatment by contributing to the pool in advance and drawing on it when needed. This is not only efficient, because it smoothens out consumption between periods of ill health and good health, but also considered to be equitable, as it helps vulnerable households who would otherwise suffer hardships (Blomqvist, 2011). Without risk pooling, households would pay for the expenses from out of pocket, which is highly regressive as it allows access only to those with the ability to pay for care.

Designing an effective health financing tool is however a challenging task. Paying for all medical expenditures from the pool would be prohibitively expensive but deciding which services to include or exclude raises difficult political and financial questions. The path-dependent nature of health systems also means that adjustments in response to technological and demographic changes are hard to introduce due to opposition from those benefiting from current arrangements. More importantly, not all pooling arrangements are alike, as some forms provide better protection than others, and at very different costs.

However, health financing pools are, and should be, more than cash registers to pay for health care, as they entail strong potential for actively managing the sector. The payers, be it government or insurance, have levers and opportunities to impose conditions that reward desired behaviour and curb undesired behaviour on the part of users and providers. Financing pools also allow the managers' opportunities to use their immense purchasing power to bargain with providers for better prices and quality.

While all financing mechanisms offer this opportunity, their relative usefulness varies considerably. Social and private insurance funds, for instance, have the potential to use their purchasing power to negotiate better deals with providers but they have no incentive to do so, as it is easier for them to reimburse claims and raise premiums, if necessary, to cover the extra costs. Medical savings accounts (MSA) are even less useful due to individual buyers' lack of capacity to negotiate with providers. Government expenditures, in contrast, give policymakers a direct tool to intervene in the sector to shape behaviour. Thus, for example, if the government wants to expand public health services or a particular health service, it can simply spend more money on them rather than require or incentivize preferred behaviour indirectly. Thus, contrary to the commonly held view among public finance economists (for a more detailed discussion, see Blomqvist, 2011), not all pooling mechanisms are functionally equivalent if seen from a broader public management perspective.

Social insurance is one of the most widely used healthcare financing mechanisms in the world. It works by collecting a set premium from members on a regular basis and, in return, entitling them to access health care as necessary. To avoid the problem of adverse selection, it is almost always mandatory. It is considered to be equitable as members contribute according to ability (premium is usually set as a percentage of income) but receive similar benefits when they fall ill. However, it also has several disadvantages. Since it depends on collecting regular premiums, usually through payroll, it is ill-suited for informal workers, whose numbers are particularly large in developing countries (Hsiao and Shaw, 2007). Social insurance programmes are also difficult to administer because they need to diligently collect contributions, manage finances, and contract with hospital networks to provide care. They also have to contend with the challenges of moral hazard typically associated with insurance programmes whereby users demand more services than they need and providers are willing to meet the demand because the payment is by a third party.

Private insurance works on the same principle as social insurance except that it is voluntary. Its voluntary nature makes it attractive to those who are or expect to be sick as well as the rich who want to access more comfortable services. The net result of their attractiveness to particular population segments is high premiums, which makes it poor value for money for the healthy, thus further narrowing the pool (Roberts et al., 2003). They are also expensive to administer, as plentiful experience from the USA suggests (Hsiao and Shaw, 2007). It is therefore unsurprising that voluntary private insurance covers only a small share of the population in most countries.

Medical savings accounts (MSA) is a unique mechanism under which individuals 'save' for their future healthcare needs. Each person has their own account which can be accessed to pay only for their own health care, subject to specified conditions. Only Singapore has mandatory MSA, though many countries subsidize voluntary contributions to MSA. It is efficient at encouraging users to make conscious decisions when accessing health care, but in most instances, patients are poorly equipped to assess the quality of the service they are purchasing. Its greatest limitation, however, is that it is of little help for those with no or little income because their income is not large enough to accumulate sufficient balance to pay for major illnesses; and that it requires to be layered with another mechanism such as insurance to pay for more expensive treatments (Asher and Nandy, 2006).

Direct government expenditure is the oldest and most widely used health financing mechanism wherein the government uses funds from its general revenues to pay for its residents' health care. It is an efficient mechanism because the larger nation-wide pooling of finances by the government makes for lower unit costs. It is also an equitable mechanism if taxes are progressive, because everyone contributes according to ability but benefits according to need.

In addition, direct government spending affords the government levers to change providers' and users' behaviour in desired ways by selectively paying for priority services. For example, governments can attach conditions specifying standards of care or costs to the allocation of funds to hospitals. If public spending is substantial, the government can use its buying power to negotiate better prices and quality through 'strategic purchasing' (Tangcharoensathien et al., 2015). Moreover, in exchange for public funding, the government can require healthcare providers to share information on costs, clinical outcomes, waiting times, and so on that doctors loathe to share as it reduces their control over the system (Pauly, 2009). While private insurers can also perform these functions, they do not face the same incentives to lower costs (and pass on those savings in the form of lower premia across their membership pool).

Critics of public expenditures are quick to highlight the limited fiscal capacities of many governments to pay for health care from their budget. For such governments, it is necessary to mobilize the funds that are currently spent out of pocket and channel them to public exchequer for spending on health care and other programmes. Large efficiency gains can be expected from elimination of OOP and diversion of funds to public exchequer and public insurance. Most countries in Asia that have succeeded in reducing out of pocket spending over the past two decades have achieved it through enhanced public expenditures (World Bank, 2018). This is corroborated by evidence from Asia, where the better managed healthcare systems rely heavily on direct public spending, especially for hospital care, as we shall see in this book. To realize the full potential of risk pooling tools, however, they need to be matched by appropriate payment tools discussed in the following section.

1.3.4 Payment: Prospective and Retrospective

After revenues have been collected and funds have been pooled, the providers must be paid for their services. How the providers are paid – that is, how their services are 'purchased' – has a vital impact on their behaviour and outcomes. The core goal of such a system is to ensure that providers remain responsive to patients' needs while ensuring that the cost of the services remains affordable to households and the society. This is not a straightforward task as the market for healthcare services is shrouded in information asymmetry, involves navigating multiple lopsided principal–agent relationships, and is mired in moral hazard, adverse selection, and other forms of market and government failures malfeasance (Blomqvist, 2011; McGuire, 2000).

First, healthcare providers are deeply guided by the incentives they face, regardless of the effects of their action on overall costs and welfare (McGuire, 2000; Cutler and Zeckhauser, 2000; Christianson and Conrad, 2011). As Roberts et al. (2003) document, if patients are willing to pay more for new technologies, hospitals are quick to acquire them, and doctors employed at these hospitals are quick to recruit patients to use these services. Similarly, if doctors are paid relatively more for certain types of interventions, such as caesarean sections, they tend to prescribe more of such services (Foo et al., 2017; Oster and McClelland, 2019) and less if the price differential is reduced (Kozhimannil et al., 2018). The providers' acute awareness of their own self-interest is paralleled by apathy on the part of users when a third party such as government or insurance is paying for the services. The result is higher expenditures which are ultimately borne by the society as a whole.

Second, healthcare providers enjoy a relative information advantage over other stakeholders, particularly users and third-party payers. Insurance companies, governments, and users usually do not know what drugs and services individual patients need or how to assess their appropriateness or costs and hence depend entirely on providers for information. This opens up vast opportunities for providers to exploit their information advantage to raise their income (Pauly, 2009). The problem is aggravated by users' indifference when a third party is paying for the services.

Third, changing provider payment arrangements is an inherently political exercise because it touches on the core financial interests of the contradictory interests of the different actors in the sector (Fuchs, 2011; Tuohy and Glied, 2011). Healthcare providers want payments to be as high as possible so they can maximize their income unencumbered from any oversight of government, regulators, or even payers, while payers want their own contributions to be as low as possible (Roberts et al., 2003). Third-party payers, for their part, are largely agnostic so long as sufficient revenues continue to stream in. These relationships are complicated by the information advantage and political power that providers wield over other stakeholders which gives them unparalleled influence over the health system (Fuchs, 2011). They use their power to resist any policy effort that erodes their material interests and undermines their control over the health system (Tuohy, 2018; Tangcharoensathien et al., 2019; Sparkes et al., 2019). It is for these reasons that the core goals of a payment system of being responsive to patients and being affordable to households and society are difficult to realize.

At the broadest level, there are two types of payment tools: retrospective and prospective. Under retrospective arrangements, of which FFS is a prime example, providers are paid after a service has been delivered. In contrast, under prospective arrangements such as capitation and global budgets, healthcare providers and payers agree in advance on prices and services to be provided. Traditionally, FFS has been the main form of payment in the private health sector while global or line-item budget has been the main payment mechanism in the public sector. In recent years, there have been efforts to move away from FFS and towards prospective payments following the realization that the latter is more effective at containing costs, which has been a major concern for reformers worldwide (Preker and Langenbrunner, 2005).

FFS payment mechanisms are easy to establish and operate and offer many advantages. Its greatest strength is that by directly tying providers' income to users' visits, it motivates them to focus on users' needs. By the same token, it allows providers' opportunities to increase the volume of services they provide with the objective of raising their own income (McGuire, 2000; Langenbrunner and Liu, 2005). Furthermore, providers may take advantage of information asymmetries to prescribe treatments and drugs that offer them the largest profits rather than therapeutic benefits. The combined effect of sensitivity to users' needs guided by the motivation to maximize income through oversupply of services adds to the payers' and eventually national health expenditures. This is confirmed by the finding of rising health expenditures in countries relying on FFS payment after they expanded public financing for health care (Wagstaff and Lindelow, 2008). FFS payments are particularly problematic when they are paid for by third-party payers because it is convenient for insurers to passively approve claims than to use their authority to monitor and constrain providers. In recent years, however, there has been immense pressure on insurers to curb FFS payments by shifting to or layering FFS with other payment mechanisms.

Prospective payments, which pay a set amount based on some agreed formula, impose an overall ceiling on expenditures and are thus effective at containing cost. On the downside, by imposing a cap on payment, the arrangement encourages providers to skimp on quality and quantity of services to preserve and maintain their income (Ellis and McGuire, 1990). Thus, while prospective payments are useful for dampening health expenditures, they also create systemic incentives to under-provide services both in terms of quality and quantity (Kutzin et al., 2016).

There are different forms of prospective payments, including line-item and global budgets, capitation, and case-based payment. Global budget and line-item budget are the oldest forms of payment and are still commonly used. Under global budget, an overall amount based on some formula is transferred to providers for all operating expenses within a year. Under line-item budget, on the other hand, providers are paid fixed annual amount for different 'items' such as salaries, capital expenditure, pharmaceuticals, and so on. The former allows managers more financial discretion while remaining within the overall limit, whereas the latter ensures that managers spend only on designated items. The latter makes for a rigid system and there is a move away from using it (Langenbrunner and Liu, 2005). Under fixed budget systems, doctors are often on fixed salaries which removes incentives for over-servicing associated with FFS but, conversely, it also undermines providers' responsiveness to patients. Moreover, fixed budgets offer continuing incentives to healthcare providers to recommend cost-effective treatments (Ensor and Langenbrunner, 2002).

Another common prospective payment tool, especially for outpatient care, is capitation. These are fixed annual payments to providers for a given menu of services for a designated population group within a catchment area. The fixed amount per capita that providers receive is meant to incentivize them to lower costs and provide preventive care. When layered with choice of providers, it can encourage providers to remain responsive to patients (Blomqvist, 2011). While capitation payment helps lower overall expenditures, it disincentivizes providers to serve patients who require high-cost interventions. In recent years, this has been addressed through risk-adjusted or age-adjusted capitation rates or even bundling out expensive services and paying from them separately. Calibrating capitation rates to accommodate variations in risk profiles requires extensive epidemiological and demographic data which increase the administrative and transaction costs of managing provider payment systems.

Under case-based payment, providers are paid according to the cases treated rather than per person or per service or per bed days. Case payment can be based on a single flat rate per case, but in most cases it is based on case diagnosis, called diagnosis-related groups (DRGs). In principle, risk-adjusted DRGs are cost-effective, technically efficient, and retain incentives for providers to improve quality. However, they require extensive data and monitoring to ensure that providers do not systematically code patients into more severe cases. Payers have started to address this challenge by layering DRGs with a global budget. This in turn transfers the financial risk to healthcare providers, which makes them understandably resistant to it. DRGs also require extensive analytical capabilities to calibrate and adjust weights and monitor provider behaviour.

Overall, there is no single payment tool – retrospective or prospective – that simultaneously encourages providers to be responsive to patients' needs, improve service quality, and be cost-conscious at the same time. Most effective payment systems therefore use a combination of tools to create the appropriate set of incentives and restraints. However, managing the combination of payment systems increases administration costs and requires more design work in calibrating different sets of payment tools (Roberts et al., 2003).

Health systems with a single or a dominant payer – such as in Canada, the UK, and Hong Kong – enjoy unique potential to shape providers' and users' behaviour through payment arrangements without resorting to subsidy or regulation. In what is called 'strategic purchasing', large buyers (i.e. payers) use their purchasing power as a lever to bargain for lower prices or higher quality vis-à-vis suppliers (Tangcharoensathien et al., 2015; Kutzin et al., 2016). Their leverage over providers is directly proportional to the share of revenue or population they represent (Kutzin et al., 2016). By the same token, the strategic purchasing power is highest when the government is also the dominant provider, because its bargaining is backed not only by market power but the sovereign authority of the state as well. (Ramesh, 2008). This is in sharp contrast to health systems with multiple providers and purchasers, such as the United States where insurers suffer from high administration costs and weak bargaining power (Tollen et al., 2020). Notwithstanding the lower administrative and transaction costs of integrated health financing systems, contemporary health reform literature insists on separating purchasing from service delivery, known as purchaser–provider split, on the assumption that competition is necessary for efficiency and patient responsiveness.

Ultimately, it is upon governments to design payment arrangements that contain costs while meeting the needs of patients. In public health systems, in which providers are typically paid on a prospective basis, the challenge is to ensure that users receive the services they need in a timely manner. In contrast, the main challenge in private systems, which usually work on FFS basis, is to curb providers' profit-seeking instincts without undermining user focus. What countries choose depends heavily on contextual conditions. Systems dominated by private providers and insurance or OOP financing will find it difficult to develop arrangements that serve the society's rather than the providers' interests.

1.3.5 Setting Standards: Regulations

Users of health services are a vulnerable group that requires assistance in ensuring that the medications and treatments they are being served are safe. This protection is typically provided through regulations, which are prescriptions and proscriptions issued by governments (Musgrave and World Bank, 1996; Roberts et al., 2003). Regulations are legally binding commands and failure to comply with them usually involves penalty. They take various forms, including rules, standards, permits, prohibitions, laws, and executive orders (Howlett et al., 2020). While health regulations are mostly designed and enforced by government departments with responsibility for health-related matters, some tasks are subject to 'self-regulation' whereby professional bodies regulate the conduct of their own members (White, 2014). Thus, different associations of doctors, dentists, nurses, and the like develop licencing procedures and codes of ethics for their members and ensure that they are complied with.

While there are many reasons for using regulations as a policy tool (see Mitnick, 1981), two are of particular significance to health care. The first is equity: different individuals have different income and varying needs for health care and, if left on their own, many will not be able to afford the care they need. Regulation is a powerful way to ensure affordable prices, including compelling payers and providers to cross-subsidize needy individuals. The second reason is a variety of market failures that afflict health care, particularly information asymmetries, externalities, and monopoly power. A lot of what goes on in the production and consumption of health care is intangible, making it hard to assess quality and cost. It is particularly difficult for users to know what they need and how much it costs and of what quality, giving providers a powerful advantage. Health care also has numerous externality effects, both positive and negative, which may be promoted or constrained by regulations. Thus, users may be compelled to take vaccination shots and providers may be required to maintain minimum standards of hygiene and sanitation. Monopoly is also a potential problem, especially in sparsely populated regions and poorer neighbourhoods where there are few providers, that governments address through regulatory tools (Roberts et al., 2003).

In health care, there are two broad targets of regulation: quality and safety of medical products and services, and the healthcare market. Medical interventions affect human bodies and, as such, ensuring patients' safety is of vital importance. Accordingly, all countries have regulations to ensure the professional credentials of medical personnel and safety of drugs, therapies, and clinical practices. However, the stringency and, especially, enforcement vary substantially across jurisdictions, based on economic, social, and political contexts. While there is a broad societal and political consensus on the need for safety of health products and services, there are considerable controversies over acceptable level of risk and cost compared to benefits, especially with respect to new drugs and therapies.

Regulations of healthcare markets, in comparison, are targeted at altering the behaviour of producers and insurers with the purpose of protecting users from medical and financial hardships. Regulations are particularly needed with respect to private providers, whose core mission is to maximize profit, the pursuit of which incentivizes them to cut corners on safety issues and over-service and overcharge users. Users, for their part, are unable to protect themselves because of their lack of knowledge on what and how much they are being prescribed. Public providers are less susceptible to these shortcomings, though they suffer from other flaws arising from public ownership (Konisky and Teodoro, 2015).

There is strong evidence that private providers induce patient demand for unnecessary goods and services with the purpose of raising their own income (Yip, 1998). Many countries tackle this through ceilings on medical manpower and investments in new medical facilities. This is easier in systems dominated by public provision and financing (as in the UK) where decisions on new technology and drugs can be made through normal planning processes. Private systems, on the other hand, require government regulations to determine investment decisions (Roberts et al., 2003). The effectiveness of the regulations varies substantially across governments, depending on their political and administrative capabilities.

Similarly, private insurers are characterized by proclivities to behave in undesirable ways that require regulatory intervention. For instance, they face incentives to engage in adverse selection, cream skimming, and denial of claims that need to be stymied or eliminated through regulations. This is a technically and politically difficult task, however, evident in the weak insurance regulations that exist in most developing and many developed countries (Roberts et al., 2003; Malhotra et al., 2018). It is unsurprising that private insurance products in weakly regulated environments offer a poor value for money and are widely avoided by the population, unless subsidized by the government. While regulations are necessary for social insurance as well, many of the critical issues can be tackled through the funds' internal decision processes rather than regulatory intervention.

In addition to their intrinsic value as command and control tools, regulations also play a vital role in supplementing and complementing other tools. For example, social insurance schemes can work only if there is a regulation requiring all members to contribute or else adverse selection will lead to spiralling costs and eventually bankruptcy. By the same token, regulations are more effective when they are accompanied by incentive and information tools (Laffont and Tirole, 1993).

Notwithstanding their many usefulness as tools for shaping providers' and insurers' behaviour, regulations are a difficult tool to use. First of all, they are not self-implementing, which means that they need to be actively enforced on the target subjects who seek to resist and evade them (Landy et al., 1990). Writing rules and enforcing them therefore require considerable technical, administrative, and political capacities that are not always easy for governments to muster (Ayres and Braithwaite, 1992). A lot of the expertise and information required to regulate health care are held by private firms, which make it difficult for even otherwise capable governments to develop effective regulations.

Before resorting to regulations, governments should consider if other tools are more appropriate for the purpose. Thus, while governments can command providers to provide free or cheaper services to the poor, it may be more effective to achieve the same objective through public provision or public subsidy because they are easier to design and cause fewer market distortions (Carmichael and Pomerleano, 2002). Similarly, instead of reducing drug prices by fixing lower prices, which are either easily evaded or cause other distortions, governments may consider smarter ways of achieving the same objective. Many public providers and social insurance funds, for example, use their vast buying powers to bargain with producers for lower drug prices or set benchmarks.

1.4 Case Selection and Background

In this book, we study health policy design of six diverse governments in Asia: China, India, South Korea, Thailand, Hong Kong, and Singapore. Together they display sufficient similarities as well as differences to shed light on the workings and effects of different health policy tools. The group includes both public and private healthcare systems, offering opportunities for gaining insights into their different modes of functioning and effects.

Our cases include two of the largest countries in the world by population – China and India – as well as two small city states – Hong Kong and Singapore. In between, there are two middle-sized countries: South Korea and Thailand. The study includes economies at very different levels of development, with India being the poorest (though still lower middle income) and Hong Kong and Singapore being some of the wealthiest places on earth. It also includes South Korea whose economy moved from a low- to high-income country within a generation. The case of Thailand is instructive because it has succeeded in building a highly effective health system over the past two decades despite sluggish economic growth. All six societies have relatively youthful population, though all (except India) are ageing rapidly.

There are significant patterns of similarities and differences in the historical legacies of these governments and their health policies. Three of the health systems studied here – Hong Kong, India, and Singapore – trace their history to British colonial rule and share many features with the traditional British National Health Service, though India has deviated substantially from the public model it inherited. The historical legacy of the Chinese health system is rooted more in central planning than the earlier colonial rule which was relatively short and confined largely to coastal areas. The rapid liberalization of the Chinese economy in the 1990s introduced *laissez-faire* elements that sit uncomfortably with the basic health architecture built on rigid controls. The contemporary health systems in Korea and Thailand are unique in that the military governments in both countries largely neglected health care until the 1970s, except for public employees and military personnel (Haggard and Kaufman, 2008). But when they did decide to reform their health systems, Korea was influenced by policies in Japan while Thailand was influenced more by practices in Britain. The occupation and region-based health insurance originally adopted in Korea (until their merger into a single-payer system in 2000) had parallels in similar schemes in Japan (Jeong and Niki, 2012). While Thailand was never colonized, it has maintained close political ties with Britain for more than a century and has been heavily influenced by its health policy thinking. Specifically, a large number of Thai students trained in Britain in the 1990s and later went on to assume senior positions in the health ministry where they introduced British policy ideas (Pitayarangsarit and Tangcharoensathien, 2009)

The health status of these societies and the efficiency with which they use their resources vary substantially. As Table 1.2 shows, there is no strong relationship between total health expenditures and the results they produce.

Korea (and USA) spends the largest share of GDP on health but its outcome measured in terms of life expectancy is behind Hong Kong and Singapore which spend considerably less. The case of Thailand is notable in that it has a respectable average life expectancy despite being a middle-income country with the lowest health expenditures in the group.

Countries in our study fare less well when measured in terms of access to affordable health care, another vital goal of health policy. The share of population facing 'catastrophic' health expenditures – defined as health expenditure exceeding 10 per cent of total household expenditures, the threshold at which households begin to reduce other essential expenditures (Russell, 2004) – is comparatively high in all countries in the region except Thailand and Hong Kong. The shares of population facing catastrophic expenditures in Korea (21 per cent),

Table 1.2 *Healthcare system performance*

	Total health expenditure, % of GDP, 2015	Life expectancy at birth, years, 2015	Population with household health expenditures exceeding 10% of total household expenditure or income, %, circa 2013*	Bloomberg healthcare efficiency score, 2018 ~
China	4.9	76.4	19.7	54.6 (20th)
Hong Kong	6.1	84.2	5.9^	87.3 (1st)
India	3.6	68.8	17.3	44.9#
Korea	7.0	82.7	21.0	67.4 (5th)
Singapore	4.2	82.9	9.0	85.6 (2nd)
Thailand	3.7	75.5	2.2	51.9 (27th)
USA	16.8	79.0	4.8	29.6 (54th)

* WHO n.d. Global Health Observatory data repository. https://apps.who.int/gho/data/view.main.UHCFINANCIALPROTECTION01v Accessed April 2020.

~ Bloomberg efficiency score 2018. www.bloombergquint.com/global-economics/u-s-near-bottom-of-health-index-hong-kong-and-singapore-at-top. Numbers in parenthesis indicate rank out of 56 countries studied.

\# Authors' calculation.

^ Van Doorslaer et al., 2007

China (20 per cent), and India (17 per cent) are very high compared to the global mean of 9.2 per cent across 133 countries (Wagstaff et al., 2018). China and Korea have high catastrophic expenditures despite universal health insurance because their programmes exclude many medical services and impose significant co-payment requirements. In Singapore, the high catastrophic expenditures are the result of the government's explicit policy to make households bear a part of their healthcare costs. In Hong Kong, the incidence of catastrophic expenditures is low because inpatient care in the territory is provided nearly free of charge by public hospitals. The star performer here is Thailand, with one of the lowest rates of catastrophic health expenditures in the world.

The Bloomberg Health Efficiency Index seeks to estimate the efficiency of healthcare systems by taking both life expectancy and health expenditures into account. The index shows that the region enjoys high efficiency in health spending in comparison to other parts of the world. Indeed, Hong Kong, Singapore, and Korea rank among the top five in the world, with other countries also displaying respectable performance.

1.5 Conclusion

This book describes the health policy design of six governments across Asia. Specifically, it explores how these governments have designed and deployed a common set of policy tools – stewardship and coordination, ownership and management, risk pooling, provider payments, and regulations – in their efforts to achieve universal coverage. This introductory chapter establishes the theoretical context for the study and situates the findings within the scholarship on policy design and UHC. In the chapters that follow, we will show that the largely positive healthcare trends in the region, albeit with some notable shortcomings, are the result of concerted efforts by governments to improve the choice and use of health policy tools.

One of the strongest improvements has been with respect to *stewardship and coordination* of health policy. All governments in the region have recognized the importance of stewardship and coordination in health policy, including firmer guidance of the implementation process. Even in China and India, two vast countries with high degree of decentralization, the central governments have begun to play a more assertive role in delivery of health care in recent years. Governance of health care in Hong Kong, Singapore, Korea, and Thailand has traditionally been highly centralized and state-driven and this continues despite some decentralization in operational matters. In addition to vertical coordination, all governments have established central agencies to promote horizontal coordination of different health policy functions spread across agencies.

One area in which deficiencies remain is the integration between primary and hospital care despite avowed commitment to integrate them. Integration is hard to achieve due to the large financial risks that hospital care poses to households, leading governments to prioritize it. The fact that primary care in the region is provided largely by the private sector and paid heavily out of pocket while hospital care is provided mostly by the public sector (except in India and Korea) and paid through third party also make integration difficult. Yet all governments are trying, with China and Singapore going about it most systematically in recent years. Given the highly privatized structure of health care in India and Korea, the chances of integration are slim unless a fundamental reorganization of the sector is undertaken.

Public ownership and management of healthcare facilities, especially hospitals, are much higher in the region than elsewhere. Indeed, hospitals are overwhelmingly in the public sector in all countries in the region except India and Korea. Governments in Hong Kong, Singapore, and Thailand have a long record of running public hospitals efficiently and recent management reforms have further improved their performance. However, public hospitals in China and especially India are unable to live up to their public service mission due to the managers' misaligned incentives. The traditional rules and norms of civil service have also hindered hospital reforms, especially in China and India.

Governments in the region have increasingly recognized the scourge of OOP payments and expanded *risk pooling* through increased spending from both budget and social insurance. Indeed, the speed at which China and Thailand, and India more recently, expanded the share of public funding has few parallels anywhere. Yet, OOP accounts for more than one-third of total health spending in all countries in the region except Thailand. Direct government spending has been more effective at shaping the behaviour of providers and users. Social insurance funds, in comparison, have viewed their role merely as a financing vehicle rather than as active purchasers who use their immense purchasing power to bargain with providers for better prices and quality.

All countries studied here are in the process of moving towards prospective *payment for providers* with the purpose of containing health expenditures. The need for it has been less pressing in countries with publicly provided and financed hospital systems – such as Hong Kong, Singapore, and Thailand – because they have always operated under line-item or global budget. In contrast, providers in China, India, and Korea have traditionally worked on FFS basis and are in greatest need for payment reforms. But it is these latter countries that have encountered the greatest difficulties in curbing FFS due to opposition from providers who understandably see the reforms as a threat to their income and professional autonomy. Political solutions to overcome their resistance need to be found if the governments are to mitigate the debilitating effects of FFS.

Regulations are the bedrock of market economy, vital for efficient interaction between buyers and sellers without imposing adverse costs on others. In health care, this is particularly important when the providers and third-party payers are private entities who, in their drive for profit, take actions that distort competition and erode the interests of users. It is the government's responsibility in such situations to establish regulations that promote competition as well as protect users. The need for such regulations is less pressing in systems dominated by public providers because the same goals can be achieved more easily through internal governmental directives. Given that private provision is most prevalent in India and Korea, these two are most in need for a strong regulatory framework but have made little headway in building it.

Ultimately, governments have to work with the policy tools that they have at their disposal. Seldom do governments have access to 'new' tools to fashion designs *de novo* (Mukherjee and Bali, 2018). Rather, their choice of policy tools is shaped by historical legacies, the interests of elites, and dominant ideas. Governments in many instances are constrained by issues related to legitimacy and societal acceptance and are therefore forced to work with instruments that are not ideal for the purpose (Capano and Lippi, 2017). This is particularly the case in health policy. The entrenched interests of elites, particularly healthcare providers, and ossified attitudes on the role of governments in the healthcare sector constrain options that health policymakers can contemplate. However, as we will see in the following chapters, Asian governments have been adept at working within the existing constraints in their efforts to achieve universal coverage.

2 *The Economic and Social Context of Health Care in Asia*

China, Hong Kong, India, South Korea, Singapore, and Thailand are home to 2.9 billion people forming more than one-third of the world population of 7.6 billion in 2018. Together they account for nearly 9 per cent of global GDP in nominal US$ terms and substantially higher when adjusted for purchasing power. While they share many economic, social, and political traits in common, they diverge even more along many other dimensions. The patterns of divergence and similarity are epitomized in the Human Development Index. On the one hand, we have Hong Kong, Singapore, and Korea ranking high in the index, which is unsurprising given that economic prosperity is a key though not only driver of social development. On the other hand, there are China and Thailand ranking in the middle of the index with India far behind. The purpose of this chapter is to provide economic and demographic background to help understand the context within which health policy exists. The key socio-economic indicators described here affect and are affected by health policies and are the subject of this chapter.

2.1 Economy

The economy and income levels of the countries studied here have risen spectacularly over the decades but deep differences remain due to different starting points and different growth rates. The city states of Hong Kong and Singapore were already high-income places by the 1970s but continued to grow rapidly in the following decades, rising to become some of the wealthiest places on earth. South Korea, on the other hand, started at a relatively low-income level and remains at the lower rungs of high-income group – its per capita GDP is less than half of Singapore, for instance – despite enjoying some of the highest growth rates in the world between the 1970s and the 1990s.

Table 2.1 *Human Development Index*

	2001		2017	
	Index	Rank	Index	Rank
Hong Kong	0.84	26	0.93	7
Singapore	0.82	28	0.93	9
Korea	0.82	30	0.90	22
Thailand	0.66	74	0.76	83
China	0.60	104	0.75	86
India	0.50	127	0.64	130

Source: World Bank 2020a

Among the developing countries studied here, China has the most impressive record in economic growth. It started late, from a low starting point, but grew at some of the fastest rates in the world. Its economy quickly surpassed Japan and many of its middle-income peers in a short period of time. In contrast, rapid economic growth in Thailand started early but stalled in the wake of the 1997 economic crisis and never regained its old strides. India, on the other hand, started late from a low base but its growth rate picked up in the late 1990s and has remained relatively high since, although, not high enough to catch up with its peers.

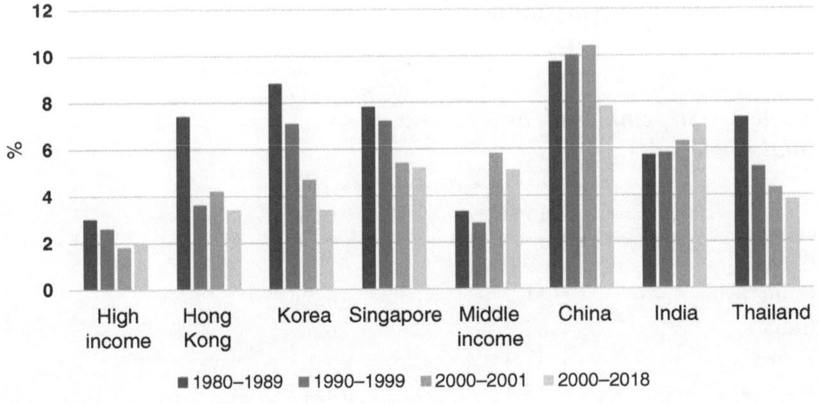

Figure 2.1 *GDP, annual average growth rate*
Source: World Bank, 2020

Table 2.2 *GDP per capita, PPP (current international $)*

	1990	2000	2010	2018
Singapore	23,852	43,864	75,348	98,827
Hong Kong	18,270	28,253	49,185	62,727
High income	18,314	27,931	39,405	50,411
Korea	8,273	18,083	30,365	39,661
Thailand	4,312	7,303	13,195	18,482
China	982	2,921	9,254	15,376
Middle income	2,714	3,995	8,230	11,685
India	1,202	2,096	4,237	6,697

Source: World Bank, 2020

The rapid economic growth in these regions has offered ample economic opportunities to their populace and unemployment has generally been low since the 1960s, even during periods of economic slowdown. However, underemployment is significant in these countries, especially in India and Thailand. This is partially a function of the high self-employment in these countries where large number of people remain on the margins of the formal economy. Informal employment is about 80 per cent of non-agricultural employment in India and 50 per cent in Thailand; the share is considerably higher as a percentage of the workforce. Such households are hard to reach through contributory social protection programmes.

Table 2.3 *Self-employed, as a percentage of total employed*

	1991	2018
China	69.0	47.0
Hong Kong	11.7	8.5
India	85.7	76.5
Korea	37.3	25.1
Singapore	13.4	14.0
Thailand	69.3	51.5

2.2 Poverty and Income Inequality

There is a growing consensus that high levels of poverty and inequality hamper the pace and sustainability of economic development. The World Bank in The East Asian Miracle (1993) argued that East Asia was remarkable in promoting rapid economic growth without increasing inequality due to equitable distribution of the benefits of growth. This is not corroborated by evidence, however, as the region contains both very equal (for example, Korea and Japan) and unequal societies (such as Hong Kong, Singapore, and Thailand). There is a clearer trend with regard to poverty, as all fast-growing countries in the region saw rapid decline in poverty, though significant pockets of poverty persist.

Poverty rates are context-specific and hence hard to compare across societies. Absolute monetary poverty is relevant to low and lower middle-income countries such as China, India, and Thailand. To compare poverty rates in middle-income countries, the number of people living on $3.20 a day adjusted for purchasing power parity (PPP) in 2011 constant prices is a useful indicator. Measured either way, poverty has declined by a wide margin in China, India, and Thailand in recent decades, as shown in Table 2.4. Poverty in higher income societies – such as Hong Kong, Singapore, and Thailand – is usually measured in relative terms, typically as percentage of households living

Table 2.4 ***Poverty headcount ratio, as a percentage of population***

	at $3.20 a day (2011 PPP)		at national poverty lines		at 50% of median household income
	2010	**2018**	**2010**	**2018**	**2018 circa**
China	28.5	5.4	5.7 (2015)	1.7	..
India	69.9	NA	29.8	6.9	..
Thailand	2.5	0.5	16.4	9.9	..
Hong Kong	NA	NA	NA	NA	20.4
Korea	0.7	NA	NA	NA	17.4
Singapore	NA	NA	NA	NA	26.6

Source: World Bank, 2020. (Ng, 2018) www.censtatd.gov.hk/hkstat/sub/so461.jsp https://en.wikipedia.org/wiki/Poverty_in_Indiahttps://data.oecd.org/inequality/poverty-rate.htm

below 50 per cent of the median household income. However, it is worth noting that relative poverty measurements reveal more about the extent of disparity in a society rather than monetary deprivations.

In the early 1980s, almost the entire Chinese population lived in poverty defined as living under $3.10 a day (2011 PPP) but by 2018 the share had declined to around 5 per cent, representing one of the fastest poverty reduction trends in human history. Using official poverty line of RMB 2,300/year, which is very low and relevant only for measuring extreme poverty, 1.7 per cent of the population was living in poverty in 2018. Poverty is largely a rural phenomenon in China, as extreme poverty has been almost entirely eradicated in urban areas. However, a large percentage of the non-poor live only slightly above the poverty line and are highly vulnerable to slipping into poverty during economic downturn. On 28 May 2020, Prime Minister Li Keqiang noted that "China has over 600 million people whose monthly income is barely 1,000 yuan (USD 140) and their lives have further been affected by the coronavirus pandemic" (Poverty in China, 2020).

India too achieved an impressive reduction in poverty rate, which when measured in terms of the PPP$ 1.90 (to account for its considerably lower per capita income) declined from 49 per cent of households in 1993 to 21 per cent in 2011. Measured by national poverty line – currently defined as US$0.36 a day for rural areas and US$0.45 a day for urban areas – poverty declined sharply from 30 per cent in 2010 to 7 per cent in 2018. But this definition of poverty is unduly low and a more realistic estimate would yield considerably higher level of poverty still prevailing in India (Drishti IAS, 2019).

In Thailand, poverty, measured by national poverty line, declined from 65 per cent in 1988 to 7.2 per cent in 2015 but increased over the following three years to 9.8 per cent in 2018. The increase was largely due to decline in consumption and income growth among the bottom 40 per cent of the population due to stagnation of wages and decline in farm and business incomes over 2015–2017 ('Thailand's Poverty on the Rise', 2020).

In Hong Kong, measured by official definition of poverty – income less than half the city's median monthly household income, that is, HK$4,000 for one person, HK$10,000 for a two-person household, and HK$16,500 for a three-person household – poverty rate stood at 20.4 per cent of the population in 2018, which was slightly higher than in the preceding several years. Poverty is particularly concentrated among the unemployed, the elderly, single-parent households, and new-arrival families. About 44 per cent of the elderly population is deemed poor, which is twice the share of the general population. The government claims that after cash (such as Comprehensive Social Security Allowance and Old Age Living Allowance) and in-kind benefits are calculated, poverty rate was only 10.6 per cent, indicating that income support programme in the territory is highly progressive ('Poverty Situation', 2019).

While more than half of the Korean population was affected by absolute poverty in the mid-1950s, the share had declined to below 2 per cent by the mid-1990s. Measured in relative terms, poverty increased from about 8 per cent in the early 1990s to 15 per cent as of 2012, partly reflecting the increase in income inequality ('Poverty in South Korea', 2020). Poverty in Korea is heavily concentrated among the aged, half of whom live in relative poverty, which is the highest proportion among Organisation for Economic Co-operation and Development (OECD) countries.

Very little is known about poverty rates in Singapore because the city state does not have an official poverty line and poverty-related data are sparsely available. According to one estimate, 11.8 per cent of all households and 7.5 per cent of working households in the country lived in absolute poverty in 2017 (Ng, 2018). The same study estimates that 26.6 per cent of the population lived in households with income below half the national median income in 2017.

Higher levels of poverty amidst rapid economic growth partially reflect skewed income distribution patterns, with increasing proportions of income accruing to upper income classes. As shown in Figure 2.2, South Korea has been more successful in ensuring an equal distribution of income compared to other countries. Notably, inequality has grown markedly in China and India and slightly in Hong Kong and Singapore. Only in Thailand has the Gini coefficient declined, albeit slightly.

The distribution of wealth is understandably far more unequal than the distribution of income. In Korea and Thailand, the bottom 40 per cent of the income households possess less than 5 per cent of their countries' private wealth while in China and India they hold only around 2 per cent. Conversely, the top 20 per cent of the households own more than three-quarters of the wealth in China and India and around two-thirds in Korea and Thailand. While comparable data are not available for Hong Kong, the distribution of wealth in the city state is likely to be worse than their regional counterparts.

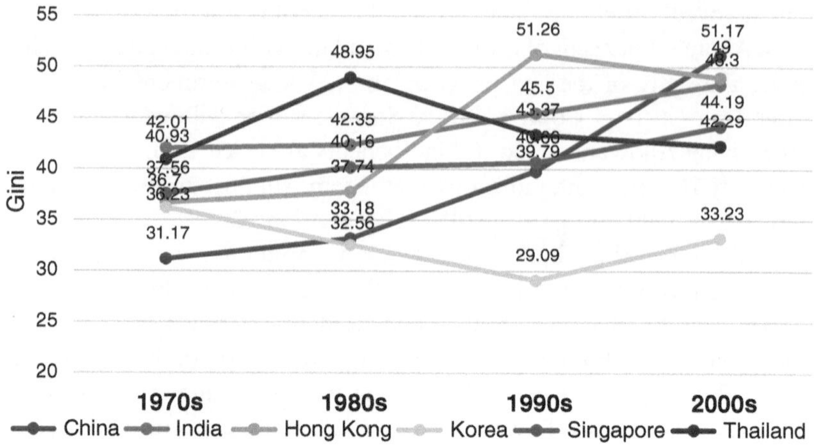

Figure 2.2 *Income inequality, Gini coefficient*

Source: The Standardized World Income Inequality Database (Solt, 2019)

Table 2.5 ***Distribution of income and wealth, by income groups***

	Net income	Net wealth		
	Gini Index 2018	Gini Index 2018	Share of bottom 20%	Share of top 20%
China (2010)	51.0	78.9	0.3	78.4
India (2012)	47.9	83.0	0.9	77.6
Korea (2011)	30.7	70.0	0.1	63.9
Singapore (2018)	39.8	73.3	–	–
Thailand (2006)	43.7	85.1	0.5	69.5

Source: (Credit Suisse, 2016; Samans et al., 2018).

The existence of multigenerational and extended family structure in the region goes some ways in mitigating the ill-effects of poverty. In such families, individuals without income (youth, women, and the elderly) tend to live with those who do have income, resulting in redistribution of income through private transfers (Jacobs, 2000).

Income and wealth inequalities are of course not the only disparities that erode household and social well-being. No less important are regional, gender, and occupational disparities that reflect and aggravate income inequalities. The rural–urban and cross-regional disparities have a particularly adverse effect on access to economic opportunities, education, and health care. Korea has relatively small interregional disparities, with the richest region having per capita income that is only 29 per cent higher than the poorest region (OECD, 2018). Bangkok and the surrounding provinces have grown as much as three times faster than the northeast and southern provinces. Poverty in Thailand is a heavily rural phenomenon, with more than half of the rural population in the bottom 40 per cent of the income distribution, compared to just one-quarter of the urban population. The regional inequalities are reflected in the election results which see parties affiliated to former Prime Minister Thaksin repeatedly winning elections that are resented and overturned by parties affiliated to the elites in the central provinces (Ariyapruchya, 2019).

In China, rapid economic growth in coastal areas has led to substantial inequalities between coastal and interior regions. The share of output of the eastern region rose from 43.8 per cent in 1980 to 55.6 per cent in 2005 while other regions' shares declined correspondingly (Li and Xu, 2008). By the late 2000s, per capita GDP in the inland regions was less than half of that in the coastal regions (Fan et al., 2011). Interprovincial differences are even larger: per capita GDP in Shanghai is 10 times as large as in Guizhou. The rural–urban gap in per capita income is about three times across the country. Income disparities are also reflected in health and other indicators: infant mortality rate (IMR) in the coastal provinces is half of the rate in inland provinces. In both coastal and inland areas, rural IMR was about twice the urban IMR.

In India, regional disparities are even more pronounced than in China. Poverty rates range from 5 per cent of the population in Goa to 40 per cent in Chhattisgarh and 37 per cent in Jharkhand (Jose, 2019). The patterns of inequality have persisted despite faster rate of economic growth in recent years in some of these poorest states. The income inequalities are reflected in education, health care, infrastructure, and other socio-economic indicators of well-being. On socio-economic indicators, the southern states perform significantly better than their northern and eastern counterparts (Kumar and Rani, 2019).

2.3 Public Finance

Many Asian economies have had an enviable reputation for managing public finances, maintaining both low taxes and expenditures and a healthy budget balance.

Government revenues and expenditures as a proportion of GDP in these countries are enormously diverse, being highest in Korea and lowest in India, which is to be expected as it is broadly correlated to income. However, the revenues of Hong Kong and Singapore governments are unusually low for high-income economies. Public revenues and expenditures as percent of GDP have grown significantly in China despite large expansion of the economy itself. Thailand too has experienced growth in public finances, though in the context of slow economic growth rate. In India, on the other hand, revenues' and expenditures' share of GDP have been flat amidst rapidly growing economy.

The modest tax revenues in these economies are confirmed by their low-income tax rates, except for South Korea where the rates are comparable to the developed economies. The tax rates in Hong Kong are by far the lowest, with individuals paying a flat rate of 17 per cent on income over US$1,778/month and corporations paying at the rate of 16.5 or 15 per cent depending on whether they are incorporated or not. However, due to the high tax threshold, 60 per cent of Hong Kong households do not pay any income tax. In Korea, the tax rate for

Table 2.6 ***Public finance, percent of GDP, annual averages***

	Government expenditures		Government revenues		Surplus (deficit)*	
	2000–2009	2010–2018	2000–2009	2010–2018	2000–2009	2010–2017
China*	18.0	21.0	15.3	19.0	1.5	2.4
Hong Kong*	18.9	19.7	18.9	23.3	0.0	3.7
India	15.7	15.5	12.3	12.5	4.8	4.5
Korea	18.0	23.9	21.1	26.1	1.0	2.6
Singapore	14.7	13.7	20.2	18.0	8.0	7.1
Thailand	16.2	18.7	17.8	19.7	1.10	0.8

* Asian Development Bank, Key Indicators for Asia and the Pacific 2019.
Source: World Bank, 2020

individuals is 42 per cent on income over US$38,000/month and for corporations, it is 28 per cent on income over US$88,420/month. In Singapore, the maximum tax for individuals is 28 per cent on income over US$141,744/year while the corporate income tax rate is 26 per cent. In comparison, the corporate income tax rate is 31 per cent in the UK and 15–35 per cent in the USA.

Remarkably, for some economies in the region, governments' revenues are substantially higher than their tax revenues, which allows the tax rates to be kept low. This is the result of significant earnings from land sales, user charges, and profits of state-owned enterprises. In Hong Kong and Singapore, as much as two-fifths of the revenues have traditionally come from non-tax sources, though the share has declined to one-quarter in recent years. In Korea, on the other hand, the non-tax sources' share in total revenues has grown from one-fifth to two-fifths in recent years. China, India, and Thailand have a more typical pattern, with only about one-fifth of total revenues coming from non-tax sources.

Hong Kong and Singapore have had some of the largest budget surpluses in the world, though it has reduced in recent years. China,

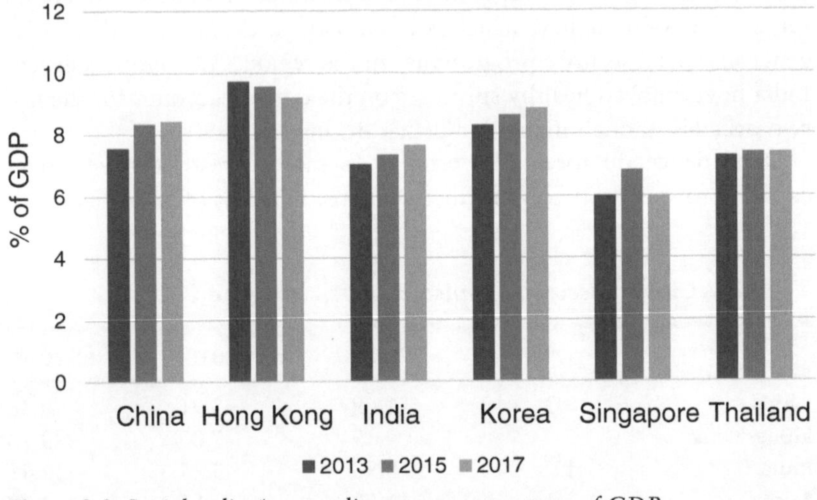

Figure 2.3 *Social policy* expenditure, as a percentage of GDP*
Note: Includes government expenditures on education, health, social protection
Source: ADB Key Indicators 2019

India, Korea, and Thailand have had moderate budget deficits – 4.2, 3.4, 1.5, and 2.5 per cent of GDP, respectively – which are within acceptable range for developing countries with large development needs.

Countries in the region are known as frugal spenders on social policy (indicated here by education, health, and social protection) and this is partially reflected in the data. While they all spend less than the OECD welfare states, there are significant variations among them. Hong Kong is the largest spender on social policy (9.2 per cent of GDP), belying its self-professed commitment to *laissez-faire*. Singapore, in contrast, is the smallest spender (6 per cent of GDP), confirming the government's self-declared antagonism towards welfare state. China and India's self-declaration to socialism and equality are not reflected in their social policies as their spending is similar to other countries in the group.

2.4 International Trade and Finance

The effects of a country's international financial position are said to increasingly impact their social policies. Large current account deficit is believed to undermine investors' confidence in a country, pressuring governments to pursue prudent and generally tight fiscal policy. On this account, these Asian economies have a tremendous reputation for sound foreign economic policy, which is often used to explain the absence of generous social security programmes in the region. All countries except India have enjoyed healthy surpluses on the current account for the last two decades, though many had deficit during the 1990s.

Dependence on foreign investment is said to restrict government capacity to spend on non-productive items, which is of great relevance

Table 2.7 *Current account balance, as a percentage of GDP*

	1990	2000	2010	2018
China	3.3	1.69	3.91	0.36
Hong Kong		4.39	7.00	3.73
India	1.0	0.98	3.25	2.41
Korea	2.2	1.81	2.55	4.78
Singapore	8.6	11.07	22.93	17.61
Thailand	8.5	7.37	3.37	5.64

Source: World Bank, 2020

Table 2.8 *Foreign direct investment, net inflows, as a percentage of GDP*

	1990	2000	2010	2018
High income	1.17	5.11	2.76	1.33
Hong Kong	4.26	41.07	36.17	26.76
Korea	0.37	2.05	0.87	0.75
Singapore	15.42	16.15	23.07	25.00
Middle income	0.66	2.50	2.97	1.99
China	0.97	3.48	4.00	1.73
India	0.07	0.77	1.64	1.55
Thailand	2.86	2.66	4.32	2.61

Source: World Bank, 2020

to these economies because some of them are highly dependent on foreign investments. Foreign direct investments (FDI) as percent of GDP is high in all countries, except South Korea, though Hong Kong and Singapore are in a class of their own with annual FDI inflow forming more than a quarter of the GDP. Inflow to India used to be low but has grown exponentially in recent decades. In absolute terms, it is China that attracts the largest sum of FDI. The high dependence on FDI makes policymakers in these places sensitive to the perceptions and interests of foreign investors.

To the extent dependence on trade is deemed to restrict social policy options, these economies would be expected to be severely restricted because of their very high degree of trade dependence. Hong Kong and Singapore are exceptional in this dimension in that trade amounts to more than 300 per cent of their GDP. This is not the case of other economies in the region, but they too have high trade dependency. Even very large economies are heavily trade-dependent, with international trade forming 28 per cent of GDP in China and 43 per cent in India in 2018.

Overall, these economies have a much higher dependence on international trade than the average for high- or middle-income countries. In this sense, one would expect these economies to be far more cautious with public spending commitments than other countries. In reality, however, there is little relationship between exposure to international trade and social spending: the governments in Hong Kong and South

Table 2.9 *Exports and trade as a percentage of GDP*

	Exports		Trade (exports + imports)	
	2000	2018	2000	2018
High income	25.84	31.90	51.58	62.65
Hong Kong	126.03	188.28	247.65	376.50
Korea	35.01	44.01	67.95	83.00
Singapore	188.35	176.38	364.36	326.19
Middle income	26.53	25.61	50.05	51.26
China	20.89	19.51	39.41	38.25
India	13.00	19.74	26.90	43.38
Thailand	64.84	66.82	121.30	123.31

Source: World Bank, 2020

Korea devote similar levels of GDP to social policy despite their very different levels of dependence on international trade.

2.5 Population and Demography

Demographic trends vitally affect the demand for different social and economic services which, in turn, affect what the government does to address them. Many Asian economies, including those studied here, have gone through a deep demographic transition in recent decades and this has, as we shall see in subsequent chapters, decisively shaped their health and other social policies.

The size of the total population is very diverse in these societies. China, Hong Kong, India, South Korea, Singapore, and Thailand together are home to almost half of the world population. China and India separately had more than 1.3 billion people each in 2018 and they together account for 36 per cent of the world population. Thailand and South Korea are medium-size countries with 69 and 52 million population, respectively, while Hong Kong and Singapore are city states with population of approximately 7 and 6 million, respectively.

The population in these economies grew rapidly from the 1950s to the 1970s but then the rate declined abruptly and now have some of the slowest population growth rates in the world. The declining population growth rate was the result of declining fertility rate in recent

Table 2.10 *Population trends*

	Fertility rate (births per woman)		Population growth (annual %)		Population aged > 65 years (% of total)	
	1980	2018	1980	2018	1980	2018
High income	1.97	1.60	0.83	0.47	11.3	17.9
Hong Kong	2.05	1.07	2.67	0.80	6.4	16.9
Korea	2.82	0.98	1.56	0.48	4.1	14.4
Singapore	1.82	1.14	1.27	0.47	4.7	11.5
Middle income	4.02	2.32	1.96	1.06	4.6	7.6
China	2.61	1.69	1.25	0.46	4.7	10.9
India	4.83	2.22	2.31	1.04	3.6	6.2
Thailand	3.39	1.53	2.07	0.32	3.8	11.9

Source: World Bank, 2020

decades, as shown in Table 2.10. Indeed Hong Kong, Korea, and Singapore, along with Japan, have the lowest fertility rates in the world, considerably lower than their high-income peers. The fertility rates in China, India, and Thailand too are lower than the average for middle-income countries. In fact, the fertility rates in all these countries except India are less than half of what is needed to maintain the current population size.

Declining fertility is gradually reflected in population growth rate, which has declined sharply over the last two decades. Population growth in Hong Kong and Singapore would have been lower still, were it not for high immigration in both territories.

Declining fertility and population growth gradually translate into ageing of the population as the number of children and working adults decline and elderly increases. The ageing of the population is eventually reflected in higher costs of pension, health care, and elderly services which the elderly themselves are not mostly able to afford on their own and hence the costs must be borne by the society as a whole. The ageing trend and the associated higher social expenditure needs are compounded by increasing life span of the population as aged live longer due to improving nutrition, lifestyles, and health care.

In the past five decades, these Asian economies have gone through one of the most significant demographic transitions in the world and the trend is likely to continue over the coming decades. Remarkably, countries in the region have aged much faster than their Western counterparts. It took twenty-six years in Europe and fifty years in USA for the old-age dependency ratio to increase from 15 to 20 per cent, but the same occurred in China, Korea, Thailand, and Singapore in less than ten years. The rapid pace gives Asian countries less time to prepare for the fiscal and infrastructure needs of the aged population. The aged dependency ratio will rise faster in future than in the past due to swift drop in the birth rate during the 1960s and 1970s, the age cohort that will be retiring in the near future. Supporting the social protection and especially healthcare needs of the age cohort retiring in the 2020s will be a major fiscal challenge for all these countries. Hong Kong and Singapore have been able to mitigate the trend by maintaining high immigration but this is not the case in other countries. Countries such as China, India, and Thailand will become 'old' before they become 'rich', which will be unlike the past when only rich countries had large aged population.

The magnitude of the problem is greater than what the above figures suggest because many social security programmes in these economies provide benefits at the age of 60 years rather than 65, which is the practice in OECD countries. Moreover, women will form a disproportionately large share of the aged population. Given that women live

Table 2.11 *Old-age dependency ratio, estimate 2000–2050*

	2000	2010	2020	2030	2040	2050
Japan	25.2	36.0	48.3	53.1	63.8	70.9
Hong Kong	15.3	17.2	26.5	43.7	55.6	64.6
Korea	10.2	15.0	22.2	37.6	54.4	65.8
Singapore	10.3	12.2	21.4	36.5	51.1	61.6
Thailand	9.5	12.4	18.4	29.2	42.3	52.5
China	9.7	11.1	17.1	25.3	39.6	46.7
India	7.2	8.0	9.8	12.4	15.8	20.6

Note: The old-age dependency ratio indicates the size of the population 65 years of age and older as a share of the working-age population (15–64 years old)
Source: Salgado, 2017

longer, have less savings, and suffer from higher incidence of debilitating diseases, the problems will only increase. Heller (1999: 52) estimates that even without the addition of new programmes or expansion of existing programmes, public expenditures on income maintenance, health, and education as a percentage of GDP will grow continuously in the coming decades. This increase will take place despite the decline in education expenditures due to declining birth rate.

Another distinguishing feature of these economies is the high level of urbanization of their societies although with variations. Almost the entire population in Hong Kong and Singapore and four-fifths of the population in Korea live in urban areas. On the other hand, China, Thailand, and India have 41, 50, and 66 per cent of the population, respectively, living in rural areas (World Development Indicators, 2019). A trend of particular concern from healthcare perspective is the ageing of the population in rural areas where health care is more difficult and expensive to deliver.

2.6 Healthcare System and Performance

International comparisons of health policies often dwell on expenditures on the one hand and some outcome indicator – such as IMR or poverty induced by health expenditures – on the other. While convenient and easy to construct, these are of limited help because a lot more than expenditures goes into shaping outcomes. More importantly, highlighting correlations between two sets of data generates unclear or misleading policy implications. What would be more useful for both policymakers and scholars is to gain insights into the different elements that contribute to better overall health outcomes. It is for this reason that this book provides a detailed examination of the governance institutions and processes and mix of policy tools in six Asian countries. The following discussion offers comparative summary of key health-related indicators in the region as a reference point for subsequent discussions.

2.6.1 Expenditures

At the global level, health expenditures tend to be strongly co-related to income levels, but not in the countries studied here, all of which spend less on health than their high- and middle-income peers. As Table 2.12

Table 2.12 *Health expenditure*

	Total health expenditure, % of GDP		Government health expenditure, % of total government expenditure	
	2000	2015	2000	2015
High income	9.4	12.4	14.7	23.3
Hong Kong	5.2	6.1	12.3	12.4
Korea	4.0	7.0	8.0	12.9
Singapore	3.4	4.2	6.7	12.4
China	4.5	4.9	6.1	9.4
India	4.0	3.6	3.3	3.1
Thailand	3.1	3.7	13.3	15.3

Source: World Bank, 2020

shows, Korea is the largest spender on health in the region, but its total spending of 7 per cent of GDP is nearly half of the average of 12.4 per cent for rich countries. In contrast, Hong Kong and Singapore, which are among the richest economies on earth, spend only 6.1 per cent and 4.2 per cent of GDP, respectively. China is the largest spender among middle-income countries, with India and Thailand far behind. What is remarkable is the similar level of spending in India and Thailand though the latter has substantially superior outcomes, as we will see later. Between 2000 and 2015, there was a sharp increase in health expenditure in Korea, similar to its high-income peers, but not in Hong Kong and Singapore, where the increase was comparatively small though still substantial. Remarkably, expenditures declined in India and increased only slightly in China and Thailand despite massive expansion of health financing programmes.

In terms of share of government expenditure on health as a percentage of total government expenditures, we see a different picture. Thailand is the largest spender from this perspective, spending 15.3 per cent of total government expenditures on health. It is followed by Hong Kong, Korea, and Singapore, where government spending forms about 12 per cent of their government's total spending, which is almost half of the average of 23.3 per cent in high-income countries.

There is a vast difference in the share of total health expenditures accounted for by public and private sources. Public sources (comprising both government budget and social insurance) form approximately one-half of total health expenditures in Hong Kong and Singapore and three-fifths in Korea, which is considerably lower than the OECD average of 71 per cent. Public sources form a higher percentage in China (60 per cent), and especially Thailand (77 per cent), and markedly lower in India (26 per cent) (World Bank, 2020).

Conversely, OOP payments form around one-half of total spending in Hong Kong, Korea, Singapore, and China. In Thailand, in contrast, OOP's share is only 12 per cent while in India it is astonishingly 65 per cent. The respective shares have been relatively stable in Hong Kong, Korea, and India but OOP's shares have declined substantially in China and Thailand and to a lesser extent in Singapore. Private insurance forms about 15 per cent of total expenditure in Hong Kong and Singapore and less than 10 per cent in the other countries. However, its share of total expenditures has been increasing gradually in all countries in recent years, often with government support, as in China, Hong Kong, and India, as we will see later in the book.

The number of doctors and nurses in a society are co-related to society's income levels and this is largely reflected in Table 2.13. However, Hong Kong has somewhat fewer doctors than expected while India and Thailand have far fewer doctors than expected. The trends are similar for nurses per capita. The case of Thailand is notable given its high health outcomes and suggests that the number of medical personnel are not strongly correlated to outcomes.

Table 2.13 ***Supply of hospital beds, doctors, and nurses, per 10,000 persons, 2019***

	Hospital beds	Medical doctors	Nurses
Hong Kong	50	19.0	70.0
Korea	115	23.7	69.7
Singapore	24	23.1	72.1
China	10	17.9	23.1
India	7	7.8	21.1
Thailand	21	8.1	29.6

Source: World Bank, 2020

The average length of stay (LOS) in hospitals is a measure of efficiency of use of medical resources in a society. It is an especially important measure in ageing societies because the aged are often allowed to stay in hospitals, despite high costs, due to lack of long-term care facilities. Hospitals also prefer longer stay because of the higher profit margins it offers. Over the years, policymakers have been trying to reduce LOS without compromising care and many have succeeded.

Table 2.14 shows vast differences in the length of hospital stays in the region: 18.5 days in Korea and 10.2 days in China compared to 4.6 in Hong Kong, 4.8 in Singapore, and 5.0 in Thailand. It is notable that the countries with the longest LOS (China and Korea) also have commercially driven health system in which providers are paid on FFS basis and OOP forms a large share of expenditures. This is unlike Hong Kong, Singapore, and Thailand, where the bulk of hospital care is provided by the public sector which is not motivated by revenues or profits. LOS has a major effect on overall costs and expenditures, as one study in Hong Kong found that declining LOS from 6.8 days in 2000 to 4.6 in 2010 (despite rising hospitalization rate) was the main reason for slower growth of health expenditures in the territory (Yip et al., 2014).

2.6.2 Outcomes

Health status indicators in the region differ widely depending significantly on income levels. Thus, Hong Kong, Korea, and Singapore along with Japan are some of the richest places in the world and also have the

Table 2.14 ***Hospital length of stay, days***

OECD (2018)	7.7
Hong Kong (2010)	4.6
Korea (2018)	18.5
Singapore (2018)	4.8
China (Shanghai) (2018)	10.2
India (2014)	5.0
Thailand (2009)	5.0

Sources: various. Statista, 2018b; SingHealth, 2018; OECD, 2019

Table 2.15 *Health status indicators, 2015*

	Under-five mortality rate	Life expectancy at birth (years)			Healthy life expectancy at birth (years)		
	per 1000 live births	**Male**	**Female**	**Both**	**Male**	**Female**	**Both**
Hong Kong	2	81.9	87.3	84.2	69.3*	72.4*	
Korea	3	79.5	85.6	82.7	70.7	75.1	73.0
Singapore	3	80.8	85.0	82.9	74.7	77.6	76.2
China	9	75.0	77.9	76.4	68.0	69.3	68.7
India	39	67.4	70.3	68.8	58.7	59.9	59.3
Thailand	10	71.8	79.3	75.5	64.0	69.8	66.8

* 2000.
Sources: World Health Indicators, 2019. (Law and Yip, 2003)

best health status indicators. Child (under five years old) mortality rate, which in many ways encapsulates a whole range of health system related indicators, is around 3 in Hong Kong, Korea, and Singapore, which is among the lowest in the world and indeed half the OECD average of 7.5. Similarly, the child mortality rate in China and Thailand is around 10, which is substantially below the average of 39.8 in middle-income countries while India's rate of 39 is about average.

Population of countries covered in this study enjoy some of the highest rate of life expectancy in the world. Indeed, Hong Kong has the highest life expectancy in the world, followed by Singapore at number three and Korea at number 10 out of 186 countries. Thailand and China also fare well, ranking 53 and 59, respectively, with India far behind at 130. Women in the region live particularly long, especially in Hong Kong.

A better indicator of health system performance is Health Adjusted Life Expectancy (HALE) which subtracts the years of life spent in disability from life expectancy. By this measure, Singapore ranks number one in the world and Korea number twelve. This is remarkable because as recently as 1997 the two countries ranked number 30 and 51, respectively ('Healthy Life', n.d.). While no recent data is available for Hong Kong, it is likely to have remained as one of the top performers. While women live longer, they also experience higher incidence of disability which is reflected in their comparatively lower HALE. Singaporean women rank number one in the world on HALE while Korean women rank number four. China and Thailand too perform well on both life expectancy and HALE when their middle-income level is considered.

Health status is a partia reflection of access to health services which the WHO tries to estimate through its Universal Healthcare Service Coverage Index (UHSCI), which measures population's coverage for essential health services on a scale of 0 to 100. UHSCI is high for Korea and Singapore and respectable for China and Thailand, especially considering their middle-income status. While data is not available for Hong Kong, it can be reasonably assumed to enjoy high service coverage given its publicly provided and financed universal healthcare system.

Table 2.16 *UHC service coverage index, 2017*

Korea	86
Singapore	86
Thailand	80
China	79
India	55

Source: World Bank, 2020

Efforts to assess the performance of the entire health systems of countries and rank them paint a mixed but largely positive picture of the Asian countries studied here. According to Bloomberg Health Efficiency Index, which measures efficiency by comparing health expenditures and life expectancy, Hong Kong, Singapore, and Korea have three of the top five health systems in the world. Remarkably, USA has the second worst system among the 55 countries on the list. Note that the index makes no judgement on quality of care or equity in access.

Economist Intelligence Unit's index titled 'Health outcomes and cost: a 166-country comparison', which also looks primarily at outcomes and spending, shows similar results (EIU, 2014). In this comparison, Japan ranks first in terms of outcomes, followed by Singapore at second, Korea at fourteenth, and Thailand at sixty-two. What is remarkable is the vast

Table 2.17 ***Bloomberg Health Efficiency Index, 2018***

Rank (Total 55)	Economy	Efficiency score	Life expectancy years	Relative cost % of GDP	Absolute cost US$
1	Hong Kong	87.3	84.3	5.7	2,222
2	Singapore	85.6	82.7	4.3	2,280
5	Korea	67.4	82.0	7.4	2,013
20	China	54.6	76.1	5.3	426
27	Thailand	51.9	75.1	3.8	219
54	USA	29.6	78.7	16.8	9,536

Source: Bloomberg, 2018

differences in costs. Singapore and Korea (and most likely Hong Kong, if it were included in the study) produce outcomes at nearly half the cost of Japan, which itself compares well with other high-income countries. The developing countries have a lower rank in terms of outcomes although they also have much lower costs.

2.7 Conclusion

The chapter shows that the region has, in most years, experienced impressive economic growth rates which made the task of addressing social problems easier but also posed new policy challenges. Steep increases in real income brought down poverty levels and, by corollary, the need for government poverty alleviation measures. The only blemish on the otherwise excellent economic situation was the high-income disparity in all societies studied here, except Korea. As a result, all indicators based on 'average' must be interpreted with caution as they may be masking large standard deviations.

Economic growth also had salutary effects on government finances as it meant increasing revenues without increase in tax. Government revenues in both absolute terms and as a percent of GDP have risen continuously in all these economies. In terms of public spending, Hong Kong, India, and Singapore are small spenders, whereas China and Korea are comparatively large spenders, though considerably smaller than their high-income counterparts elsewhere.

The declining birth rate and increasing life expectancy are having a significant impact on social policy in these economies, and the trend will accentuate further in the coming decades. Less will be needed for education and a lot more on health and social security, not just because the aged will exceed children in numbers but also because the protection requirements of the aged tend to be more expensive to meet.

All countries in our sample, except India, enjoy some of the best health statuses in the world. Remarkably, they accomplished this at modest cost, as they all spend around one-half or less of the average total health expenditures in high-income countries. The case of Thailand is the most remarkable because it spends less than one-third of the high-income average. In terms of share of total government expenditure, however, Thailand is the largest spender, followed closely by Korea, Hong Kong, and Singapore. What this shows is that high government expenditure does not necessarily lead to larger total expenditures, as is often suggested by commentators who claim that government expenditures are inefficient. A yet more beneficial effect of high government expenditures is that it is associated with low OOP expenditures, which is inefficient, inequitable, and a major source of poverty.

3 *China*

Rebuilding the Health Care System

The healthcare system in China has gone through more frequent and deeper changes than any country studied here, or perhaps anywhere in the world. It started as a typical centrally planned system which was working effectively in providing basic health care until it was replaced by a freewheeling system amidst opening up of the economy that started in the 1980s. The system began to change again in the mid-2000s and we can now see outlines of a modern system emerging that uses a variety of policy tools to meet the functional imperatives of a modern healthcare system. The recent reforms have targeted all key aspects of the healthcare system, though critical deficiencies in management and regulation of hospitals remain.

The scale of health reforms in China is vast and the scholarship on it burgeoning. However, efforts to understand the reforms have not kept up with the speed and scope of the reforms, especially given that most scholarly works are detailed empirical analyses that miss out on the bigger picture. The studies are also not grounded in public policy which prevents them from focusing on the underlying problems and how they are addressed by the reforms. In this chapter, we will focus on the policy tools that the Chinese government has used to rebuild the healthcare system and assess the extent to which the efforts have ameliorated the problematic conditions.

The Chinese healthcare system until the beginning of reforms in the late 1980s was simple and remarkably coherent, relying mostly on organizational tools backed by straightforward fiscal transfers, social insurance, and simple regulatory controls. It was a highly centralized system with services delivered through public hospitals and health centres. This was followed by, in the 1990s, a de facto market-based system which saw massive increase in costs and decrease in access. The government is currently in the process of building a modern healthcare system relying on a variety of policy tools to steer and manage the sector, as we shall see in the following discussion.

3.1 Policy History

3.1.1 1950–1980

Before reforms began in the 1990s, China's healthcare system was a part of the centrally planned economy in which the population had almost universal access to basic primary and hospital care as well as pharmaceuticals. It was a relatively simple system characterized by public provision and financing and fixed provider payment. Nearly all healthcare services were provided by the public sector and financed through government-sponsored health insurance.

In addition to an extensive network of public hospitals, health centres, and village clinics that offered healthcare services, government agencies and medium and large state-owned enterprises (SOEs) usually operated their own clinics, which provided free primary care to their employees. Hospitals and other health facilities were owned by the government and prices were usually set below cost. Basic primary care and public health services in rural areas (where more than four-fifths of the population lived) were provided virtually free by 'barefoot doctors', who were mostly middle school graduates. All medical staff at all levels were salaried employees on fixed income.

Nearly the entire population was covered by one of the three social insurance schemes: Government Insurance Scheme (GIS) for government employees, Labour Insurance Scheme (LIS) for the urban population, and Cooperative Medical Scheme (CMS) for the rural population. GIS was financed from government budget and covered all current, retired, and disabled government employees as well as teachers and university students. Members paid neither premiums nor charges except for certain excluded services and drugs. LIS was financed by SOEs out of their welfare fund equivalent to 11–14 per cent of total wages. Its members received health services for free, and their dependents could claim reimbursement of up to 50 per cent of their medical costs. CMS was operated at the village level, and financed through rural welfare funds and member contributions. The scheme provided 90 per cent of rural population with access to basic care and covered 50–70 per cent of their total medical expenditure (Liu and Yi, 2004).

3.1.2 1981–1996

The institutional and economic foundations of China's health system began to unravel with the launch of economic reform in the 1980s. First, the public healthcare facilities faced acute financial difficulties due to rising costs (caused by relaxation of various price controls) and declining subsidies. Government budget used to contribute nearly 50 per cent of public hospitals' revenues until the 1980s, but by 2000 the share had declined to less than 10 per cent (Wang, 2004). Second, economic reforms severely undermined the capacity of many SOEs to fund health care for their employees. As many SOEs folded across the country, their current and retired employees and their families were left uncovered by any health insurance. By 1998, only 45 per cent of the urban population and much lower share of the rural population were covered by any health insurance. Of those who remained covered, the benefits were often reduced through higher co-payment requirement. Third, the dismantling of collective farms during the 1980s led to the demise of the CMS, leaving the vast majority of rural population without any form of healthcare coverage. At the end of the 1970s, 90 per cent of the rural population had access to reasonable quality care and some protection against catastrophic expenses (Ooi, 2005), but by the end of the 1980s the proportion had shrunk to 5 per cent.

Amidst the great financial hardships that both providers and users were experiencing, the government continued to reduce healthcare subsidies, which rapidly declined to less than 10 per cent of the providers' expenses by the early 1990s (Eggleston, 2012). The reduction was not inadvertent but a part of deliberate policy to promote marketization of the sector by pressuring providers to earn fees directly from users to recover their costs. In addition to reducing fiscal burden on the government, this was expected to sharpen the providers' market instincts which were believed to have been dulled by central planning. Adopting the strategy used to reform SOEs, the government reduced subsidies for hospitals and at the same time allowed them autonomy to earn income from sales of services and drugs (Wang and Wang, 2007).

The hospitals, for their part, responded enthusiastically to the government's new policy. They not only began to charge fees but also incentivized the managers and physicians to earn more by tying their compensation to the revenues they generated. Aggressive prescription of expensive drugs and diagnostics rapidly became widespread (Liu and Hsiao, 1995; Xu et al., 2010). In rural areas, 'barefoot doctors', who had formed the backbone of the rural healthcare system, became independent entrepreneurs who earned their livelihood by selling drugs.

The simple policy measure of reducing subsidies and encouraging providers to collect fees from users had the effect of establishing a nearly free market in health care in one quick swoop. The results on the ground were swift and dramatic: OOP spending and THE rose exponentially while access declined sharply. While the increase in OOP was by policy design, the rapid growth of THE and reduction in access were not.

In an effort to help users access basic services, the government set artificially low prices for primary care, essential drugs, and standard diagnostic tests and surgeries while allowing hospitals to set their own charges for the rest. But this did little to address the problem, as hospitals increased their supply of services not covered by the restrictions, such as prescribing expensive diagnosis and pharmaceuticals and longer hospital stays (Liu et al., 2000; Meng and Eggleston, 2008). According to a conservative estimate, 20–30 per cent of China's overall healthcare expenditures were spent on services and drugs that were unreasonable or unnecessary (Zhong, 2001).

In the meanwhile, insurance coverage continued to shrink, from 70 per cent of the population in 1981 to 20 per cent in 2003 (Development Research Center of the State Council, 2005). At the same time as insurance coverage was declining, the government's share of THE halved, from 40 per cent to fewer than 20 per cent. Correspondingly, OOP payments increased to fill the gap, from 20 per cent of THE in 1978 to 58 per cent in 2005. Remarkably, though not surprisingly given the lack of restraints on providers, THE rose from 3 per cent of GDP before the reforms to 4.4 per cent in 1998, at a time when China witnessed spectacular increases in its national income.

3.1.3 1997–2008

Given the grave problems of rising costs, unaffordability, and waste, the government began to respond to the problems in the late 1990s. It introduced various social insurance schemes, starting with Urban Employee Basic Medical Insurance (UEBMI) for formal sector workers in 1998, followed by the introduction of the New Cooperative Medical Scheme (NCMS) for the rural population in 2003, and the Urban Resident Basic Medical Insurance (URBMI) for informal sector workers, children, and the elderly in urban areas in 2007. The launch of social insurance programmes without reform of the provision system centred on revenue maximization by providers showed lack of understanding of the causes of the unsatisfactory situation that had triggered the reforms.

The providers' continued focus on revenue generation meant that the additional resources afforded by insurance were used to augment their own income rather than ease the users' financial burden. As a result, insurance did little to reduce total or private expenditures or improve access. The rising OOP spending placed significant burden on households. In 2003, the poorest income quintile in urban areas spent about 11 per cent of their income on health care while their rural counterparts spent 27 per cent of their income (Yip and Mahal, 2008). Unsurprisingly, 50 per cent of the poorest households did not seek care when ill due to financial reasons (Yip and Mahal, 2008). Of those who did obtain treatment often fell into poverty: 30 per cent of poor *(pinkun)* households in 2003 said that healthcare costs were the reason for their poverty. In 2008, 13 per cent of Chinese households experienced catastrophic health expenditure (Wei et al., 2018).

In addition to increasing expenditures and declining access, hospitals suffered from inefficiencies, reflected in low bed occupancy rate (60 per cent) and long duration of hospital stay. Moreover, patients tended to be treated in costly ways through excessive use of drugs and diagnostics (Wagstaff et al., 2009). Launch of insurance programmes without reform of payment arrangements is believed to have aggravated the problems.

3.1.4 2009–Present

By the mid-2000s, the government realized that its policy was causing immense hardships for households, which was undermining the population's trust in government. The Development Research Center of the State Council (2005) was blunt in its assessment of the 1990s reforms, concluding that they were 'basically unsuccessful'. After extensive consultation, in 2009, the government published a report, 'Opinions on Deepening Pharmaceutical and Health care System Reform', which offered a road map to establishing a universal healthcare system by 2020. To attain the goal, the government committed RMB 850 billion (US$125 billion) to 'four beams' of reform: health insurance for all, public health as the foundation of medical services, essential drugs of reasonable quality and cost, and an efficient hospital system focused on care rather than profits. These were to be complemented by 'eight pillars' comprising mechanisms related to management, operations, financing, pricing, supervision, human resources, regulations, and information. The reform momentum continued with the issue of 'guidelines on the reform of public hospitals' in February 2010. In March 2011, the State Council issued the 'reform of public hospital working arrangements', which called for separations of management and supervision, administration and operations, treatment and medication, and profitable and non-profitable services in public hospitals.

The 12th Five-Year Health Plan of China (2011–2015) set a target to reduce the percentage of OOP payment in total health expenditure to below 30 per cent by 2015. The target was achieved, as OOP's share of THE declined to 29 per cent in 2016 as the share contributed by government and social insurance increased (Wei et al., 2018). In 2015, the government announced that it would establish a tiered healthcare system and use payment structure to incentivize secondary and tertiary hospitals to refer patients with defined diagnoses and stable conditions to primary care facilities.

The 13th Five-Year Plan (2016–2020) is more ambitious and sets out a road map for establishing a viable healthcare system for both urban and rural residents. It calls on local governments to restrain increase in medical expenses, with average growth rates kept below 10 per cent for public hospitals. The plan also calls for integration of the different insurance schemes. The 13th plan was followed by the launch of 'Healthy China 2030 Plan' in October 2016 which specifies the goal of achieving health status of high-income countries by 2030. Moving away from FFS towards prospective payment arrangements such as DRG and capitation is a prominent part of the strategy. The macro-governance of the healthcare system has also been altered with the establishment of the National Healthcare Security Administration (NHSA) in 2018 with the overall responsibility for financing and payment matters. Correspondingly, the National Health Commission (NHC) has been established with the responsibility for managing the provision and delivery of health care.

3.2 Health System Performance

Under the publicly provided and financed arrangements that existed until the early 1980s, the healthcare status of the Chinese population improved at one of the fastest rates in history. Between 1952 and 1982, infant mortality rate (IMR) declined from 200 to 34 per 1,000 live births, while life expectancy increased from 35 to 68 years (Blumenthal and Hsiao, 2005). More remarkably, these achievements were reached at a relatively low cost, as total health expenditures formed less than 3 per cent of GDP throughout the 1960s and 1970s.

The high health status that China had already achieved by the 1970s allowed it to ride out the policy lapses of the 1980s and early 1990s. Following decentralization and marketization of the sector during the period, the rate of improvement in population's health slowed down but did not deteriorate. This was partly a result of economic boom and the associated reduction in poverty rate, which enabled households to devote increasing resources to better sanitation and nutrition. Yet, this was an unremarkable achievement considering the vast and increasing resources that were spent on health care after the 'reforms' that began in the 1980s. In the global ranking of health care constructed by the World Bank in 2000, China fared lowly 144 out of 191 countries (WHO, 2000), which was a sharp decline for a country once admired as a leader in the developing world.

Heath protection in China improved tremendously after the launch and expansion of social insurance schemes which now cover around 95 per cent of the population. Yet, OOP expenditures remain high at 29 per cent of total expenditures in 2016, despite massive decline from 60 per cent of total in 2011. However, unlike urban employees who experienced a significant decline in health expenditures due to insurance cover, rural residents' average health expenditures actually increased (Liu et al., 2017). One study found that majority of New Rural Cooperative Medical Scheme (NRCMS) members continued to face catastrophic expenditures even after insurance reimbursement due to increased expenditure arising from the use of higher level facilities, more expensive treatments, and longer length of stay (World Bank and WHO, 2019). The situation has improved in recent years, with the incidence of catastrophic health expenditure declining markedly (Yip et al., 2019). Yet, 28 per cent of respondents in a 2017 nationally representative survey reported that health care was the largest financial burden on their household expenses (Yip et al., 2019).

Total health expenditures in China are large and growing – from 4.5 per cent of GDP in 2000 to 6.6 per cent in 2018 – despite rapid expansion of the economy during the period. Looking closer, about 70 per cent of the increase in health expenditures was due to increase in volume of services, and the remaining 30 per cent due to increase in charges per admission or visit (Yip et al., 2019).

The rise in total health expenditure has been driven mostly by sharp growth in spending from public sources, especially social insurance. More significantly, the expenditures are projected to continue to grow in the coming decades. Based on current trends, it is estimated that total health expenditure in China would increase from 5.4 per cent of GDP in 2015 to 9.1 per cent in 2035, an average increase of 8.4 per cent per year in real terms, of which more than 60 per cent would be on inpatient services (World Bank and WHO, 2019). It is estimated that savings equivalent to 3 per cent of GDP can be achieved by lowering treatment costs alone, highlighting the importance of efficiency (World Bank and WHO, 2019).

The large increase in healthcare costs in China has been driven not just by demand-side forces rooted in population's rising expectations and epidemiological and demographic shifts but also by providers' drive to generate revenues from patients. The supply-side pressures promoting expenditures is most evident in trends related to hospitalization and prescription of drugs and diagnostics. In China, hospitalization rates are around 15 per cent, while the average length of hospital stay is 9.8 days, compared to the OECD average of 7.3 days. (World Bank and WHO, 2019). Similarly, drug sales in China account for 40 per cent of overall health expenditure, which is more than twice the OECD average of 16 per cent (World Bank and WHO, 2019). Alarmingly, majority of hospitalization and drug prescriptions are estimated to be unnecessary for the treatment (Qian, 2017). It is unsurprising that the income generation method of hospitals has been a major focus of reforms in China over the past decade.

The Chinese healthcare system is highly hospital-centric: hospitals account for 54 per cent of national health expenditures, compared to the OECD average of 38 per cent (World Bank and WHO, 2019). There are 4.1 hospital beds per 1,000 population in China, which is higher than the OECD average. Average number of visits to hospitals increased by 93 per cent between 2008 and 2017 compared to an increase of 50 per cent to primary care facilities. Indeed, hospitals are playing an increasingly large role even in outpatient services, expanding their share of outpatient visits from one-third in 2010 to more than three-fifths in recent years (World Bank and WHO, 2019). The government is concerned about the hospital-centric system and is taking steps to build a tiered system of care in which primary care providers will be the foundation pillar and the first point of contact for users.

Patients report a generally high and improving level of satisfaction based on their experience with hospitals and clinics. In a nationwide survey, more than 80 per cent of the respondents viewed their care experiences as positive, especially with the clinical aspects of care (Hu et al., 2019). Patients at larger and specialist hospitals are especially satisfied with their services (Hu et al., 2020). Healthcare managers are harsher in their assessment. Nearly half of all health insurance managers surveyed in 2015 gave low rating for financial protection, equity, efficiency, and fund management and expressed pessimism about achieving UHC (Shan et al., 2017).

3.3 Stewardship and Coordination

Two of the most notable features of the Chinese healthcare system are vertical decentralization and horizontal fragmentation. These institutional features have affected the government's ability to set direction for the sector and shape the sector's functioning and outcomes. Finally, in the mid-2000s, the government began to realign the governance arrangements so as to assert central direction while leaving operational matters to local governments and hospital managers.

There are five levels of government in China. In addition to the national and 31 provincial governments (including 4 municipalities and 5 autonomous regions), there are more than 300 prefectures and prefecture-level cities, 2,800 counties and districts, and nearly 40,000 towns and townships in China ('Administrative Divisions of China', n.d.). All jurisdictions have their own legal, personnel, and financial resources which they use to design and implement a broad range of programmes. The central government is meant to set broad policies while provincial governments design the programme details and local governments implement them. Financing responsibilities are also divided by level of government: provincial governments finance provincial hospitals, county governments finance county hospitals and preventive institutes, while township governments finance township health centres. There is only limited transfer of funds from higher to lower levels of government, though the central government's share has increased somewhat in recent years.

China has a centralized political system with highly decentralized system for designing and delivering healthcare and other public services. Health legislation is the responsibility of the National People's Congress, though major health policies and reforms are often initiated by the State Council and the Central Committee of the Communist Party. Provincial governments develop the actual health programmes which are then implemented by local governments.

Notwithstanding the efforts of the central government to assert its role, the health sector in China remains highly decentralized in some respects but not others. Broadly, the providers enjoy immense autonomy in financial matters, a legacy of loosening of financial reigns in the 1980s and 1990s with the objective of encouraging them to be more entrepreneurial. On the other hand, the providers are subject to numerous rules and regulations, especially pertaining to personnel matters, rooted in the central planning days that severely limit hospital manager's operational autonomy.

Similar to provision, governance of healthcare financing in China is also decentralized and fragmented, though this has changed recently following consolidation of all insurance funds under NHSA. Earlier, UEBMI and URBMI funds were pooled at the municipal level, while NRCMS was pooled at the county level. As a result, there were 7,450 pooling units in the country which not only made coordination difficult but also limited their bargaining power vis-à-vis providers. It also limited the scope for cross-subsidizing from healthier to sicker members and from richer to poorer regions (Liu, 2009; McIntyre, 2007).

In addition to vertical divisions, there are horizontal fragmentations wherein particular functions are spread across several agencies at the same level. Thus, there are about a dozen agencies with overlapping responsibilities in health care. Central government agencies often maintain parallel provincial and local branches with different levels of coordination. The involvement of variety of agencies with their own missions and interests fosters competition and rivalry, which in turn necessitates additional coordination which has not been forthcoming (Qian, 2015).

Different aspects of health care come under different government departments and levels of government, each with its policy agenda and capacity. Local health bureaus are responsible for public hospitals, while local social security bureaus are responsible for health insurance, and local commerce bureaus for distribution of drugs. As a result, implementation and monitoring of programmes are sporadic and disjointed (World Bank and WHO, 2019).

To overcome the governance shortcomings of a decentralized and fragmented system, in 2018 the government reorganized the entire macro-governance structure of the health sector. The National Health and Family Planning Commission (NHFPC), which had replaced the Ministry of Health in 2013, was relaunched as the National Health Commission (NHC) with overall responsibility for provision of health care, including planning, regulation, and administration. At the same time, a parallel new agency called National Healthcare Security Administration (NHSA) was established with responsibility for healthcare financing and payment. NHSA subsumed the responsibilities of administering UEBMI, the Urban–Rural Resident Basic Medical Insurance programme (which had emerged from the merger of urban and rural resident schemes), and the Medical Assistance programme. NHSA is also responsible for drug pricing and procurement. As a purchaser of services on behalf of 95 per cent of the population, NHSA has immense purchasing power to drive change in the sector (Yip et al., 2019).

3.4 Ownership and Management

Consistent with its legacy of central planning, health care in China is characterized by the dominance of public providers. Thus, public hospitals account for more than 80 per cent of hospital beds, 64 per cent of licensed doctors, 82 per cent of inpatient admissions, and 40 per cent of ambulatory visits (Xu et al., 2019). However, they are driven by the same financial imperatives as their private counterparts, which is unlike traditional public hospitals. Similar to private providers, public hospitals rely heavily on paying patients, face hard budget constraints, and are guided by incentives centred on their own pecuniary benefits (Wagstaff et al., 2009). The market concerns stfile public hospitals' public service spirit and distort their accountability to the governments that own them.

Although public ownership of hospitals is the norm, the Chinese government has been promoting private hospitals since the 1980s in the expectation that competition would improve efficiency and fill gaps in the public sector's offerings. In 2012, the government declared the goal of increasing private health care's share of hospital beds to rise to 20 per cent, a target reached in 2015. Private providers play a larger role in outpatient care, in which they account for nearly 50 per cent of all providers, though still only 10 per cent of outpatient visits. Weak and inconsistent regulatory environment coupled with aggressive competition from public providers is believed to be the main reason why private providers remain small. The government has ramped up its efforts in recent years, including additional tax concessions for not-for-profit private hospitals. At the same time, the government is promoting private health insurance through income tax rebate of up to 1,080 yuan (around $160) a year for the purchase of commercial medical insurance. In 2017, private hospitals accounted for 25 per cent of all hospital beds, 18 per cent of outpatient visits, and 14 per cent of inpatient visits (Yip et al., 2019).

Chinese healthcare providers enjoy unparalleled level of autonomy in financial matters (Bloom, 1998), though this has diminished somewhat in recent years. They have broad discretion in financial and asset management, launch of new services, acquisition of equipment, and borrowing. Most significantly, they are able to retain and spend accumulated surplus earnings, which gives hospital managers both means and motives to be entrepreneurial (World Bank and WHO, 2019). The lack of effective accountability mechanisms – with multiple agencies at different levels of government issuing ambiguous and even contradictory directives – further expands their scope of discretion.

At the same time, Chinese public hospitals have little discretion in matters related to human resources management. Despite relaxations in recent years, key decisions on human resource management, compensation, and service pricing remain with government agencies. Hospitals have no control over the size of the allocated staff quota or their salary, thus creating mismatches between hospitals' requirements and the allotted staff. Quotas are specific to institutions which aggravate the problem by restricting mobility across hospitals and localities. To get around the rigidities of the quota, hospitals resort to hiring contract staff, who are often paid lower salary and benefits. Lack of control over these costly and crucial operational matters constrains managers' ability to plan the use of resources and manage their facility efficiently.

The government realizes that the governance arrangements for providers are misaligned – in that the government has the responsibility for many operational matters while hospitals decide on policy matters when the opposite is the preferred arrangement – and has been trying to reform them since 2009 (Qian, 2017). The reforms seek to rejig the arrangements so as to allow providers' autonomy in personnel and financial management but also steer them to work in public rather than their own interest.

To enhance accountability, some local governments have appointed independent governing boards to oversee and supervise the operation of public health facilities. A number of cities have established Hospital Management Council (HMC) for supervising hospitals within their locality. Although they are meant to provide independent supervision, in reality they are often too closely tied to their respective governments to live up to expectations. HMCs in Shanghai and Zhenjiang, for example, are staffed by civil servants who reflect government policies rather than provide independent supervision. The government is aware of the problem and has directed relevant authorities to diversify the membership of governing bodies (World Bank and WHO, 2019).

The primary health care (PHC) system in China has been weak and fragmented since the 1980s and massive efforts are underway to revive them. There are separate primary care systems for rural and urban areas. In the former, it is delivered mainly through village doctors and health workers in rural clinics and general practitioners (GPs) in rural township, whereas in urban areas it is delivered through community, secondary and tertiary hospitals. More than 90 per cent of all urban community health centres and township health centres and 63 per cent of village clinics are publicly owned. Village clinics function as for-profit entities with revenue generated primarily from government subsidies, paid mainly through the health insurance programmes and public health service programmes (Li, 2017).

Following the 2011 publication of national guidelines for establishing a GP system in China, the total number of practising doctors and registered nurses in primary care increased by more than a quarter between 2010 and 2017. In addition, the average annual salary for

PHC professionals increased about two and half times over the same period. The government also increased subsidies for PHC providers. Subsidies' share of the revenues of rural health facilities rose from 23 per cent in 2010 to 37 per cent in 2017, whereas for urban facilities the share increased from 25 to 45 per cent over the same period (Ma, 2019). Yet the quality of primary care remains low in China. In 2014, only 6 per cent of all physicians were licensed GPs (Fang, undated). In PHC facilities outside urban hospitals, 31 per cent of doctors do not have appropriate educational qualifications and the share is even smaller in rural clinics (Li, 2017). A 2017 study showed that township health centres provided the correct treatment only 38 per cent of the time and village clinics only 28 per cent of the time (Yip et al., 2019).

In a bid to guide patients to use more appropriate facilities, the government has been promoting regional healthcare alliances connecting tertiary, secondary, and primary care in a regional area through training and advisory support. However, such networks are specific to the insurance scheme, normally at the prefecture level for urban schemes and at the county level for rural schemes. People can use out-of-network health services, but must pay higher co-payments. The government has set a target of universal registration with a family doctor by 2020. Such family doctors will serve as the first point of contact for accessing higher level services within the alliance. However, the progress to date remains limited with only minimal forms of cooperation, which is understandable given the alliance members' mutual competition for revenues paid on FFS basis. The incomes of hospital managers and physicians are still tied to hospital profits and it is not in their interest to encourage shift from curative care to health promotion (Yip et al., 2019).

3.5 Financial Risk Pooling

Similar to many other countries, there are three main sources of healthcare financing in China – government budget, social insurance, and OOP payment – with private insurance playing a small but growing role. However, as mentioned in the introductory chapter, it is budget spending that gives governments the most direct way of intervening in the sector and shaping its behaviour. Insurance schemes too offer the potential to play a vital role in shaping providers' behaviour when they are large enough to give the buyer monopsonic powers. However, although insurance reimbursements form more than half of hospitals' total revenues, public insurers in China have not used their purchasing power to shape the providers' behaviour (Qian, 2017), though this may be changing gradually with the establishment of NHSA.

Before reforms began in the 1980s, the Chinese government channelled funds to healthcare providers in the form of direct subsidies. Since the providers' income was not tied to service volume or user fees, there was little incentive for overcharging or over-servicing. But the lack of linkage between providers' income and users' satisfaction also meant that the former had little incentive to heed patients' preferences (Liu et al., 2017). This changed with reforms launched in the late 1980s under which government drastically reduced its subsidies and encouraged providers to compensate for the loss by earning revenues directly from users. In the absence of effective controls on their behaviour, providers engaged in aggressive and often unethical practices to earn income, including excessive and unnecessary prescription of tests and drugs. By reducing subsidy, the government inadvertently lost a key policy tool for ensuring access to health care.

Recognizing the limitations of its earlier reforms, in the late 2000s the government began to increase both budget and insurance financing for health care. Currently, nearly half of expenditures from government budget are on subsidies for insurance premia, mostly for rural and urban residents' schemes, whose members earn too little to contribute sufficiently towards their insurance. The remaining government funds provide subsidies to PHC facilities to build infrastructure and deliver free or subsidized care. Central and local governments also fund the public assistance programme for the poor, called the Medical Financial Assistance Scheme (MFAS). Government subsidies formed 36 per cent of PHCs' income in 2017, a large increase from 12 per cent in 2009 (Yip et al., 2019).

Public insurance re-emerged as a major source of health financing in the mid-1990s, starting with the launch of UEBMI for formal employees, followed by schemes for informal urban workers (URBMI) and rural dwellers (NRCMS). Even though enrolment in the schemes was almost universal, they offered little security against large co-payments that the insurance schemes required. It was only in the late 2000s that the insurance schemes, especially URBMI and NRCMS, began to offer meaningful protection to members, after the government increased subsidies and expanded insurance coverage. Insurance funding is mostly for hospital-based services which have the unintended consequence of discouraging visits to primary care clinics.

Currently, there are three main public health insurance schemes in China: Urban Employee Basic Medical Insurance (UEBMI) launched in 1998, New Rural Cooperative Medical Scheme (NRCMS) launched in 2003, and the Urban Resident Basic Medical Insurance (URBMI) launched in 2007. There is also the Government Insurance Scheme (GIS), which offers health insurance to government employees in addition to other social benefits. Finally, there is the Medical Financial Assistance Scheme (MFAS), established in 2003, which provides means-tested support for health insurance premium and OOP payment.

UEBMI is a mandatory insurance scheme for formal sector workers (excluding non-employed family members) in urban areas. After two decades of existence, membership stalled at around 71 per cent, indicating the difficulties that mandatory programmes face in a vast country with large migratory population. Insurance premium is 8 per cent of a member's salary, paid 6 per cent by employers and 2 per cent by employees. However, local governments are allowed flexibility, which has resulted in different rates, with employers' contribution ranging from 10 per cent in developed eastern cities to 4 per cent in poorer north-western cities.

For the rural population, the NRCMS was established in 2003. Although voluntary, nearly the entire target population enrolled in the scheme within a few years of launch. NRCMS was followed by the launch of URBMI in 2007 for migrant workers as well as the non-employed (children, students, elderly, and disabled) living in urban locations. It was a voluntary programme that enrolled only 17–18 per cent of the target population (Zheng and Qian, 2019), which is unsurprising given the various uncertainties faced by migrant workers. Approximately 75 per cent of the premium for both NRCMS and URBMI is paid by central and local governments while the insured individuals themselves contribute the rest. Both schemes offer similar benefits and contain individual and pooled accounts, with the former for critical outpatient care and the latter for inpatient care.

To overcome coverage gaps for rural and urban residents in the informal economy and to promote economy of scale, in 2016 the government began the process of merging the three insurance schemes. As of 2020, NRCMS and URBMI schemes have largely been amalgamated, but UEBMI remains separate due to its separate structure though the government has affirmed its commitment to amalgamation. Under the integrated scheme that has emerged, all members pay the same premium and enjoy the same benefits, regardless of their location (*hukou*). Informal urban workers and rural dwellers pay only 20 per cent of their total premium, with the remainder paid jointly by central and local governments, shared 60:40, respectively.

China has had a separate financing scheme – the Government Insurance Scheme or GIS – for government employees, communist party officials, and public service units (such as schools and hospitals). The scheme was launched in 1952, and extended to members' children in 1955, retired GIS members in 1956, and university students in 1989. The scheme used to be administered by the central government, but in 1989 the responsibility was transferred to local governments. In 1998, the scheme was merged with UEBMI, though the merger is yet to be fully completed. GIS used to be funded entirely from government budget and paid for 100 per cent of the medical costs for the active members and 50 per cent of the costs for children and retirees. Although the contribution and reimbursements are meant to be the same for all UEBMI members in a locality, in reality the former GIS members continue to receive additional benefits in the form of government contribution to their individual health accounts. The amount of such transfers increased from RMB 7.5 billion in 2011 to RMB 15.4 billion in 2016, though as a percentage of total UEBMI expenditures the share has remained stable at 2 per cent (Ministry of Finance, Budget report, various years).

In addition to promoting social insurance, the Chinese government has been promoting private health insurance in recent years with income tax rebate annually. Private insurance is purchased mostly by high-income people to cover deductibles, co-payments, and coverage gaps in public health insurance schemes. Private health insurance has been growing rapidly, with premium already amounting to 6 per cent of total health expenditures despite narrow and shallow coverage (Fang, undated). Expansion of private insurance and the moral hazards associated with it are likely to, if left unchecked, raise costs and undermine the improvements achieved through expansion of budget and public insurance financing.

Individuals living below the poverty line, which is determined by the relevant local government, can apply for assistance under MFAS, established in 2003. The scheme is funded jointly by central, provincial, and local governments. The available assistance is used mainly to pay health insurance premium or the co-payment required at the point of service. In 2018, 76.7 million people (approximately 5.5 per cent of the population) received support for insurance premium payment while 53.6 million people (3.8 per cent of the population) received support for direct health expenses. (Hai, 2020).

3.6 Provider Payment

Fee-for-Service (FFS) is the norm for paying providers in China despite efforts to curb it. The government realizes that FFS and the moral hazards it entails are at the root of a lot of what ails the Chinese healthcare system and it is trying to contain its use. Accordingly, different levels of government have adopted different payment arrangements such as global budget, Disease-Based Payment (DBP), Diagnosis-Related Groups (DRGs), and payment for performance (P4P). Governments are also experimenting with capitation payment for integrated healthcare networks.

In 2012, the central government introduced global budget in the form of annual reimbursement cap for each health facility nationwide. Some governments have experimented with a prospective global budget for integrated healthcare alliances comprising primary care facilities which were paid a fixed annual amount. The objective is to incentivize hospitals to reduce their expenditures by collaborating with primary care facilities (Xu et al., 2019).

About two-thirds of hospitals in China have implemented DBP for conditions with well-defined admission criteria and treatment procedures. Under this system, hospitals receive fixed payment amount for treating certain diseases and there is extra payment if the treatment complies with standard clinical pathways. Some local governments have also adopted payment based on DRG, which distinguishes treatments by their clinical procedures, costs, complexities, and so on. (Xu et al., 2019). However, DRGs are complex to adopt and have had only limited uptake so far. Many provincial governments have also adopted performance-based salary to discourage providers' incentives for over-servicing. The performance-based salary consists of a basic salary forming 60–70 per cent of the total salary and a performance component of 30–40 per cent (Ma et al., 2019).

Excessive prescription of drugs and overcharging for them are major problems in China and have been key targets of government's reforms. Until 2015, expenditures on drugs accounted for around 40 per cent of total health expenditures, which is one of the largest in the world. For hospitals, profits from sale of drugs formed 47 per cent of outpatient revenues and 38 per cent of inpatient revenues in 2014, compared to 3.2 per cent of total revenues that came from government subsidy (Fang, undated). To reduce expenditure, in 2015 the government required providers to sell drugs at cost, with no markup over procurement price, which within two years saw decline in drug expenditures by as much as one-quarter. To offset the loss of a major source of income for providers, prices for medical services such as physician visits and nursing have been increased. Many provincial governments have also introduced a 'general consultation fee' for each patient visit or admission rather than the treatments they offer. Local governments are also being encouraged to raise subsidy for providers to discourage them from seeking ways to earn additional revenues from users.

Provincial governments have also rejigged financial tools to weaken incentives that lead providers to concentrate on revenue-generating activities. Many local governments have ended or reduced bonuses for physicians who were linked to the revenues they generated and instead put them on fixed salaries with a small component of their remuneration tied to performance. Some provincial governments, notably Shanghai, have placed a hard budget constraint on total personnel spending (World Bank and WHO, 2019).

Other changes to payment systems include reduction in reimbursement for hospital treatments without referral from a primary care provider. Relatedly, reimbursement rates for primary care services at hospitals have been reduced to discourage hospitals from offering such services. However, the difference in reimbursement rates is not large enough to overcome users' preference for tertiary-level facilities, which are perceived to provide higher quality care. It also matters that tertiary and secondary hospitals work aggressively to attract patients to their facilities for even basic care.

3.7 Regulations

The nature and sophistication of health regulations in China have evolved over the years. In the beginning, to address the unaffordability issues caused by market-based financial tools – FFS, OOP, and competition for patients – the government imposed price controls on essential drugs and services. Little changed in reality, however, as providers began to prescribe expensive drugs that were outside the regulated list. Many went as far as receiving kickbacks from the suppliers for prescribing their products (Yip and Hsiao, 2009: 614). To reign in rampant price gouging for drugs, the government ramped up the controls in 2009. It published a list of 307 basic medicines that were to be available at all public health centres at purchase price, without markup. Subsequently, an auction system for purchasing medicines was adopted to introduce transparency and reduce prices through open competition. The drugs purchased through auction are then sold at cost without markup.

Price schedules in China have traditionally favoured drugs and high-technology procedures at hospitals rather than primary care and this is being reformed. There are pilot programmes to set fees based on true costs, unlike the past when fees were below cost, which led providers to oversupply services that were outside of controls. Local governments have taken measures to reduce hospitals' ability to profit from prescription and sale of diagnostics and especially drugs, including limitations on what physicians can prescribe. Similarly, treatment protocols have been introduced to standardize diagnosis and treatment.

China has also updated and tightened the licencing requirements for hospitals, clinics, doctors, nurses, pharmacists, and the like. One notable recent change is the change in regulation allowing physicians to work in more than one facility. Previously, they could work only in one facility and most stayed in their job for much of their working life. This had the advantage of blunting financial motives but was inconsistent with efforts to promote private hospitals, which were not always in a position to hire full time.

3.8 Conclusion

China achieved the goal of providing necessary health care to the entire population in the 1960s, earlier than most developing countries. During the central planning era, the government controlled most aspects of production and financing and used its dominance to ensure that the entire population received basic health care. It was a simple system relying mostly on organizational (state ownership and social insurance) and financial tools (subsidy) targeted at the supply side. Medical personnel were paid fixed salaries which offered no incentive to overcharge or over-service, though it also meant impersonal and frugal services.

Reforms introduced in the 1980s and 1990s gutted the existing system and replaced it with ill-conceived, market-based institutions and practices that raised costs and reduced access. The government responded by expanding social insurance in the late 1990s, but this did little to improve affordability or reduce costs as the extra resources were siphoned off by providers who increased their charges and volume of services. It was at this point in the mid-2000s the government realized that piecemeal reforms were insufficient and that it needed to use a comprehensive range of tools to address the pervasive problems. The ensuing reforms have involved the use of a mix of policy tools, targeting all key health policy problems.

China's reforms over the past decade have emphasized reassertion of central government stewardship in setting clear policy direction for the country and ensuring that they are followed by all levels of government. The central government has also introduced organizational reforms to promote coordination of activities of the various agencies in the sector. In parallel, great efforts have been made to improve the performance of public hospitals, the dominant providers in China. Chinese public providers have had exceptional degree of autonomy, especially in financial matters, which they used to advance their own financial interests. To guide them towards focusing more on serving the public interest, the government has set limits and instituted various accountability mechanisms. At the same time, hospital managers have been given more autonomy in operational matters, especially with regard to personnel management issues. The government is also promoting the role of private providers to foster competition but the progress remains limited due to stiff competition posed by public providers.

Chinese policymakers are also paying more nuanced attention to how health care is financed in the country, following the realization from experience that health insurance is unsuited for population segments that do not have income regular or sufficient enough to pay premium. The government has consequently increased the subsidy towards premium for informal workers and rural residents. It has also increased subsidies for primary care, which is typically under-provided when hospitals are driven by financial concerns, as is the case in China. In addition, the government has taken a swathe of measures to curb FFS payment and move towards capped payment arrangements. Provincial and local governments have launched dozens of different experiments with the purpose of reducing the incentives without extinguishing them altogether. However, regulations controlling the quality and safety of drugs and behaviour of providers remain weak, which is a critical shortcoming given the finance-driven orientation of public hospitals in China.

The healthcare system in China proved its mettle in the fight against Covid-19 pandemic. After initial confusion, the province of Hubei with the help of the central government marshalled resources at an unprecedented speed and largely beat the pandemic in two months (Mei, 2020). While much of the success was due to the country's strict social distancing measures, hospitals were highly successful in treating infected patients, with a death rate of 4.9 per cent of total confirmed cases in Wuhan and 2.1 per cent nationwide.[1] The death rate was high initially due to lack of understanding of the new disease, though the low priority given to public health (compared to hospital care) also contributed to overburdening of hospitals that raised death rates. The government's efforts were aided by the public ownership of most hospital beds which were mobilized rapidly for Covid-19 treatment. The newly established NHC also played a key role in coordinating nationwide efforts against the pandemic. However, while the overall response was largely effective, it highlighted the weaknesses of the public health system in China.[2]

[1] www.worldometers.info/coronavirus/coronavirus-death-rate/.
[2] www.thinkchina.sg/chinas-public-healthcare-system-needs-revamp.

4 India
Reforming a Private and Fragmented Health Care System

India lags behind other countries at similar levels of economic development on most conventional health indicators. More worryingly, an overwhelming share of healthcare expenditure is still financed through OOP payments, which is a major cause of hardship for bulk of the population (Dreze and Sen, 2011; World Health Organization, 2019). The poor healthcare performance has traditionally not posed a political problem for successive governments, allowing them to neglect the sector.

Recently, however, health care and social policy more broadly have become a central feature of the national political discourse and the country is amidst implementing its most ambitious healthcare programme. The *Pradhan Mantri Jan Arogya Yojana* (PM-JAY) programme announced in 2018 and rolled out in 2019 aims to provide healthcare coverage to half a billion Indian citizens. It is already the world's largest single-payer health system and offers hope that it will reduce the population's financial sufferings.

The central objective of this chapter is to assess the evolution of the healthcare system in India and examine the policy tools in use to understand the country's preparation for achieving universal health coverage (UHC). In this chapter, we will see that healthcare system continues to be handicapped by a weak public sector and an inadequately regulated private sector, which together form an inhospitable context for publicly financed programmes to succeed.

4.1 Policy History

India was a British colony until 1947 and the colonial government had little interest in health care for the population other than its employees. After independence, the government focused primarily on industrial and agricultural development whilst neglecting social sectors such as education and health care. India's early health policy documents (for example, the Bhore Committee Report, 1946) envisioned the delivery of healthcare services through a network of government-owned and operated facilities: primary care through subcentres and primary healthcare centres (PHCs), secondary care through community health-care centres (CHCs), and advanced medical care through specialist hospitals in cities.

Starved of funds and with limited accountability, the health centres have been poorly run, suffer from chronic absenteeism, are often out of essential medical supplies and equipment, and avoided by patients other than those without an option (Rao, 2017). A government report estimated that nearly 25 per cent of PHCs and 40 per cent of CHCs suffered from inadequate infrastructure (Government of India, 2012). In 2017, it was estimated that CHCs faced a shortfall of 82 per cent of specialists compared to the stipulated manpower requirements (Government of India, 2018).

For most people seeking affordable healthcare services, district and speciality hospitals in urban areas have been the only option. They cater not only to the urban population but also to rural patients who flock to these hospitals bypassing the PHCs and CHCs. Unable to keep pace with the growing demand, public hospitals can only provide inadequate and low-quality services, thus creating a void that has been filled by private providers. The latter of course offer little relief to the vast majority of the population which has limited means to pay for private services.

However, unlike the neglect of the general population, the government has offered reasonably good medical care to those employed in the formal sectors, especially in the public sector. The 1950s saw the launch of new programmes offering a comprehensive range of medical services to those in the civil service and armed forces and their dependents. Expenditure on these schemes increased gradually to form as much as 10 per cent of total government expenditure on health care (Gupta and Chowdhury, 2014). The government also established mandatory Employees' State Insurance Scheme (ESIS) in 1952 for those employed in the formal private sector and their dependents. The ESIS, supervised by the Ministry of Labour, functions through a network of hospitals it owns as well as recognized private hospitals.

By the 1960s, India had in place a barely functioning health system for the majority of the population and a slightly better system for the public sector and formal private sector employees and their families. The neglect of medical care for the vast majority of the population allowed the private sector to grow and dominate healthcare delivery in the country.

Accession to the Alma Ata declaration in 1978 refocused the government's attention on health care and led to the publication of the National Health Plan (NHP) in 1983, the first policy document on the subject since independence. The document set ambitious health outcome targets but contained few details on how it intended to achieve them. Indeed, of the eight healthcare components of the plan (nutrition, prevention of food adulteration, maintenance of drug quality, water supply and sanitation, environmental protection, immunization, maternal and child health services, school health and occupational health), none dealt with curative services, the most expensive component of health care. In the meanwhile, the private sector continued to expand and thrive. Unsurprisingly, all NHP goals were missed (Rao, 2017).

The macroeconomic crisis in 1991–1992 and fiscal stringency conditions associated with the loan from the International Monetary Fund (IMF) resulted in significant reduction in health budgets across state governments. The scaling back of the public programmes coupled with a weak regulatory framework continued to fuel the growth of private healthcare providers. This was accompanied by a decline in the share of public spending on health care from 1.4 per cent of GDP in the mid-1980s to 0.9 per cent of GDP in 2001 (Rao, 2017).

After nearly two decades of policy drift aggravated by the economic crisis, the government announced a new NHP in 2002, which was reaffirmed in the Tenth Five-Year Plan (2002–2007). In addition to the familiar proclamations of commitment to equity, affordability, primary care, and the need for increased government spending, NHP 2002 heralded two new departures. For the first time, the government stated that it would promote decentralization and explicitly stated its intention to expand the role of the private sector in health care. The shift was followed by unprecedented devolution of responsibility for health care to local government coupled with the exponential growth of private delivery. The marriage of privatization and decentralization promoted by the new policy resulted in further fragmentation of the health system which only aggravated the dismal state of the healthcare system (Bali and Ramesh, 2015a).

In 2005, soon after winning the election on 'common man' platform, the Congress-led government launched the National Rural Health

Mission (NRHM) and later the National Health Mission (NHM) with a three-fold increase in budget primarily aimed at strengthening maternal and child health, disease prevention, and primary care for the rural population. The central government provided funding to state governments on a matching basis for delivering the NRHM programmes. Many states did not have the absorptive capacity to spend the expanded resources, and even within states, the better-off districts were able to claim more funds, thus widening disparities (Rao, 2019). Furthermore, the increased central funding did not necessarily result in increased overall spending as many state governments reduced their own spending on the sector (Rao and Choudhury, 2012; Bowser et al., 2019). However, the programme did contribute to the strengthening of maternal and infant health (Rao, 2017: Table 5.1), but did little to reduce OOP expenditure on health.

India took the first serious step towards establishing national health insurance for those in the informal sector in 2007 with the launch of *Rashtriya Swasthya Bima Yojana* (RSBY). The scheme provided free inpatient services, to a maximum of INR 30,000 (US$400) per annum to recognized poor households in designated private and public facilities. While RSBY was a major improvement over the existing programmes, it offered relatively low benefits and was fraught with implementation challenges relating to contractual opportunism (Maurya, 2015). The programme gradually covered 41 million families but did little to reduce OOP spending on health care (Mauyra and Ramesh, 2019; Karan et al., 2014).

Following the realization of RSBY's deficiencies, in 2010 the government established the High Level Expert Group (HLEG) which reported in the following year. The report for the first time since independence proposed detailed measures for the government to take to reform the sector comprehensively. It recommended universal access to essential services through integration of all health programmes under the central government's direction and a massive increase in government funding from 1.2 per cent of GDP in 2011 to 3.0 per cent in 2022. However, little was done to implement the recommendations of the HLEG in the following years.

In 2014, a newly elected government increased the benefit level of the RSBY to INR 100,000 but left the programme design intact and its shortcomings unaddressed (Rao, 2017; Maurya and Ramesh, 2019). However, it introduced a range of other reforms aimed at improving public health outcomes which touched on health care. This included the launch of a national sanitation programme (Swachh Bharat Mission) and a programme (*Pradhan Mantri Ujjwala Yojana*) to provide subsidized cooking gas connections to poor families to reduce reliance on wood-fire stoves (Bhaskar, 2019).

In 2017, the central government announced its most ambitious healthcare programme, *Ayushman Bharat Yojana*, later renamed as *Pradhan Mantri Jan Arogya Yojana* (PM-JAY). The programme subsumes the RSBY, is paid for by central and state governments, and implemented by state governments. A newly created agency, the National Health Authority (NHA) was established to provide policy oversight and stewardship. The programme was rolled out in 2018, a year before the general elections.

After decades of policy neglect, the launch of PM-JAY represented a massive increase in the government's involvement in the country's feeble and fragmented healthcare system. The reform is anchored in two policy documents – the report of the High Level Expert Group (2011) and the NHP (2017) – which called for doubling of public spending on health care and engaging the private sector more strategically in delivering health services. There is little admission in the documents that the same private system which it sees as the solution is the key cause of the problem that the government is trying to address.

4.2 Health System Performance

India has not performed well in international rankings of health systems. In the WHO's 2000 ranking of 192 countries, India ranked 112 on health system performance and 121 on goal attainment (WHO, 2000). Despite improvements in recent performance, it still ranks 145 out of 195 countries in quality and accessibility of health care in the recent Global Health Access and Quality Index (2018) (PTI, 2018). *The Economist* in its ranking of 166 economies in 2014 placed India in the fifth tier (out of six) on health outcomes and costs. India's poor showing on international indices is for a large part shaped by its overwhelming reliance on OOP payments and consequently the inadequate financial protection its health system offers.

Over the past two decades, about two-thirds of all healthcare expenditure in India have been financed through OOP payments, which is a grave indictment of the risk pooling mechanisms that exist

Table 4.1 *India's health indicators in a comparative perspective*

	Life expectancy at birth, total (years)		Maternal mortality ratio (per 100,000 live births)		Mortality rate, under-5 (per 1,000 live births)	
	1995	**2015**	**1995**	**2015**	**1995**	**2015**
China	69.9	75.8	72	27	47.5	10.7
India	60.4	68.0	471	174	108.7	47.7
Indonesia	65.0	68.9	326	126	66.7	27.2
Vietnam	71.9	75.6	107	54	41.2	21.7
Low middle-income countries	61.3	67.3	470	251	106.3	52.8

Source: World Bank, 2018

in India. It is not uncommon for patients and their families to borrow or sell assets to pay hospital bills, which temporarily or permanently pushes them into poverty (Mishra and Mohanty, 2019; Sangar et al., 2019). Keane and Thakur (2018) estimate that OOP payments increased poverty rates by 11 per cent in 1999–2000, 12 per cent in 2004–2005, and 18 per cent in 2010–2011. Their finding implies that 50 million people in 2011 fell into poverty due to the burden of OOP payments.

Although India's healthcare performance is similar to the global average for low middle-income countries, it lags behind countries at similar levels of economic development in Southeast Asia (Table 4.1). On a positive note, maternal and child mortality rates have declined notably in recent years which reflects the increased policy focus and government spending on maternal health (Mishra and Mohanty, 2019; Rao, 2017).

India's headline numbers nonetheless mask wide variations in the performance across states, which stem from different levels of economic development, state capacity, and demographic profiles (Rao, 2015; Table 4.2). The performance of many of the eastern and north-eastern states such as Odisha, Assam, and the populous Uttar Pradesh have been unequivocally poor. These states also have lower levels of economic development, and lower fiscal and state capacity to address these challenges (Rao, 2019). For example, Kerala has a life

Table 4.2 ***Regional disparities in health indicators in India***

	High performing states	Poorly performing states
Infant mortality per 1,000 live births	Kerala (12), Tamil Nadu (21), Delhi (24), Maharashtra (24)	Madhya Pradesh (54), Assam (54), Orissa (51), Rajasthan (47)
Maternal mortality per 100,000 live births	Kerala (66), Maharashtra (87), Tamil Nadu (90)	Assam (328), Uttar Pradesh/ Uttarakhand (292), Rajasthan (255), Odisha (235)

Source: Rao 2015

expectancy at birth of 74.2 years – 12 years more than in Madhya Pradesh (62.4); and the risk of death during pregnancy or levels of infant mortality in Kerala is fourfold lesser than in Madhya Pradesh (Rao, 2015). Similarly, the share of births attended by skilled personnel ranges from 20–30 per cent in Uttar Pradesh, Jharkhand, Chattisgarh, and Assam to 80–90 per cent in Tamil Nadu and Kerala.

The data on the years of life lost (YLL) in India reflect similar disparities. While the top causes of premature death – diarrheal diseases, lower respiratory infections, and pre-term birth complications (Institute for Health Metrics and Evaluation, 2010) – at the national level have not changed significantly over the past two decades, the relatively developed southern states face a greater share of YLL caused by 'lifestyle diseases' such as diabetes and cardiac and pulmonary diseases. India, therefore, faces multiple challenges in managing its disease burden: the lesser developed eastern and north-eastern states bear the brunt of conventional communicable disease and public health challenges, while the more developed southern and western states are increasingly experiencing higher burden of lifestyle diseases (Rao, 2015). Wide variations in health status make it increasingly difficult for the central government to develop policies that are equally effective across the country.

The healthcare system in India is characterized by severe shortcomings along several critical dimensions. First, public health infrastructure is inadequate and dilapidated, unable to cope with the increasing demand for services (Radwan, 2005; Bali and Ramesh, 2015a). Most public spending is focused on operational needs and a negligible share is devoted to infrastructure or capacity building (Gupta and Chowdhury,

2014). Second, the distribution of healthcare resources and consequently outcomes across rural and urban areas and across states are highly inequitable (Rao and Choudhury, 2012; Dreze and Sen, 2011). Third, the healthcare workforce in the public sector is highly inadequate, a point noted in successive government heath infrastructure surveys. The lack of capacity in public hospitals is a key reason for the poor quality of services (Mohanan et al., 2016; Bali and Ramesh, 2021; Anand, 2017). Fourth, existing healthcare risk pools are small and shallow and offer perverse incentives to healthcare providers (Bali and Ramesh, 2015a; Prinja et al., 2019). Schemes such as the ESIS and the Central Government Health Scheme (CGHS) cover a small fraction of India's population in the formal sector, while those in the informal sector are covered through programmes that offer relatively small benefits or require treatment at poorly functioning public hospital system. These challenges are amplified by low public expenditure on health which has remained stagnant despite repeated promises to increase it.

India's healthcare infrastructure has also not kept pace with the growing demand for medical services which adversely affects healthcare performance. While the number of physicians per 1,000 patients has risen over the past two decades from 0.5 in 2000 to 0.82 in 2018, it falls short of the WHO recommended guidelines of 1 per 1,000 patients (World Bank, 2020a). In addition, there are about 3 million nursing personnel and 0.8 million AYUSH (Ayurveda, Yoga & Naturopathy, Unani, Siddha, and Homoeopathy) doctors serving a population of 1.3 billion. The 0.7 hospital beds per 1,000 persons have been constant since the 1980s (World Bank, 2020a). The shortage of beds is aggravated by wide variations in the distribution of hospital beds (Singh et al., 2020).

4.3 Stewardship and Coordination

Successive governments since independence have done little to steer the sector, provide stewardship, and coordinate different levels of government and various agencies involved in health care (Bali and Ramesh, 2015b). By sheer neglect, the government promoted private provision and financing, making India one of the most privatized healthcare systems. The thoughtless development of private provision had other adverse collateral effects, including lack of efforts to establish risk pools to pay for private care and regulations to govern the behaviour of private providers. These failings resulted in the emergence of a system with health care delivered at poorly regulated private hospitals paid for by households out of pocket, which compromised the population's physical as well as financial well-being.

The problems afflicting the Indian health system are crucially rooted in lack of committed and capable stewardship (Maurya and Ramesh, 2018; Virani 2019). K. Sujata Rao, a retired secretary in the Ministry of Health and Family Welfare notes that "the attention of health ministers [in state governments] has disproportionately focused on the procurement and transfer of doctors and nurses. Most have sought to maintain the status quo and been wary of fundamental changes, discouraged out of the box thinking or radical solutions . . . [and there is] a lack of an appreciation of the complexities of the health sector" (Rao, 2017:125). This leadership vacuum is compounded by frequent transfers of senior bureaucrats who advise the minister. For example, the central Ministry of Health and Family Welfare experienced a turnover of seven secretaries (the highest ranked administrative official) during the eight years between 2011 and 2019. Often, the officials move on to the next assignment before they have had a chance to familiarize themselves with the nuances of the sector. Instead of addressing systemic problems, successive governments have resorted to making marginal changes to existing programmes or introducing new programmes in an ad hoc and patchy manner.

The problems are aggravated by the overwhelming dominance of the private sector which understandably serves only those with the means to pay for it. This in practice creates two tracks in India's health systems: a feeble public system unable to cope with the demand and a burgeoning private sector accessible only to those with the means to pay. Recent reforms, reflected in the launch of RSBY and then PM-JAY, represent the government's admission of failure of the public system and acceptance of the reality that the private sector is the only viable option for delivering health services. But buying healthcare services from private providers is not a panacea, as it requires even firmer direction due to its innate focus on profits rather than public service (Langenbrunner and Liu, 2005). Addressing these challenges requires strong and active stewardship on the part of the government which has not been forthcoming, though the situation has improved somewhat in recent years.

The problems are compounded by weak vertical integration due to India's federal division of responsibilities. Not only is the public spending on health low, as much as 85 per cent of the total is by state governments and only 15 per cent by the national Ministry of Health and Family Welfare. The negligible public spending denies the government meaningful opportunities to set policy direction and steer the sector (Maurya and Ramesh, 2019; Virani, 2019). Further, most of the Ministry of Health (MoH) spending is directed at operations rather than infrastructure and capacity building (Gupta and Chowdhury, 2014). Many state governments suffer from weak fiscal capacity and depend heavily on transfers from the central government, but the transfers have been insufficient to meet the enormous needs of the poorer states (Rao & Choudhury, 2012; Rao & Singh, 2005; Rao, 2019).

The central government responded to these challenges in the late 2000s through increased funding under NHM but that made little difference as many state governments simply reduced their own health spending in response. While the NHM spending improved maternal and child health outcomes, it did little to reduce OOP payments for the general population (Rao, 2017). Eventually, in 2008 the central government enhanced its financing for the sector through the introduction of RSBY and subsequently PM-JAY. The PM-JAY has lowered the co-funding requirement for the poorest states to as low as 10 per cent. While the financing boost has helped, the same states also suffer from poor governance which makes it difficult for them to implement the PM-JAY, which is a technically complex programme. The programme is largely designed, funded, and managed by the central government, but implemented by state governments who in turn have to work with a network of insurance companies and hospitals to deliver the programme. While some of the southern states have extensive experience in managing health financing programmes, for others it is a new experience (Forgia and Nagpal, 2012).

In addition to vertical fragmentation associated with federalism, the governance of health care in India is marred by horizontal divisions. There are separate agencies at both central and state levels performing similar functions with no overarching mechanism to coordinate them. More worryingly, the agencies are often awarded functions in areas in which they have little expertise. RSBY and ESIS, for instance, are managed by the Ministry of Labour, which has limited expertise in either health policy or health insurance (Maurya and Mintrom, 2020; Bali and Ramesh, 2015b). Central government spending is allocated across a range of government agencies that run their own healthcare

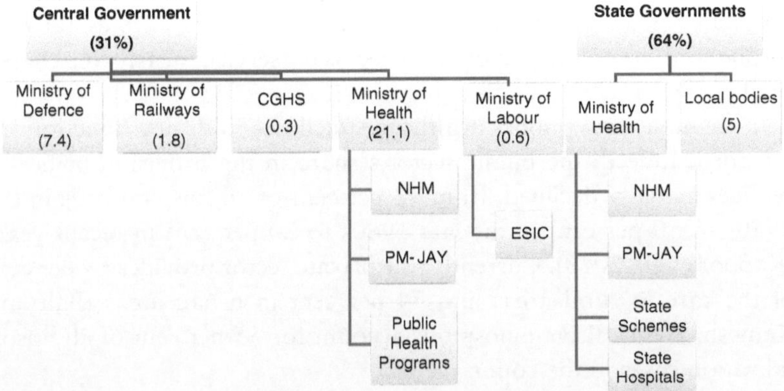

Figure 4.1 *Key government agencies involved in health care in India*
Note: Numbers within parenthesis refer to share of total public spending on health care in 2014. (Gupta and Chowdhury, 2014).

programme with limited coordination among them (Figure 4.1). This has resulted in multiple occupation-based healthcare programmes covering defence, railways, central government, and other employees, often with similar design features.

The problem of horizontal fragmentation is being addressed, to some extent, with the establishment of NHA, the agency responsible for implementing PM-JAY. NHA's establishment followed the NHP 2017 which had recommended that the government play a more strategic role in managing the sector. NHA is not only trying to coordinate the efforts of the different agencies active in the sector but also developing relationships with external private providers and payers, something in which the government has little experience.

4.4 Ownership and Management

There is a deep disjuncture between what the Indian policymakers envision and the tools they use to realize it. Since early days of independence, the government's policy documents – particularly those by the Bhore Committee in 1946, the Murdaliar Committee in 1956, and the Kartar Singh Committee in 1967 – have espoused publicly provided health care. They particularly called for a network of government facilities providing primary, secondary, and tertiary care throughout the country. Accordingly, health sub-centres, centres, and hospitals owned and operated by state governments were established throughout the country to offer all necessary services nearly free charge to all.

In reality, the public healthcare facilities did not function as intended. Indeed, the public sector's share in the provision of health services actually declined, from 92 per cent of all hospital beds in the 1950s to 50 per cent in the late 1960s to 30 per cent in recent years (Kapoor et al., 2020). Currently, the private sector provides 69 per cent of the care in rural areas and 74 per cent in urban areas (Bali and Ramesh, 2021). Private hospitals account for 55 per cent of all hospitalization cases in the country.

In the public sector, each health sub-centre is designed to cater to a population of 3,000–5,000, primary health centre (PHC) a population of 20,000–30,000, and community health centre (CHC) a population of 80,000–120,000. Each sub-centre has an auxiliary nurse/midwife and two health workers but no doctor. Their purpose is to provide basic child and maternal care, and counsel families on immunization, nutrition, and other programmes. PHCs are the first point in the health system for accessing curative care. They are staffed by one medical doctor and fourteen paramedical staff and are equipped with rudimentary medical infrastructure. CHCs are staffed by four medical doctors and twenty-one paramedical staff and contain approximately thirty hospital beds with essential diagnostic facilities such as X-ray machines and laboratories. In March 2019, there were 160,713 sub-centres (157,411 rural and 3,302 urban), 24,855 PHCs, and 5,335 CHCs in the country.

While public health centres physically exist as intended, they do not always provide the intended services due to lack of essential personnel, supplies, and utilities. For example, a quarter of sub-centres and a tenth of PHCs do not have access to regular water supply and electricity. The health centres without basic facilities are concentrated in Uttar Pradesh, Uttarkhand, West Bengal, Assam, and Arunachal Pradesh (Government of India, 2015: Table 36b). While there has been progress over the past decade, overall the status of rural health infrastructure remains woefully inadequate. Studies show that public facilities are often out of essential medical supplies and equipment, and avoided by patients other than those without an option (Radwan, 2005). They also face acute manpower and infrastructure shortages:

20 per cent of the sanctioned positions at the sub-centres, 24 per cent of the sanctioned doctors' positions at PHCs, and 66 per cent of the sanctioned specialist positions at CHCs were vacant in March 2019 (Bali and Ramesh, 2021). Without the necessary supplies and personnel, the public healthcare facilities are unable to function as intended and, understandably, avoided by users.

Absenteeism in public hospitals is also a common problem in India. Nearly 40 per cent of health service providers were found to be absent on a given day in public healthcare facilities (Bajpai, 2014; Panagariya, 2008a). Not only do users find the staff at public health facilities uncaring, the staff themselves feel handicapped and stifled as routine management decisions require approval from the state health authorities (Rao, 2017). Medical staff in the public sector are permanent employees of the state government with ironclad job security and limited incentives to improve performance. The government has no practical tool available to apply accountability or discipline to improve performance. Furthermore, the frequent transfers of medical personnel and supervisory staff within district offices do little to make the rural healthcare system more responsive to patients' needs. This, in turn, leads those needing care to flock to hospitals in urban areas (Das et al., 2012).

Figure 4.2 outlines broad governance relationships that are associated with public healthcare schemes.

As we can see in the organization chart, fund flows do not often correspond with the lines of accountability. The central Ministry of Health and Family Welfare, for example, has no direct control over public hospitals because the latter are owned and operated by state governments. More significantly, even state governments lack meaningful mechanisms for enforcing accountability on the managerial and medical staff of the facilities they own and operate.

Management of public hospitals rests with senior physicians and nurses with only minimal management training. Indeed, superintendents of public hospitals are practising physicians for whom management is not the primary job responsibility. Due to civil service rules that require frequent transfers across agencies and districts, officials do not have the opportunity to develop expertise in health care or any particular sector. The combination of inadequate funding, and the relatively poor quality of services provided by public hospitals pushes patients towards private providers, who readily welcome them.

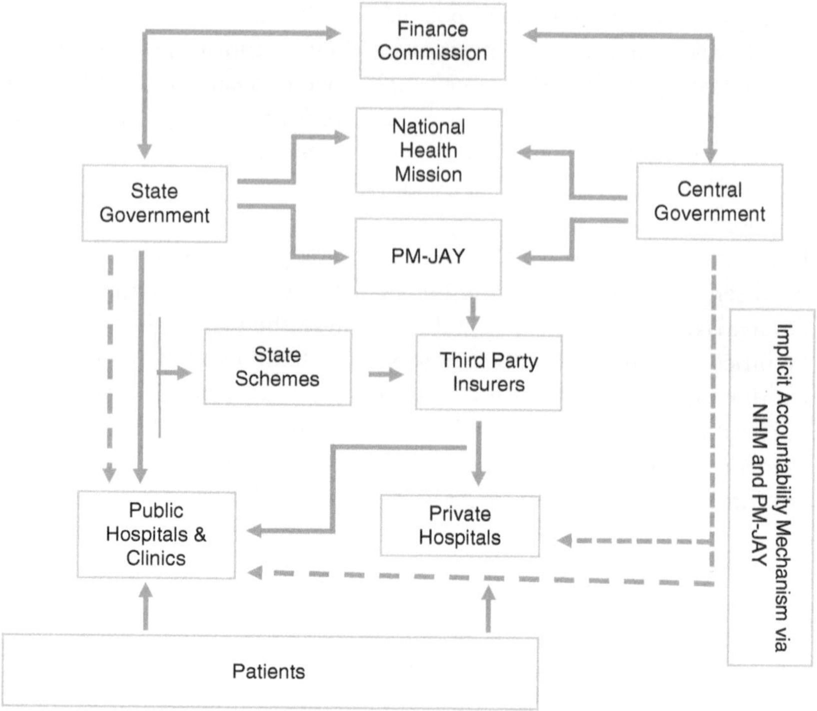

Figure 4.2 *Governance relationships in India's health system*
Note: Solid arrows highlight the flow of funds; dashed lines highlight accountability relationships.
Source: Authors'

The government also imposes little oversight or accountability over private healthcare providers or insurers. Private providers have unfettered autonomy and are only subject to minimal licensing requirements. The lack of effective governance mechanisms has profound implications given the dominance of private providers. The combination of a feeble public hospital system that does not inspire confidence among patients and a weakly governed private health system driven mainly by profit makes for a dysfunctional healthcare system, the results of which are evident in the sector's poor performance.

The National Health Mission, RSBY, and now PM-JAY have opened new opportunities for improving the sector's governance. The NHM programmes, under which some of the funds flow directly to hospitals and require them to track and report performance, have given the government a direct window into the hospitals' performance and a leverage for strengthening accountability. For example, some NHM funds are directed to patient welfare committees (known as *Rogi Kalyan Samitis*) with the aim to improve accountability and responsiveness in public hospitals and health centres. The PM-JAY programme offers similar opportunities for improving hospitals' performance. To be empanelled by the programme, both public and private hospitals must meet and maintain certain minimum conditions regarding facilities and performance. Moreover, to be reimbursed by PM-JAY, public hospitals – nearly half of empanelled hospitals are public – must compete with private hospitals to attract and retain patients. This has now given public hospitals incentives to focus on users' needs and preferences, as a part of the additional income that hospitals generate from PM-JAY may be utilized at the discretion of the hospital managers. For example, hospitals that receive complete accreditation from the National Accreditation Board are paid at a higher level (up to 15 per cent more) for the services they provide. As PM-JAY benefits are portable, private and especially public providers are incentivized to heed to patients' needs. These financial incentives also serve as a tool for the central government to monitor and improve the performance of public hospitals.

4.5 Financial Risk Pooling

India spends around 4 per cent of GDP on health care, which is, contrary to widespread perceptions among Indian commentators, not low for a lower middle-income country with a very young population. What is remarkable though is that public spending accounts for less than a quarter of total health expenditures, which makes it one of the world's most privately financed health systems (Bali and Ramesh, 2015a). The small public spending severely limits the government's ability to intervene in the sector and improve the population's health.

Much of the private financing for health is in the form of OOP payment, accounting for 60 per cent of total health expenditures. Private insurance forms only about 10 per cent of total expenditures due to their relatively high costs for a variety of reasons (Ellis et al., 2000). In terms of public spending, state governments account for about two-thirds and the central government about a third of total health expenditure. Each level has a range of programmes run by different agencies for their own employees or for different population segments as depicted in Figure 4.1.

Total public spending on health has been stagnant at about 1 per cent of GDP since the late 1990s, only recently increasing to 1.3 per cent. Much of the state governments' expenditures are in the form of block grants to public hospitals, CHCs, PHCs, and subclinics. Block grant is also the mode of financing NHM, which is funded by central and state governments on a matching 75:25 basis. Since many states lack both administrative and fiscal capacities to implement the programme, less than half the funds allocated by the national government remain unspent (Rao, 2019; Rao and Choudhury, 2012).

India also has several other public health insurance programmes: the Employees' State Insurance Scheme (ESIS) established in 1952, the Central Government Health Scheme (CGHS) established in 1954, *Rashtriya Swasthya Bima Yojana* (RSBY) established in 2008, and the *Pradhan Mantri Jan Arogya Yojana* (PM-JAY), which in 2018 subsumed the RSBY. In addition, many state governments have established their own public insurance schemes that complement PM-JAY.

The ESIS, implemented in 1952, is a social insurance programme for formal private sector employees and their dependents. The ESIS is administered by the Employees' State Insurance Corporation (ESIC) and is supervised by the Ministry of Labour and Employment. The ESIS functions through a network of hospitals it owns and operates, as well as recognized private and public hospitals throughout the country. Employers contribute 4.75 per cent of the monthly wages while the employees contribute 1.75 per cent, subject to a monthly wage ceiling of INR 15,000. The ESIS offers a generous benefit package without annual or lifetime limits and does not require any form of cost-sharing or user fees for treatment at ESIS hospitals. The programme covers about 133 million individuals (ESIS, 2018), including dependents, forming approximately 10 per cent of the population. The low coverage is unsurprising given that more than 80 per cent of the labour force in the country is employed in informal sectors. ESIS hospitals generally have low occupancy of around 51 per cent and a high average length of stay (La Forgia and Nagpal, 2012). In recent years, the ESIS has generated significant surpluses because higher wage earners often elect to use private services instead of ESIS hospitals (La Forgia and Nagpal, 2012).

The CGHS was initially designed to provide outpatient services to employees of the central government in New Delhi but has since been expanded to provide comprehensive services in most major cities in India. The design features of the CGHS are important as many schemes for employees of state governments and state enterprises are modelled after it. The scheme covers current and retired employees of the central government as well as their dependent family members. The contribution rate for the programme is based on the member's pay grade and varies between INR 50 and 500 per month. Its members enjoy generous benefits with no co-payments, deductibles, or monetary cap on yearly or lifetime coverage or exclusions for pre-existing conditions. CGHS beneficiaries are required to register with an accredited primary care provider nearest to their residence. However, members are issued smart cards which allow them to access services at designated public and private facilities anywhere in the country. The CGHS uses a combination of its own facilities for basic primary care (i.e. wellness centres) and purchases inpatient and diagnostics services from public and private health facilities.

Rashtriya Swasthya Bima Yojana until its replacement entitled poor households to access free inpatient and outpatient services at designated private and public facilities. The scheme is implemented through private insurance companies funded from premium paid by central and state governments in 75:25 ratio. Many states have introduced their own more generous versions of the scheme. RSBY and the cognate state programmes covered 300 million individuals, or more than a quarter of the country's population, in 2010.

In 2018, PM-JAY replaced RSBY in an expanded form. Unlike its predecessor, which concentrated only on hospital care, PM-JAY includes both primary and hospital care. The primary care component targets the much-needed strengthening of primary health centres in the country. The palliative care component covers services at designated public or private hospitals for about 1,400 types of medical interventions up to a benefit level of INR 500,000 (US$7,000) per year, per family. The programme is meant for the 40 per cent of India's lowest income population who amount to 110 million families. The premiums are borne by central and state governments in a 60:40 ratio except for three poorest states where the ratio is 90:10. While 50 per cent of the empanelled hospitals are public, they account for only 37 per cent of hospitalizations and 25 per cent of expenditure (Government of India, 2019b) indicating users' preference for private providers.

Table 4.3 *Key programme features of RSBY and PM-JAY*

	PM-JAY	ESIS	CGHS
Enrolment	110 million families (no ceiling on the number of members) identified using 2011 socio-economic census	59 million	4.0 million
Funding source	Premiums mostly shared 60:40 between central & state governments, but 90:10 in three states	Payroll contributions	50–500 INR per month; Subsidized by MoH
Risk pooling	Benefit ceiling of $7,000 per year; risk pooling by states or insurance companies	Comprehensive	Comprehensive
Provision	Public and private hospitals empanelled after accreditation	At ESIS-owned hospitals	At CGHS-empanelled hospitals
Purchasing	Competitive bidding process	At ESIS hospitals or FFS for others	Competitive bidding process
Provider payment	Per diem payment. Uniform rates across the country; some discretion to state governments; Prospective price; retrospective payment	Fixed Salary	Package rates that providers have to bid for
Benefit package	1,400 defined medical service packages available at any empanelled hospital. No co-payment.	At ESI hospitals or FFS	FFS

Source: Authors

4.6 Provider Payment

The prevailing payment arrangements in India over-incentivize supply in the private sector and under-incentivize it in the public sector. This is because the incomes of doctors and managers in private hospitals and clinics are partially or entirely tied to the revenues they generate. In contrast, employees of public healthcare facilities receive fixed emolument unrelated to performance, which gives them no reason other than individual motivation to do more than the minimum required. Indeed, the terms of employment in the public sector are such that there is little fear of any penalty if they do not meet even the minimum requirements. The opposite is true in the private sector where the doctors' income is directly tied to the revenues they generate (most of them are paid on an FFS basis), thus incentivizing them to oversupply services and even overcharge for them (Langenbrunner and Liu, 2005). Correspondingly, private providers are keenly focused on attracting and pleasing patients while there is no such concern in the public sector. This latter point is perhaps the main reason for users' poor perception of public facilities and the motivation for preferring private providers.

Most physicians (especially senior physicians and surgeons) in the private sectors are affiliated to, but not employed by, hospitals and are paid on an FFS basis. They run their practice at hospitals which receive a share of the total income they generate. Those employed in public hospitals, on the other hand, are permanent employees of state governments and receive a monthly salary and are subject to personnel conditions applicable to civil servants. Physicians in public hospitals are allowed to engage in private practice only during after-hours though the restriction is commonly ignored. It is therefore understandable why the same physicians are more attentive to their work in private practice than in their primary place of employment.

PM-JAY seeks to curb the ill-effects of FFS by capping the amount it pays the providers. Its payment design – emphasizing prospective payments – represents a marked departure from existing practices in India's health system. The NHA has developed a standard benefit list of treatments for which it pays providers preset amounts that are uniform across the country. Similar to Thailand, there is provision for additional payments for high-risk cases. In principle, patients are not supposed to make any payment for treatment, though in reality, they continue to make informal payments. In a recent study, 27 per cent of patients in Madhya Pradesh and 11 per cent in Gujarat incurred OOP costs, especially in private hospitals (Trivedi et al., 2020). The average size of OOP payment in Madhya Pradesh was INR 41,000 or 28 per cent of per capita income in 2019. The continued need to pay out of pocket despite PM-JAY insurance is often due to poorly conceived conditions imposed by state governments. For example, while the programme covers cancer treatment at private hospitals in Madhya Pradesh, they are not reimbursed for chemotherapy, which is reserved for public hospitals which the latter do not always offer (Trivedi et al., 2020).

4.7 Regulations

Healthcare systems with extensive public provision and financing allow the government many levers to intervene in the sector. In contrast, systems with only limited public provision and financing, as is the case in Korea and especially India, require a strong regulatory framework for correcting market failures and broadly guiding the behaviour of the key actors seeking to take advantage of the failures. The central and state governments in India have been unable to rise to the challenge of establishing an appropriate regulatory framework for the sector. Since the private sector grew and proliferated as a result of government neglect and oversight rather than deliberate policy, insufficient attention was paid to its development. Eventually, the private sector became simply too large to be easily brought under government regulations.

While professional councils such as the Medical Council of India (1956), Dental Council of India (1948), Nursing Council of India (1947), and Indian Medicine Central Council (1970) provide statutory guidelines and standards for medical care, they play a relatively passive role. Similarly, while India has the necessary regulations for ensuring the safety of medical services, drugs, and devices, they have limited resources and their enforcement is often lax and patchy (Bali and Ramesh, 2021).

In India, there is almost no regulation controlling the number of hospitals, distribution of hospitals and hospital beds, and the types of high-technology diagnostic services. This, in turn, has resulted in the proliferation of privately owned diagnostic centres, pharmacies, and clinics of uneven quality. For example, it is common for diagnostic centres and pharmacies to pay commissions to doctors who refer patients to them (Nagral and Nundy, 2017), although the practice violates the 2012 Code of Ethics of the Medical Council of India (Vaidya, 2018). Moreover, doctors in public hospitals in most state governments are not allowed to practice in the private sectors, but they often do. Similarly, scheduled pharmaceuticals require medical prescriptions but are commonly available over the counter across pharmacies in violation of the Drugs and Cosmetics Act.

Yet another challenge stems from the limited regulatory oversight of the insurance industry in India which proliferated during the 2000s. Since 1999, nine dedicated health insurance companies have been established. However, the poor claims ratio of 58 per cent in 2016 in the private sector raises concerns about their adequacy and consumer protection. On the other hand, government-financed health insurance schemes, which include RSBY and PM-JAY, have a claim ratio of 117 per cent (2016), suggesting that they are financially unviable (Malhotra et al., 2018; Patnaik et al., 2018). This is the result of gaps in regulation such as the absence of a requirement to refund in case of poor claims ratios, restrictions on the practice of rejecting claims without providing a reason, and missing vital information in the policy document. Moreover, regulations that do exist are often poorly enforced, as there are no penalties for violation. The flawed manner in which ombudsman is appointed further undermines the effectiveness of the regulatory regime governing the healthcare sector in India.

The establishment of NHA is a promising step towards improving the regulatory architecture to support India's largely private health system. The NHA has established detailed guidelines and protocols for empanelling hospitals, accrediting them, and monitoring their performance. It relies on a system of incentives and controls to shape provider behaviour. For instance, providers found to have engaged in malfeasance are de-empanelled and their infringements are publicly displayed on PM-JAY's website (PTI, 2019a). Similarly, hospitals are publicly lauded for improving quality and being responsive to patients (ANI, 2019). NHA's success in developing an effective regulatory architecture would go some way in improving the services that residents receive from private providers. If NHA lives up to the promises, it will be following the footsteps of the NHSO in Thailand which has played a critical role in shaping the Thai health sector (Tangcharoensathien et al., 2015).

4.8 Conclusion

Under pervasive neglect of successive governments, the health system in India performs poorly by just about every measure, even by the standards of middle-income countries. The situation has improved significantly in recent years after the central and some state governments began to pay greater attention to expanding health protection. Many governments have made health care a key component of their election platform and launched expensive programmes providing tax-funded health care to the population. They are, however, yet to come to grips with the fundamental shortcomings of the system which limit the protection that the programmes offer to the population and stand in the way of achieving UHC.

After decades of policy drift, health policy in India was rejuvenated with the launch of NRHM, followed by RSBY and later PM-JAY. For the first time, the government did not simply declare grandiose goals or announce extra funds but offered details on governance structures and policy tools to be employed to implement health programmes. The programmes indicate the central government's assumption of stewardship of the sector, notwithstanding the constitution which vests the state governments with the responsibility for health care. The recent establishment of NHA is also a notable effort to strengthen the sector's governance.

India has a nationwide network of public healthcare facilities at the village, block, district, and city levels, but they have been starved of funds and, more importantly, managed poorly, leading all but the poorest population to avoid them. In response, instead of building the public sector's capacity, in recent years the government has accepted the private providers' dominance and decided to work with them. At the same time, it has taken measures to improve the performance of public providers. For instance, public hospitals have been given not only additional funds through national programmes but their managers have also been granted more flexibility in the use of funds. However, the measures are too limited to meet the challenges of improving the performance of public healthcare facilities, still a crucial provider of services to the poor.

Traditionally, the pooling of risk for financing health care has been dismal in India and a major cause of poverty in the country. Recent programmes, notably RSBY and its successor PM-JAY, indicate that the government has begun to recognize the gravity of the problem and is making serious efforts to address it. Financing the programme from tax revenues rather than contributions and insurance premiums recognizes the reality of informal economy which makes a regular collection of contribution almost impossible. The programmes not only provide much-needed funding from the general budget but also allow the government levers to intervene in the sector and shape the providers' behaviour. However, the programmes do little to protect the non-poor not covered by the programme.

Fee-for-service by which private providers are paid is one of the key reasons why the healthcare system in India performs poorly. Recent programmes such as NHM, RSBY, and PM-JAY seek to address the problem by introducing capped payments. However, as the experience from Singapore and Thailand suggests, capped payments require continuous calibrations and extensive stakeholder consultations, which is a challenging task for the weak healthcare bureaucracy that exists in India. The OOP payment that PM-JAY members continue to make despite prohibition highlights the poor state of health policy implementation in the country.

A strong regulatory framework is essential for healthcare systems dominated by private provision and financing, but this has not been forthcoming in the case of India. While the Indian government has established regulatory frameworks for the pension (The Pension Fund Regulatory and Development Authority) and insurance sectors (The Insurance Regulatory Development Authority), they have failed to develop one for the health sector. Existing regulations that govern hospitals, pharmacies, and providers are weak, patchy, and often ignored. Recent policy documents recognize the lacunae, but no concrete policy action has emerged to address them.

The Covid-19 pandemic hit India hard, despite early and hard lockdown, which thoroughly tested the country's health system. India has had a relatively low rate of case fatality, though this reflects the country's young and rural demography rather than the capabilities of the health system. Public hospitals were quickly stretched to their limits while the private system played only a marginal role at best in the fight against the pandemic. For example, by the end of May, 99 per cent of ICU beds were occupied in the greater Mumbai region (The Print, 2020). This pushed patients to private hospitals, many of whom profiteered by charging exorbitant prices. Many state governments responded by invoking the Disaster Management Act (2005) and introducing price ceilings on Covid-19 related services at private hospitals. While most private hospitals complied with these regulations, some denied treatment to patients on the grounds that they were functioning at capacity (Raghavan et al., 2020). The absence of reliable data on the number of hospital beds – particularly in intensive and critical care – made it difficult for governments to coordinate responses across hospitals. This subsequently forced many state governments (including Maharashtra and New Delhi) to introduce regulations to 'take over' hospital beds in the private sector. In the national capital of New Delhi, regulations were introduced requiring private hospitals to prominently advertise existing hospital capacity at the entrance. Many state governments have had to thus focus their attention on monitoring and managing private providers rather than controlling infections. The pandemic has served as a timely reminder to policymakers to pay more attention to management and regulatory issues in both public and private health facilities.

5 South Korea

Universal Health Care in the Context of Private Provision

South Korea (simply Korea, hereafter) is the only OECD country included in this book and it offers fascinating insights into health care in Asia compared to its Western counterparts. The health status of its population is among the best in the world and total healthcare expenditures are modest, though the highest among the six economies studied in this book. It has a classic social insurance system for financing, coupled with privately provided health care paid on FFS basis, a combination known for driving up healthcare costs.

Korea has one of the most extnensively studied health systems in Asia, possibly only after Japan and China. There are studies covering almost every aspect of Korean health care, including cultural, social, and political contexts, in addition to its medical and economic aspects. Yet there is no systematic study comprehensively examining specific policy problems that exist in the country, the policy tools that are used to address them, and the outcomes they produce. The objective of this chapter is to bridge this vital gap.

In this chapter, we will see that the Korean government has gone to great lengths to establish mechanisms and processes to provide health care to all while containing financial burden on both households and the government. The root cause of the high OOP payments in Korea is the FFS mode of paying providers, which incentivizes oversupply of services that generate higher returns for providers. Unable to replace FFS with capped payments due to political opposition from providers, the government has had to resort to controlling fees and volume of services and requiring substantial co-payment from patients. It has also established detailed decision and monitoring processes to curb oversupply and overcharging, which have shown only limited success. The financing and payment arrangements and weak regulations coupled with political power of the vested interests make it very difficult to reduce the burden of OOP expenditures on households without shifting the burden to the government, a burden that the latter is unwilling to shoulder.

5.1 Policy History

The Korean health system shares many features with Japan, which in turn was inspired by the German system. The end of Japanese rule was followed by an increasing influence of American medical practices, mainly in the form of increasing role of private provision and financing. The first major health policy development in contemporary Korea is the adoption of Medical Insurance Act in 1963 by the military government that had come to office through coup d'etat. The Act provided for a voluntary insurance scheme financed from equal contribution by employees and their employers, with the government paying for the administrative expenses. Without incentives or coercion, it is not surprising that the take-up rate was too small to make much difference. A private member's bill establishing compulsory health insurance was passed by the legislature in 1970, but it was never implemented due to the government's opposition. The health minister argued that there needed to be major improvements in national income and state's finances, along with medical resources, before Korea could have national health insurance.

The situation changed with the introduction of means-tested Medical Aid program (Medicaid) for the poor in 1977. In the same year, National Health Insurance (NHI) was established following legislative amendments making health insurance compulsory for firms employing 500 or more employees, which was subsequently extended to all public sector employees and school employees in 1979. Health insurance coverage to other workers was expanded in phases: membership threshold for private firms was reduced to 300 employees in 1979, 100 in 1981, 16 in 1983, and 5 in 1988. The motivation for the narrow coverage initially and gradual expansion thereafter was to start with population segments that were the easiest to reach and enrol, as it is hard to enrol and collect premium from employees of small firms and the self-employed.

The final expansion of NHI took place when the rural self-employed were covered in 1988 and the urban self-employed in 1989. The schemes for the rural and urban self-employed included anyone not covered by other health insurance schemes or Medicaid. Since it is hard to assess the income of those outside formal workforce, separate insurance pools were established for them. Large corporations had their own firm-level insurance societies while small and medium-sized firms and the self-employed belonged to insurance society in their geographic area. The rationale for different funds for the self-employed was to account for their different sources, levels, and forms of income which were often informal and hence difficult to assess and collect.

The extension of insurance to the self-employed nearly doubled the coverage rate, but at great cost to the government as the programme required large subsidies to remain financially viable. The massive expansion occurred against the backdrop of the 1987 presidential election – the first free national election in the country in nearly two decades – in which the ruling party candidate (Roh Tae Woo) sought votes and legitimacy by offering universal health care along with national pension scheme and minimum wages. The booming economy generated the resources to finance these massive social programmes.

However, the health insurance scheme began to show deep financial problems by the early 1990s. The problems were rooted in insurance premium that had been set unsustainably low in order to launch the programme quickly. To cover the gap between low insurance revenues and substantially more expenditures, NHI required users to make co-payments out of pocket, but this was insufficient to cover the entire gap, resulting in growing deficit. The high administrative costs of running small insurance pools aggravated the deficits, in addition to promoting regional inequities (Na and Kwon, 2015).

These concerns triggered another round of healthcare reforms. In 1996, the government established a high-powered Health Care Reform Committee chaired by the prime minister, which submitted its report in the following year. The committee recommended integration of the 350 separate health insurance societies into a single national scheme, reform of pricing for medical equipment and pharmaceuticals, and separation of prescription and dispensing of drugs. The proposed changes were both technically and politically challenging, but the government moved to adopt them immediately.

The recommendation to establish a single-payer insurance by merging hundreds of insurance societies for the self-employed faced opposition from formal employees who feared that the merger will lead to increase in their contribution rates. The government was eventually able to overcome their opposition with the support of other stakeholders, particularly rural health societies and pharmacists. A new National Health Insurance Corporation (NHIC) responsible for the entire insured population became operational in 2000 and a uniform premium structure came into effect in 2002. The merger of diverse societies with different revenue and expenditure patterns was helped by a centralized claim review system, uniform statutory benefits coverage, and a uniform fee schedule that already existed (Kwon, 2009).

The establishment of NHIC was accompanied by the formation of an independent Health Insurance Review Agency (HIRA) with the responsibility for reviewing reimbursement claims, evaluating provider remuneration system and the benefit package, and conducting quality assessments for providers. HIRA's establishment was in response to the demands of providers who complained that NHI focused excessively on revenues rather than meeting patients' or providers' needs. The government, for its part, wanted to establish a specialist agency with expertise in reviewing reimbursement claims, managing quality, and assessing the appropriateness of care (Kwon, 2009).

Efforts to separate prescription of drugs from their sale turned out to be the most difficult component of the reforms due to opposition from doctors, who depended on drug sales for a substantial portion of their income. The medical profession called a national strike to which the government responded with arrest of strike leaders. Faced with widespread disruptions and angst among the population, the government agreed to increase doctors' fees by 25 per cent in phases (at the cost of US$1.9 billion over two years), reduce admission quotas at medical schools, and increase wages for medical interns and residents. The government also agreed to defer reform of the payment system for providers.

The increased payment for providers in the context of sluggish economic conditions strained NHIC's finances, as its annual operating deficit increased and accumulated balance declined from a surplus of KRW 9 trillion in 2000 to a deficit of KRW 18 trillion in 2001 (Na and Kwon, 2015). To tackle the resulting financial crisis, the government froze physicians' fees, decreased drug prices, increased co-payments, and suspended the planned expansion of benefits. In addition, the government temporarily increased the subsidy to the self-employed from 28 to 50 per cent funded from revenues from taxation of tobacco products. The measures quickly returned the fund to surplus which lasted for several years.

In the meanwhile, HIRA launched many programmes to improve the sector's efficiency. In 2007, it launched a new pay for performance programme called the HIRA-Value Incentive Program (HIRA-VIP) on a pilot basis. The programme rewards high performers with financial incentives amounting to 1 per cent of NHIC reimbursements, while penalizing performance below the 2007 baseline. The Medical Association and the Hospital Association fiercely opposed the programme because they believed it would reduce physicians' clinical autonomy. Most hospitals and doctors preferred voluntary participation (Lee et al., 2012) and the programme is yet to be fully rolled out.

HIRA also launched a new programme mandating economic evaluation before drugs or treatments are included under reimbursement list, making Korea the first country in Asia to impose this requirement. Its mandate is rather narrow, however, focused only on higher price drugs compared to their alternatives, unlike most other countries that use it for determining coverage of all new drugs (Bae et al., 2018). It also initiated 'risk sharing' for new expensive drugs in 2013 which were excluded from NHI reimbursement. Under the programme, patients can apply for a life-threatening, anti-cancer drug or orphan drug without any substitutable alternatives to be partially reimbursed by NHI, with the remainder paid by the patients themselves. The programme has been used only to a limited extent due to the large share of costs that must be borne by patients (Bae et al., 2018).

To reduce the problem of high OOP payment that had festered for decades, the newly elected President Moon Jae-in announced major healthcare reforms in August 2018. The goal of his reform is to reduce the potentially heavy financial burden on patients without imposing unsustainable fiscal burden on the government. It seeks to do that by increasing reimbursement for expensive treatments while requiring higher co-payments for simple treatments. As a part of the programme, the government plans to expand NHI scheme coverage to additional 3,800 medical services and raise the overall average reimbursement rate from 63 to 70 per cent at the cost of 30.6 trillion won ($28.1 billion) over five years. Furthermore, rather than using a positive list whereby coverage is available only for designated treatments and drugs, the government intends to have a negative list which would allow coverage of all treatments that are not explicitly excluded. The increased coverage would then be managed by regulating prices, treatment methods, and number of treatments (Yulchon, 2018). The government plans to increase insurance premiums to pay for the expansion, though the planned increase is estimated to be insufficient and is expected to add to the deficit. For example, health insurance premiums in Korea are 6.1 per cent, compared to 10 per cent in Japan and 14.6 per cent in Germany.

Medical doctors have vehemently opposed President Moon's proposal, with thousands demonstrating on the streets. The main reason for their opposition is that the expansion of service coverage will lead to additional regulatory controls over what and how much they can charge. Currently, treatments excluded from NHI allow doctors the freedom to set their own prices for uncovered services which will no longer be the case once a particular service is covered. Physicians claim that they lose money on covered services, a loss they recover from providing uncovered (and hence unregulated) services. By mid-2020, two years after the launch of the reforms, the measures were yet to be fully implemented.

5.2 Performance

The health status of the population in Korea has improved continuously over the last several decades. Life expectancy at birth has increased from 66 years in 1980 to 82 years in 2015. Similarly, child mortality rate (under 5 years of age) declined from 36.2 in 1980 to 3.4 in 2015 which is one of the lowest in the world (World Bank, 2019). Non-communicable diseases are the main causes of mortality, especially cancer, cardiovascular diseases, and cerebrovascular diseases, with rate of death from diabetes rising sharply. Yet, healthy life expectancy, which adjusts for years of disability, stands at 73 years, indicating that an average person spends only 9 out of 82 years of their life, mostly during old age, with disability.

The improving health status partially reflects the country's increasing health resources. The number of hospitals increased from 590 in 1990 to 1,949 in 2007, while the number of modern medical clinics increased from 11,172 to 26,145 over the same period (Mathauer and Xu, 2009). At the same time, the number of primary care clinics declined slightly, from 3,592 to 3,445. Korea has the highest number of hospital beds per capita in OECD: their numbers increased from 17 per 10,000 persons in 1981 to 47 in 2000 to 110 in 2014. The increase in hospitals and hospital beds has been accompanied by increase in hospitalization rates and length of stay. The number of hospital admissions per capita doubled from 0.06 to 0.12 over the same period (Kwon, 2018). Correspondingly, the number of days per hospital admission increased from 12 in 1990 to 16.5 in recent times, which is more than double the OECD average of 8.1 days (OECD, 2016). Advanced medical equipment, such as MRI scanners, are used more widely in Korea than in other OECD countries.

The number of physicians per 10,000 persons in Korea increased from 5 in 1981 to 13 in 2000 to 22 in 2014. Remarkably, 93 per cent of Korean doctors are specialists. More remarkably, 70 per cent of doctors even in outpatient clinics are specialists. Conversely, only 8 per cent of doctors in the country specialize in family medicine. It is arguable that the motivation for specialization and advanced technology is rooted largely in the higher income they offer.

Expanding medical resources in Korea have been accompanied by rising expenditures following the launch of NHI. Total health expenditures (THE) formed 3.7 per cent of GDP in 1990 (the first year after the launch of NHI) to 4.0 per cent in 2000 (the year NHI funds were integrated) to 7.3 per cent in 2016. While this is lower than the OECD average of 8.8 per cent of GDP, it is higher than the countries studied here. Similarly, per capita spending on health increased from US$726 in 2000 to US$2,678 in 2018 which is, again, lower than the OECD average of US$4,928, despite rising at a much faster rate (OECD Health Statistics).

Over the period 1980–2016, public sources' (including government budget and NHI spending) share of THE nearly tripled, from 20 to 59 per cent. At the same time, OOP's share of THE decreased from 66 to 34 per cent (OECD Health Statistics). Of the total spending, outpatient care forms approximately 42 per cent of THE, followed by 34 per cent for inpatient care and 25 per cent for pharmaceuticals. The share of expenditures on outpatient care and pharmaceutical drugs in Korea is notable for their large size. In terms of expenditures by types of facilities, hospitals receive 46 per cent of total health expenditures, while clinics receive 29 per cent, and pharmacies 25 per cent.

Public health insurance in Korea offers broad coverage of services but only shallow protection, as a significant percent of the population spends more than a quarter of its income on health care. In this regard, Korea is similar to China in that nearly entire population is covered by health insurance and yet 4–5 per cent of the population spend more than a quarter of their household expenditure on healthcare service (Chu et al., 2019). As a result of the large OOP payments, one out of twenty-five households in Korea is vulnerable to catastrophic health expenditure and even bankruptcy (Ho Kyeong, 2018).

Catastrophic expenditures are closely related to age: 10 per cent of pensioners' households face catastrophic health expenditure compared to 2 per cent of other households (Mathauer and Xu, 2009). Hospitalization is also a major cause of impoverishment: more than 30 per cent of households with a hospitalized member encountered catastrophic expenditure in 2007, compared to an average of 3 per cent of all households. Instances of OOP payment for outpatient services and medicines causing financial catastrophe are uncommon (Mathauer and Xu, 2009).

The medical aid scheme mitigates the adverse effects of OOP to some extent. Empirical evidence shows that lower income Koreans covered by medical aid were 2.26 times more likely to receive inpatient care and 1.23 times more outpatient care than those who were covered by NHI (Sohn and Jung, 2016). While this largely reflects the fact that Medicaid recipients disproportionately suffer from multiple chronic conditions, it also shows that Medicaid is widely used by the indigent population (Mathauer and Xu, 2009).

There is evidence of substantial unnecessary use of expensive hospital care and prescription of drugs in Korea. Spending on health prevention and promotion accounts for only 1.9 per cent (2007) of total health expenditures, compared to OECD average of 2.6 per cent. Avoidable hospital admissions for chronic conditions (such as asthma and diabetes) are significantly above the OECD average (OECD, 2016). Similarly, caesarean delivery rate of 36 per cent of live births (2013) is the highest among OECD countries (average of 27.6 per cent) (Kim et al., 2017). The Korean healthcare system is also heavily dependent on sale and consumption of pharmaceutical drugs (both prescribed and over the counter), attracting around 22 per cent of total current health expenditures. After more than two decades of persistent efforts, there is evidence of some success, as drug expenditures' share of THE has declined from around 29 per cent in the early 2000s to around 25 per cent in recent years.

Overall, however, Koreans are largely satisfied with their healthcare system. A survey in 2017 showed that respondents gave the healthcare system an average score of 71.8 out of 100. This is at odds with the fact that nearly 87 per cent of the respondents also had private health insurance plans at the average monthly cost of 147,000 won ($136.50) in addition to NHI coverage (Ho Kyeong, 2018).

5.3 Stewardship and Coordination

Korea has a highly centralized political and administrative system and this is reflected in the healthcare sector (Chun et al., 2009). The Ministry of Health and Welfare (MOHW) and specialized agencies created by it are in charge of all important aspects of making and implementing health policy in the country. However, the centralization is partially offset by horizontal division of governance functions across several agencies. More importantly, the overwhelming dominance of private providers in both primary and hospital care substantially limits the government's capacity to shape the sector. To offset these countervailing and conflicting forces, the government has established complex decision structures to make joint decisions. While the arrangements have been successful in producing the desired results on routine matters, they have succumbed to issues that threaten the providers' fundamental interests.

MOHW is the lead agency in the sector and has the primary responsibility for managing the healthcare system. Its central position makes for strong coordination among government agencies involved in health care. However, it has limited tools to intervene in the sector in the context of limited public ownership, unlike some other countries covered in this study – China, Hong Kong, Singapore, and Thailand – where widespread public ownership affords the government an array of direct policy tools.

The Ministry of Food and Drug Safety's (MFDS) role in health policy arises from its responsibility for ensuring the safety of foods, pharmaceuticals, medical devices, and cosmetics. MFDS is responsible for assessing and granting licenses for drugs, medical devices, and diagnostics, and formulating and enforcing standards and regulations for such products. The Ministry of Strategy and Finance (MSF) also plays an important role by virtue of its responsibilities for overall financial management and its membership of most high-level committees. It is authorized by law to monitor all public agencies' performance, which gives it leverage to control budget spending, personnel management, and major projects (Chun et al., 2009).

MOHW houses the powerful Health Insurance Policy Deliberation Committee (HIPDC) which serves as the apex body for approving major changes related to health insurance, including contribution rates, benefit packages, and service pricing. One of its key objectives is to align insurance premium and providers' fees, which in the past were dealt with separately and were frequently out of sync (Na and Kwon, 2015). The committee consists of 24 members chaired by vice minister of MOHW and meets on a monthly basis. The members are nominated by the organizations representing the key stakeholders: 8 members by payers (labour, employers, civic groups, patient groups, etc.), 8 by providers (physicians, hospitals, dentists, pharmacists, nurses, traditional physician association, etc.), 4 by key government agencies (MOHW, MSF, NHIS, and HIRA), and 4 independent experts appointed by the government. The composition of membership is intended to provide a platform for bringing together conflicting interests to sort out differences and arrive at joint decisions. The four expert members have a disproportionate say in the committee because the payers' and providers' representatives are evenly matched in number and usually take self-interested opposing positions which require resolution by experts (Kwon, 2018).

There are two semi-autonomous agencies with responsibilities for implementing the healthcare insurance programme: the National Health Insurance Service (NHIS) and the Health Insurance Review and Assessment Service (HIRA). The former is responsible for the financing aspects of NHI: verifying eligibility, collecting and managing premium, and reimbursing providers. HIRA, on the other hand, is responsible for purchasing and payment matters: reviewing medical claims, assessing quality of medical services provided, and setting rules and standards for benefits.

Since the consolidation of more than 350 autonomous insurance pools into a single insurance pool in 2000, NHIS has been the sole manager and buyer of public health insurance in the country. It is responsible for levying and collecting health insurance contributions and other fees, administration of insurance benefits, and negotiation of medical fees with healthcare service providers. After the establishment of a single pool, administration costs declined from 8 per cent of NHI expenditures to 4 per cent (Chun et al., 2009). As the largest buyer of health services in the country, NHIS wields vast leverage over the sector. Yet, its hands are tied because there are large parts of private hospitals' operations related to services that are not covered by insurance. The vast political powers of hospitals and doctors also dilute their monopsonic bargaining powers. Doctors and hospitals, for example, have for decades successfully resisted their efforts to reform the payment system (Kwon, 2009).

The Health Insurance Review and Assessment Service (HIRA) is responsible for reviewing medical fee claims, assessing the appropriateness of health care provided, and determining medical fee schedules and drug prices. Its establishment was partly motivated by concerns that NHIS was overly focused on financial matters which needed to be tempered by consideration for quality and appropriateness of health care. Within HIRA, there exists a range of autonomous specialist committees for conducting technical evaluation of proposals for new services, drugs, and medical devices. The Medical Procedure Expert Evaluation Committee is responsible for clinical services, while the Medical Device Expert Evaluation Committee is in charge of medical devices and diagnostics. Cost-effectiveness, in addition to safety and clinical effectiveness, is the key evaluation criterion for these committees. Both committees consist of 20 members, each of whom is randomly selected for every committee meeting from a pool of 300 experts and meet monthly to make decisions on manufacturers' reimbursement applications and appeals (MOHW, n.d.).

The division of insurance responsibilities between NHIS and HIRA raises questions of effectiveness. On the one hand, the division promotes specialization and potential checks and balances. HIRA, especially, has raised the standards of technical evaluation and improved the quality of purchasing decisions (Kwon, 2018). On the other hand, the functions handled by the two agencies are often intertwined that causes duplication and confusion. That both report to the same MOHW, however, mitigates some of the problems resulting from division of responsibilities.

To counter the providers' vast political powers and relative information advantages, the government has created specialist committees including technical experts to promote non-political decision-making. Thus, the National Evidence-based Collaborating Agency (NECA) is an autonomous body responsible for assessing the safety and clinical effectiveness of new medical services, technologies, and health products for use by consumers, providers, and policymakers (EIU, 2017). The Health Insurance Policy Committee under the MOH is in charge of reviewing and deliberating on health insurance, including healthcare benefit standards and costs, contribution rates, and fees. The Health Insurance Finance Committee of NHIC, on the other hand, is in charge of reviewing matters relating to health insurance finance and the remuneration contracts with providers, as well as reviewing and adjusting contribution rates. Similarly, the Health Care Review and Evaluation Committee of HIRA is in charge of reviewing medical reimbursement claims. The membership of these committees comprises of all key stakeholders – representatives of provider associations, trade unions, consumer associations, government agencies, and experts – who are expected not only to contribute technical expertise but also to promote political compromises among contending interests.

There are also unusually detailed conditions on what NHIC will or will not reimburse and the processes by which such decisions are made. Since 2007, NHIC has only reimbursed drugs and medical devices that are on the positive list. Manufacturers are required to submit both medical and economic evidence to HIRA in support of their application for inclusion. HIRA's evaluation is based on analysis of the Drug Reimbursement Evaluation Committee (DREC) and Medical Device Expert Evaluation Committee (MDEEC), both of which are independent advisory committees composed of specialists and representatives of the Korea Food and Drug Administration (KFDA), consumers, and HIRA (Yoo et al., 2019). The elaborate evaluation procedures for approving new drugs and treatments result in low application rates and even lower success rate by OECD standards. As much as two-thirds of applications for reimbursement of new products and one-quarter of applications for new pharmaceutical products are rejected (EIU, 2017). While this is effective for lowering expenditures, it is at the expense of patients' access to new technology and medicines, except for those who can afford to pay out of pocket (EIU, 2017).

Once a medical service, device, or drug has been approved for use, prices are determined by negotiations between the NHIS and each relevant provider association such as the Korean Medical Association, Korean Hospital Association, and so on. Similar mechanisms are in place for determining the prices of pharmaceuticals and medical devices, though the negotiations are with the products' manufacturers rather than provider associations. The inclusion of provider groups in price negotiations impedes decision-making because their mutual differences make agreement difficult (Kwon, 2018). The result is preservation of status quo which has the salutary effect of holding expenditures down but also has the adverse effect of stalling reforms.

NHIS is handicapped in its negotiations with producer groups because of difficulties in obtaining reliable costing data held by private providers, who do not share information readily except when it suits them (Kwon, 2018). To ascertain better insights into hospitals' operations, their costs and to test new ideas, NHIC runs its own hospitals, but this is no match to the data held by private firms (Mathauer and Xu, 2009).

Korea also has a highly formal process for verifying insurance claims, which sets out detailed guidelines like the number of times a service can be provided. All healthcare providers send their reimbursement claims to HIRA on a monthly basis. After a computerized check for errors and omissions, an indicator-based review is conducted. At the end of the review, HIRA sends the results to the NHIS, which then pays the providers. The entire process takes less than two weeks (Na and Kwon, 2015). It is arguable that the elaborate process for payment for insured services incentivizes providers to expand supply of non-covered services because they do not require HIRA's verification.

One aspect that has largely escaped reform in Korea is the lack of integration between primary and hospital care. Healthcare facilities in Korea are not clearly divided by levels and patients can access any level they choose for their first and subsequent visits. There is no gatekeeping for visits to tertiary and higher level facilities, though additional co-payments are required for visits without referral. Patients tend to choose larger hospitals regardless of their needs due to widespread perceptions that the best doctors and equipment are located there. Hospitals account for approximately 15 per cent of all outpatient visits and 29 per cent of outpatient expenditures in the country (Kim et al., 2018). Hospitals encourage this behaviour because of the extra revenues such patients bring, claiming that this allows them to offset the loss from providing inpatient services provided under NHI contract. To compete with hospitals, primary care clinics have had to transform themselves into mini-hospitals providing a broad range of surgical procedures due to the higher income opportunities they offer. The government has made repeated efforts to establish a stratified healthcare delivery system with measures such as referral slip for visiting tertiary hospitals and higher co-payment for outpatient services at hospitals but has had little success in altering patients' or hospitals' behaviour (Kwon, 2018).

The government hopes to improve sector-wide integration by building an extensive information technology infrastructure. Its Drug Utilization Review system, launched and operated by HIRA, is deemed as one of the most extensive for monitoring prescription of drugs in OECD. The system tracks the prescription and dispensation of all drugs through all outlets at the individual level to check for duplications and conflicts in medication (OECD, 2012). HIRA also publishes vast information on the performance of healthcare providers to help users make informed choices. Moreover, it gathers and shares information on prescription rates of antibiotics and injectables, number of medicines per prescription, and cost of medicines prescribed. Similarly, it provides comparative performance information to help healthcare providers improve their performance by comparing themselves to their peers (Na and Kwon, 2015).

5.4 Ownership and Management

The Korean government makes light use of public ownership as a policy tool, as only 7 per cent of hospitals and 12 per cent of hospital beds are in the public sector, which is significantly lower than its OECD or Asian peers (Yeo et al., 2016). Similar to India, the private sector's dominance is so pervasive that it poses major obstacles to the government's efforts to pursue the goal of providing universal health care at affordable costs.

The public hospitals' share of all hospitals in the country decreased from 35 per cent in 1975 to 5 per cent in 1994 and has stayed at that level while the private hospitals' share increased correspondingly from 65 to 95 per cent. In parallel, the public hospital beds' share of all beds in the country decreased from 53 per cent in 1977 to 13 per cent in 1997. The situation is similar with primary care facilities, approximately 10 per cent of which are in the public sector while 90 per cent are run privately. The decline of public provision was not the result of deliberate policy decision but the result of the government's neglect of health and other social policy matters due to its focus on economic development drawing remarkable parallels with the evolution of India's health system.

Public facilities are concentrated in rural hospitals, public health clinics, and military hospitals. While public hospitals are few in number, they perform disproportionately large social service functions.

The 33 local public hospitals – most dating back to Japanese colonial rule – provide medical services at a lower cost, serve more low-income people, and provide fewer non-covered services compared to their private counterparts (Yeo et al., 2016). All these factors lead to financial losses which make them dependent on government subsidies in order to survive.

All private hospitals in Korea are required to operate on not-for-profit basis under the Medical Act. Nevertheless, they compete vigorously for revenues and surplus which they use to expand facilities, acquire new technology, and pay doctors. Indeed, even public hospitals act more like their private counterparts due to the competition they face in attracting doctors and managers steeped in the dominant norms and ethos of the private sector. However, for those services not included in the NHI benefits package, which allows providers to levy charges as they wish, public hospitals tend to charge lower prices than private hospitals (Kwon, 2018).

Private hospitals enjoy vast legal autonomy in both strategic and operational matters, including decisions to build, buy, sell, or close down hospitals without restrictions. Competition among providers is fierce which leads them to focus on pleasing users, which promotes user focus but also leads to unnecessary services and higher overall expenditures. The intense competition also undermines cooperation between primary and secondary providers and stymies policy coordination. For instance, laboratory tests done in one facility are often not used by others even though this is required by law. Medical records are also not shared among providers. These critical shortcomings call for stronger regulation of private providers which has not been forthcoming, as we will see shortly.

5.5 Pooling of Financial Risk

Similar to other countries studied in this book, health care in Korea is financed from multiple public and private sources. Of all the sources, social insurance and OOP constitute nearly 80 per cent of THE, while the government budget accounts for only a small share. The small direct spending by the government gives it only limited leverage to shape the sector which compounds the limitations arising from small public sector in provision of health care.

Public health expenditures (including both government budget and social insurance) rose steadily from 39 per cent of the total in 1990 to

49 per cent in 2000 to 54 per cent in 2014 (WHO National Health Accounts). Among public sources, NHI forms four-fifths, while the government budget forms only one-fifth. More than half of the budget financing is spent on medical aid, with much of the remainder spent on public health and subsidy for NHI premium for the self-employed (Mathauer and Xu, 2009). Subsidy for NHI premium was high at the outset – as much as 44 per cent of total premium in 1988 – but has gradually declined to less than 22 per cent over the years (Mathauer and Xu, 2009). The government also pays for healthcare services for prisoners and military personnel from its budget. Moreover, it offers tax relief for OOP payments exceeding 3 per cent of households' disposable income and private health insurance premiums of up to a ceiling (Chun et al., 2009). Overall, however, the amount of government transfers is too small to serve as a strong tool for shaping the providers' or users' behaviour.

NHI, mandatory for the entire population since 1989, is the largest source of healthcare financing in Korea. The premium for the formally employed is based on monthly wages while the premium for the self-employed takes into account a range of factors, including property, income, age, gender, and household size. The premium is split evenly between employers and employees, except that the government contributes 20 per cent of the premium for private school teachers (the employer pays 30 per cent), 46 per cent for the rural self-employed, and 34 per cent for the urban self-employed. The government also contributes partial or complete premium for those on Medicaid benefit.

NHI premiums have grown steadily over the years, rising from 3.9 per cent of monthly income in 2002 to 6.5 per cent in 2019 and are expected to continue rising in the foreseeable future (Yonhap, 2019). However, the premium is still low by OECD standards and also lower than the maximum of 8 per cent set by the NHI Act. In comparison, the premium is 10 per cent of income in Japan, 14 per cent in Germany, and 13 per cent in France. The comparatively low premium limits NHIS's ability to expand or deepen insurance coverage and reduce OOP expenditures.

NHI covers full range of inpatient and outpatient services, including drugs, nursing services, and traditional medical therapy. However, NHI requires substantial co-payments even for insured services: 20 per cent of the bill for inpatient services, 30–55 per cent for outpatient services, and 30–40 per cent for drugs, in addition to a flat fee of about US$4 for each unit of service. Furthermore, co-payments for approved drugs are kept at a maximum of 5 per cent of the total bill for cancer patients, 10 per cent for rare diseases, and 30 per cent for infectious diseases and other prescription medicines, to keep them affordable to patients. There are also ceilings (cumulative over six months) on co-payments for insured services but not uninsured services. Finally, patients covered by Medicaid are exempt from co-payments for insured services.

NHI users need to make full payment for uninsured services, that is, services not included in the benefits package. It is believed that much of the expenditures on non-covered services are for new technology and branded drugs of unclear medical benefits (Kwon, 2018). Providers actively induce demand for uncovered services and medicines because there are no price controls on them, which allows them to set higher prices. There is scant information on prescription and use of uncovered drugs and services because providers are not required to report it to NHIS or HIRA.

Medicaid benefits are available to persons and households living below the poverty line as defined by the government. There are two types of beneficiaries depending on capacity for work. Class 1 beneficiaries are people who are unable to work while Class 2 beneficiaries are those who are able to work independently but are poor. The former are not required to make any payment at the point of service while the latter pay a reduced amount of the designated co-payment. Majority of the beneficiaries fall under Category 1 (Chun et al., 2009).

Much of private health spending in Korea is out of pocket, which formed 57 per cent of THE in 1990, 41 per cent in 2000, and 36 per cent in 2014. OOP payments take two forms in Korea: for covered and non-covered services. OOP payments for covered services are controlled by the government and typically amount to approximately 20 per cent of the total treatment bills borne directly by users. Payments for non-covered services, on the other hand, are entirely uncontrolled and form the bulk of OOP in Korea. OOP is a weak policy tool, useful only for restraining users' demand with little effect on the demand induced by providers. OOP also has huge implications for equity, hitting the lower middle-income households particularly hard, because the poor are covered by Medicaid.

Private insurance accounts for small but growing share of THE: 1.2 per cent in 1990, 5 per cent in 2000, and 6.2 per cent in 2014 (Kwon,

2018; WHO National Health Accounts). This is a very small share, especially considering that 77 per cent of Koreans purchase private insurance (Osental, 2018). Unsurprisingly, the wealthy and healthy have higher tendency to buy private health insurance and are significantly more likely to utilize inpatient care than people who do not (Jeon and Kwon, 2013).

5.6 Provider Payment

Fee-for-service (FFS) is the mode of payment for 93 per cent of the NHI benefits expenditure and even a higher percentage of OOP expenditures. As mentioned earlier, all outpatient and inpatient visits involve payment of some fees, though there are exemptions and concessions for certain designated groups and treatments. The requirement for fee payment in nearly all instances is a central part of the Korean healthcare system and overwhelmingly shapes the behaviour of users and especially providers. Specifically, FFS encourages providers to focus on satisfying users but it also incentivizes them to increase the volume and intensity of services, especially services with larger profit margin. To mute FFS's distortionary effects, MOHW has developed a formula for calculating fees – called Resource-Based Relative Value Score (RBRVS) – which takes into account the amount of work and resources (such as manpower, facilities, equipment, etc.) involved in a particular medical treatment. Significantly, the RBRVS rates do not apply to services not covered by NHI.

The fee schedule for specific medical procedures in Korea is believed to be lower than in comparable OECD countries, though strict comparison is difficult. The reason for the low fees goes back to the time of the launch of NHI when fees were set lower than the prevailing rates in order to keep insurance premium affordable. Efforts to contain the rate of fee increase have persisted, except for 1999–2001 when the government raised fees to compensate for physicians' loss of income resulting from restriction on sale of drugs. While tight controls over fees have kept costs and insurance premium low, it has also encouraged providers to find ways to prescribe services that are outside the controls. Their efforts are facilitated by insurance rules that allow providers to offer both covered and non-covered services during the same episode of illness, which offers opportunities for substituting non-covered services for covered ones. Providers argue that they use income from non-covered services to cross-subsidize for lower reimbursement rates for covered services. However, the claim cannot be verified due to lack of systematic information on the volumes and types of non-covered services (Mathauer and Xu, 2009).

In light of the distortionary effects of FFS, the government has been trying to promote alternative payment mechanisms for more than two decades. It launched a DRG pilot programme on a voluntary basis in 1997. Following findings of positive results – such as reduction in length of stay, average number of tests, and use of antibiotics without a negative effect on quality of care (Kwon, 2018) – the pilot was scaled up to a regular but still voluntary programme in 2002. Under the Korean DRG classification system, which is adapted from the US Medicare system, healthcare providers have reimbursed hospitalization costs based on a fixed fee schedule for an illness regardless of the type or volume of healthcare services. This is supplemented by an adjustment mechanism for the type of hospital, exceeded length of stay, and so on. In 2012, DRG payments became mandatory for all providers, but only for seven types of surgeries, which is less than 1 per cent of all DRGs that exist on the books. DRG payments form only a small part of total inpatient expenditures.

The continued reliance on FFS for all but seven DRGs, and that too on a voluntary basis, is due to the political opposition from provider groups who oppose it for fear of reduction in income and loss of autonomy. For hospitals, DRGs are simply not attractive enough for the more complex and varied cases that they treat (Mathauer and Xu, 2009). It is arguable that policymakers are unrealistic in their hope that widespread adoption of DRG, even if possible, will strengthen cost controls in the absence of volume control or a budget cap.

In addition to DRG, the government has promoted payment for performance (P4P). In 2004, it launched a three-year pilot project for acute myocardial infarction and caesarean delivery in tertiary and general hospitals. The programme's objective is to incentivize physicians and healthcare providers to improve quality of care and/or reduce costs. In 2011, the programme was expanded to acute stroke, pre-preventive antibiotics in surgery, high blood pressure, and diabetes. Performance measures for C-section, for example, include the difference between actual and risk-adjusted rates. Financial incentives, amounting to as much as 1 per cent of total health insurance reimbursement, are available to hospitals in the highest performance band.

Correspondingly, hospitals in the lowest performing band are subject to 1 per cent reduction in reimbursement. The government is considering expansion of the programme to other areas and smaller hospitals and including other performance measures but has made little headway due to providers' opposition.

Finally, daily per diem is the mode of payment for long-term stay in convalescent and psychiatric hospitals. This is to ensure that providers do not oversupply (as might be the case under FFS) or undersupply necessary services (as is the incentive under fixed prepayment) at long-term care hospitals. Per diem payments accounted for 7.5 per cent of total inpatient expenditures in 2013.

5.7 Regulations

A strong regulatory framework is necessary to mitigate the vast market failures in health care which help providers to pursue their self-interests at the expense of societal welfare. This is especially the case in Korea where weak government presence in provision allows private providers a dominant role. Insurance payment is a potentially powerful tool for shaping the behaviour of providers and users but in Korea it is limited by the widespread prevalence of OOP for uncovered services paid on FFS basis. Regulations in Korea have come a long way in addressing critical problems in health care but have fallen short of the enormous challenges posed by private providers.

Korea has all the standard licencing requirements for medical practitioners and facilities expected in modern healthcare systems. Similarly, it has the necessary regulations for ensuring the safety of medical services, drugs, and devices. Additionally, as mentioned earlier, all medical care providers must participate in NHI scheme and cannot distribute profit to shareholders by law. Moreover, under Article 15 of the Medical Act, providers cannot deny necessary medical services 'without legitimate cause'. The Emergency Medical Act even more clearly states the right to access emergency care in all hospitals (Chun et al., 2009).

Overall, health care in Korea is lightly regulated by OECD standards, even though the regulations have strengthened over the years. The light regulatory regime in Korea arguably reflects the industrial policy mindset of policymakers centred around growth and competition rather than serving social purpose. The regulatory and legal frameworks for health care focus on delivery of insured services rather than citizens' access to them (OECD, 2012). Thus, there is almost no regulation controlling the number of hospitals, location and size of hospitals and hospital beds, and the number and distribution of high-technology diagnostic services. There is also no compulsory inspection of hospitals with fewer than 300 beds, which may explain the prevalence of a large number of small hospitals in Korea. Moreover, fees for non-covered services, special treatment, and room charges are outside regulation despite the fact that their absence is a major reason for over-provision of health care in Korea. Indeed, the government does not even know about the volume and costs of these charges because the providers do not need to report them. Enforcement of existing regulations is also weak. For example, patient referral system exists but is enforced only for Medicaid users (Mathauer and Xu, 2009).

In some respects, however, regulations are unusually detailed and include what are otherwise managerial functions, such as cost-sharing schedule. HIRA's process for reviewing insurance claims and evaluation of insurance benefits too is remarkably detailed (Mathauer and Xu, 2009). There are also patches of strict regulations to ensure quality of medical services, such as requirement that all medical professionals must undergo eight hours of relevant professional training each year to maintain licence.

5.8 Conclusion

Korea has built a world-class healthcare system over a relatively short period of time since the 1980s. The population's health status is one of the best in the world while total expenditures are modest compared to its OECD peers, though higher than in the countries studied here. Yet the system suffers from deep flaws. On the supply side, there is a heavy reliance on advanced hospital care, specialist outpatient services, and expensive medicines. For users, health insurance does not cover substantial portion of healthcare expenditures which they must pay for out of pocket.

The government has displayed strong stewardship in building governance structures to suit the local context characterized by private provision, insurance financing, and FFS payment. This is a difficult policy environment, as policymakers in the United States have repeatedly learnt from experience in trying to reform their healthcare system.

But unlike the United States, Korea has the advantage of a highly centralized administrative system with immense experience in designing and executing major policy reforms. The government has established a health governance system that is highly coordinated, both vertically and horizontally. To draw in outside expertise and support, there is an extensive advisory system comprising key stakeholders as well as technical experts.

The system for providing health care in Korea is remarkably simple: most outpatient as well as inpatient care is almost entirely provided by the private sector with minimal restrictions. The provision system is hospital-centric, separate from primary care, and dominated by specialists. Private providers are guided heavily by financial concerns, reflected in long hospital stay, and heavy use of advanced diagnostic and expensive prescription drugs. The providers are normally paid from national health insurance combined with co-payment, except for non-covered services which are paid entirely through OOP payment. Fee-for-service is the main mechanism for paying providers. Over the years, the government has made intermittent attempts to curb FFS, through measures such as close scrutiny of insurance claims, DRG, and Pay for Performance. However, the efforts have run up against opposition from providers who have been united in opposing efforts that curb their income or autonomy.

Determined to promote universal access to health care while containing costs in a healthcare context characterized by private provision, insurance and OOP financing, and FFS payment, the Korean government has had to employ elaborate institutional processes to shape the behaviour of users and providers. There are detailed processes for deciding on services covered, FFS rates, benefit standards, cost-sharing rates for insured services, claims management, payment schedule, and so on. This is unlike the situation in Hong Kong, Singapore, and Thailand, where governments achieve the same objectives directly through the exercise of ownership rights and attaching conditions to fiscal transfers.

Korea has probably come as far as it is possible to address the hurdles to universal care at affordable costs through administrative solutions. Only a political solution can solve the problem of high OOP but how to develop and implement it has eluded successive governments, including the current president. In many ways, Korea's problems are similar to China, in that they are centred on skewed providers' incentives but its democratic context makes solutions harder to enforce. That providers in Korea are mostly private, unlike China where they are mostly public, also makes reforms more difficult.

The Korean government's response to the sudden outbreak of Covid-19 in February 2020 has been widely commended for speed and comprehensiveness (see Lee et al., 2020). Crucial to the success was a hospital triage system that sorted patients according to severity of symptom and assigning them to hospitals or community centres accordingly. Its experience in dealing with MERS in 2015 helped with the swift response (Kwon, 2020). Universal NHI coverage, complemented by waiver of co-payment for treatment of infectious diseases, was instrumental to the success because it allowed provision of care to everyone who needed it. The response to the pandemic also highlighted that it is public hospitals that are the first line of defence in a national emergency, as they looked after 90 per cent of patients even though they constitute only 10 per cent of all hospital beds in the country. Private hospitals, on the other hand, tend to be reluctant to invest in prevention and treatment of infectious diseases due to the low-profit margin they offer.

6 *Thailand*

Universal Health Care under Government Stewardship

Thailand is celebrated in the health policy and development literature as a poster child for achieving universal healthcare coverage at relatively low costs. This is contrary to the reactions at the time of adoption of major reforms in 2002 when they were widely criticized as populist and profligate. Instead of rising rapidly as widely alleged and feared, total health expenditures increased modestly while the population's access to health care improved vastly. In addition to extending coverage to the entire population, the reforms introduced sophisticated policy tools to pay healthcare providers, which afforded the government new levers in managing the sector. By almost any indicator, Thailand's health policy indicators are better today than they were two decades ago.

What accounts for Thailand's success in providing health care to its population? In this chapter, we will show that Thailand has a relatively simple health system: most services are provided at publicly owned and operated hospitals, and paid for by the government through tax revenues. Layered to the combination of public ownership and funding is a purchaser–provider split and frequent changes in policy settings based on data on utilization and costs. The combination of these tools enables the government to monitor quality of services while maintaining a watchful eye on costs. Importantly, Thailand has invested in developing policy capacity over the past two decades. Policymakers in the Ministry of Public Health (MOPH) and other statutory agencies are supported by an ecosystem of trained policy professionals in designing and implementing prospective payment programmes.

While Thailand's policy choices to organize and manage the health system may appear simple, what is remarkable is that the tools it uses – a combination of stewardship, ownership, funding, and capped payments – complement and reinforce each other. The only tool that Thailand does not use extensively is regulation which, as we will see in this chapter, is not necessarily damaging due to the other tools in place. Public ownership and funding offers sufficient levers to the government to create incentives for healthcare providers to control costs, promote responsiveness, and improve quality of services without the need for extensive regulations.

Thailand's health reforms offer valuable lessons for other countries seeking universal health coverage. As analysts have pointed out, capped payments and the expertise of its administrative agencies, coupled with strong backing of Prime Minister Thaksin have been crucial for its success. A vital factor for the reforms' success that has been overlooked by observers is the concerted use of multiple policy tools to intervene in the sector. Thailand's case also highlights the power of vested interests to stall reforms affecting them adversely, evident in civil servants' ability to retain their tax-funded scheme offering far more generous benefits than available to the general population.

6.1 Policy History

Thailand's health schemes have developed in a disjointed and piecemeal fashion since the early 1970s. The Workman's Compensation Scheme (WCS), introduced in 1974, was the first to be implemented and covers work-related injuries. The next in line was the Low Income Card (LIC) scheme launched in 1975. The scheme provided informal workers medical services at public hospitals which were reimbursed on capitation basis. However, the capitation rates were low and hence hospitals avoided serving LIC members. Yet, the scheme's coverage increased and covered a third of the population by 2000, indicating the population's demand for health security (Kongsawat, 2000).

In 1983, the government launched the Voluntary Health Card Scheme (VHCS) to cover the non-poor who were uninsured. The scheme was initially implemented in only selected rural areas and expanded nationwide in 1991. The VHCS operated as an insurance scheme, with contributions from its voluntary members matched by the government. Similar to the LIC scheme, providers had little incentive to serve VHCS members as they were reimbursed at rates below their cost (Hsiao and Shaw, 2007). Furthermore, as the scheme had no cost-sharing feature and was voluntary, it was plagued with adverse selection and excessive utilization (Ramesh, 2009; Kongsawat, 2000). At its peak, the VHCS covered a fifth of Thailand's population, and was disbanded after the Universal Coverage Scheme (UCS) was implemented in 2002.

The Civil Servants Medical Benefit Scheme (CSMBS) adopted in 1980 covers civil servants, military personnel, and their dependents and is administered by the Ministry of Finance. It is financed entirely from government budget and offers the most generous benefits among programmes existing in Thailand, covering both ambulatory and inpatient care, with no exclusion. The programme has been fraught with over-servicing and inflated billing, as the Ministry of Finance has little capacity in health insurance matters and its officials face little incentive to prevent misuse or reign in spending. Outpatient expenses in the programme are approximately four times higher than for other health financing programmes in Thailand.

The Social Health Insurance (SHI) was launched in 1990 to cover formal private sector employees following the adoption of the Social Security Act which also introduced pension and other social programmes. The scheme is supervised by the Ministry of Labour and is financed from matching contributions from workers, their employers, and the government. Similar to India's ESIC programme, the main limitation facing SHI in Thailand is that by definition it cannot cover the vast majority of the population that are in informal employment.

The Traffic Victim Protection Act passed in 1992 requires all vehicle owners to participate in an insurance scheme for injuries resulting from motor vehicle accidents. Since most motorists are covered by some other health financing scheme, the scheme does not offer practical benefits to the population.

Collectively, these health financing programmes covered about 70 per cent of the population in the late 1990s despite several rounds of expansion (Table 6.1). However, apart from the SHI and CSMBS, the schemes were unable to cope with the fiscal pressures of the Asian Financial Crisis in 1997 which left many households without access to adequate health cover (Ramesh, 2009).

In the aftermath of the 1997–1998 economic crisis, the political conditions were right for the launch of a universal health financing programme. The newly formed Thai Rak Thai Party (TRTP) made health care the centrepiece of its election manifesto and won by a landslide. (Ramesh, 2009; Tangcharoensathien et al., 2004). It followed up on its promise and set about to establish a universal health programme immediately. Officials at the MOPH worked with

Table 6.1 *Key health programmes in Thailand*

Pre-2002	Population covered, %	Post-2002	Population covered, % (2018)
Social/Medical Welfare Scheme	38	Merged into UCS	75
Voluntary Health Card Scheme	12		
Uninsured	~30		
CSMBS	11	CSMBS	10
Social Health Insurance	9	Social Health Insurance	15
Private Insurance	~9	Private Insurance	< 10 (supplementary cover)

Source: Authors

researchers from the Health Systems Research Institute (HSRI) and International Health Policy Program (IHPP) to generate proposals for the government's consideration. Learning from the mistakes of the contribution-based SHI and the open-ended CSMBS programmes, it was decided early on that any new programme would be non-contributory and require providers to work within a strict budget (Tangcharoensathien et al., 2007). The resulting Universal Coverage Scheme (UCS) popularly called the '30-Baht Scheme' (as it initially required a co-payment of 30 baht or US$1 in 2002 prices) was thus non-contributory, financed from government revenues, and delivered at contracted hospitals that were paid using a combination of payment modes. The UCS covers all citizens not covered by SHI or CSMBS, thus closing the coverage gap that existed at the time of its adoption.

6.2 Health System Performance

Thailand achieved most of the health-related millennium development goals (MDGs) before 2015, and is on track for achieving the revised Sustainable Development Goals (SDGs). These achievements are attributable largely to two broad reasons: sustained expansion of health infrastructure, especially primary health centres in poorer regions, and expansion of public health financing programmes (Evans et al., 2012; Hanvoravongchai, 2013). Improvements in health care in Thailand are evident in its international rankings. It ranked 47 out of 192 in the World Health Report 2000. In the 2018 Bloomberg Health Efficiency Index, however, it ranked 27 out of 55 countries, ahead of Netherlands, Denmark, the United Kingdom, and Germany.

Thailand's life expectancy at birth has increased from 70.6 in 2000 to 76.9 in 2018. Its current under-five child mortality rate of 9.1 and infant mortality rate of 7.8 per 1000 live births are lower than the averages for upper middle-income countries of 12.6 and 10.8, respectively (World Bank, 2020a). Similarly, the life expectancy at age 65 (in 2017) of 21.7 for females and 18.5 for males is comparable to high-income countries (21.9 and 18.6 for females and males, respectively). Its health-adjusted life expectancy of 16.3 for females and 14.2 for males at age 65 is higher than high-income countries (16.2 and 13.8 for females and males, respectively) (Kyu et al., 2018). Despite its significant strides in improving conventional health outcomes, Thailand still faces a significant burden from both infectious and non-communicable diseases.

What is most impressive about Thailand's healthcare performance is its relatively low levels of spending. Total healthcare expenditure in Thailand has hovered between 3 and 4 per cent over the past two decades. In 2018, it spent 3.8 per cent of GDP on health care or about two-thirds of what typical upper middle-income countries spend. Most of the healthcare spending is paid for by the government. The government's share in total health spending has increased from 58 per cent in 2000 to about 80 per cent in recent years. Public spending on health care consists of tax revenues (70 per cent) and mandatory social insurance contributions (10 per cent).

The increased public spending has been associated with a decline in OOP payments which reduced from 34 per cent of total health expenditures in 2000 to 9 per cent in 2018. The remarkable decline in OOP spending has salutary implications for health equity and even efficiency. For example, using 10 per cent of average household income as threshold, the incidence of catastrophic health expenditure in Thailand declined from 4 per cent of population in 2002 to 2 per cent in 2015. More remarkably, the incidence of poverty reduced from 38.5 per cent in 2000 to 6.6 per cent in 2015, when spending on health care declined (Tangcharoensathien et al., 2020). These achievements are the result of a multipronged approach to health policy design, as we will see in the following discussion.

6.3 Stewardship and Coordination

Health policy and administration in Thailand is highly centralized, despite the commitments to decentralization and devolution outlined in the 1998 Constitution and the Decentralization Act of 1999 which required financing and management of all healthcare facilities to be devolved to local governments. However, less than 1 per cent of health centres and hospitals are devolved and local governments play a relatively small role in financing and delivering healthcare services (Huntington and Hort, 2015). The reason for this is the local governments' self-described lack of 'preparedness' to shoulder responsibility for running healthcare facilities (MOPH, 2019), and reluctance to change arrangements that are believed to be already working well.

The MOPH is the apex agency that oversees the healthcare sector in collaboration with other agencies, including the Health Systems Research Institute (HSRI), the Thai Health Promotion Foundation (THPF), the National Health Security Office (NHSO), and the National Health Commission Office (NHCO). In addition, there are research groups within the MOPH – such as the International Health Policy Program (IHPP) set up in 2001 and the Health Intervention and Technology Assessment Program (HITAP) established in 2007 – that produce data and analysis on the healthcare system.

The HSRI, established in the early 1990s, concentrates on health system research and plays a key role in strengthening health policy capacity. It is common for policy officials across agencies in the MOPH, including the HSRI, to receive training in health economics and policy, pursue research programmes, and disseminate findings in international journals (Pitayarangsarit and Tangcharoensathien, 2009). Although the governing board of HSRI is chaired by the Minister of Public Health, it enjoys operational autonomy and its researchers play an important role in informing the design of key health programmes (Tangcharoensathien, 2004). The Thai Health Promotion Foundation, established in 2001, provides technical support and financing to reduce risk factors for non-communicable diseases. It is funded through an ear-marked 2 per cent levy on excise taxes on tobacco and alcohol. The National Health Security Office

(NHSO), established in 2002 as a part of universal coverage reforms, is tasked with responsibilities for purchasing healthcare services for UCS enrollees. NHSO reports to the National Health Security Board (NHSB) which is chaired by the Minister of Public Health. The NHCO convenes the annual National Health Assembly (NHA) in part to engage non-state actors in policy deliberations. While these agencies enjoy operational autonomy, there is significant coordination among them, not the least because they all ultimately report to the Minister of Public Health (Evans et al., 2012).

In terms of vertical coordination, the Thai health system is organized along regional- and district-based networks. There are 13 regional management offices responsible for coordinating and managing resources; a district health system (DHS) that coordinates activities within a district and across different agencies involved in health care; and a primary care cluster (PCC) that focuses on promoting better health outcomes and providing palliative care. The DHS and PCC also work with agencies in the Ministry of Interior that provide sanitation and disease control services, and the Ministry of Natural Resources and Environment. In addition, healthcare providers in Thailand are clustered in what are known as 'contractor networks' that serve the population in their catchment area. For example, the UCS requires primary care to be delivered through a contractor network comprising of a health promotion centre and a hospital (WHO, 2017).

The PCC is the first point of contact for primary care, and the DHS for those seeking hospital care. Each district has a community hospital with 30–120 beds, and is supported by 10–15 subdistrict (*tambon*) health centres with 3–5 medical staff. The district hospital also serves as a gatekeeper to accessing specialist care at regional and provincial hospitals. The share of district and subdistrict health facilities in primary care is about 90 per cent (Phudit Tejativaddhana et al., 2018).

The launch of UCS in 2002 was followed by separation of provider–purchaser functions, with MOPH being responsible for provision while the newly established NHSO was given the responsibility for purchasing. The resulting provider–purchaser split has made for a more fragmented structure but is in line with modern thinking that separating the two promotes efficiency, transparency, and accountability (Evans et al., 2012). Following the split, the NHSO contracts with MOPH hospitals to deliver services for members enrolled in the UCS.

Unlike Singapore and Hong Kong, and similar to China and India, multiple agencies are involved in paying for and delivering healthcare services. However, as we will see in this chapter, MOPH maintains a magisterial command over the system. The broad governance relationships in the Thai health system are illustrated in Figure 6.1. MOPH has two distinct sets of responsibilities: making and implementing health policies as well as owning and operating all public hospitals. There are different governance structures for different agencies implementing the health financing schemes: NHSO reports to MOPH, the Social Security Office (SSO) to the Ministry of Labour, and the CSMBS to the Ministry of Finance. These agencies in turn contract with public hospitals and/or hospital networks to provide care for their respective members.

As the largest purchaser of healthcare services in the country, NHSO serves as the informal steward for the sector even though the Ministry of Finance and Ministry of Labour zealously guard their respective SHI

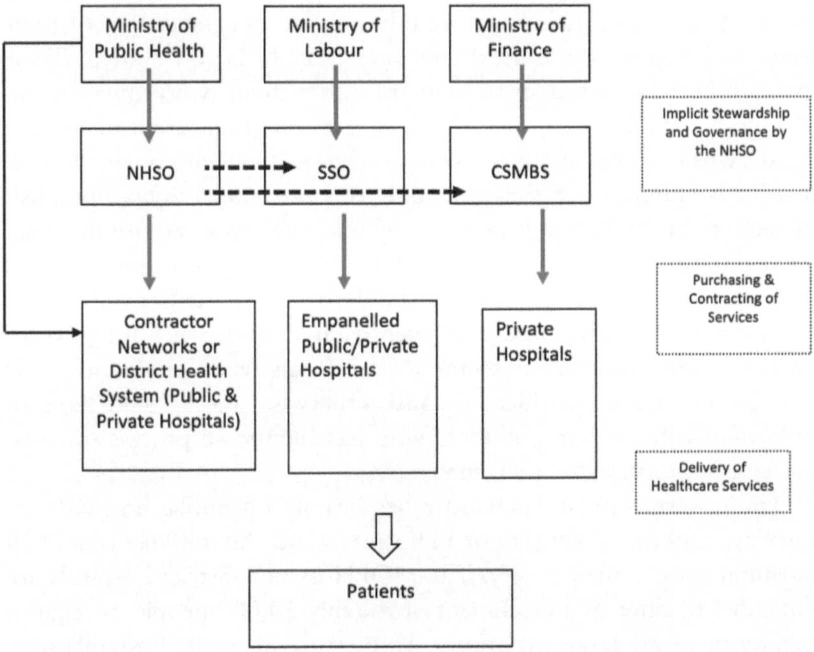

Figure 6.1 *Governance relationships between patients, providers, and payers in Thailand.*

and CSMBS programmes. The vast amount of information it possesses allows NHSO to highlight anomalies even in programmes sponsored by other ministries. Its relatively high analytical capabilities allow it to shape the implementation of SHI and CSMBS programmes even though they lie outside its formal area of responsibility. Thus, for example, the comparative data on costs and utilization produced by NHSO highlighted the inefficiencies of CSMBS which triggered the reforms of FFS payment that were the root cause of its high expenditures (Tangcharoensathien et al., 2015b). These reforms also introduced changes in how providers within the SHI were paid bringing it in line with arrangements in place in the UCS.

6.4 Ownership and Management

In health economics, public hospitals are viewed as prone to vast government failures despite their potential to correct market failures (Preker and Harding, 2003). In this conception, government monopoly and lack of hard budget constraints make them inherently undesirable in most instances. This is an unduly narrow understanding of their role, and betrays the critical role they play in health system. Their performance is affected by factors other than who owns them (Ramesh, 2008b; Shen et al., 2007; Eggleston et al., 2008b). Indeed, who owns hospitals matters less than what is done with ownership, as public hospitals offer the potential to more effectively accomplish objectives in the health system that would otherwise require financial or regulatory tools (Ramesh, 2008b). For instance, public hospitals can be used to set price and quality standards for the entire industry if managed strategically and competently. It is also easier for governments to introduce major sector-wide reforms when public hospitals are the dominant providers because otherwise, they would have to negotiate with private providers who have immense powers to resist measures that threaten their interests.

The government in Thailand relies heavily on public hospitals, as they account for 75 per cent of all hospitals and about 80 per cent of all hospital beds in the country. The MOPH-owned district hospitals are intended to cater to a catchment of roughly 50,000 people, though in reality there are large variations. An increase in public hospitals over the years has reduced the population per hospital bed from about 700 in 1980s to about 400 in the past decade. They account for nearly all the hospital beds in many poorer areas in the Northeast and Southern regions, unlike private hospitals that are concentred mostly in cities. In remote, sparsely populated areas, MOPH hospitals are often the sole providers.

Not only does the public sector own most of the hospitals, most of the personnel involved in the delivery of healthcare services are public employees. The most recent data suggest that around 80 per cent of doctors, 90 per cent of dentists, 85 per cent of pharmacists and professional nurses work in the public sector, mostly at MOPH facilities (Jongudomsuk et al., 2015; MOPH, 2014). All hospital staff – except for some contract workers – are paid salaries but may receive additional financial incentives such as 'hardship allowance, non-private practice incentives, and specific pay for performance, such as pay based on the number of surgical cases provided at all levels' (Tangcharoensathien et al., 2015a:158).

There are 51,308 primary health centres at the village level, and 10,120 health centres in urban areas, each with 3–5 health workers. There are about a thousand public hospitals in the country, of which 75 per cent are district hospitals, 10 per cent are regional hospitals, and the rest are specialist hospitals that serve as regional referral centres (MOPH, 2017). All of these facilities are owned and operated by the MOPH.

Thailand also has a small but vigorous private sector in hospital care serving both local and foreign patients. The number of private hospitals especially grew following the launch of SHI in 1991 which offered insurance cover for treatment at private hospitals to formal private-sector workers. It is estimated that 65 per cent of SHI members are registered with private hospitals. Following rapid proliferation in the mid-1990s, a large number of private hospitals closed down in the wake of the 1997 economic crisis. There are about 322 private hospitals in the country, 30 per cent of which are located in Bangkok, serving mostly rich Thais and foreigners (Wibulpolprasert et al., 2011; Tangcharoensathien et al., 2015a). They find it hard to compete with public hospitals which offer sparse but good services involving no OOP payment. This is reflected in their average occupancy of 50 per cent, compared to 80 per cent for public hospitals.

Hospitals, including public hospitals, must meet accreditation standards in order to be empanelled by the NHSO. The Hospital

Accreditation Programme was transformed into Healthcare Accreditation Institute (HAI) in 2009 with the purpose of promoting higher standards of services increasingly demanded by NHSO. To be accredited, hospitals must demonstrate systems for risk prevention and quality maintenance. International standards including quality management systems (ISO 9001), environmental management systems (ISO 14001), and quality management systems of medical laboratories (ISO 15189) are also encouraged by HAI. The close relationship between MOPH and NHSO, and the fact that most hospitals are publicly owned and largely publicly financed ensures that hospitals meet standards. There are financial incentives available for seeking accreditation but they are insufficient to alter behaviour and so what makes the difference is the direct leverage that MOPH exerts over them (Hanvoravongchai, 2013).

While public hospitals enjoy broad operational autonomy in management of resources, procurement of medicines and supplies, and human resource matters, they are not legal entities – unlike Singapore – and are not allowed to raise debt or borrow money. Indeed, the decision-making structure at public hospitals is centralized, with the ministry appointing general and regional hospital directors. Moreover, it is MOPH and not hospitals that are signatory to the contract with NHSO, giving the ministry complete control over prices and other contractual conditions.

Hospitals are subject to financial audits by purchaser organizations including the SSO for SHI scheme, the Comptroller General's Department for the CSMBS, and the NHSO for the UCS. The NHSO also randomly audits 1 per cent of hospital claims. Currently, NHSO filters out anomalies in claims using specific criteria (e.g. significant deviations from averages, etc.) and then follows up with onsite audits. In 2012, the NHSO clawed back US$6.0 million in over-reimbursements to hospitals. There are also other ways in which NHSO contains expenditures. For instance, to reduce the resources devoted to litigating adverse medical outcomes, UCS beneficiaries may be compensated up to a maximum of 200,000 baht (about US$5,700) without the formalities of a judicial review (NHSO, 2014).

6.5 Pooling of Financial Risks

A pressing goal of modern healthcare systems is to promote risk pooling in healthcare financing and, correspondingly, reduce OOP spending. At a time when many governments in the developing world, including China and India, are pursuing this goal through insurance-based and contributory programmes, Thailand realized as early as 2002 that such programmes were difficult to implement in societies with a large informal sector and hence opted for non-contributory programmes. The decision was informed by decades of experience with implementing different contributory programmes which left about a third of the population uncovered despite concerted efforts for universal coverage. Similarly, the design of the UCS benefit package and arrangement for paying providers were based on the lessons it had learnt from past programmes, especially the need to avoid open-ended FFS payments (Tangcharoensathien, 2019). The success of the strategy is evident in steep decline in OOP spending over the past two decades.

There are three health finance pooling programmes in Thailand: Civil Servants Medical Benefit Scheme (CSMBS), Social Health Insurance (SHI), and Universal Coverage Scheme (UCS). CSMBS covers civil servants and their dependants who together form about 10 per cent of the population, while SHI covers formal sector workers who form about 10 per cent of the population. The CSMBS is a non-contributory programme funded entirely from government budget and is administered by the Ministry of Finance. SHI, on the other hand, is a contributory programme funded from monthly contribution of 1.5 per cent of wages each by the employees, their employers, and the government, subject to wage ceiling of 15,000 baht. The scheme is administered by the Social Security Office (SSO), whose administration costs are subsidized by the government.

UCS covers 80 per cent of the population and is paid for entirely from government revenues. It requires no payment from users at the point of service, except when patients visit hospitals without appropriate referral from their primary care provider. The programme is administered by MOPH which gives it extensive leverage and command over the sector. In exercise of its purchasing functions on behalf of UCS members, MOPH is able to set standards for clinical care, treatment, and pricing that serve as a benchmark for the entire sector. It is thus able to achieve objectives that would otherwise require regulations.

Benefits and treatments covered under the UCS are decided by the Committee on Benefits Package convened by the National Health Security Board (NHSB). The committee's evaluation is largely based on a cost-effective assessment though there is leeway in making decisions based on public interest (Youngkong et al., 2012). Over time, CSMB and SHI benefits have covered towards what is offered under UCS.

6.6 Provider Payment

Thailand has broad experience with different provider payment mechanisms. The oldest programme, the CSMBS, has traditionally reimbursed healthcare expenditure on an open-ended FFS basis with limited oversight, resulting in considerably higher expenditures than other financing programmes in the country. Recent studies have found that health expenditure per member in the CSMBS is about four times that of the UCS. The SHI, on the other hand, relied on capitation payments to reimburse providers, which is effective in containing costs but offers inadequate incentives for providers to be responsive to users.

Based on the experience with CSMBS and SHI, the UCS largely avoids FFS, and instead relies on three different payment mechanisms. First, it uses an age-adjusted capitation rate per person registered with the contractor network. Second, it uses case-based payment rate for inpatient care that is layered with a strict global budget. Third, there are set fees for high-cost and priority treatments. These mechanisms are meant to incentivize providers to be cost-conscious and remain within their allocated budget. This layering and packaging of prospective and retrospective payment tools is the result of sustained efforts over the past decade to calibrate different payment tools to create the appropriate incentives and disincentives.

The SHI programme initially used an inclusive capitation rate. While this was effective in controlling costs, it promoted undersupply of necessary but expensive medical procedures (Hsiao and Shaw, 2007). The problem was in part addressed through extra payments for high-cost treatments, and in part through quality reviews and hospital accreditation standards. Recent payment reforms to the SHI programme have brought it in line with the UCS programme: outpatient care is paid for by capitation and inpatient care through DRG is accompanied with a global budget.

The CSMBS in the past paid healthcare providers on FFS basis through reimbursement claims submitted by members. To rein in expenditures caused by open-ended FFS payments, the scheme was reformed to cover cost of hospitalization only in public hospitals, and only ambulatory care continues to be paid on an FFS basis. Inpatient care is now reimbursed on DRG basis, although global budgets are not applied, which limits its usefulness in containing costs. Pharmaceuticals used for ambulatory care are reimbursed on the basis of set fee schedule while drugs for inpatient care are paid on DRG basis. However, even after the reforms, data show that CSMBS pays more for the same drugs (Thit et al., 2016) and has higher payment rates for the same clinical condition than other schemes (Patcharanarumol et al., 2018). Efforts to introduce global budgets in the CSMBS programme have been unsuccessful due to fierce opposition from civil servants who benefit from the status quo.

Under UCS, a single capitation rate is applied throughout the country for both public and private providers. The capitation is paid to a contractor network which allocates the funds to primary care centres and smaller hospitals within the network based on their utilization rates. Secondary care requires a referral from the primary care doctor and was initially reimbursed on the basis of a national global budget, but a regional global budget is currently employed. This is estimated based on expected admission utilization rate and unit cost per admission at different levels of hospitals (Tangcharoensathien et al., 2015b).

In terms of pharmaceuticals, contracted provider networks are required to purchase from the Government Pharmaceutical Organization (GPO), a statutory body under the MOPH that manufactures and retails drugs (Hollaway, 2012). The GPO is the largest purchaser of pharmaceuticals in Thailand which allows it to exploit its monopsony power. It is also the largest producer of drugs in the country which gives it monopoly power in selling to contracted providers. Hospitals are not allowed to charge UCS patients for medicines, but can charge a markup of 10–30 per cent on drugs dispensed to CSMBS and SHI patients, for which they are later reimbursed on the basis of agreed fee schedules. Although there are no national guidelines for use of generic drugs, the capitation payments for pharmaceuticals under SHI and UCS encourage their use.

The focus on cost-effectiveness within the NHSO largely stems from the National Health Security Act, 2002 which mandates the agency to act as a 'strategic purchaser' of healthcare services. This is in sharp contrast to the CSMBS administered by the Ministry of Finance, which is responsible only for overseeing reimbursements for medical expenditure. The different statutory roles of the agencies have shaped their approach to designing provider payments, particularly in the use of DRGs (Patcharanarumol et al., 2018).

DRG is a key innovation in Thailand's capped payment system. Indeed, it was one of the first developing countries to implement a prospective payment system for most of its population. There are currently two base DRG rates: a regional rate with global budget and a national rate for referrals. Higher cost treatments have been 'bundled out' from the DRG and are now reimbursed on a fee schedule. This is layered with routine audits to ensure that patients are not assigned to more expensive diagnostic groups.

Thailand's experience with capped payments highlights the complexity in designing provider payments (Langenbrunner et al., 2009). No single arrangement – FFS, capitation, or DRGs – can create appropriate incentives for healthcare providers. Governments need to rely on a combination of payment tools (Wu and Ramesh, 2014) and be prepared to continually adjust them in response to underlying data on costs and utilization. Such calibrations require extensive data and investments in analytical capacity to understand and respond to variations in utilization of high-cost services. Thailand has built extensive capabilities in compiling disaggregated data on costs, utilization, and regional morbidity patterns and epidemiological profiles. The information is then used by the NHSO to calibrate capitation rates and subsidies that it pays to contractor networks (Patcharanarumol et al., 2018). Providers in turn use data on costs and utilization in their negotiations with the contractor network and the NHSO. While healthcare providers resist sharing of data (Pauly, 2009; Fuchs, 1978), this has not been the case in Thailand because the government owns and finances most hospitals and is able to directly shape their behaviour.

The experience of introducing DRG in CSMBS highlights the institutional and political constraints that circumscribe healthcare reforms (Preker et al., 2013; Fox and Reich, 2017). Elites within the Ministry of Finance resisted all efforts to adopt DRGs until their resistance became untenable due to the overwhelming evidence suggesting the programme's wastefulness. However, even when DRG was adopted, it was unaccompanied by a global budget (Tangcharoensathien et al., 2015b).

6.7 Regulations

The health sector in Thailand is relatively lightly regulated which is not unexpected given that public sector's dominance in provision and financing of care reduces the need for extensive regulatory tools necessary in systems dominated by private provision and financing. Its regulatory landscape includes the Medical Registration Division (MRD) of the MOPH which licenses clinics and private hospitals to operate, and the Food and Drug Administration (FDA) which regulates the access and availability of medical products and drugs. Professional associations such as Medical Council, Pharmacy Council, and Dental Council, have some authority for self-regulation. Prices for over-the-counter medicines are regulated by the Ministry of Commerce and through the national formulary for prescription medicines.

Unlike India, the absence of a regulatory framework that governs the distribution of private providers or the supply of diagnostic equipment has not posed major challenges given the overwhelming presence of public hospitals. Thailand also uses regulatory guidelines to address the shortage of manpower across the sector. For instance, all physicians and nurses must work in community hospitals for three and four years, respectively, after they graduate (Teerawattananon et al., 2003).

To the extent regulatory tools are used, they are in the form of rules and directives for citizens and healthcare providers to manage some of the malfeasance associated with third-party payments. For example, all members of the UCS must register with a contractor network (or DHS). This directive is to ensure that there is no adverse selection as healthcare providers are paid an age-adjusted capitation (for inpatient care) for members registered within their catchment area. Current rules allow a member to change contractor networks twice a year but in practice this is rarely used. Similarly, patients cannot access hospital care without a referral under the UCS unless in the case of emergencies. Strict gatekeeping rules implicitly encourage those with means to opt out of the public system and pay for their care at private clinics and hospitals.

To control costs, the SSO and NHSO require medicines to be prescribed using the national formulary. The National List of Essential Medicines (NLEM) was adopted in 2007, which is revised annually and includes about 900 chemical entries classified on the basis of cost-effectiveness. The listing guides the prescription and dispensing behaviour for all hospitals paid through UCS, SHI, and CSMBS. However, hospital networks have the autonomy to develop their own formularies specifying essential and non-essential drugs and many have done so.

Hospitals procure non-essential drugs from the revenue they generate as well as UCS allocations from NHSO. The funds allocated by NHSO must be used to purchase pharmaceuticals in line with an approved procurement plan. To ensure competition, any purchase more than 100,000 baht must be tendered – but most contractors overcome this by purchasing for lesser value but more frequently (Hollaway, 2012).

To improve monitoring and evaluation, NHSO is required to publish an annual report and its Secretary General is held accountable through annual performance reviews. This practice is however not followed for the SSO and the CSMBS programmes. There are also extensive guidelines that require health technology assessments and cost-effectiveness projections for new interventions or changes to benefit packages under the UCS. A committee under the NHSB of 31 members (including stakeholders from NGOs) helps in these deliberations. These conditions do not however apply to the CSMBS which does not have established criteria for including new interventions and only relies on a Technical Advisory Committee of medical specialists.

6.8 Conclusion

Thailand's health system displays exceptional performance in terms of archiving universal access and high health status at modest cost. There are few parallels anywhere in the world where the government was able to so drastically reduce OOP payments without compromising quality or financial sustainability over such a short period of time. This is the result of concerted efforts covering all critical aspects of health system.

The Thai health system is remarkably simple in form and essence. It is a highly centralized system – despite decentralization required by the constitution – with the Ministry of Public Health (MOPH) providing clear stewardship to all segments of the sector. An excellent primary care system owned and operated by MOPH spanning the entire country forms the bulwark of the system. Similarly, hospital care is dominated by MOPH owned and operated public hospitals that provide no-frills but clinically sound necessary services to the entire population, except for the small segment of the local population (and foreign tourists) that prefers the comfort of private hospitals. The public hospitals are tightly integrated with primary care providers in their district through various financial and administrative arrangements. The performance of public hospitals and clinics is maintained through quality standards and accreditation procedures rather than financial or other rewards.

The funding for both primary and public care has come increasingly from government budget, mostly as subsidy for the Universal Coverage Scheme, with OOP payment forming a small and shrinking share of health expenditures. Healthcare providers are paid prospectively for the most part: on capitation basis for primary care and DRG basis for hospital care, with small FFS component for priority services. Regulations in Thailand are light, partly due to the dominance of public providers who can be controlled through ownership and financing tools. The government's policy to promote medical tourism in the country also requires liberal regulatory regime so as to allow private providers to offer innovative products at costs the (international) market will bear.

The government's success in establishing sophisticated healthcare system over a relatively short period of time was helped by public ownership of most hospitals and a nationwide network of primary care clinics and its position as the largest buyer of healthcare services. In the absence of these conditions, as is the case in India, the Thai government would have had to negotiate with private providers and insurers and overcome their political power and information advantage.

While the popular support for UCS remains strong, this cannot be taken for granted going forward. The upper and upper middle class finds the services provided by the public sector too frugal and relies on private services, except for high-cost medical treatment procedures. This small but powerful section of the society has pressured the government to expand subsidy for private insurance and, relatedly, private hospitals. Public providers will need to work hard to meet the rising expectations of the population if it is to retain their political support.

The mettle of Thailand's health infrastructure was tested by the global pandemic in 2020. The relatively low number of infections and case fatalities in 2020 did not pose a challenge for Thailand's health system. Early responses to the pandemic and a strict regimen of surveillance, reporting, quarantining, and contact tracing resulted in containing widespread transmission of the virus across the country. Thailand's past experience of the H1N1 pandemic and its reflexive practice of policy learning and capacity building helped health authorities in responding to the pandemic. Sustained investment in primary care, community engagement through over a million volunteers working in villages across Thailand, and oversight through a newly established Center for COVID-19 Situation Administration – all central tenets of Thailand's health system – were also key features of the government's response to the pandemic.

7 Hong Kong

Keeping Health Care Simple

Hong Kong has one of the best healthcare systems in the world, noted for its low costs and high equity. It is a rather simple system, with hierarchical governance structures and sparse policy tools centred on public ownership and financing that have undergone only minimal changes since the 1960s when they were established by the colonial government heavily influenced by policy thinking in Britain at the time. The territory has been spared of much of the modern healthcare 'reforms' that have been tried out in other countries.

Notwithstanding the stellar results, the Hong Kong government itself has not been happy with its health system and made repeated efforts to change it, without much success due to political obstacles. It has expressed concerns about high and rising costs and cast doubt on the system's long-term financial viability. It is lost on the government that rising healthcare expenditures are not unique to Hong Kong and nor are projections of severe fiscal strains in the future due to ageing population, growing demand, and rising prices. It also does not fully appreciate that many of its coveted reforms promoting larger role for private provision and financing are known drivers of costs that will aggravate the very problems it seeks to solve. At the root of the government's desire for change lies its fundamental discomfort with a state-dominated system, which it finds inconsistent with its self-professed preference for markets and private sector.

The purpose of the chapter is to describe the development and functioning of the healthcare system in Hong Kong and examine the policy tools that underlie it. The case shows that it is possible to achieve universal health care through traditional organizational and fiscal policy tools. The case also serves as a cautionary lesson for health policy and international consultants proposing complex mix of policy tools in health care when simpler tools used effectively can achieve superior results.

7.1 Policy History

Hong Kong was a British colony until 1997 and the colonial legacy is evident in its healthcare system to this day. During the first century of its rule, the colonial government paid little attention to the population's health, except for those working for the government. The inadequacies of the existing system were starkly exposed with the influx of refugees from the mainland following communist victory in 1959. It was estimated that four-fifths of the residents could not afford health care, which was provided largely by the private sector at the time.

To address the scarcity, the government began to establish and expand hospitals and clinics that provided medical services nearly free of charge (Gauld and Gould, 2002). The authorities believed that public facilities would be used only by the poor as those with the means to pay would be turned off by austere services and long waiting times. This attitude was reflected in a 1964 White Paper titled *Development of Medical Services in Hong Kong*, the first policy document on health care to be published by the colonial government. The report stated that the government was committed to "providing, directly or indirectly, low cost or free medical and personal health services to the large section of the community which is unable to seek medical attention from other sources." To this day, the government continues to adhere to the notion that its role is to provide services to those unable to pay from other sources.

In reality, public hospitals came to be used not just by the poor but increasingly the middle class which did not mind the inconveniences for free services. The result was that public hospitals came to gradually replace private hospitals as the dominant supplier. Continuous improvements in services at public hospitals further undermined private hospitals, which were not only expensive but were also viewed as providing inferior clinical services due to concentration of experienced medical personnel in the public sector. The resulting rapid increase in government expenditure on hospitals did not raise serious fiscal concerns as the economy grew even more rapidly and generated ample public revenues to pay for the increased expenditures.

However, as the population's income grew and expectations rose, it expressed displeasures over overcrowded facilities providing austere services. The hospital managers too were frustrated with the rigid rules imposed by the Medical and Health Department. The government released a White Paper in 1974 to respond to the complaints. The document called for regionalization of public hospitals whereby each region was to have one large hospital supported by a network of smaller hospitals. This did little to ease the overcrowding problems at large regional hospitals because users preferred them to smaller hospitals. Unable to meet the growing demand for hospital care and address the general dissatisfaction with the disparities between regional and non-regional hospitals, and between public and subvented hospitals, the government began to explore ways to restructure the entire hospital system.

An Australian management consultant was commissioned to inquire into the system and the report, popularly known as the Scott Report, was published in 1985. The Scott Report was surprisingly ahead of its time and featured a line of thinking which came to be later described as New Public Management (NPM). It recommended corporatization of public hospitals under an autonomous entity – called the Hospital Authority – run by professional managers. Doctors and nurses at public hospitals opposed the proposal to grant more autonomy to hospitals because they thought they would lose out by being outside the civil service. Staff at subvented hospitals, on the other hand, strongly supported the proposal to put them on the same footing as their public counterparts. To build support for the proposed reforms, the government offered public hospital staff salaries and conditions similar to the civil service. Eventually, the recommendations were adopted and the Provisional Hospital Authority was established in 1989.

The Provisional Hospital Authority published a report on financing issues in the first year of its existence. The report proposed recovery of 15–20 per cent of the costs of services from patients, instead of the less than 3 per cent that was the case at the time. The proposal made little headway in the context of a booming economy, which yielded increasing government revenues, and public apathy towards what was not yet a significant problem.

The Hospital Authority (HA) was eventually established in 1990 and began functioning in 1991. It was to be an independent statutory body with control over personnel and resource allocation matters, in line with the objective of freeing public hospitals from rigid civil service rules. Individual public hospitals, in turn, were allocated fixed budgets by the HA but were otherwise given broad autonomy in routine personnel and purchasing matters. The enhanced autonomy for HA and hospitals was expected to increase the efficiency of hospitals as well as the quality of services.

Having dealt with management of public hospitals, the government turned to reforming financing arrangements. The first formal attempt to reform healthcare financing began with the publication of the consultation paper *Towards Better Health* in 1993. The paper listed five options for reducing the government's healthcare financing burden: (1) cost recovery of 5–15 per cent of operating cost; (2) semi-private rooms in public hospitals for extra charges and itemized billing; (3) government-coordinated voluntary and private health insurance; (4) compulsory public insurance; and (5) the Oregon approach of giving the community a say on how public healthcare resources are allocated.

The 1993 discussion paper drew more than 500 submissions, most of which supported option 2 (establishing semi-private rooms) or option 3 (voluntary private insurance). In late 1993, the government decided to support options 2 and 3 favoured by the submissions (Bennett et al., 1996). However, it did not follow up with the necessary legislative or regulatory changes on the grounds that none of the options enjoyed sufficiently broad support in the community (Legislative Council, 2005). There were speculations at the time that the government backed off in an effort to not further complicate the rocky ongoing negotiations with China over the future of Hong Kong (Woo, 2006).

The appointment of Shane Solomon, who had management rather than medical background, as the head of HA indicated the government's renewed focus on financial and management issues. In a speech to the Hospital Authority Convention in 2006, Solomon highlighted the importance of more professional management of hospitals and the need for reforms to make high-income earners pay a greater percentage of the costs. He may have had in mind the Australian system where high-income earners are induced to buy private insurance despite the availability of universal access to public hospitals. The Secretary of Health Bureau at the same HA convention hinted at establishing private clinics in public hospitals.

The second attempt at healthcare financing started with the publication of *Improving Hong Kong's Health Care System: Why and For Whom?*, popularly known as the Harvard Report, prepared by the Harvard Team of consultants appointed by the government (Harvard

Team, 1999). The report concluded that while the existing system was equitable as well as efficient, it needed to be put on sounder fiscal footing because otherwise health care would be taking up 20–23 per cent of the total government budget by 2016. The centrepiece of its reform proposal was the establishment of two new financing schemes: Hospital Security Plan (HSP) and Medisage. The former was a low-cost insurance scheme for specified chronic illnesses while the latter was a compulsory individual medical savings account plan which could be drawn upon only during old age. Furthermore, it proposed cost recovery of up to 30 per cent of the costs of services at public hospitals through user charges.

The government did not express its own views on the Harvard Report's recommendations and the general public did not understand all the data and technical arguments presented in the report. Of those who did participate in the discussion, the majority strongly opposed the proposals for higher user charges and compulsory contributions to HSP and Medisage. Remarkably, doctors launched an unprecedented campaign against the proposal for centralized purchase of outpatient services by HA. With the public opposed to contributing to HSP and Medisage, private doctors opposed to central purchase by the government, and the HA opposed to erosion of its role, the Harvard Report had no realistic chance of being adopted.

The government's response to the Harvard Report was in the form of a consultation paper *Lifelong Investment in Health* published in 2000. The paper rejected the Harvard Report's proposed HSP on the grounds that a compulsory insurance scheme would increase labour costs, promote overuse, and be prone to deficits. It instead proposed higher user charges for non-essential services and a compulsory medical savings account scheme called Health Protection Account (HPA) to which everyone would contribute 1–2 per cent of their income during the ages of 40–60 years and draw on it after the age of 65 years. The proposal was met with widespread opposition from the general population which disliked the idea of compulsory contribution to individual medical savings or paying higher user charges. Business groups opposed the proposal on the grounds that compulsory employer contribution would raise labour costs and thus reducing their international competitiveness. The proposal gradually disappeared from the agenda in the face of widespread opposition to its core features.

The HA submitted a counterproposal to the Harvard Report recommending that only essential healthcare services be fully subsidized by the government, with the costs of other services partially or fully recovered through user charges. It recommended establishment of a medical savings scheme to help patients pay for non-subsidized services. It also proposed expanding its services to primary care provided through its network of hospitals and clinics and funded on a capitation basis. Its thinking was subsequently reflected in the consultation document 'Lifelong Investment in Health' issued by the Health and Welfare Bureau in 2000. Economic recession and weak government at the time made adoption of deep reforms nearly impossible. Unsurprisingly, the public deeply opposed any measures that imposed additional costs on them, including user charges, insurance, or savings. At the same time, the medical association and private doctors opposed the proposal to expand HA's services to primary care.

After repeated failures in reforming financing arrangements, the government returned to reforming provision of health care, including hospital services. It published a consultation document titled 'Building a Healthy Tomorrow' in 2005 which recommended deep changes in the delivery of health care. The document recommended that the HA hospitals concentrate only on (1) acute and emergency care; (2) services for low-income and underprivileged groups; (3) illnesses that entail high cost, advanced technology, and multidisciplinary teamwork; and (4) training of healthcare professionals. If put into effect, all non-emergency and non-acute cases would have shifted to general practitioners. The proposal confronted too many powerful vested interests to succeed politically and died by neglect.

The fourth effort at financing reform started with the publication of the consultation document *Healthcare Service and Financing Reform* in 2008. The document merely laid out different financing options without making a recommendation, though the accompanying commentary indicated preference for compulsory medical savings account (MSA). The ensuing public consultation showed severe public opposition to any scheme requiring mandatory contributions, be it to medical savings account or insurance. Correspondingly, the population indicated strong preference for voluntary schemes heavily subsidized by the government.

Public consultations over the 2008 document were followed by the publication of another consultation paper called *My Health, My*

Choice in 2010. The document was based on the feedback received on the 2008 paper and a government-commissioned consultancy report (Milliman Limited, 2010a). The cornerstone of the document was the proposal for a private voluntary health insurance scheme called Health Protection Scheme (HPS). It is notable that the proposal mentioned choice for users as the main objective, and not financial sustainability of the healthcare system, which was the stated reason for seeking reforms. HPS sought to promote private health insurance and private health care by subsidizing premium for the young (who tend to avoid insurance) and the aged (whose premium tend to be higher). Subsequent announcements by the government indicated that the subsidies would cost the government up to HK$50 billion (Food and Health Bureau, 2010).

Following public discussion on the HPS proposal – which unsurprisingly faced little opposition given that it offered subsidy and, in addition, was voluntary (Ramesh, 2012) – the government passed the necessary legislations in 2018 and the programme came into effect in April 2019. The scheme was named Voluntary Health Insurance (VHIS) to emphasize its voluntary nature. The stated objectives of the scheme are (1) Enhance the protection level of hospital insurance products; (2) Provide the public with an additional choice of using private healthcare services through hospital insurance; and (3) Relieve the pressure on the public healthcare system in the long run (Food and Health Bureau, 2019). These are modest objectives, far short of the lofty original goal of improving the healthcare system's sustainability. In retrospect, it is apparent why a government lacking popular legitimacy could only launch an unremarkable programme of uncertain value that offended no powerful group (Ramesh, 2012).

7.2 Performance

The healthcare system in Hong Kong performs exceptionally well and indeed has been ranked number one in the world by Bloomberg Health Efficiency Index (Miller and Lu, 2018). The index shows that Hong Kong population lives a long life while spending remains relatively low.

The territory's average lifespan of 82 years for men and 88 years for women are some of the longest in the world whereas the infant mortality rate of 1.5 per 1,000 live births is one of the lowest (Government of Hong Kong SAR, n.d.). Similarly, its average Health Adjusted Life Expectancy (HALE) at birth of 70 years for men and 76 years for women is the third and second highest, respectively, in the world (Law and Yip, 2003).

Remarkably, Hong Kong's fine healthcare status has been achieved at rather modest expenses, as its Total Health Expenditure (THE) forms only 6.2 per cent of GDP which is considerably lower than the OECD average of over 9 per cent. Hong Kong's THE is estimated to be 18 per cent below the rate 'predicted' by its income level (Wagstaff, 2005). No less remarkably, the population enjoys a highly equal access to health care due to the availability of hospital and primary care nearly free of charge. This is in the low incidence of catastrophic healthcare expenditures, as only 5 per cent of the population faces healthcare expenditures exceeding 15 per cent of the total annual per capita household spending (Wagstaff, 2005). By another measure, in 2004 less than 1 per cent of the population faced catastrophic healthcare spending, defined as more than 40 per cent of non-food household expenditure (Wong and Tin, 2015).

Much of the credit for Hong Kong healthcare system's superior performance in terms of expenditures and access is attributable to the efficiency of the public system. Public hospitals maintain an average occupancy rate of 90 per cent. The duration of hospital stays has declined over the years, from 6.8 days in 2000 to 4.6 days in 2010, despite increase in the number of visits per patient from 1.9 to 2.1, and the increase in hospitalization rate from 87.4 to 94.5, due to ageing of the population. The decline in the mean duration of hospital stay has been shown to be the primary reason for containing increase in health expenditure in Hong Kong (Yip et al., 2014). In comparison, the performance of private providers and insurers is discernibly sub-par. The average annual premium (US$6,432 in 2018) for private health insurance is the second highest (after the USA) in the world. It is notable that most of the private insurance is for services at private hospitals, which are many times more expensive than their public counterparts (Pacific Prime, 2018).

Drug expenditures in Hong Kong are comparatively modest, forming only 10 per cent of total expenditures, despite rising approximately 7 per cent per annum between 2005 and 2015. In 2017, drug expenditures amounted to US$17 billion, of which the Hospital Authority accounted for roughly 54 per cent (Mercurio, 2019). As a public provider operating under fixed budget constraints, HA is highly cost-conscious which helps with keeping a lid on expenditures. Although not legally required, HA facilities largely use generic drugs which are substantially cheaper (Lee et al., 2018). This is unlike private providers who largely use branded drugs that yield them extra profit. There is no separation of prescription from dispensing of drugs, which is not a problem at public facilities due to absence of financial incentives on their part but causes overprescription at private facilities.

Notwithstanding the good performance at low cost, there have been widespread concerns in Hong Kong over growth of THE, which grew at an average annual rate of 5.8 per cent in real terms between 1989 and 2016 compared to average economic growth rate of 3.8 per cent. The increase was largely driven by government health expenditures, which increased by 490 per cent cumulatively during the period, compared to 277 per cent for private expenditures. As a result, the public share in total health expenditure rose from 40 per cent of THE in 1989 to 52 per cent in 2016 (Food and Health Bureau, n.d.).The government's share of spending was especially large in inpatient care, 73 per cent of total inpatient spending in 2016.

The government has since the mid-1980s feared that its large and growing expenditures on inpatient care are fiscally unsustainable in the long run. Health care already formed over 16 per cent of the total government expenditures and was projected to rise to 20 per cent in 2020 and 27 per cent in 2033. While this was small in comparison to other rich countries, the government has claimed that it is large for the territory's low-tax environment (Food and Health Bureau, 2008). The government's professed commitment to 'big market, small government' also comes in the way of accepting responsibility for government-provided free health care regardless of the technical merits of the arrangement (Tsang, 2006).

The Hong Kong population too is not as satisfied with the healthcare system as one might think based on the positive performance indicators. Its greatest reason for dissatisfaction is the long waiting times at public hospitals. The average time, for example, is 3 hours for emergency, 3 months for CT scan for liver tumour, 5 months for liver cancer treatment, 31 months for cataract surgery, and 31 weeks for new cases for specialist outpatient services (*Sing Tao Daily*, 22 April 2009; *The Sun*, 8 March 2008; *Hong Kong Daily News*, 15 June 2010; Food and Health Bureau, 2008). While no one is denied immediate treatment in life-threatening situations, people would prefer quicker access to medical services according to their own perceived need rather than as determined by the public provider. They do not appreciate that one of the secrets of low THE in Hong Kong is its reliance on fixed global budget for public hospitals which necessarily involves rationing for non-urgent services. Provision of healthcare services on demand, as is the case in FFS system, would in all likelihood expand supply and reduce waiting time but will also increase expenditures (McPherson et al., 1981; Wu and Ramesh, 2009).

Notwithstanding the anxieties over long waiting periods, an independently commissioned survey in 2017 found an overall positive experience with inpatient service at HA hospitals, with 91 per cent of respondents rating 7 or above on a 10-point scale, with mean score of 8.1 (Government of Hong Kong SAR, 2018). This was an improvement over the 2015 survey in which 86 per cent of the respondents gave positive rating for overall experience with the mean score of 7.8. Other surveys, including survey by He (2018), have found similarly high public satisfaction with public hospitals.

7.3 Stewardship and Coordination

In the early years, healthcare governance in Hong Kong bore semblance to the arrangements that existed in Britain at the time. The expansion of the government's role in health care in the 1960s was led by the Medical and Health Department which owned and operated its own facilities, similar to the arrangement in Britain. It was a large organization employing 13 per cent of the entire civil service and receiving 9 per cent of the government's total expenditures. It also indirectly controlled the non-profit private hospitals through subventions that formed a bulk of their revenues.

The top-down healthcare system that existed was an effective mechanism for achieving the government's goal of providing basic services to everyone who needed them. All aspects of provision and financing were internal to the government, thus improving coordination and reducing transaction costs. However, the command and control tools by which it worked did not contain mechanisms for responding to users' needs. It was also a rigid system that frustrated both hospital managers and users. Managers were dissatisfied with the detailed approval processes for routine matters regarding appointments and purchases. The users too were displeased with the lack of choice of treatment or doctors. To address these concerns, the government established the Hospital Authority (HA) in 1990 as an autonomous corporate entity to professionally manage all government-owned health facilities in the territory.

Although there has been some devolution of operational functions, the healthcare system in Hong Kong remains centralized with clear hierarchical relationships and well-defined roles. The Food and Health Bureau (FHB) sets healthcare policy and allocates resources. The Department of Health (DH), on the other hand, is the implementing agency, which includes regulating and licencing providers and insurers. The latter's role expanded in 2019 with the launch of the Voluntary Health Insurance Scheme (VHIS) which it administers. DH also provides disease prevention, health protection, treatment, and rehabilitation services and runs public health campaigns.

The responsibility for operating public hospitals and outpatient clinics rests with the Hospital Authority which answers to the Department of Health. To promote flexibility and responsiveness, HA has increasingly decentralized its operations and created governing boards at every level, each answering to the board above it in the hierarchy. The territory is divided into three 'regions' – Hong Kong Island, Kowloon, and New Territories – each with its own governing board and advisory committee. The regional boards play a largely advisory role, rather than the operational role proposed by the Scott Report.

The public healthcare facilities are further divided into 'clusters': Hong Kong Island and New Territories have two clusters each while Kowloon has three clusters. Each cluster includes a broad range of outpatient, inpatient, and convalescence facilities. The clusters are intended to improve coordination of hospital services at the local level and eliminate duplications and redundancies among nearby hospitals. Since 2000, each cluster has had its own chief operating officer responsible for allocating budgets to hospitals within their clusters and overseeing their performance.

Each public hospital has a Hospital Governing Committee which receives regular management reports from the hospital chief executive, monitors operational and financial performance of the hospitals, participates in human resources and procurement matters, and engages with the community. Also, doctors have remained dominant in the management of hospitals, contrary to earlier expectations that HA would be run by professional managers. The continued dominance of doctors arguably had the effect of muting revenue motives characteristic of those trained in business management.

To further address persistent concerns and criticisms of how public hospitals were managed, the government established a Steering Committee on Review of Hospital Authority in 2013 which submitted its report in July 2015 (Food and Health Bureau, 2015). The committee recommended that the HA Board should play a more active role in managing the HA, coordinating services, setting standards and targets, and closely monitoring performance. It also recommended that the HA should establish effective mechanism to take patients' feedback into account. The government accepted the report though the status of its implementation is unclear.

7.4 Ownership and Management

In Hong Kong, direct provision by the government is the tool of choice for providing hospital care while private provision is the main tool for delivering outpatient care. Some 77 per cent of all hospitals (54) and 86 per cent of all hospital beds (32,973) are in the public sector which together accounts for 90 per cent of all hospital admissions (Food and Health Bureau, 2017). Notably, the shares have not changed despite efforts to expand private hospitals in recent years.

The Hospital Authority currently operates 42 hospitals (of which 14 provide acute care) and 74 general outpatient and 45 specialist outpatient clinics. Acute care hospitals are strategically located in densely populated areas and are connected to a network of non-acute hospitals in neighbouring areas. The facilities are organized into seven regional clusters, each offering a comprehensive range of acute and convalescent services.

Primary care, in contrast to hospital care, is dominated by private providers who account for approximately 70 per cent of all outpatient visits while HA-affiliated clinics account for the rest. Private clinics are spread throughout the island and there is intense competition among them leading to patient-focused services at reasonable prices.

There is no meaningful vertical referral system in Hong Kong, as users can visit any provider they choose without restriction or penalty. At public hospitals, the specific type and volume of service to be provided are determined by the medical staff. At private clinics, on the other hand, what services patients receive depends not only on their medical needs but also on their ability to pay. In fact, it would be difficult to establish a vertical referral system in Hong Kong given that outpatient care is dominated by private hospitals and inpatient care by public hospitals and there is no platform for collaboration between them.

There is competition among public hospitals in the sense that users are free to visit any hospital they wish, but this does not substantially alter hospitals' behaviour because fixed transfers from the government are the source of vast majority of their income. Private clinics, on the other hand, compete vigorously with each other and public clinics. Since government facilities are heavily subsidized, private providers must make up for it by providing user-focused services at prices that most households can afford. The government has recently sought to expand the role of private providers in hospital care too, but success has been limited because it is hard to compete against public hospitals that provide high-quality services at highly subsidized prices. A small number of private hospitals are able to survive largely due to long waiting period for non-urgent services at public hospitals.

In Hong Kong, there is no effort to promote efficiency in public hospitals through competitions, as is the case in Singapore. The establishment of Hospital Authority as a corporatized entity in Hong Kong was not accompanied by new incentives or regulations to promote better performance on the part of managers. Indeed, HA's establishment reduced competition because it created one large organization controlling most of the hospital beds in the colony that provided comprehensive range of services nearly free of charge. Its establishment was also not accompanied by any change in funding mechanism, as the government continued to fund it through block grants as before. Moreover, there was no reward for higher productivity or consumer satisfaction and nor was there any penalty for not meeting performance targets (Preker and Harding, 2003: 14–15).

To compensate for Hospital Authority's near-monopoly status and the resulting lack of competition as a mechanism for improving service quality in hospitals, the government employs a broad range of procedural tools to promote and maintain standards and contain costs. Elaborate 'Patients' Charter', quality standards, and 'Professional Code of Practice' for public hospitals are in place to improve service quality. Each hospital has arrangements for maintaining quality and monitoring clinical and management performance. Hospitals have also established community advisory bodies and considerable efforts are made to make them work.

There are also various accountability measures in place to ensure performance. Thus, all administrative and service units within HA are subject to annual review of performance with respect to inputs such as number of beds, clinical protocols, and information system (Yip and Hsiao, 2003: 405). To further enhance accountability, there exists elaborate series of contracts between different levels of HA administration. "The Government first contracts the public health care services to HA; HA then contracts the resource management to the clusters, then to the public hospitals, then to the department heads and then to each specialty." (Lam, 2015) Such contracts are meant to promote vertically integrated accountability but critics argue that in reality they only promote extra paperwork.

The lack of competition in the healthcare sector and the absence of performance incentives are, however, not as disastrous as argued by mainstream health economists and New Public Management protagonists. Indeed, it is arguable that the absence of market competition has lowered marketing and management costs, which are reflected in public hospitals' relatively low administration costs in Hong Kong. The public provision of services to all residents is also highly equitable, reflected in low incidence of catastrophic expenditures.

7.5 Financial Risk Pooling

Of total health expenditures (THE), approximately half comes from government budget, making it by far the largest source of financing. In comparison, 34 per cent comes from OOP, while employer-provided insurance constitutes 7 per cent, and self-purchased private insurance 8 per cent (2016). The government's share of THE increased while the OOP's share declined during the 1980s and 1990s and their respective shares have remained stable since then (Food and Health Bureau, 2017). Government expenditures are mostly incurred at public hospitals while OOP expenditures are mostly on primary care, which is provided privately for the most part (Food and Health Bureau, n.d.). By funding majority of the territory's health expenditures from its budget, the government is able to directly shape the character and functioning of the sector.

Financing for public providers consists largely of fixed annual budgets provided by the government to the Department of Health, which transfers the funds to HA, which in turn transfers to clusters down to individual facilities. The size of the transfer is determined largely by historical precedence with some adjustment for demographic shifts. The lump sum payment imposes a global limit to what the HA and individual clusters and public hospitals can spend. While this has the virtue of containing costs, it also necessarily involves rationing of the volume of services, especially non-urgent elective treatments, which leads to long waiting time promoting public dissatisfaction and anguish.

Over 90 per cent of the public hospitals' revenues come from government subventions and only 5 per cent from user charges. This is unsurprising given the low hospital charges, ranging between HK$100–120 (US$12.90–15.50) per day for inpatient care and HK$180 (US$23.22) per Accidents & Emergency visit when the average operating cost is approximately HK$3,300 per hospital bed per day (Yip et al., 2014). In contrast, private providers, most of whom are in primary care, are paid largely from private sources (OOP and insurance).

Transfer of funds through subventions not only establishes a clear line of accountability but also allows the government to directly set conditions regarding production and delivery of medical services in the territory, thus offsetting the need for detailed regulations. Furthermore, public financing for hospital care promotes both efficiency and equity by protecting households from potentially catastrophic bills for hospital treatment. Primary care, on the other hand, does not pose the same level of risk because the cost of individual episode of treatment is usually small and within the means of most household. However, the cumulative burden of outpatient care may be heavy, especially for households with older members, and public clinics are a great help for such households.

Notwithstanding the advantages of public financing for hospital care, the government has been apprehensive about its open-ended commitment in the context of an ageing society and has been promoting private financing as a supplementary source of funds. However, all alternative private sources of revenues – higher user charges at public facilities, compulsory public insurance, and/or compulsory individual medical savings – are politically difficult because they require households and/or employers to make additional payments for what has been free of charge. It is unsurprising therefore that all financing reform efforts over two decades failed, except for the most recent reform establishing VHIS. The latter succeeded because it provides subsidy at cost to the government rather than impose costs on users or employers.

Going forward, the role of private insurance will expand due to subsidies available under VHIS. The programme offers tax deduction for the amount paid for health insurance premium, amount not exceeding HK$8,000 a year is available, to the taxpayers and their dependents. Notable benefits offered by VHIS- certified insurance plan include: coverage for unknown pre-existing and congenital conditions, guaranteed renewal up to the age of 100, and no 'lifetime benefit limit'. However, insurers are allowed to reject applications for insurance and to set their own premium, including premium loading for age.

VHIS sits uneasily with the existing healthcare architecture in Hong Kong built on public provision and financing of hospital care. With subsidized insurance, the insured will be able to buy private services in addition to enjoying the right to use public services (Yin and He, 2018). While the government argues that VHIS will reduce the burden on public hospitals, there are reasons to be sceptical. According to one study, those who purchase VHIS are likely to be existing users of private hospitals who will use private services more unhesitatingly (Institute for Health Policy and Systems Research, 2015). Moreover, the various caps and deductibles imposed by private insurance will deter acute patients from using private hospitals and they will continue to rely on public hospitals. The most optimistic estimates show that VHIS will reduce waiting time at public hospitals rather than reduce demand for their services, thus making no or little impact on the latter's expenditures (Milliman Limited, 2010b).

There are valid reasons to fear that expansion of VHIS will lead to increase rather than decrease in expenditures due to the private sector's higher cost structure. According to one estimate, "the cost per inpatient discharge (including deaths) in the private sector is 2.42 times higher than that in the public sector" (Institute for Health Policy and Systems Research, 2015). That private insurance pays private hospitals on an FFS basis, which is a known fuel for cost escalation, will further raise expenditures. It would foster a political economy that galvanizes beneficiaries to band together to demand higher subsidies and to cover for increased premium resulting from higher spending due to FFS. This is different from the current global budget method of financing public hospitals, which presents no incentive for maximizing revenues or profits.

It is arguable that Hong Kong already has a highly cost-effective healthcare financing system. Hospitals are allocated fixed global budgets which undermine any incentive for providers to engage in over-servicing and overcharging, the perineal problem in health care. Global budgets are an effective mechanism for containing costs because the sum allocated to each hospital forms an overall ceiling within which managers must remain. Even though there is no cap on demand, it exists at the point the providers' funds are exhausted. A strong accountability arrangement is necessary to ensure that funds are not wasted.

7.6 Provider Payment

As mentioned earlier, public hospitals and clinics are paid lump sum amounts by the government on an annual basis. There are user charges, but they are small and based on the number of days of hospitalization rather than services received. The revenues they generate are too small to incentivize public hospitals to over-service users for income.

Initially, HA allocated funding to hospital clusters based on historical budget allocation but in 2009 it introduced a 'Pay for Performance' funding model whereby hospital clusters are given additional funding for service growth in areas of greatest need, improvement in patient safety and quality, staff development, and technology advancement. A Diagnosis Related Group (DRG) based casemix system for acute inpatient services has been developed to measure hospital performance and determine baseline budget. Following the principle of 'same service, same funding', patients are classified into different DRGs based on the number of cases treated by the hospitals, adjusted for the complexity of cases (Legislative Council, 2014)

Private providers, both outpatient and inpatient, are paid mostly on FFS basis directly by users and/or private insurance. However, the role of FFS will expand in the future as a result of expansion of visits to private hospitals fostered by subsidized voluntary insurance.

Individual physicians and other medical personnel in both public and private sectors are mostly on standard salary package with a small component tied to performance. They therefore have no significant financial incentive to oversupply services or overcharge for them. This is arguably a strong reason for the low-cost healthcare system that exists in Hong Kong.

7.7 Regulations

Given the dominance of public provision and financing of health care which allows the government's vast scope for shaping providers' and users' behaviour through internal administrative rules, there is relatively less reliance on regulatory tools in Hong Kong. However, there does exist a small private hospital sector and a large private outpatient services sector which need to be regulated. There are a range of regulations for licensing of doctors, hospitals, nurses, pharmacists, and so on, similar to those that exist in other jurisdictions. For the most part, private providers in Hong Kong are lightly regulated in order to promote competition among them.

Information as a tool has also not been heavily used to alter providers' or users' behaviour, except for public health where the government has been active since the 1960s. Information asymmetry is not a major issue in healthcare systems dominated by public provision and financing of hospitals. However, this is changing with the launch of VHIS with explicit objective of promoting private provision and financing. To protect users, new rules have to be devised to ensure information disclosure and consumer awareness to prevent abuse.

7.8 Conclusion

The government-dominated health care in Hong Kong developed without much deliberation or planning in the context of dire healthcare needs amidst economic boom in the 1970s and 1980s (Ramesh, 2004; Tang, 1998). The public coffers were overflowing and the contradictions between the government's avowed commitment to *laissez-faire* and expansion of public provision and financing of health care (as well as education and housing) were easy to overlook. The circumstances changed when public expenditures on health care rose exponentially and policymakers became aware of the fiscal implications of ageing population. The privatization wave sweeping the world in the 1990s also affected Hong Kong (Mok and Lau, 2002) as the government made intermittent attempts to expand private financing but they made little headway due to the population's opposition.

There are several reasons why Hong Kong has one of the most inexpensive and equitable healthcare systems in the world (Harvard Team, 1999). It has a tightly organized administrative machinery for health care that offers firm stewardship and coordination to the sector. The government is helped by the territory's small size and population, further accentuated by the administration's relative autonomy from politics. The public ownership of most hospitals, their financing from government budget, and paying them on a global basis make for a solid health system from a design perspective, as outlined in Chapter 1. In such a heavily public system, the need for regulations is confined mostly to setting safety standards for drugs and therapies rather than market competition. This will, however, change if the government succeeds in its ongoing efforts to expand private hospitals, which will need to be appropriately regulated.

The publicly provided and financed hospital system in which providers are paid on a global basis makes for an equitable as well as efficient system. The public system, while open to all, is used more heavily by low-income households which enhances equity. Remarkably, the comprehensive provision of equitable and accessible hospital services does not impose undue burden on the state, as public hospitals and clinics provide only necessary services due to lack of incentives for oversupplying, unlike their private counterparts. The long waiting list for elective services, while understandably resented by those affected, is an integral part of public hospital system working under the constrain of a global budget which often unavoidably involves some rationing of services.

Notwithstanding the comparatively low costs and superior outcomes of Hong Kong's healthcare system by international standards, the government has been worried about its financial sustainability in the long run and made persistent efforts to reduce its share. However, its efforts to get users and/or employers to contribute more towards health care have been frustrated by opposition from those who are used to paying virtually nothing for hospital care. As a last resort, it has turned to subsidizing private insurance which will likely increase fiscal burden on the government in the future but has the short-term advantage of appeasing the middle class which is frustrated by the long waiting time for elective procedures in public hospitals.

Hong Kong has displayed a remarkable record in tackling the ongoing Covid-19 pandemic. While much of the success is due to the population's exemplary mask-wearing and hand-washing efforts which limited the spread of the virus and infection (only 4,995 cases out of population of 7.5 million as of 18 September 2020), public health facilities played their part too (Hartley and Jarvis, 2020). The public health system was able to quickly expand capacity and attend to those who needed critical medical help, evident in case mortality rate of under 2 per cent.

The relatively simple healthcare system in Hong Kong offers useful lessons for other countries desiring to deliver basic health care to their population. The publicly provided and financed system can be highly cost-effective and equitable if run well. On the downside, such a system lacks mechanisms for promoting user focus which is a reason for dissatisfaction among middle-class households desiring more personalized services but unable or unwilling to pay for it. Singapore, as we will see in the next chapter, addresses this problem by promoting competition among hospitals and making users bear a larger portion of the costs which, while making services more plentiful, has adverse implication for equity.

8 *Singapore*

Synergizing the State and Market in Health Care

Singapore's healthcare system ranks among the best in the world in terms of various commonly used criteria, prompting *BusinessWeek* magazine to rank it as the healthiest country on the planet (Flynn, 2016). What is even more remarkable is that the fine outcomes are accomplished at less than half the costs in countries at comparable level of economic development. Policy reformers around the world have taken note of the achievements in recent years and have explored emulating it, though most observers admit that the social and political conditions in the island state are too different to allow emulation (see for example, Carroll, 2019).

As we will see in this chapter, there is no grand health 'policy' in Singapore, other than broad commitment to make basic services available at affordable costs to households without unduly burdening the government. It has pursued these goals with a range of policy tools targeting specific problems that work in tandem and are fine-tuned constantly in response to policy learning and changing circumstances. The arrangements have been commended by the Secretary General of the World Health Organization, Dr Margaret Chan, for balancing the "advantages of competitiveness and other market forces with the need for state intervention to steer these forces in the right direction" (quoted in Paul et al., 2015).

Singapore has unusually elaborate governance arrangements and relies on an array of complementary policy tools to manage its healthcare sector. To make the complex combination of governance arrangements and policy tools work effectively, the government has had to target and guide the behaviour of all key stakeholders – providers, payers, and users – as discussed in this chapter (Bali and Ramesh, 2017; Wu and Ramesh, 2014).

The findings of this chapter are at odds with the dominant narrative in the literature which attributes its success to the government's ideology of self-reliance, conspicuously embodied in the individual medical savings scheme called Medisave. For example, according to the international consulting firm Watson Wyatt, "The key to Singapore's efficient health care system is in its emphasis on the individual to make a significant contribution towards their own health care costs." (Tucci, 2010). While the government has indeed been ideologically committed to individual responsibility, this has not come in the way to resorting to deep and widespread interventions in the sector to fundamentally alter its operation.

The island state has an inordinately complex health system comprising a broad range of policy tools and it is their combined working, and not that of any one tool, that explains the system-level performance. As we shall see in this chapter, by focusing on simply on one tool, observers miss the bigger picture as well as the details of the healthcare system in the country. And yet, as we shall see, policymakers in Singapore are stymied by ideological blind spots that leave crucial problems insufficiently addressed.

8.1 Policy History

Health care was a low priority for the colonial government and it built only basic facilities for the local population. In the final years of colonial rule, however, the government became increasingly involved in direct provision of health care, partly reflecting the development of the National Health Service in Britain. The People's Action Party (PAP) government that took office in 1959 retained and gradually expanded the arrangements while introducing modest reforms (Haseltine, 2013). It established new public health clinics and expanded School Health Services and compulsory free immunization against a range of diseases. Government spending on health care rose from $53 million in 1970 to $154 million in 1975, with public hospitals receiving two-thirds of the total (Paul et al., 2015). At a broader level, the government worked on improving sanitation, housing, and education, which together with expanded public health measures played a major role in improving the population's health. It is arguable that these public health and social policy interventions contributed the most to improving the population's health then and for decades later.

The expansion of the government's role in health care was, however, against the grains of its belief that health was an individual responsibility. The attitude was clearly expressed by the then Health Minister and future Prime Minister Goh Chok Tong in May 1981: "There is no place for a cradle-to-grave welfare health system in Singapore. Such a system is politically motivated and disregards the basic truth that resources are finite in terms of funds, doctors, nurses and other supporting staff" (Quoted in Paul et al., 2015). The scepticism towards publicly provided and financed health deepened with the increasing awareness of the financial implications of publicly financed health care in the context of population ageing and epidemiological transition (Aw and Low, 1997; Phua, 1990).

It was within the context of these broader changes that the government announced the National Health Care Plan in 1983 outlining the strategy for meeting the demands of a growing, ageing, and increasingly affluent population. The 1983 Plan firmly rejected social insurance on the grounds that third-party payment would lead to oversupply by producers and overuse by consumers. It instead proposed a range of measures to improve the efficiency of hospitals through competition coupled with higher user charges for patients and new financing schemes to pay for them.

In 1984, the government initiated efforts to make public hospitals more efficient by restructuring them as autonomous corporate entities though still fully owned by the government, starting with NUS Hospital. The newly corporatized public hospitals were encouraged to compete for paying patients as this was expected to enhance efficiency. To enable them to compete, it granted them broad autonomy in operational matters. It also launched efforts to establish new funding schemes to pay for health care.

Medisave, established in 1984, was the first of many new healthcare funding initiatives to be launched during the period. The scheme required the entire workforce, including the self-employed, to save 6–8 per cent (currently, 8.5–10 per cent) of one's monthly income (subject to a ceiling) and age to pay for hospital care of the account holder and his/her immediate family except siblings. However, Medisave imposes severe exclusions and ceilings intended to allow accumulation of funds to pay for expensive inpatient care when needed. Since savings in one's Medisave is one's own money, it is expected that people would spend it cautiously.

In 1990, the government established a publicly organized but privately financed voluntary health insurance scheme called Medishield. It covered hospitalization expenses for surgery and outpatient treatment for specified 'serious' illnesses. Its establishment, despite the government's principled opposition to social insurance, was in response to the realization that most Medisave accounts were insufficient for treatment of major illnesses. For those seeking more comprehensive coverage than offered by Medishield, it established Medishield Plus in 1994 as a voluntary scheme for treatment in private hospitals and higher class wards in public hospitals.

The realization that there was a segment of the population that could not afford even highly subsidized health care led the government to establish a public assistance scheme called Medifund in 1993. It is an endowment trust fund built on an initial US$59 million contribution from the government. To prevent depletion of the fund, only income from the fund is used to pay bills of those unable to afford hospital care.

The results of the reforms – increased autonomy and competition among hospitals, and establishment of Medisave, Medishield, and Medifund – were not what government had expected. While the service quality improved, fees increased even more markedly. The restructured hospitals shut down subsidized wards, raised doctors' salaries, and acquired latest technology in order to attract high-spending patients. The government also came to realize that Medisave, Medishield, and Medifund – collectively described as 3M – together paid for less than 10 per cent of total health expenditures, imposing significant hardship on households (Asher and Nandy, 2006).

Faced with widespread concerns, the government set up a Review Committee on National Health Policies in 1991 which submitted two reports over two years. It subsequently established a high-level Ministerial Committee which published its report, Affordable Health Care: A White Paper in 1993 (MOH Singapore, 1993). This was an important turning point as it acknowledged the limitations of the market and the need for direct government intervention to steer its operation. The White Paper stated that Singapore's health system should neither be "a totally regulated national health service, nor a *laissez-faire* system where providers have full freedom to organise and to price". It recommended that public hospitals (which received government subsidies) be subject to controls in the key areas of pricing and operations while private providers would be subject to minimal controls.

The White Paper set out a number of 'principles' for governing the health sector. First, it stated that health was an 'individual responsibility' and reliance on government budget or social insurance was to be avoided. Second, it called for competition in the sector to improve service and raise efficiency. Third, and at the same time, it proposed direct government intervention when 'the market fails to keep health-care costs down'. The combination of market competition and enhanced state intervention spelled out in the White Paper was audacious and ahead of its time. Contrary to the dominant thinking in favour of the market at the time, it asserted that competition should be promoted under strong government stewardship. "Market forces alone will not suffice to hold down medical costs to the minimum.... The government has to intervene directly to structure and regulate the health system" (MOH Singapore, 1993: 3). It called for direct measures to control costs while simultaneously promoting competition. At the same time, the government made it clear that it had no intention to set up a publicly funded health system. The 1998 Budget statement stated bluntly: "We believe that extensive welfare programmes damage the fabric of our society as they discourage individual responsibility, self-reliance, community support and the work ethic." (cited in Tang, 2000, 42).

In the following years, the government introduced price caps on medical services, imposed controls on introduction of new technology and specialist disciplines in government hospitals, and tightened overall supply of doctors (Barr, 2001: 714). The direction set by the White Paper continues to frame the government's policy to this date, with reforms consisting largely of changing the specific combinations of market competition and state control. However, the commitment to requiring households to share costs has remained unchanged.

In 2011, the government launched the Health care 2020 Masterplan to guide building 'an inclusive healthcare system for the future'. The Masterplan focused on developing the long-term care sector with a variety of new measures to expand capacity and raise standards of care. This was understandable given that existing arrangements for providing and financing long-term care services had been stretched to the limits (Chin and Phua, 2016). It also clarified the role of polyclinics as focusing on lower income Singaporeans while leaving the rest to be served by private clinics. Moreover, it proposed expansion of nursing homes through public–private partnerships whereby the government would build and own the facilities but tender out the operations to voluntary organizations. The government would also operate three or four nursing homes to maintain a window on the sector.

The government came under enormous pressure to increase financial support for the healthcare sector after the 2011 General Elections in which the ruling party received the lowest share of votes in fifty years. To ease financial burden of OOP payment, the government progressively relaxed the conditions for accessing Medisave, Medishield, and Medifund. It has also increased public funding for hospitals, offered subsidies towards Medishield premium, and injected S$3 billion into the Medifund endowment fund. Medishield was made mandatory – under the new name of Medishield Life – and expanded to cover pre-existing illnesses without age limit, unlike its predecessor. Moreover, the Community Health Assist Scheme (CHAS) was launched in 2012 as a means-tested programme for the aged to receive subsidized services at participating outpatient and dental clinics. The changes undertaken over the 2010–2017 period together resulted in a dramatic 250 per cent increase in the government's healthcare spending (Straits Times, 2017).

Notwithstanding its core belief that health care is primarily the responsibility of households, the government has come to gradually accept that this was impractical for ageing population and, as a result, politically untenable. It has also accepted that public expenditures on health and other social policies will continue to rise markedly in the future. To prepare for the eventuality, the government has indicated its intention to raise Goods and Services Tax to 9 per cent in 2021 to generate the extra revenues needed for increased expenditures.

8.2 Performance

The international community first took notice of Singapore healthcare system's performance in 2000 when WHO ranked it sixth out of 191 countries for 'overall health system performance' largely due to high health status indicators and low expenditures (WHO, 2000). Such accolades became more frequent in the following decade. The *Bloomberg Health Efficiency Index* 2014 ranked Singapore second out of 55 countries (Hong Kong was first). Similarly, the Economist Intelligence Unit's comparison of 166 countries showed that while Singapore ranked second in the world (behind Japan) in terms of health outcomes, it achieved them at considerably lower costs (EIU, 2014).

Singapore's infant mortality rate of 2.1 per 1,000 live births in 2015 ranks sixth best in the world (behind Iceland, Slovenia,

Finland, Japan, and Luxembourg) while its life expectancy of 83 years at birth ranks the eighth highest in the world (behind Hong Kong, Japan, Macao, Italy Spain, Switzerland, Iceland, and France) (World Bank, 2020). Between 2003 and 2013, life expectancy increased by 3.5 years for males and 2.9 years for females, compared to the average increase of 2.5 and 2.0 years, respectively, in OECD (MOH Singapore, 2016). The gain in life expectancy has been driven by declining mortality rates in cardiovascular diseases among those aged 50 years and older.

Singaporeans are not just living longer, but are also enjoying healthier lives, evident in rising Health Adjusted Life Expectancy (HALE). According to 2013 *Global Burden of Disease* study (GBD) covering 188 countries, males in Singapore spend the equivalent of 70.75 years of their life free of disease and/or injury, which is the second highest (after Japan) in the world. The female HALE at birth, at 73.35 years, ranks third highest in the world (behind Japan and Andorra) (Murray et al., 2015). The *Bloomberg Global Health Index* 2017 of 163 countries ranked Singapore as the fourth healthiest country in the world (behind Italy, Iceland, and Switzerland) based on variables such as life expectancy, causes of death, and health risks.

While Singapore's health status indicators are good, it is the low cost at which they are achieved that makes its performance outstanding. Its total healthcare expenditures form 5 per cent of GDP, which is less than half of the average for high-income countries. Another remarkable feature is that public spending forms only a small share of the total health spending: 42 per cent in Singapore, compared to the average of 62 per cent for high-income countries.

One of the reasons for the low total expenditures in Singapore is low utilization as well as low cost of services. Hospital admissions rate in Singapore is 90 per 1,000 persons, compared to OECD average of 143 admissions. Similarly, the average length of stay (ALOS) in Singapore is 5.8 days which is considerably lower than the OECD average of 8.5 days. While the low admission rate and short ALOS may indicate high quality of services, it may also be due to hesitation to seek admission at hospitals and staying there until fully recovered because of the fear of OOP payment required (DBS, 2016a).

Another positive indicator is low cost of medical services, indicating higher level of efficiency. Major surgeries in Singapore cost 62–92 per cent less compared to other high-income countries. According to one report, a heart bypass surgery that would cost S$130,000 in the United States costs just $18,000 in Singapore (Flynn, 2016). The low costs can be largely attributed to the active management of the healthcare facilities by the government and its persistent efforts to keep the costs down.

However, OOP payments form 32 per cent of total health expenditures in Singapore, which imposes heavy burden on household finances. Although no published data is available to estimate equity effects of health expenditure in the country, it is likely to be significantly inequitable. This is indicated by estimates that 9 per cent of Singaporean households in 2012 spent more than 10 per cent of their expenditures on health care (WHO, n.d.). The increase in the private insurance's share of total health expenditures is also likely to worsen equity in healthcare financing. The rising costs at private hospitals, increasingly financed from private insurance, is also a matter of concern: bill sizes at private hospitals increased at an average annual rate of 9.7 per cent between 2012 and 2014, compared to 0.7 per cent at public hospitals (Health Insurance Task Force, 2016).

8.3 Stewardship and Coordination

The government, ruled continuously by the People's Action Party (PAP) since 1959, has been remarkably clear and consistent about its health policy vision: an affordable healthcare system financed partially by households. Realizing this vision has been difficult, however, as affordability is hard to maintain when health care is funded privately. As a result, the government has had to establish an elaborate governance framework to align market competition and private responsibility with the objective of providing health care to all at affordable costs.

The PAP government began to put its stamp on the healthcare system in the 1980s following the realization that the existing system centred on public provision and financing was too rigid to respond to the needs of users and increasingly expensive for the government. It sought to address both problems by promoting competition among providers to improve efficiency and larger payments by users to reduce burden on the government. The results were not what the government anticipated as competition enhanced responsiveness to users but it also, worryingly, raised expenditures. This unleashed a series of reforms over three decades with the purpose of simultaneously promoting responsiveness and affordability. While the specific measures have changed, sometimes drastically, the broad objective has remained unchanged.

A unique feature of Singapore is the strong stewardship role played by the Ministry of Health Holdings Private Limited (MOHH), a government-owned holding company that owns and manages all government healthcare assets, including public hospitals and clinics. Given its role as an owner and manager, MOHH is in a unique position to provide system-level direction, oversight, and coordination to the sector. It is the de facto lead agency for nationwide coordination of manpower and talent development, information technology, and infrastructure development (Quah and Neo Boon, 2016). It has played the lead role in developing the National Electronic Health Record (NEHR) platform, which it later turned over to newly formed Integrated Health Information Systems Private Limited, also fully owned by MOHH, for management.

Being a compact urban city with excellent transport and communication infrastructure, vertical coordination has not been a significant issue in Singapore. Yet, one of the unintended effects of autonomy for hospitals and competition among them was fragmentation of the healthcare system that made resource planning difficult. To promote a better balance between collaboration and competition, in 2000, the government established two vertical 'clusters' of public hospitals: National Healthcare Group (NHG) and Singapore Health Services (SingHealth). Each cluster consisted of a network of hospitals and clinics that together offered full spectrum of primary hospitals and specialist centres. As a result of the clustering, the competition was restricted to two large government-owned network of healthcare facilities rather than among individual hospitals, as the latter was believed to promote duplications and waste.

Health clusters are legally incorporated entities owned entirely by the MOHH. They in turn own the hospitals within their clusters, making for a vertically integrated system. The strategic intent of the regional clusters is to promote vertical integration of services, establish a more effective patient referral system, and reap economies of scale while at the same time maintaining a modicum of competition in the sector. To highlight their social purpose, the clusters and individual hospitals are designated as 'not-for-profit' entities.

After several years of experience with clusters, the government concluded that two clusters were not enough to produce the desired results. Accordingly, in 2012, the two clusters were reorganized into six Regional Health Systems (RHS). The aim of the reorganization was to offer integrated care to patients closer to home while at the same time offering them more choice to visit facilities in nearby regions. After six years, in 2018, the government concluded that six regional clusters were too many for a small country and the number of clusters was reduced to three: National Healthcare Group (NHG), SingHealth Group, and NUHS Group. The speed with which the government introduced major changes is remarkable as is the government's determination to fine-tune the system to achieve the desired balance between centralization and decentralization.

Horizontal coordination of health policy in Singapore is also strong with the Ministry of Health at the helm. MOH has a broad mandate to make health policies, manage public funding for health care, and regulate all health care facilities in the country. Its goal, as it states modestly, is "to ensure that good and affordable basic medical services are available to all Singaporeans. We achieve this by providing subsidised medical services while promoting individual responsibility for the costs of health care services." The statement contains all the key phrases that characterize the healthcare policy of the government: 'basic medical services', 'subsidized' but not free services, and 'individual responsibility'.

MOH pursues its policy goals through its roles as policymaker, financier, and regulator. In addition, it actively uses public ownership as a tool to shape the sector, as we shall see in the following section. In all these roles, cost-effectiveness is a paramount concern. At one point, it even described its mission as building the 'world's most cost-effective health system'.

Singapore also has many specialized agencies reporting to MOH to deal with specific aspects of health care. The Health Promotion Board is responsible for promoting public health awareness and practices in the country. Second, the Health Sciences Authority, established in 2001, is a statutory board responsible for regulating sale and use of pharmaceuticals. Third, the Agency for Integrated Care (AIC), established in 2009, is responsible for promoting patient-focused integration of primary, intermediate, and long-term care. It fosters coordination of service delivery, such as facilitating patient transition from general acute hospitals to community hospitals, nursing homes, and community-based facilities such as day centres and home care service

(Liu and Haseltine, 2015). Fourth, the Group Purchasing Office established in 2000 is responsible for bulk purchase of drugs, medical supplies, equipment, and IT services at the national level. Fifth, the Agency for Care Effectiveness (ACE) was established in 2015 to assess and advise on new drugs and medical technology. It is staffed by skilled analysts specializing in assessing the costs and benefits of different medical procedures, devices, and drugs.

One crucial area in which coordination is lacking is the delivery of primary care and its linkage to hospital care. Primary care is provided largely by private clinics, except for the small number of public polyclinics. Most primary care providers are small clinics which make for intense competition among them. The primary care system in Singapore is particularly deficient in continuity of care, as diagnostic test results and medication records stay with the clinic and not shared with other clinics (Khoo et al., 2014). The recently established compulsory Electronic Health Record platform is expected to improve continuity of care only in the long run.

To promote greater coordination in primary care, the government has been promoting Community Health Centres and Family Medicine Clinics. Recently, it launched the Primary Care Networks (PCNs) scheme that encourages GPs to come together and form networks to provide coordinated care to chronic care patients. Under the scheme, patients receive care through a multidisciplinary team ("Primary Care Networks," undated). To encourage formation of PCNs, MOH provides funding, administrative support, and information and training to GPs when they form such networks.

8.4 Ownership and Management

Singapore is an enthusiastic user of public enterprises as a policy tool, notwithstanding its free market reputation. However, its public enterprises are unique in the way they are incorporated and managed, as we shall see in this section. Essentially, public enterprises operate as autonomous private companies over which the government exercises close controls as an owner and regulator. Ownership allows the government to intervene in pointed ways not possible through other policy tools.

In primary care, in contrast to hospital care, the government uses public ownership sparingly but strategically. There are only 20 public clinics – called polyclinics – compared to 1,700 private general practice (GP) clinics that are mostly run as solo or small group practices. ("Primary Care Networks," undated). Unlike private clinics that receive no subsidy, polyclinics are heavily subsidized: 75 per cent concessions in fees for children and aged and 50 per cent concession for other Singapore citizens. In addition to being subsidized, the polyclinics offer complete range of diagnostic and treatment services. As a result, they are heavily used by chronic disease patients despite longer waiting times. Thus, while polyclinics account for only 20 per cent of all primary care visits, they account for 45 per cent of acute care visits (Khoo et al., 2014). In fact, they play a more important role than suggested by these raw numbers as they set a price and quality benchmark for the private clinics to follow.

Hospital care in Singapore is thoroughly dominated by public hospitals which account for approximately 75 per cent of all hospital beds and 80 per cent of inpatient admissions in the country. The public hospitals are used heavily, evident in occupancy rates of between 85 and 93 per cent (Singapore Business Review, 2019). Private hospitals play an ancillary role overall, serving mostly local residents seeking non-urgent care in comfortable setting. They also play vital role in promoting medical tourism to the country.

What sets Singapore's public hospitals apart from their counterparts elsewhere is that they are registered as private firms even though they are entirely owned by the government. This is not unique to public hospitals, as such firms are found in all leading sectors in the country, including Singapore Airlines in aviation, Neptune Orient Lines in shipping, and DBS Bank in development finance. They are run professionally and indeed profitably while helping to achieve policy goals. The combination of public ownership and the organizational structure of a private company allows the government to achieve its goals in ways not possible with either alone.

Singapore has set up an elaborate management system to overcome the problem of inefficiencies and impersonal services that often characterize public hospitals. Its most important tool in this regard is the ownership of public hospitals by MOHH, a holding company fully owned by the Ministry of Health but registered as a private company. It owns and manages all government healthcare assets, including all public hospitals, polyclinics, and specialist medical facilities. Its portfolio includes National Healthcare Group (NHG), National University

Health System (NUHS), Singapore Health Services (SingHealth), and agencies such as Agency for Integrated Care (AIC) and Integrated Health Information Systems (IHiS). The three healthcare clusters (NHG, NUHS, and Singhealth) in turn own public hospitals which are also individually incorporated as separate entities. The hierarchical ownership structure makes for a highly integrated system, notwithstanding their legal autonomy.

Although MOHH and the clusters are technically autonomous legal entities, MOH is deeply embedded within them through various procedural mechanisms. The MOH appoints the chairman and members of the cluster boards, which include senior officials from the ministry. Furthermore, MOH's approval is required for the appointment of the chief executives and chairmen of the public healthcare institutions. These mechanisms provide the government direct presence in the operation of public hospitals.

Public ownership of what are legally private firms allows them the autonomy they need to operate in a competitive environment yet remain within the government's direct reach (Ramesh, 2008). Public ownership also affords the government a direct channel for receiving market feedback. Moreover, the government as an owner can shape hospitals' behaviour without resorting to onerous regulations or purchase negotiations that would be necessary if they were truly private firms. Public ownership has been particularly useful, for instance, for controlling user charges, physicians' remunerations, and the number of hospital beds in different ward classes that have been instrumental in keeping costs of inpatient care low in Singapore (Lai, 2016). Other benefits of public ownership include "distribution of resources across subsidised and non-subsidised patients and greater opportunity to influence practice to ensure a focus on cost-effective care. Centralization of aspects of hospital management – for example, drug procurement – also allows savings to be reaped. Highly specialised services are restricted to designated centres, and purchase of expensive equipment centrally scrutinised, taking into account national capacity already available" (Lai, 2016). Indeed the public hospitals' performance affects also their private counterparts. As Lai (2016) summarizes, "If subsidised and quality care in the public sector is reasonable and available alternative, private providers will need to ensure that they do not price themselves out of the market."

Public hospitals in Singapore are exposed to competition despite government ownership and overwhelming dominance of the market because the policymakers view competition as necessary for promoting efficiency and customer focus. Patients are free to choose any hospital they wish and government subsidies largely follow patients, exerting pressure on providers to attract patients by offering quality services at competitive prices. To allow public hospitals to devise their own competitiveness strategy, managers have been given broad autonomy in operational matters within the framework established by MOH and MOHH.

There are numerous procedural mechanisms by which the Ministry of Health oversees, coordinates, guides, and directs public healthcare facilities that are legally autonomous and operate in a competitive environment (Quah and Neo Boon, 2016). First, a 'Standing Policy Agreement' sets out the public mission of the clusters and demarcates their respective roles and responsibilities. Second, 'Service Level Agreements' (SLAs) between MOH and clusters set out performance targets and objectives in return for government subsidy. SLAs are reviewed and renewed at regular intervals on the basis of performance measured by Public Hospital Scorecard. Third, there is the annual 'Statement of Priorities' which states the priorities and key performance indicators for the following year. The stated priorities are used for assessing the performance of hospitals and clusters.

In addition, there are performance management and measurement mechanisms in place for public healthcare facilities. In 2008, the Ministry developed the 'National Standards for Health Care' (NSHC) for public institutions. Scorecards are used to assess all public hospitals and clusters' performance against the national standards which form the basis for the annual Public Hospital Performance Report. The comparative performance data contained in the report are designed to drive continuous improvement and innovation. The report also forms the basis for drawing up SLA and Standard Operating Procedure (SOP).

An innovative tool the Singapore government has employed to improve service quality and contain prices is to collect and disseminate information on hospital charges and clinical outcomes. As the Minister of Health put it, "For economics and markets to work, we must make sure that the conditions for market competition exist. That is why I published the bill sizes for the common medical treatments .... When competition is brought to bear on these services, we will then have the right incentives for the health care providers to do the right thing, to raise standards even as they reduce cost." (Khaw, 2005) Since 2003, public hospitals have been required to publish their total mean and median bill sizes (including charges for room, treatment, surgery, laboratory test, etc.) for different common conditions and procedures. The collated data on 65 common surgical procedures are subsequently published on the MoH's webpage.

The MoH maintains online calculators which patients can use to estimate the cost of treatment of particular episodes of illness and calculate the share of the bill that may be financed by Medisave and ML. Importantly, they can make informed choices on the particular hospital and ward class in which they seek treatment. It is likely that the information made available on the web page is used more widely by hospitals (rather than patients) who use the information to benchmark their prices and outcomes against their competitors'.

In future, National Electronic Health Record (NEHR) – which was established in 2011 but became mandatory in 2018 – will be used increasingly to improve patient care and hospital management. All providers are required to contribute data regarding hospital admission, visit, and discharge history; laboratory test and radiology results; history of medication, surgeries, and procedures; allergies and adverse drug reactions; and so on. Only authorized healthcare professionals who are providing care to the specific person can access NEHR. Individuals, however, have the option to opt out of the platform.

Finally, MOH conducts annual survey of patient satisfaction and expectations which indicates high favourability rating. In the 2015 survey, 86 per cent of respondents rated their satisfaction levels with the public healthcare institutions as 'excellent' or 'good', compared with 79 per cent in 2014 (MOH Singapore, 2015).

8.5 Financial Risk Pooling

To simultaneously rely on both competition and government controls to pursue health policy goals, Singapore employs a range of financial policy tools: subsidy for providers and users; compulsory savings; insurance; and OOP payment. Of these, some (government budget, Medishield Life, Medifund, and Eldershield) are 'pooled' sources while others (Medisave and OOP payments) are sourced individually by households. The different tools are meant to accomplish different purposes and yet work in concert.

Government budget is the source of 42 per cent of total health spending in Singapore, mostly in the form of subsidies to public hospitals, for whom government transfers are the source of half of total revenues. The subsidy amounts are based on the number of patients and treatments performed in subsidized wards in public hospitals. There are four classes of wards in public hospitals – A, B1, B2, and C – providing different levels of privacy, amenities, and choice. Patients in 'C' class wards are subsidized up to 80 per cent of the costs while those in 'A' ward receive no subsidy.

There are also other schemes financed from the government's general budget. For instance, Community Health Assist Scheme (CHAS) launched in 2014 is a means-tested scheme that offers additional grants to low-income household (below S$1,800 per capita per month) for subsidized outpatient and dental treatment. Similarly, ad hoc schemes such as the Merdeka Generation Package and the Pioneer Generation Package subsidize insurance premiums as well as co-payments for elderly Singaporeans born before 1950. There are also extensive public health campaigns fully financed by the government.

Subsidies for drugs are also provided to eligible patients through the Standard Drug List (SDL) and Medication Assistance Fund (MAF). Drugs on the SDL are subsidized at 50 per cent for all Singapore citizens who are patients in a public healthcare institution while patients from lower to middle-income households can receive a higher subsidy of up to 75 per cent. Prescription of subsidized drugs in public hospitals is restricted to "clinically relevant and cost-effective drugs considered as basic therapies that are essential for the management of common diseases afflicting the majority of our population" (Pwee, 2009).

Subsidies in Singapore are meant not only to make health care affordable but also to allow levers for enhancing cost-effectiveness. Accordingly, subsidies for public hospitals and government clinics are only for basic medical services that are cost-effective and of proven value. Moreover, subsidy allows MOH to shape providers' behaviour through service and other conditions in return for receiving funds. Subsidy also allows the government to set different levels of fiscal support for different purposes and target groups. Thus, maximum subsidy is available for lower class no-fringe beds in public hospitals. Similarly, different levels of subsidy are available to different age and income groups at outpatient clinics.

However, government subsidies do not cover the full costs that users are billed, not even in the most subsidized wards and nothing at all in the non-subsidized wards. To help users with paying for the rest, the government launched a limited insurance scheme called Medishield (later renamed as Medishield Life or ML) for financially 'catastrophic' serious illnesses. To keep costs low, ML covers only a limited range of serious illnesses (such as cancer and heart diseases) and imposes hefty deductibles and co-insurance. ML premium ranges from $16 a month for a 29-year-old in 2019 to $68 a month for a 69-year-old, without subsidies. Deductibles vary by age and ward class, and range from $1,500 to $3,000 while co-insurance ranges from 3 per cent to 20 per cent depending on the size of the medical bill. ML is intended to cover the cost of hospitalization in a Class B2 or C ward: those choosing higher class wards are expected to pay out of pocket or through private insurance. Given the restrictions, it is unsurprising that Medishield claims form only 2 per cent of total health expenditures (DBS, 2016). These are however expected to increase in the future.

The government realizes that a large number of people prefer services at facilities more comfortable than what is offered in Class C or B2 wards and it therefore allows the purchase of private insurance to cater for their needs. Indeed, about 40 per cent of patients opt for private ward classes (Class B1 and Class A) in the public hospitals or private hospitals even though there is no subsidy (Lai, 2016). To cover the part of the bill that must be paid out of pocket, users may buy Integrated Shield Plans (IPs). Nearly two-thirds of Singaporeans have purchased IPs. Incidence of IP claims has been growing at 9 per cent per annum while the overall average bill size increased by, between 0.6 and 8.7 per cent per annum, depending on the medical provider (Health Insurance Task Force, 2016). The government is aware of the moral hazards of IP and has taken measures to curb it. Since 2020, all IP policyholders have been required to make a co-payment of at least 5 per cent of the total bill.

Another pooled source of healthcare financing is Medifund which was launched in 1993. It is a means-tested public assistance programme for those who cannot afford to pay user charges even in subsidized wards through their Medisave and Medishield entitlements and private savings. Under Medifund, patients in lowest class wards at public hospitals and outpatients may apply for complete or partial waiver of their bills. The scheme is funded from earnings from a government endowment fund (amounting to S$4 billion in 2018) rather than its operating budget. Medifund accounts for a negligible share of total healthcare expenditure: it paid out an average amount of S$170 to 766,000 applicants in 2014 (Kathleen, 2019).

Finally, the ElderCare scheme, launched in 2000, provides grants to intermediate and long-term care facilities to subsidize care for low- and middle-income patients (Liu and Haseltine, 2015). ElderShield makes monthly direct cash payments for designated services, including nursing home, facility- and home-based health care. It is funded from a separate endowment fund amounting to S$3 billion. The scheme is still small but likely to grow as the population ages rapidly in the coming years which will lead an increasingly larger number of aged without sufficient family support or financial resources to access private aged care facilities.

In contrast to the above schemes, Medisave is a compulsory but non-pooled private financing programme to pay for health care. Indeed, when launched in 1984 it was the first and remains the only mandatory national-level Medical Savings Account scheme in the world. Medisave members contribute 8–10.5 per cent of income, depending on age, to their own *individual* account. Members can withdraw funds from their account to pay for their own and their dependents' approved healthcare expenses and pay for ML premium. The average balance in Medisave account was S$27,900 in 2019 (GovTech, 2020). Notably, Medisave accounts for only 5 per cent of the total healthcare spending in the country (DBS, 2016).

Notwithstanding the large number of financing programmes that exist, OOP payment is the second largest source of healthcare financing, accounting for 32 per cent of total health expenditure in Singapore. OOP in Singapore is a deliberate policy tool, not an unintended consequence as in many countries, for achieving the goal of making households personally responsible for their own and their families' health care (Asher and Nandy, 2006; Barr, 2001). All healthcare services in Singapore levy charges that reflect the costs of the services. Even when the charges are heavily reduced (as in Class C wards) due to government subsidy, patients are expected to contribute at least a small share of it (Thomas, 2016). The adverse effects of OOP are in part mitigated by lower prices at public hospitals and polyclinics which receive government subsidies and offer cost-effective services.

However, the amount of OOP expenditure for hospitalization is relatively small, on average forming 4 per cent of the total cost of hospitalization in Class C wards in public hospitals (Lai, 2016). It is of course considerably higher for the 40 per cent of users who choose treatment in higher class wards or private hospitals. It is also higher for outpatient services, which are generally less subsidized except for the aged.

8.6 Provider Payment

The normal mode for paying providers in Singapore is FFS in both public and private sectors. Visits to outpatient clinics, whether public or private, involve some payment for services and medication. Similarly, treatment at both public and private hospitals involve payment of itemized charges. However, at public hospitals, in most instances the user pays only a part of the charges directly, with the remainder paid through third-party sources such as government subsidy, Medisave, and Medishield. In private hospitals, on the other hand, users pay higher fees not only because of considerably higher charges but also due to the absence of government subsidy and ceiling on ML and IP reimbursements.

Medical and other staff at public hospitals and clinics are on fixed monthly salary, though they are also usually entitled to bonus tied to performance measured by a range of criteria. Staff at private facilities, on the other hand, receive income that is more closely tied to the revenues they generate, which is a known driver of healthcare expenditures. To contain prices at private healthcare facilities, the government has published benchmark prices for a range of common medical procedures based on the Report of the Fee Benchmarks Advisory Committee published in 2018 (Ministry of Health, 2018). While not binding, the benchmarks serve as strong signals to both providers and users as to what they should charge or pay for services.

The arrangements for how the government transfers to public hospitals have changed over the years, moving increasingly in the direction of prospective payment. Until the 1980s, public hospitals were allocated block grants based on historical precedence and number of beds. In the 1990s, the formula was changed to payment based on the volume of services which, predictably, saw an increase in the quantity of services provided. To curb the resulting increase in expenditures, in 1999 the government introduced diagnosis-related group (DRG) based funding for inpatient and day surgery and piece rates for outpatient services. DRG formula pays a standard averaged rate for treatment of particular diseases regardless of the intensity of treatment or acuity of the individual patient. Clinical outcomes are closely monitored to prevent under-servicing, a problem innate to capped payment arrangements. To ensure providers do not take advantage of users, who bear significant co-payment that they pay out of pocket in Singapore, cost-recovery ratios for different types of services are monitored to ensure that patients pay only the target rate.

8.7 Regulations

Regulatory interventions in the healthcare sector in Singapore are, on appearance, not too onerous as they mostly consist of the usual restrictions related to licencing and safety. This partially reflects the fact that overwhelming majority of hospitals are in the public sector which can be directed to behave in a particular manner without the need for regulation. As an owner of public facilities, the government can oversee and direct all significant aspects of their operations: the types and volume of specialized clinical services they provide, the fees they charge, the salaries they pay, and the equipment they purchase.

While public institutions have always been subject to MOH rules, private facilities were largely unregulated until 1993 when the Private Hospitals and Medical Clinics Act (PHMCA) came into effect (Tan et al., 2016). Even then, the regulations applied only to modern mainstream medical services. Public hospitals and clinics were brought under the ambit of the Act after their corporatization in the 1980s and 1990s. The PHMCA and its subsidiary legislation promulgated the minimum standards for maintaining a healthcare service. The relevant subsidiary legislations include Sale of Drugs Act, The Poisons Act, and The Medicines (Advertisement and Sale) Act. To address changes in the sector, PHMCA was replaced by the Healthcare Services Act (HCSA) in 2019. Under the new Act, licences are granted based on the type of healthcare service provided rather than the physical premises of the provider, with licensing conditions tailored to each type of service ('Changes to Health Care Regulation in Singapore', 2018). Remarkably, Singapore has no controls on drug prices, as anyone who can afford to pay has access to drugs that are commercially available.

Regulation of private providers must contend with the policy goal of promoting competition in the sector. Private providers are seen as crucial to the objective of attracting medical tourists to the country and enhancing the quality of public hospitals by posing competition to them. Public providers receive substantial subsidies from the government, and notwithstanding autonomy in operational matters, are subject to strict controls regarding the number of hospital beds and their distribution by wards, fees they charge, acquisition of technology, and expansion of specialties. Private providers, on the other hand, are lightly regulated but receive no or little subsidy. Through different forms of interventions for public and private providers, the government is able to simultaneously pursue different policy goals.

There is also self-regulation in the form of practice guidelines and codes of ethics and conduct, by professional bodies such as the Singapore Medical Council, Singapore Dental Council, Singapore Nursing Board, and Singapore Pharmacy Board. The Councils play an active role in stewarding the sector and ensuring that costs are 'competitively priced'. There have been instances where the Medical Council has censured providers for 'overcharging' patients. Such controls on provider behaviour reduce their monopoly power in the health sector and reduce room to engage in maleficent behaviour.

New medical drugs and devices must be approved by Agency for Care Effectiveness (ACE) before they can be made available to users. Its Health Technology Assessment (HTA) compares health technologies with alternative treatments to identify interventions that are clinically effective and offer the best value for patients. Its reports are considered by the Drug Advisory Committee of Ministry of Health which is responsible for the final decision on procurement and reimbursement.

Chronic Disease Management Programme (CDMP) requires that providers must be in compliance with prescribed disease management protocols for users to be able to pay through Medisave. The requirement gives Medisave a lever to impose standard of practice on providers, which not only discourages non-essential treatment but also specifies treatment that ought to be provided (Lai, 2016).

In 2018, MOH introduced benchmarks for medical fees jointly developed by doctors, hospitals, and other stakeholders. Public providers are required to adhere to the benchmark fees, though compliance is voluntary for private providers. Such benchmarks are expected to give doctors a yardstick for measuring their own prices while giving patients a clearer idea of comparative costs.

8.8 Conclusion

What distinguishes Singapore's health policy after two decades of reforms is the clarity of the government's goals and the large selection of tools by which they are pursued. The government fervently believes in the philosophy of 'individual responsibility' wherein people look after their own and their family's needs. It works assiduously to put this idea into practice, despite epidemiological and demographic factors that make large-scale government intervention unavoidable. Its interventions are guided by efforts to contain costs so as to keep the services affordable, not just to households but also to the government. The government has built an elaborate governance structure and provides active stewardship to the sector to achieve its goals.

The core of the system is built on public provision of hospital care, which poses the greatest financial risk to households. Public hospitals are paid largely through government budget and paid mostly on prospective basis. The government uses a mind-boggling range of policy tools to ensure that public hospitals are cost-conscious as well as user-focused: competition, public ownership combined with private management practices, benchmarks, contracting, disclosure of billing information, and so on. The policy tools governing provision are complemented by a vast array of financing tools – government subsidy, Medisave, Medishield Life, Medifund, OOP payment, and private insurance – each with their own terms and conditions specifying their use in great detail. In contrast to public hospitals, private hospitals and clinics (which provide vast majority of primary care in the country) are neither provided financial support nor subjected to onerous regulations.

A distinctive feature of the health policy tools in Singapore is the effort it devotes to developing specific tools to target specific behaviours while maintaining coherence and complementarities among them. The government constantly fine-tunes the tools' mix and settings to enhance their effectiveness. In this, it is helped by its agnostic attitude towards planning and market (Wu and Ramesh, 2019). The approach is buttressed by willingness to take risk with new tools and reversing choices when the effects are found to be undesirable. This is experimental governance – typically a forte of local governments (Sabel and Zeitlin, 2012) – writ large at the national level. However, few governments in the world have the administrative capacity or political autonomy to enjoy such a free hand to design and employ policy tools as they wish, severely limiting the transferability and relevance of policy lessons from Singapore.

The healthcare system in Singapore has held up well against the ongoing Covid-19 pandemic which saw the island state experiencing one of the highest infection rates in the world, with 57,543 confirmed cases out of population of 5.6 million as of 18 September 2020. Yet only 27 cases were fatal, or a case fatality rate of less than 0.1, compared to global average of 3.2 (Global Change Data Lab, n.d.). While a part of the reason is the demographics of the infected population (mostly young and healthy workers, only a small percentage of whom exhibited severe symptoms), the health system managed the cases with great efficiency (Woo, 2020). All Covid-19 patients above the age of 45 years or with underlying conditions or with severe symptoms were cared for in hospitals while the remainder were treated in community isolation facilities ('Explainer: Why Singapore's Covid-19 death rate is the world's lowest'). This proactive hospitalization led to lower ICU admission rate of 0.3 per cent of cases (Goh and Chong, 2020), compared to 41 per cent ICU admission rate in the United States (Lagasse, 2020). Experience in dealing with SARS epidemic in 2003 and well-functioning public hospitals proved crucial to dealing effectively with the pandemic.

9 *Conclusion*
Health Policy Design in Asia

While providing health care to the entire population has been an aspiration of most governments since the mid-twentieth century, it is only over the past decade that they have made concerted efforts to make it a reality. Following the adoption of the Sustainable Development Goals (SDGs) by the United Nations in 2015, members of the WHO committed themselves to achieving Universal Health Coverage (UHC), defined as ensuring that everyone has access to health care without suffering financial hardships. In the ensuing years, international and national efforts concentrated heavily on establishing new health financing programmes and mobilizing new resources to fund them. Subsequently, scholars and commentators began to highlight the importance of 'governance' of health, emphasizing transparency, participation, and accountability in how policies are made and implemented. While the broadening of the discussion was an improvement, the discussions were largely abstract and overlooked the specific policy tools and mechanisms by which UHC was to be achieved. The discussions particularly overlooked the importance of public ownership as a policy tool for providing affordable health care to the population.

The health policies studied in this book and the tools they employ vary considerably, as do the results they produce. Overall, all our cases have experienced good or excellent outcomes. According to Bloomberg's ranking, Hong Kong, Singapore, and Korea (along with Italy and Spain) have the top five most efficient health systems in the world. China and Thailand perform better than many high-income counterparts while India, the laggard in the region, has improved its performance in recent years. Countries in our study fare less well when measured in terms of access to affordable health care. The share of population facing 'catastrophic' health expenditures is comparatively high in all countries in the region except Thailand and Hong Kong. It is notable that these two health systems are heavily characterized by public provision and financing and prospective payment for providers. The book concludes that public ownership and financing of health care are vitally important health policy tools provided they are accompanied by governance structures that incentivize efficient management. Private hospitals and financing, on the other hand, require targeted regulation to ensure they remain affordable.

9.1 Stewardship and Coordination

9.1.1 Stewardship

After a vigorous move towards decentralized and market-led governance in the 1990s, policy discussions have turned to reasserting the role of governments, especially national governments, in ensuring that everyone has access to necessary health care. Calls for decentralization, liberalization, marketization, and new public management have given way to pleas for direction and stewardship in public policy. The change in tone and substance of the debate is particularly evident in health care, which is a highly complex sector with diverse and contradictory demands that necessitate not only mechanisms for heeding local community and individual preferences but also central direction and responsibility. While local participation and decision-making improve the quality of choices, decentralized decision-making weakens coordination, undermines economy of scale, and fosters interregional inequities. To overcome the obstacles, central governments have turned to setting minimum standards for healthcare services across the country and assisting the less developed regions while still allowing scope for local initiatives.

Contemporary health policy discussions emphasize the importance of a central authority ultimately responsible for ensuring that all residents within its territory have access to necessary health care. It is its responsibility to ensure that all critical policy functions are appropriately allocated so that the right level of government is responsible for the functions for which it is most suited and there exist mechanisms for holding them accountable. For policy effectiveness, functions and responsibilities need to be sited where they can be best discharged rather than wholesale centralization or decentralization. Such arrangements are best described as 'centralized decentralization' wherein a strong central authority is combined with opportunities for local innovation (Chua et al., 2019). The nature of the relationships between central and local authorities and the extent and form of the direction that national governments provide to their local counterpart vary considerably across nations, as in the six countries studied here.

China has had a decentralized administrative system for centuries which was put under greater central control during central planning but saw a revival in the wake of economic liberalization in the 1990s. Stretched for funds, local governments drastically reduced funding for public providers and, as compensation, gave them broad discretion to manage their affairs and earn income as they chose. The providers responded enthusiastically and were highly successful in generating income from users. Their income from user charges not only made them financially autonomous but it also deprived the government leverage to affect their behaviour. Faced with rising expenditures driven by providers' revenue-maximizing behaviour, in 2005 the central government started to offer greater policy direction along with additional funds for priority services. Provincial governments too began to assert control over their local counterparts, who in turn enhanced their supervision of public hospitals, especially in financial management matters.

Hong Kong and Singapore, on the other hand, started as centralized systems and have continued in that form. However, they recognize the benefits of decentralized decision-making and have taken measures to allow local autonomy in operational matters while maintaining central oversight and control. Thus, they have established regional clusters which enjoy considerable autonomy in operational matters while carrying out central policy directives. It is more complicated in Singapore where each regional cluster is registered as a separate company with complete operational and financial autonomy, though fully owned and closely supervised by the government. Each public hospital is also an autonomous private company with complete operational autonomy but, again, is fully owned and closely monitored by the government.

India has a federal system and, as in other federations, health care is largely a state (i.e. provincial) subject. Moreover, the overwhelming dominance of private providers, most of whom are small, makes for a highly decentralized and fragmented system. The new national programmes launched in the mid-2000s – especially, the National Rural Health Mission and the RSBY – saw the central government reasserting its role in the sector through increased public funding backed by detailed programme conditions designed to promote performance and accountability. The launch of PM-JAY in 2018 and the additional funding it offered came with strict programme conditions that further strengthened central direction. However, since it is the state governments that administer these programmes, their administrative and financial capabilities greatly affect the programmes' varied performance.

Korea has a highly centralized system for making and implementing healthcare policy. Only public health is devolved to local governments, which run a limited number of public hospitals and clinics. The Ministry of Health and Welfare (MOHW) is the clear lead agency responsible for setting policy directions and managing the sector. However, the reality of private sector's dominance severely limits the government's ability to carry out its intentions when faced with their opposition to measures that might undermine their income or autonomy. The strike by junior doctors in August 2020 during raging pandemic highlighted the political power of the medical professionals in Korea. To overcome the private providers' resistance, MOHW has established specialized agencies and participatory decision processes to balance and offset the conflicting interests with some degree of success.

Thailand too has a largely centralized healthcare system with limited local government involvement in policymaking or service delivery. This is the case despite the Decentralization Act of 1999 which mandates the transfer of the responsibility of all public services, including health and education, to local governments. However, efforts to decentralize health care were halted in 2002 following many local governments' reluctance to assume additional expensive and technically challenging responsibilities. Yet, while the overall system remains centralized, significant autonomy in operational matters has been gradually transferred to local governments. The country is divided into Health Districts, each of which has networks of health centres, small hospitals, and a large district hospital.

9.1.2 Coordination

In addition to vertical stewardship, effective functioning of health system requires horizontal coordination across agencies within governments. In recent years, Asian governments have increasingly turned to establishing elaborate governance structures in which policymaking and implementation functions are clearly delineated. At the same time, central agencies enjoying the government's strong backing have been established to lead and coordinate the sector. Governments have also increasingly established specialized agencies with responsibility for particular governance functions such as claims review, technology assessment, safety management, and so on that have had to be offset by enhanced coordination by central agencies.

As is to be expected in a vast country such as China, the challenges of coordinating the healthcare sector are gigantic. There are at least eleven ministries with overlapping responsibilities in health care which complicates decision-making (Qian, 2015). Local health bureaus are responsible for public hospital, local social security bureaus for health insurance, and local commerce bureaus for distribution of drugs, which make implementation and monitoring sporadic and disjointed. Moreover, local governments' evaluation systems emphasize the performance of individual agencies and officials rather than policy outcomes which impede collaboration and coordination.

In 2018, the Chinese government established the National Health Commission (NHC) with the responsibility for coordinating provision and delivery of health services and the National Healthcare Security Administration (NHSA) for coordinating health financing and payment functions. The importance of a central coordinating authority in healthcare delivery was recently highlighted by the NHC's role in leading China's efforts to fight Covid-19 pandemic. Similarly, NHSA has been leading the national efforts to integrate the fragmented insurance pools wherein UEBMI and URBMI were pooled at the municipal level, while NRCMS was pooled at the county level. The fragmentation made coordination difficult, limited their bargaining power vis-à-vis providers, and reduced the scope for cross-subsidizing from healthier to sicker members and from richer to poorer regions. The NHSA's ongoing efforts to amalgamate them is meant to not only enlarge insurance pools and improve management efficiencies but also bridge gaps in insurance coverage across the population.

Notwithstanding the deep political maladies that grip the territory, Hong Kong has a tightly integrated administrative structure that is well-equipped for coordinated action due to the comparatively small number of administrative agencies involved in managing the health sector. The Food and Health Bureau (FHB) is in charge of policy matters and the Department of Health is responsible for implementing policies and operating public healthcare facilities. The operational responsibility for managing public hospitals is vested in the Hospital Authority, which answers to the Department of Health. The territory does not have many specialized agencies responsible for different aspects of health care, as is the case in Singapore and Korea, for example.

It is only very recently that India prioritized health care and paid attention to coordination. India's federal structure coupled with differing administrative capacities of state governments has resulted in the emergence of multiple state-level health insurance schemes that are poorly coordinated. Governance is also undermined by poorly conceived allocation of responsibilities for administering health policy. Thus, the national health insurance scheme (RSBY) is administered by the Ministry of Labour and not the Ministry of Health and Family Welfare, which is responsible for other health financing programmes. The government's main health insurance programme (PM-JAY) features a complex design involving multiple forms of transactions among different levels of government (central, state, and district) on the one hand and insurance firms and hospitals on the other. The complex arrangements are meant to improve efficiency and accountability but inadvertently create vast scope for gaming and corruption. The establishment of the National Health Authority in 2018 with the responsibility for providing stewardship and oversight to the sector is a notable development in improving coordination of the government's healthcare activities.

The administration of health policy in Korea is highly centralized with the Ministry of Health and Welfare (MOHW) responsible for managing the sector. It is supported by two high-level agencies: National Health Insurance Service (NHIS) with responsibility for financing aspects of the health insurance, and the Health Insurance Review Agency (HIRA) responsible for purchasing and payment functions as well as technology assessment. As the single purchaser of healthcare services, NHIC has huge market power. HIRA, on the other hand, has quickly built impressive technical capacity to carry out its diverse functions.

However, the centralized management of health care in Korea is offset by the existence of a healthcare system thoroughly dominated by private providers who are able to resist and even defy government direction. The fierce competition among providers undermines cooperation, weakens integration between primary and secondary providers, and stymies policy coordination. As the largest buyer in the country, NHIS has the potential to play a larger role in shaping the sector, but is handicapped by the large additional payments that users must make out of pocket for non-covered services which are outside its control and indeed knowledge.

In Singapore, the Ministry of Health (MOH) bears the overall responsibility for steering and managing the healthcare sector, including system governance and financing, services and manpower planning. But Singapore government has multiple policy goals which make coordination challenging: it seeks to control costs and yet allows competition to improve service quality (which in health care usually raises costs). Similarly, it seeks to maintain Singaporeans' access to basic health care at affordable prices but at the same time it seeks to maintain a robust private healthcare system catering to medical tourists. In parallel, it seeks to contain the competition so that public providers do not lose sight of their public service mission. The simultaneous pursuit of different and somewhat contradictory goals makes coordination challenging but it is helped by the fact that the government owns and controls most hospital beds in the country. The existence of an autonomous Ministry of Health Holdings Private Limited (MOHH) as the owner and manager of all healthcare assets of the government is an immense help with coordination of the sector.

Thailand has had a centralized healthcare system with the public sector as the dominant provider since the early 1950s. The Ministry of Public Health (MOPH) is responsible for making and implementing health policies as well as owning and operating almost all public hospitals in the country. The launch of Universal Coverage Scheme (UCS) was accompanied by the establishment of the National Health Security Office (NHSO) to coordinate and implement the scheme. The NHSO's responsibility for managing UCS, which caters to 80 per cent of the population, gives it tremendous power to shape the sector. However, there is no centralized agency in Thailand to coordinate the different health financing programmes operated by different ministries. But this weakness is somewhat offset by the fact that most health care in the country is delivered by hospitals and clinics owned and operated by MOPH.

9.1.3 Integration of Primary and Hospital Care

The healthcare system in China, similar to other countries in the region, is hospital-centric and largely disconnected from the primary care system. The government's policy efforts to integrate the two are instructive in illustrating the difficulties that policymakers everywhere face in making primary care the foundation of a healthcare system. Before the 1980s 'reforms', China had public healthcare clinics throughout the country that provided no-frills basic care to the entire population. But the system collapsed in the 1990s as rural cooperatives crumbled and large swathes of the country were left without an adequate primary care system. Where it did exist, it was mostly provided by private clinics and township health centres. The situation changed again when hospitals expanded their outpatient services for generating additional revenues and quickly became the dominant supplier, especially in urban areas.

The Chinese government launched a comprehensive plan to rebuild primary care in 2009 and the efforts were ramped up in 2015. It is currently building new facilities, refurbishing the existing ones, and providing grants for equipment and public health measures. It has increased reimbursement for visits to lower level facilities while reducing it for primary care at hospitals. However, success has been limited because the difference in reimbursement rates is not large enough to overcome users' preference for hospitals, which are perceived to provide higher quality and technologically sophisticated care. It also matters that tertiary and secondary hospitals work aggressively to attract primary care patients to their facilities.

Hospital care in Hong Kong, similar to Singapore, is provided largely by the public sector while primary care is provided mostly by private clinics. The bifurcation of role has become increasingly problematic in the context of ageing population that frequently shifts between the two and requires continuity of care. To address their needs, the government is expanding the capacity of public clinics while also subsidizing users to use private clinics. However, that primary care is provided for profit while hospital care is provided nearly free of charge makes integration of the two very difficult.

The primary care system in India is weak and linked poorly to the hospital system. Both hospital and primary care are dominated by private providers who compete aggressively among themselves.

Private primary care providers are mostly in solo practice and do not share patient information with other providers. There is a nationwide network of government-owned and operated primary clinics at the village, district, and municipality levels but they are chronically under-resourced with reputation for poor service and, as a result, largely avoided by all but the poorest. No referral from primary care provider is required to visit public or private hospital and there is no financial penalty for going directly to any level of facility one wishes. However, the National Health Mission programme targeted at mothers and children has had a positive impact on improving the quality of primary care in the public sector. Similarly, the wellness component of the PM-JAY programme has some potential for integrating primary and secondary.

The provision of health care in Korea is even more private than in India and its hospital system is similarly disjointed from primary care. The public facilities that exist are concentrated in rural hospitals and primary care clinics. A peculiar feature of the Korean primary care system is that most providers are specialists working in private clinics that operate like mini hospitals. Hospitals too maintain their own outpatient clinics, staffed mostly by specialists. The government has been promoting visits to primary care facilities by reducing reimbursement for direct visits to hospitals and specialists without referral but the difference is not large enough to deter users who continue to visit hospitals in the first instance. Rural areas are acutely short of doctors, a problem the government is trying to address by increasing the number of doctors but is facing severe opposition from the medical profession concerned about loss of income due to increased supply.

Primary care in Singapore is dominated by private clinics that are spread throughout the island and offer convenient access at a reasonable price. Public clinics are yet cheaper due to government subsidies but often involve longer waiting time. Chronic disease patients seek treatment mainly from polyclinics due to the subsidized and broader range of services they offer despite longer waiting time. While both hospital and primary care systems work reasonably well, there is little integration between the two and this is unlikely to change in the near future due to their different ownership and incentives. However, the recent introduction of mandatory Electronic Health Record system is expected to promote collaboration through information sharing.

Thailand has the most vertically integrated healthcare system in the region with public providers tightly linked in a vertical chain centred around regional health 'districts' of about 50,000 population. The districts are the organizing unit for the Universal Coverage Scheme and the first point of contact for users. Each district has a community hospital with 30–120 beds serving as an anchor for 10–15 subdistrict (*tambon*) health centres with 3–5 paramedical staff. The district hospital also serves as the referring body for accessing specialist services at regional and provincial hospitals. The progressive strengthening of the local healthcare facilities is reflected in the fact that the district and subdistrict health facilities' share of all primary care utilized increased from 54 per cent in 1977 to 87 per cent in 2010 while the share of regional and provincial hospitals declined correspondingly (Phudit et al., 2018). Yet the subdistrict health centres have been under constant stress due to shortage of medical professionals. Shortage of general practitioners is also a deep and growing problem, as medical graduates overwhelmingly choose to pursue their career as specialists.

9.2 Ownership and Management

In health care, the organizational tools a government uses to produce and deliver services have a major impact on the sector's performance and outcomes. The form of ownership – public or private – is particularly important because of its effects on the use of other tools such as payment arrangements and regulations. Government ownership provides policymakers with mechanisms to intervene directly through internal decisions and instructions, instead of negotiation with external parties and regulations, as is the case with private ownership. However, the management of public hospitals and clinics is difficult due to innate limitations of traditional bureaucratic structures and processes in service delivery organizations. Job security for staff and lack of hard budget constraints compound the problems and challenge governments to adopt management processes that improve their performance while maintaining their public service orientation.

Overall, state ownership in the region is high in hospital care and low in primary care. This is not unexpected given that capital requirements are higher for hospitals, which is easier for governments to meet. Hospital care also poses greater financial risk to households, making a more compelling case for state intervention. Managing public hospitals efficiently is a major challenge that governments in the region have dealt with differently and with different levels of success, as we will see in the following discussion.

In China, the legacy of central planning continues with regard to ownership of hospitals, as the different levels of government together own and operate over 80 per cent of hospital beds spread throughout the country. Since 2005, the government has developed a range of tools to address concerns that public enterprises are inherently prone to inefficiency due to their monopolistic characteristic and government links. Accordingly, the responsibility for overseeing public hospitals has been increasingly transferred from government bureaus to independent hospital boards for more effective direction and oversight. A number of cities have established Hospital Management Council (HMC) for supervising hospitals within their locality. Appointment and promotion criteria and processes have also been made more transparent and meritocratic. At the same time, hospital managers have been given greater autonomy in human resource management matters, including relaxation of the employees' quota system which sets annual limits for staffing numbers. In parallel with greater management autonomy, the government has increased its controls over financial management at hospitals, which were seen as abusing their autonomy to improve their income at the expense of users. As a part of the efforts to strengthen accountability, insurance funds are strengthening their machinery for monitoring and reviewing the quality and costs of services.

In Hong Kong, where public hospitals enjoy near-monopoly in hospital care, the government has resorted to a variety of mechanisms for improving performance. To promote service quality and strengthen accountability, it requires all levels of Hospital Authority to enter into service agreements that specify detailed conditions relevant to the level. Thus, the government first enters into contract with HA to provide specified healthcare services which then contracts with regional clusters, which do the same with individual hospitals. The content of the agreements are of regulatory nature but the government implements them through contracts rather than regulation. While intended to promote accountability, it is arguable that the agreements actually undermine it by emphasizing form over substance.

In the absence of market competition, the Hong Kong government makes extensive use of surveys to ascertain users' opinion of service quality. The surveys show that while the population is largely satisfied with the services at public hospitals and clinics, there is widespread dissatisfaction with the long waiting times for non-urgent services, congested wards, and spartan quality of amenities (waiting areas, washrooms, wards, etc.). The level of dissatisfaction is particularly high among middle- and upper income households, who would prefer shorter waiting times, more choice of doctors, and higher quality amenities but without paying for them in any form. The disaffected population does not appreciate that the shortcomings are an integral part of the policy strategy to contain costs.

The provision of health care in India is overwhelmingly private, as public facilities account for less than 30 per cent of both hospital and primary care. While public health facilities providing free services exist at all levels of government, they are often severely short of personnel and supplies and the services they provide are patchy and unreliable, leading most users to avoid them when possible. The public facilities have been in need of management reforms for decades, but governments have found it difficult to motivate medical workers in the context of ironclad job security. The government's own administrative processes are rigid and ritualistic, unsuited for service industries, but the political obstacles to reform are simply too steep to allow improvements. The shortcomings of the public sector are a boon to private providers who are thriving under government neglect. They compete aggressively among themselves for patients which undermines scope for collaboration and integration. Moreover, their singular focus on generating profits leads them to engage in unethical and even illegal behaviour which compromises quality while promoting waste for the system and poverty among users. The end results of the weaknesses in both public and private sectors manifest in the country's poor healthcare outcomes.

The Korean government makes light use of public ownership, as public providers account for less than 10 per cent of inpatient and outpatient services utilized in the country. While small and working largely by the norms prevalent in the private sector, public providers in Korea are more likely to be located in poorer areas and are less likely to impose additional OOP payments for uncovered services. The private providers that overwhelmingly dominate the provision of services are intensely competitive which, as is common in health care, promotes oversupply and overcharging.

In Singapore, government-owned and operated hospitals are the dominant providers of inpatient care. However, they are subject to policy-induced conditions that are novel. The government has sought to promote efficiency and user focus in public hospitals by changing the market structure rather than through management reforms as in China. Specifically, public hospitals in Singapore are legally private firms which allows them the autonomy they need to operate in a competitive environment and yet remain within the government's direct control (Ramesh, 2008b). Exposing public hospitals to competition also affords the government a channel for receiving market feedback. Moreover, the government as an owner can shape hospitals' behaviour without resorting to onerous regulations or purchase negotiations that would be necessary if they were truly private firms. This has been particularly useful for controlling user charges, physicians' remunerations, and the number of hospital beds in different ward classes at public hospitals, for instance. Although MOHH and the individual hospitals are autonomous entities in a legal sense, MOH is deeply embedded within them through various procedural mechanisms such as appointment of the governing boards.

In Thailand, the Ministry of Public Health (MOPH) directly owns and operates a nationwide network of hospitals at every level. Furthermore, government payments are the largest source of income for the public hospitals and clinics. To improve efficiency in the context of pervasive dominance of public provision and financing, it has introduced purchaser–provider split whereby the National Health Security Office as payer enters into formal contracts with providers, most of whom are owned by MOPH. The NHSO's position as the single largest purchaser of healthcare services in the country gives it a strong bargaining power vis-à-vis providers. The clustering of hospitals in networks and the competition among networks for patients promote accountability as well as market responsiveness. At the same time, the capitation and DRG payments by which they are paid promote cost-consciousness despite public ownership. The risk of NHSO misusing its monopoly power is mitigated by the fact that it reports to the National Health Security Board which is chaired by the Minister of Public Health.

9.3 Pooling of Financial Risks

Financial tools in health care are more than a channel for providing material resources for producing and delivering health services to those in need. The tool by which the government spends money is as important as how much and on what it spends, as different tools offer different advantages and disadvantages. Social insurance, for instance, is an effective tool for reducing households' financial risks but it is worthless at containing costs due to the moral hazards of third-party financing. Similarly, OOP payment is efficient for making providers focus on users' preferences, but does not equip the latter to assess the usefulness and quality of the service they are purchasing. Direct government expenditure, on the other hand, is weak at incentivizing service improvements but the global spending limit that accompanies it is effective at controlling costs. More importantly, the government can use its spending as a lever to change providers' and users' behaviour in desired ways by attaching conditions to fiscal transfers. However, to realize the full potential of risk pooling tools, they need to be matched to payment tools discussed in the following section.

Public sources – comprising both government budget and social insurance – account for a considerably smaller share of total health expenditures in Asia than the average of 80 per cent for high-income countries. They form approximately half of total health expenditures in Hong Kong and Singapore and slightly more in Korea, as shown in Table 9.1. They form the largest share in Thailand and China – 60 and 77 per cent, respectively – and the smallest in India.

Table 9.1 ***Sources of total health expenditures, Percent***

	Government budget and social insurance*		Out of pocket		Private insurance
	2000	2015	2000	2015	2015
High income	59.9	79.7	15.9	13.7	–
Hong Kong	56.2	49.5	32.9	34.7	15
Korea	49.6	58.9	43.6	34.0	7
Singapore	36.3	52.8	48.2	32.0	15
Middle income	38.6	51.8	46.6	37.9	–
China	22.0	60.2	60.1	35.1	7
India	20.7	25.6	71.7	64.7	10
Thailand	57.9	77.4	34.2	12.5	10

* It is not advisable to separate the two sources because a large portion of budget spending (especially in China, Korea, India, and Thailand) is on subsiding compulsory social insurance.

Source: World Development Indicators. www.fhb.gov.hk/statistics/download/dha/en/table2_1718.pdf

Government budget is one of the largest sources of healthcare financing in all countries, in Asia and elsewhere. However, in China, India, Korea, a large share of government spending is towards subsidy for social insurance, which works differently and produces different results. Budget spending when targeted at producers, instead of subsidizing insurance, gives the government levers to directly alter the recipients' behaviour. Conversely, the lack of subsidy deprives governments of a crucial lever to shape the stakeholders' behaviour. This is especially the case with private providers who are otherwise difficult to control due to their information advantage and political power. Direct government spending is a substantial source of healthcare financing only in Hong Kong, Singapore, and Thailand which gives their governments strong levers to shape the sector. In other countries, government subsidy is mostly for subsidizing insurance premium and public health initiatives.

Social insurance can also serve as a powerful tool for altering the behaviour of providers and users, but only if the government actively uses its bulk purchasing function, which is not always the case. Social insurance is the largest source of health financing only in China and Korea among the countries studied here. Insurance funds in China are the country's largest purchaser of health services, but they have been fragmented which undermines their ability to negotiate with providers for better prices or quality. The ongoing amalgamation of the funds is likely to overcome this key shortcoming of the Chinese health system. The NHIS in Korea has been an active purchaser but has had to deal with the united front presented by private providers who wield enormous market and political power in the country.

India also has a national health insurance scheme – PM-JAY – but the premium is paid entirely by the government and, as such, is better described as budget-financed programme. In Thailand, only SSS, which is meant for formal private sector workers and covers 10 per cent of the population, can be said to be a true social insurance programme. Realization of the limitations of SSS in a country with large informal workforce is what led policymakers to choose budget financing for UCS in 2002. Singapore too has a limited but national health insurance programme called Medishield Life (ML), established in 2014. The objective of the ML is to help households' finance treatment for catastrophic illnesses through low-cost insurance, which imposes numerous deductible and co-payment requirements to keep the premium low.

Private insurance is a major source of healthcare financing in many countries – Switzerland and the United States, for example – but not in the countries studied here. Even when there is significant population coverage, it accounts for only a small percentage of expenditures as it is bought mostly by the rich and healthy. Thus, more than 70 per cent of the population in Korea buys private insurance, yet it covers only 7 per cent of total health expenditures. A similar situation exists in Singapore where two-thirds of the population have bought private insurance but it accounts for only 15 per cent of total expenditures. Singapore government is alarmed by the moral hazards associated with private insurance – which it fears will foster 'buffet syndrome' among users – and has imposed significant minimum co-payment requirements.

In contrast to Singapore which discourages private insurance, China, India, and Hong Kong are actively encouraging it through tax concessions. As a result, private insurance has grown rapidly in China though it still accounts for less than 7 per cent of total health expenditures. India too offers tax incentives for purchasing private health insurance which accounts for about 10 per cent of THE. The VHIS scheme in Hong Kong is still new and little is known about its effects though it is unlikely to make much difference in the short term given that high-quality care is available at public facilities nearly free of charge. Thailand allows deduction of up to baht 25,000 (US$800) from income tax for purchase of private health insurance, which accounts for 10 per cent of THE. In every country in the region, there is a section of the non-poor population who do not fully trust the public schemes and buy private insurance for additional security despite the high costs and limited protection they offer.

Out-of-pocket spending forms about one-third of total health expenditure in China, Singapore, Hong Kong, and Korea and nearly two-thirds in India. Although these shares are smaller than they were in the past, they are still larger than WHO's recommended ceiling of 20 per cent of THE. On a positive note, the OOP shares of THE have almost halved in China and Thailand since 2000. In Thailand, OOP now forms only 12 per cent of THE which is lower than the OECD average. Much of the OOP expenditures is on primary care, including over-the-counter drugs, which is for the most part is neither provided nor financed by governments in the region.

In China, direct government expenditure on health has increased vastly in recent years. Much of the expenditure is for subsiding insurance premium for rural residents and migrant workers who often do not have income or job security to contribute sufficiently and regularly. Subsidy for primary care infrastructure has also increased in recent years. The increased subsidy has not only lightened the burden on households but it has also given the government leverage to guide the providers to contain costs and pay more attention to public health. Previously, when the subsidy was low and providers were encouraged to earn directly from users, the government lacked financial levers to intervene in the sector.

In Hong Kong, the government employs subsidies for public hospitals as the main tool for financing health care and shaping the sector. Government transfers form more than 90 per cent of public hospitals' total revenues, in contrast to private hospitals that receive almost no subsidy. The lump sum payment it offers to public hospitals imposes a global limit to what individual hospital can spend, which is a powerful if blunt mechanism for containing costs. On the other hand, global budgets necessarily involve rationing of the volume of services, especially non-urgent elective treatments, which leads to lengthy waiting time disliked by the population. The recently adopted Voluntary Health Insurance Scheme increases demand for private hospital services which is likely to raise total expenditures. Widespread subscription to VHIS risks creating a two-tier health system in which the rich will use private hospitals providing services on demand while the rest of the population will be served by public hospitals subject to rationing of elective services.

In India, public spending on health has remained around 1 per cent of GDP over the past two decades. Much of the spending is in the form of subsidies for national programmes (such as NHM and PM-JAY) and delivered through state governments. Government spending on primary care and public health and sanitation has increased substantially in recent years but remains small given their importance to developing countries. Government facilities operate under a yearly budget based largely on historical trend without any condition intended to improve performance. The PM-JAY is an entirely government-funded scheme that sets out detailed conditions for providers, insurers, and users with the purpose of setting nationwide service standards. However, the implementation of the programme varies widely across the country depending on the state government's administrative capacity.

The Korean government makes heavy use of financial policy tools, including government transfers, tax relief, and social insurance. The central government provides modest subsidy to NHIS while local governments subsidize insurance premium for the self-employed and indigent and public healthcare services. Overall, the transfers are too small to serve as an effective tool for shaping the providers' behaviour. Social insurance is the largest source of healthcare financing in Korea and the primary tool by which the government intervenes in the sector. NHIS enjoys tremendous power over both users and providers through the conditions it imposes regarding contributions and benefits. However, private providers have successfully resisted NHIS's efforts to control their behaviour through purchase conditions. Users in Korea are required to make substantial co-payment for covered services and pay entirely for non-covered services. While this is somewhat effective at limiting government expenditures, it has serious adverse implications for access and equity.

Singapore government is an active user of targeted financial tools to shape the sector. It offers different levels of subsidy for different purposes and target groups. Thus, maximum subsidy is available for lower class, no-fringe beds in public hospitals. Similarly, different levels of subsidy are available to different age and income groups at outpatient clinics. In return for receiving subsidy, public providers must abide by service and other conditions set out by the Ministry of Health. Users too are provided subsidies on a targeted basis. Thus, means-tested support is available for hospital care under Medifund and for drugs purchase through Standard Drug List (SDL) and Medication Assistance Fund (MAF) schemes. The requirement to contribute to Medisave and make partial or full OOP payments for all services are a part of the government's deliberate policy to ensure that the principle of individual responsibility for health is upheld.

Government subsidies are the largest source of healthcare financing in Thailand. A large part of the funding is in the form of fiscal transfers to the Ministry of Public Health (MOPH) for operating hospitals and primary care clinics. The government also fully pays for the premium of UCS members, who form 80 per cent of the population. In addition, the government fully subsidized the cost of medical services for CSMBS members. The National Health Security Office's (NHSO) position as the single payer for UCS and SHI gives it tremendous market power to negotiate favourable prices and raise service quality. The NHSO has used its purchasing power to introduce capitation payments for primary care and DRG payment for hospital care, and to reduce prices of drugs and medical supplies through bulk buying. As a purchaser, the NHSO has been remarkably more successful than NHIS in Korea because it deals mostly with public providers while the latter deals mostly with private providers.

9.4 Payment Arrangements

Financing of health care does not end with collection of revenues and pooling of funds, as the funds must be distributed to providers. The mechanisms by which the funds are distributed – the payment arrangements – have a major impact on costs as well as outcomes. An effective payment system is one that incentivizes providers to provide only necessary services and in a cost-conscious manner.

At the broadest level, there are two types of payment methods: retrospective and prospective. Under retrospective arrangements, such as FFS, providers are paid after a service has been delivered, usually based on the volume of services. In contrast, providers are paid in advance under prospective arrangements such as global budget, case-based payment, and capitation payment. Traditionally, FFS has been the main form of payment in the private health sector while global or line item budgets have been the main payment mechanism in the public sector. The popularity of prospective payment mechanism has increased in recent decades, following the realization that it is more effective at containing costs, which has been a major concern for reformers in most countries.

While FFS is effective at making providers attend to users' needs, it also allows the former opportunities to increase the volume and price of services they provide with the objective of raising their income. Furthermore, providers may take advantage of information asymmetries to prescribe and sell treatments and drugs that offer them the largest profits. FFS is particularly problematic in the context of insurance financing because it is convenient for insurers to passively approve claims rather than to use their payment authority to reduce cost or improve quality. Prospective payments, on the other hand, set a limit on expenditures and are thus more effective at containing cost but are prone to encouraging providers to skimp on quality and quantity of services. It is upon governments therefore to design payment arrangements that restrain costs while meeting users' needs. In public health systems, in which providers are typically paid on a prospective basis, the challenge is to ensure that users receive the services they need in a timely manner. In contrast, the main challenge in the private system is to curb providers' profit-seeking instincts without undermining user focus.

China, India, and Korea have had a privately financed healthcare system in which users are paid on an FFS basis. Their governments recognize the limitations of the arrangement and have been making efforts to introduce elements of prospective payments, with varying degrees of success. In China, providers are paid on FFS basis according to fee schedule, which limits the prices but not the quantity of services. Providers complain that the set prices are unduly low, which leads them to prescribe drugs and treatments not included in the schedule. To curb FFS, provincial and local governments throughout China have launched reforms introducing capped payments centred on some version of capitation for primary care and case-based payments (such as DRG) for hospital care. Many local governments have also ended or reduced physicians' bonus linked to the revenues they generate and put them on fixed salaries with only a small component tied to performance.

The public health system in India, which is massive though small in comparison to the private sector, is paid through a combination of line item and global budgets. While there is no risk of runaway expenditures under this arrangement – indeed a large share of allocated budgets remain unspent every year in many states – there is a massive undersupply of services. The much larger private sector is paid almost exclusively on FFS basis but changes are underway, as PM-JAY pays a predetermined 'package rate' for various treatments at hospitals. However, the vast majority of outpatient and inpatient services in India are delivered outside of PM-JAY arrangement and continue to be paid on FFS basis, mostly out of pocket.

Korea has the most private health system among the countries studied here and the providers there are paid almost entirely on FFS basis. To curb expenditures, the government has set price schedules, which are determined by protracted negotiation between NHIC and providers. However, providers are allowed to charge what they wish for a large number of drugs and services not covered or only partially covered by insurance, incentivizing them to prescribe them rather than those on the covered list. The government has for years been trying to introduce capitation and DRG payments and 'pay for performance' mechanisms but made little headway due to fierce opposition from private providers.

Hong Kong has an uncomplicated payment system for health care, centred on global budgets for public hospitals and clinics and FFS in private hospitals and clinics. The size of the global budget is determined mostly on the basis of historical precedents, adjusted for demographic and other trends. The medical staff at hospitals are traditional public servants paid a fixed monthly salary. The government, through Hospital Authority, has set quality standards to ensure that users receive necessary services of acceptable quality in a timely manner. The fine performance of public hospitals in Hong Kong, measured by outcomes, is a useful reminder that it is possible for traditional payment systems to work well if managed effectively. The FFS payment prevalent in primary care, which is dominated by private providers, also works reasonably well due to the intense competition in a densely populated territory.

The payment arrangements in Singapore are similar but somewhat more complicated than in Hong Kong. Public hospitals are paid based on a combination of case mix and global budget. However, the subventions are accompanied by more detailed performance conditions and quality standards than is the case in Hong Kong. Medical and other staff in Singapore are eligible for performance bonus, but it is not tied to the revenues they help generate. The payment arrangements for primary care providers are almost identical to what exists in Hong Kong.

In Thailand, public hospitals are paid on DRG basis while primary care clinics are paid on a capitation basis. To offset the inherent limitations of DRG and capitation payment, there is a small FFS component for priority services – such as immunization and cataract surgery – that the government seeks to promote. Implementing DRG and capitation payment regimes are challenging tasks requiring immense analytical and administrative resources and it is to the credit of the Thai public managers that they have made the arrangements work and are constantly making improvements.

9.5 Regulations

Regulations are a powerful tool available to the government to shape the healthcare sector according to its goals and priorities. It is especially important in systems dominated by private provision where the government lacks organizational and financial tools to shape the providers' behaviour. Conversely, systems with large public ownership and high budget spending do not require extensive regulations because they can achieve the same objectives as owners and payers. This is the case in China, Hong Kong, Singapore, and Thailand which have high public ownership and comparatively light regulations. In contrast, a strong regulatory framework is vital in healthcare systems dominated by private providers, such as in India and Korea, though these governments have not adequately risen to the challenges.

With government as the owner of most hospitals in China and the party system deeply embedded in every healthcare organization, the government has multiple channels for guiding the stakeholders' behaviour. However, since even public providers in China are driven by financial motives, the government has had to control their behaviour through licencing, safety, and pricing regulations, supplemented by political direction. The quality of regulations has improved over the years, in line with general improvements in health governance in the country. In the past, for instance, the government sought to improve affordability by imposing price controls on essential drugs and services but the measure did not work as intended because the providers simply prescribed drugs that were outside the regulated list. In recent years, the government has bought essential drugs through auctions and then sold them to users without markup, which has been more effective in reducing prices. China has also updated its licencing requirements for hospitals, clinics, doctors, nurses, pharmacists, and the like in line with developments in the industry and the rising expectations of the population.

With near-monopoly of the public sector in hospital care in Hong Kong, the government does not need the same sort of formal regulations as in systems dominated by private hospitals. Accordingly, the government employs regulation-like procedural policy tools to promote quality and contain costs. The Hospital Authority has developed elaborate 'Patients' Charter', quality service standards, and 'Professional Code of Practice' to promote service quality in public hospitals. Hospitals have also established community advisory bodies and considerable efforts are made to make them work.

The ad hoc and haphazard development of the healthcare system in India is reflected in its regulations for the sector. Regulations governing licenses for medical professionals, healthcare providers, medical universities, and pharmacies exist, but they are often poorly designed and inconsistently applied. Without significant presence in the market as a producer, the government lacks the information needed to understand the sector's dynamics. The problem is compounded by weak administrative capacity due to bifurcation of responsibilities between central and local governments. The resulting weak regulatory framework has allowed the private sector to grow and operate without necessary oversight and control, leading to poor service quality and rapacious behaviour.

For Korean policymakers, similar to their Indian counterparts, the greatest challenge is to set up a regulatory framework for a sector thoroughly dominated by private providers. Overall, the healthcare sector is lightly regulated by OECD standards though more heavily compared to other countries in the region. For instance, the number of hospitals, location and size of hospitals and hospital beds, and the number and distribution of high-technology diagnostic services remain unregulated in Korea. Similarly, fees for non-covered services, special treatment, and room charges are outside regulation despite the fact that their absence is a major reason for over-provision of health services in Korea. Enforcement of existing regulations is also weak. For example, referral system exists but is enforced only for Medicaid users who form a very small portion of total users. In some respects, however, Korean regulations are unusually detailed and include what are otherwise managerial functions, such as cost-sharing schedule, and specific penalties for non-compliance. Moreover, there are detailed processes for determining services covered, FFS rates, benefit standards; cost-sharing rates for insured services; claims management; and payment schedule. Such details are unnecessary in Hong Kong, Singapore, and Thailand where the same can be achieved directly due to extensive public ownership and capped prospective payment.

In Singapore, regulations focus on licencing and safety issues. Under the Healthcare Services Act (HCSA) of 2019, which replaced the Private Hospitals and Medical Clinics (PHMC) Act, licences are granted based on the type of healthcare service provided rather than the physical premises of the provider as was the case earlier. As the owner and the largest source of finances for public hospitals, the government has more direct means available to issue directions to providers than regulations. The private sector is small but challenging to regulate due to the policy goal of promoting competition among hospitals. Lightly regulated private providers are also seen as crucial to the objective of promoting innovations in attracting medical tourists to the country.

Thailand has a robust regulatory framework for health care despite the dominance of the public sector in provision and financing. First, hospitals must meet accreditation standards before they can register with the NHSO for participation in the UCS system. Second, members of the UCS are required to register with their preferred GP within their district health system. They are also required to secure a referral from their GP to access advanced care at district hospitals free of charge. Third, UCS and SSS funding is tied to utilization rates which creates incentives for providers to attract users to sign up with their network. Fourth, health technology assessments and the national list of essential medicines (NLEM) govern what types of treatment are allowed under respective health insurance programmes in Thailand.

9.6 Health Policy Design in Asia: Discussion

The purpose of this chapter has been to present a comparative portrait of the policy tools employed in health care in Asia, the effects they trigger, and how they affect the achievement of universal health coverage. Examination of the design and implementation of key health policy tools in China, Hong Kong, India, Korea, Singapore, and Thailand show improvements along all main dimensions of health policy design.

All governments in the region are engaged in broadening the range of policy tools they use to build universal health coverage in their country. However, while the use of individual tools matters, what matters more is how they are used collectively, as the use of mismatched and contradictory tools does more harm than good. Each policy tool has its own internal logic which can be harnessed to promote synergy across tools. Our study shows that different governments have different levels of success at coherent use of policy tools.

Perhaps the greatest improvement in the use of policy tools has occurred in China, where health policy reforms since the late 1990s have touched every critical aspect of the healthcare sector. Health policy reforms started in earnest in 2005 when the government unequivocally stated that the central government was ultimately responsible for ensuring the population's access to health care. In the ensuing years, it vastly increased public funding for health care, mostly on subsidizing insurance premium for informal workers and rural residents but also substantially on public health infrastructure. It also took measures to curb the commercial orientation of public providers, professionalize their management, and rebuild primary care system and integrate with hospital care. This was accompanied by measures to promote prospective payment mechanism and dozens of innovative experiments are still ongoing across the country. Regulations too are being strengthened and modernized in ways that allow managers and doctors more autonomy but under greater government oversight. While it is still early to assess the impact of the vast range of reforms, initial indicators show flattening of total expenditures and reduction of OOP expenditures while improving the quality of care.

Hong Kong's healthcare system, especially the hospital system, is patterned after the traditional British National Health Service (NHS) and displays many of the latter's strengths and weaknesses. Much of the hospital care and a small part of the primary care in the territory are provided nearly free of charge at facilities owned and operated by the government using funds from the public exchequer. Medical and other staff at public hospitals and clinics are regular employees on fixed salary. Efficient management of the facilities has been maintained through meritocratic recruitment and reward and extensive surveys to ascertain users' feedback. Public funding is concentrated on public hospitals in the form of fixed prospective payment while primary care is funded mostly on FFS basis through OOP payment. The targeted public expenditure allows government to contain its expenditures while mitigating households' risks of catastrophic health expenditures. Since the government owns and directly controls the facilities, the need for shaping their behaviour through regulations rather than simple directives has been less imperative. This simple system has produced one of the best healthcare systems in the world in terms of both low costs and universal access.

Yet, policy makers in Hong Kong have fretted over public expenditures and made intermittent efforts to expand private provision and financing. Their reservations about the public system are shared by the upper middle classes who would prefer shorter waiting times, more choice of doctors, and better amenities. What is lost in the debate is the understanding of the essential features of the system – publicly owned and financed system paying providers on a global budget basis – that accounts for its effectiveness. The public system is intended to provide only essential services and, hence, the long waiting times for non-essential services and spartan amenities are a part of the design that underlies its success in containing expenditures. The recent efforts to expand private provision and financing aided by government subsidies are likely to dilute and undermine the system's essential character and underlying strength.

Singapore too inherited its health system from the British and has largely retained it, though it has added market-oriented features that make it distinct and extraordinarily complex. Singapore government has a reputation for firm stewardship and this is reflected in the way it guides and coordinates the sector. The provision of hospital care is dominated by public providers while primary care is provided largely by the private sector. A unique feature of public hospitals is that they are incorporated as private firms fully owned by the government which allows for management according to the rules and norms of the private sector but under close government supervision to ensure that they serve public service goals. There are various measures in place to ensure competition among providers, though there are limits to their autonomy in generating revenues. Government is the largest source of fund for hospital care, followed by a combination of OOP payment, Medishield Life, private insurance and Medifund. Primary care, on the other hand, is funded largely out of pocket. The government's payment to hospitals is mostly in the form of global budget adjusted for case load while households pay for both hospital and primary care on FFS basis. Regulation of private providers and insurers is light, partly to allow the private sector to be innovative in attracting medical tourists to the country. A particularly distinct feature of the health policy in the city state is the efforts to provide users with information to empower users with knowledge on service quality and prices while making purchase decisions.

The elaborate use of a vast range of health policy tools in Singapore is in sharp contrast to the local and international commentary on the country highlighting, and often extolling, its medical savings account (Medisave). As we saw in the book, Medisave accounts for a very small share of total healthcare expenditures in the country and, in all likelihood, compounds rather than offsets the inequitable effects of OOP payments. The government's preference for individual responsibility, reflected in its insistence on at least some household contribution to their own health care in the form of OOP and Medisave payments, accounts for the relatively high incidence of catastrophic health expenditures in the country. On the positive side, the country does not suffer from long waiting times for non-urgent services, as in Hong Kong, the territory with which it otherwise shares many features in common.

The roots of the Indian healthcare system also go back to the British system centred on public provision and financing. Successive governments in independent India proclaimed commitment to expanding the public system and, accordingly, established public health centres and clinics and public hospitals throughout the country. However, the state governments, which bore the primary responsibility for operating the public health facilities, provided inadequate funds, personnel, and supervision which led to deterioration of the public system to the point it was avoided by all but those without choice. The decrepit state of government-owned facilities allowed private providers to grow to fill the void and quickly become the dominant supplier. The system that emerged was severely deficient along all critical dimensions of health policy design: it lacked stewardship and coordination, was dominated by private providers, who were paid out of pocket on an FFS basis, and operated in a loose regulatory environment. The system was inefficient, expensive, and inequitable which was reflected in the population's poor health status and financial hardships.

It is only since 2005 that successive governments in India have devoted serious attention to health care, backed by additional public spending. India now has a national health insurance scheme (PM-JAY) for the poor intended to cover 500 million people, making it the world's largest health insurance scheme. The scheme is characterized by many modern design features, including enhanced central government stewardship, a central agency (National Health Authority) to coordinate the policy initiatives, central budget financing accompanied by spending guidelines, and fixed payment for services. However, primary care, which is of greatest relevance to poorer and rural communities, continued to be neglected and poorly integrated with hospital care. The low administrative and fiscal capacities of many state governments are also a major impediment to universal health coverage in India.

The Korean healthcare system is one of the top performers in the world in terms of efficiency but suffers from excessive specialist treatments and high OOP payments due to certain institutional arrangements in place. On the one hand, it has a centralized governance system with strong vertical stewardship and inter-agency coordination. National Health Insurance is the largest source of healthcare financing, followed by OOP payments. On the other hand, the provision system is hospital-centric and entirely dominated by private providers who are paid on FFS basis and operate in relatively relaxed regulatory environment. The government has for long recognized the hardships caused by OOP and made persistent efforts to reduce it through cost controls and payment reforms. However, its efforts have been repeatedly thwarted by private providers whose cooperation is essential due to virtual absence of public providers as a viable alternative. The Korean case is a telling reminder that the governance and financing reforms emphasized by recent health policy literature are insufficient unless accompanied by reform of how health care is provided and paid for in systems dominated by the private sector.

The health system in Thailand has a strong reputation for providing universal access to health care in an efficient and equitable manner. The achievement is the result of well-orchestrated use of policy tools covering all key aspects of the health system. The Ministry of Public Health is a well-respected institution that has provided active stewardship to the sector for several decades, irrespective of the particular government in office. It owns and operates a nationwide network of hospitals and clinics that provide good quality care for free or for nominal charge. Government budget forms three-quarters of total health expenditures, with only a small share coming from out of pocket and private insurance that are incurred mostly at private facilities. Prospective payment in the form of DRG for hospital care and capitation for primary care is the norm in the public sector, with FFS used only for promoting certain priority services. Private providers, on the other hand, are paid mostly on FFS basis.

Regulations are relatively light, partly because of the dominance of the public system but also to support private hospitals' efforts to attract medical tourists. The case of Thailand demonstrates that it is possible for governments in developing countries to achieve universal health care at modest cost through orchestrated use of policy tools targeting all key problematic areas.

9.7 Concluding Observations

The world has been rife with discussions about the pressing need for universal health coverage for two decades. A lot of the discussion has rightly focused on mobilizing financial resources and strengthening governance. What these discussions have overlooked, however, is that the absence of accessible health care is in essence a policy problem that requires a problem-solving approach if it is to be addressed. Accordingly, this book proposed a policy design approach highlighting problematic conditions in five crucial respects – governance, ownership and management, risk pooling, payments, and regulations – that characterize the health sector that policymakers must address. It then applied this framework to six dynamic but diverse economies in Asia to understand their efforts to achieve universal health coverage. The study found remarkable progress towards universal health coverage in all these countries, though they were also found to suffer from significant shortcomings. We argue that the achievements and lapses are attributable to their relative success in addressing the five critical health system functions.

Not all use of policy tools are of course a result of deliberate design. Many useful and less-than-helpful tools which governments have come to rely on are a combination of deliberate or careless choices in the past, when the implications of the various choices were not fully understood. Understanding these historical path dependencies and particularly those linked to colonial legacies is important for understanding the manner in which contemporary tools are used. Former British colonies, for example, inherited some form of the 'Beveridge' healthcare system found in the UK which, in retrospect, has shown greater efficacy than their Bismarckian counterparts found in other countries. History is not destiny of course, but choices made at the onset of the development of a healthcare system severely constrain subsequent policy choices.

The impact of such path dependencies and policy legacies is acute. Contemporary governments that started out with a centralized health system built on public provision and financing, for example, have had a distinct advantage in achieving universal health coverage at affordable costs. In such systems, the government owns and operates health facilities, funds them from public exchequer, and pays the providers prospectively through fixed budget allocations. Such systems have a lower need for regulatory oversight and information because all key health system functions are internal to the government. As a result, they already have in place most of the critical design elements for effective delivery of universal health care.

The Achilles heel of such government-dominated systems, however, is poor management of public facilities due to lack of controls or incentives needed to guide managerial and medical staff to serve users in an appropriate and cost-effective manner. Poor management of public facilities undermines public trust which leads to their decay and abandonment. The erosion of the public health system then promotes the proliferation of private providers which, unless regulated effectively, fosters the growth of private financing and FFS payments that are major impediments to universal access to health care. Publicly organized health systems that have avoided this fate are those with governments which have taken active measures to support their public hospitals and clinics with not only resources but also necessary management reforms. The potential and limitations of the public systems are evident in the countries studied in this book.

The contemporary health systems of all countries studied here, except Korea, started with dominant role of the state in both provision and financing. However, for different reasons, governments in China and India let their public healthcare systems decay which set them back in their struggle to establish universal health coverage. In contrast, Hong Kong, Singapore, and Thailand not only preserved their public system but adapted and finessed it to suit their evolving needs. It is these health systems, as we saw in the book, that have performed the best in providing health care to their population at sustainable costs. China, India, and Korea have struggled to build their system to meet their growing needs, though high public ownership and healthy public finance in China has given it an advantage in its rebuilding efforts. The additional public funding that Chinese,

Indian, and Korean governments have mobilized for health care offers opportunities for creatively intervening in the sector to address the deficiencies in ownership, financing, and payment arrangements that impede the achievement of universal health care. While the cards contemporary health policymakers were dealt by history are important, so is how they play their hand.

References

Administrative Divisions of China (n.d.) In Wikipedia. Retrieved 6 June 2019, from https://en.wikipedia.org/wiki/Administrative_divisions_of_China.

Alvarez-Rosete, A., Hawkins, B. and Parkhurst, J. (2013) Health system stewardship and evidence informed health policy. Working Paper No. 1. London: London School of Tropical Hygiene and Medicine, GRIP-Health Programme.

Anand, C. (2017, February 11) When hospitals infect you. Retrieved 6 August 2020, from www.thehindu.com/sci-tech/health/When-hospitals-infect-you/article17289370.ece.

ANI (2019, October 21) In a first, PMJAY rewards 3 hospitals with 'gold certificate' for quality culture. Retrieved 6 August 2020, from https://health.economictimes.indiatimes.com/news/hospitals/in-a-first-pmjay-rewards-3-hospitals-with-gold-certificate-for-quality-culture/71690374.

Ariyapruchya, K., Chua, K. C., Bandaogo, M. A. S. S., Lathapipat, D., Osornprasop, S., Yang, J. (2019) Thailand economic monitor: Inequality, opportunity and human capital. Thailand Economic Monitor Washington, DC: World Bank Group. http://documents1.worldbank.org/curated/en/154541547736805518/pdf/.

Asher, M. G. and Nandy, A. (2006) Health financing in Singapore: A case for systemic reforms. *International Social Security Review*, *59*(1), 75–92.

Asian Development Bank (2019) *Key indicators for Asia and the Pacific 2019*. doi: 10.22617/FLS190428-3.

Aw, Tar Choon and Low, Linda (1997) Health care provisions in Singapore. In Tan Teck Meng and Chew Soon Beng (Eds.), *Affordable Health Care: Issues and Prospects* (pp. 50–71). Singapore: Prentice-Hall.

Ayres, Ian and Braithwaite, John (1992) *Responsive Regulation: Transcending the Deregulation Debate*. Oxford: Oxford University Press.

Bae, E.-Y., Kim, H. J., Lee, H.-J. et al. (2018) Role of economic evidence in coverage decision-making in South Korea. *PLoS ONE*, *13*(10), e0206121.

Bajpai, V. (2014) The challenges confronting public hospitals in India, their origins, and possible solutions. *Advances in Public Health, 2014*. doi: 10.1155/2014/898502.

Balabanova, D., Mills, A., Conteh, L. et al. (2013) Good Health at Low Cost 25 years on: Lessons for the future of health systems strengthening. *Lancet, 381*, 2118–2133.

Balabanova, D., Oliveira-Cruz, V. and Hanson, K. (2009) *Health Sector Governance and Implications for the Private Sector*. Washington, DC: Results for Development Institute.

Bali, A. S. and Hannah, Adam (2021) Policy styles in healthcare: Understanding variations in health systems. In Michael Howlett and Jale Tosun (Eds.), *Handbook of Policy Styles* (pp. 105–122). Routledge.

Bali, A. S. and Ramesh, M. (2015a) Health care reforms in India: Getting it wrong. *Public Policy and Administration, 30*(3–4), 300–319.

(2015b) Mark time: India's march to universal health care coverage. *Social Policy & Administration, 49*(6), 718–737.

(2017) Designing effective health care: Matching policy tools to problems in China. *Public Administration and Development, 37*(1), 40–50.

(2018) Policy capacity: A design perspective. In Michael Howlett and Ishani Mukherjee (Eds.), *Routledge Handbook of Policy Design* (pp. 331–344). Oxfordshire: Routledge.

(2019) Assessing health reform: Studying tool appropriateness & critical capacities. *Policy and Society, 38*(1), 148–166.

(2021) Governing healthcare in India: A policy capacity perspective. *International Review of Administrative Sciences, 87*(2) 275–293.

Barbazza, Erica and Tello, Juan E. (2014) A review of health governance: Definitions, dimensions and tools to govern. *Health Policy, 116*, 1–11.

Barr, M. D. (2001) Medical savings accounts in Singapore: A critical inquiry. *Journal of Health Politics, Policy and Law, 26*(4), 709–726.

Béland, D. (2016) Ideas and institutions in social policy research. *Social Policy & Administration, 50*(6), 734–750.

Bennett, C. L., Pei, G. K. and Ultmann, J. E. (1996) Western impressions of the Hong Kong health care system. *Western Journal of Medicine, 165* (1), 37–42.

Bhaskar, Utpal (2019) NDA Ujjwala surpasses targets, provides 80.33 million LPG connections. *Live Mint*. www.livemint.com/news/india/nda-ujjwala-surpasses-targets-provides-80-33-million-lpg-connections-1568006076964.html.

Bhore, J. (1946) *Bhore Committee Report*. Delhi: Government of India.

Blomqvist, Å. (1991) The doctor as double agent: Information asymmetry, health insurance, and medical care. *Journal of Health Economics, 10*(4), 411–432.

(2011) Public sector health care financing. In S. Glied and P. Smith (Eds.), *The Oxford Handbook of Health Economics* (Vol. 1, pp. 257–284). Chippenham: Oxford University Press.

Bloom, G. (1998) Primary health care meets the market in China and Vietnam. *Health Policy*, *44*, 233–252.

Bloomberg (2018) These are the economies with the most (and least) efficient health care. Retrieved from www.bloombergquint.com/global-economics/u-s-near-bottom-of-health-index-hong-kong-and-singapore-at-top.

Bloomberg Global Health Index (2017) Retrieved from www.bloomberg.com/news/terminal/OMYWL56NKN52.

Blumenthal, D. and Hsiao, W. C. (2005) Privatization and its discontents – The evolving Chinese health care system. *New England Journal of Medicine*, *353*(11), 1165–1170.

Bowser, D. M., Jha, R., Bhawalkar, M. and Berman, P. (2019) The challenge of additionality: The impact of central grants for primary health care on state-level spending on primary health care in India. *International Journal of Health Policy and Management*, *8*(6), 329.

Boyer, L., Francois, P., Doutre, E., Weil, G. and Labarere, J. (2006) Perception and use of the results of patient satisfaction surveys by care providers in a French teaching hospital. *International Journal for Quality in Health Care*, *18*(5), 359–364.

Brinkerhoff, D. W. and Bossert, T. J. (2008) *Health Governance: Concepts, Experience, and Programming Options*. Health Systems 20/20. Washington, DC: USAID. www.healthsystems2020.org.

Brinkerhoff, D. W. and Bossert, T. J. (2014) Health governance: Principal–agent linkages and health system strengthening. *Health Policy and Planning*, *29*(6), 685–693.

Broom, A. and Doron, A. (2020) Antimicrobial resistance, politics, and practice in India. *Qualitative Health Research*, *30*(11), 1684–1696.

Campos, P. A. and Reich, M. R. (2019) Political analysis for health policy implementation. *Health Systems & Reform*, *5*(3), 224–235.

Capano, G. (2011) Government continues to do its job. A comparative study of governance shifts in the higher education sector. *Public Administration*, *89*(4), 1622–1642.

Capano, G. and Howlett, M. (2020) The knowns and unknowns of policy instrument analysis: Policy tools and the current research agenda on policy mixes. *SAGE Open*, *10*(1).

Capano, G. and Lippi, A. (2017) How policy instruments are chosen: Patterns of decision makers' choices. *Policy Sciences*, *50*(2), 269–293.

Capano, G., Howlett, M. and Ramesh, M. (2015) Bringing governments back in: Governance and governing in comparative policy analysis. *Journal of Comparative Policy Analysis: Research and Practice*, *17*(4), 311–321.

Carmichael, J. and Pomerleano, M. (2002) *The Development and Regulation of Non-Bank Financial Institutions*. Washington, DC: World Bank.

Carroll, Aaron E. (2019) What can the U.S. Health system learn from Singapore?", *New York Times*, April 22, 2019. www.nytimes.com/2019/04/22/upshot/singapore-health-system-lessons.html.

Cejudo, Guillermo M. and Michel, Cynthia L. (2017) Addressing fragmented government action: Coordination, coherence, and integration. *Policy Sciences*, *50*(4), 745–767.

Chin, C. W., and Phua, K. H. (2016) Long-term care policy: Singapore's experience. *Journal of Aging and Social Policy*, 28(2), 113–129. doi: 10.1080/08959420.2016.1145534.

Chindarkar, N., Howlett, M. and Ramesh, M. (2017) Introduction to the special issue: "Conceptualizing effective social policy design: Design spaces and capacity challenges". *Public Administration and Development*, *37*(1), 3–14.

Christianson, J. B. and Conrad, D. (2011) Provider payment and incentives. In S. Glied and P. Smith (Eds.), *The Oxford Handbook of Health Economics*. doi: 10.1093/oxfordhb/9780199238828.013.0026.

Chu, Annie, Kwon, Soonman and Cowley, Peter (2019) Health financing reforms for moving towards universal health coverage in the Western Pacific Region. *Health Systems & Reform*, *5*(1), 32–47.

Chua, P. M. H., Toh, Y., He, S., Jamaludin, A. and Hung, D. (2019) Centralised-decentralisation in Singapore education policymaking. In *Innovations in Educational Change* (pp. 3–21). Singapore: Springer.

Chun, C.-B., Kim, S.-Y., Lee, J.-Y., and Lee, S.-Y. (2009) Republic of Korea: Health system review. *Health Systems in Transition*, *11*(7), 1–184.

Credit Suisse, A. G. (2016) Global wealth databook 2016.

Culyer, A. J., Newhouse, J. P., Pauly, M. V., McGuire, T. G. and Barros, P. P. (Eds.). (2000) *Handbook of Health Economics* (Vol. 1). Amsterdam: Elsevier.

Cutler, D. M. and Zeckhauser, R. J. (2000) *The anatomy of health insurance. In Handbook of Health Economics* (Vol. 1, pp. 563–643). Amsterdam: Elsevier.

Dabla-Norris, Era, Kochhar, Kalpana, Suphaphiphat, Nujin, Ricka, Frantisek and Tsounta, Evridiki (2015) *Causes and Consequences of Income Inequality: A Global Perspective*. Washington, DC: International Monetary Fund.

Dahl, R. A. and Lindblom, Charles E. (1953) *Politics, Economics and Welfare*. New York: Harper and Brothers.

Das, J., Holla, A., Das, V., Mohanan, M., Tabak, D. and Chan, B. (2012) In urban and rural India: A standardized patient study showed low levels of provider training and huge quality gaps. *Health Affairs (Project Hope)*, *31*(12), 2774–2784. doi: 10.1377/hlthaff.2011.1356.

DBS (2016a) Asian insights SparX – Future of Singapore: Singapore Health Care www.dbs.com/insights/uploads/160704_insights_growing_old_with_you.pdf.

(2016b) Future of Singapore: Growing Old With You Group Research, DBS Asian insights number 28. www.dbs.com.sg/treasures/aics/templatedata/article/generic/data/en/GR/082016/160802_insights_future_of_singapore_growing_old_with_you.xml&cid=aio_heroblock_sg_in_02#.

de Savigny, D. and Adam T (Eds.), 2009, *Systems Thinking for Health Systems Strengthening, Geneva, Alliance for Health Policy and Systems Research*; WHO.

Development Research Center of State Council (2005) An evaluation and recommendations on the reforms of the Health System in China. China Development Review (Supplement).

Display fee, bed capacity details outside hospitals (2020) Retrieved 6 August 2020, from www.thehindu.com/news/cities/Delhi/display-fee-bed-capacity-details-outside-hospitals/article31799165.ece.

Does Singapore need a new approach to health care? Comparing SDP's health care policy with 3M health care scheme (2019) The Online Citizen. Retrieved from: www.onlinecitizenasia.com/2019/07/15/does-singapore-need-a-new-approach-to-health care-comparing-sdps-health care-policy-with-3m-health care-scheme/.

Dreze, J. and Sen, A. (2011) Putting growth in its place. Outlook, November 14, 2011.

Drishti, I. A. S. (2019, October 10) Poverty estimation in India. Retrieved 10 August 2020, from www.drishtiias.com/to-the-points/paper3/poverty-estimation-in-india.

Duggal, R. (2009) Sinking flagships and health budgets in India. *Economic and Political Weekly*.

Eggleston, K. (2012) Health care for 1.3 billion: An overview of China's Health System, Asia Health Policy Program working paper # 28.

Eggleston, K., Ling, L., Qingyue, M., Lindelow, M. and Wagstaff, A. (2008a) Health service delivery in China: A literature review. *Health Economics*, *17*(2), 149–165.

Eggleston, K., Shen, Y. C., Lau, J., Schmid, C. H. and Chan, J. (2008b). Hospital ownership and quality of care: what explains the different results in the literature?. *Health Economics*, *17*(12), 1345–1362.

Eggleston, K., Wang, J. and Rao, K. (2008c) From plan to market in the health sector? China's experience. *Journal of Asian Economics*, *19*(5–6), 400–412.

EIU (2014) Health outcomes and cost: A 166-country comparison. Economist Intelligence Unit. https://stateofreform.com/wp-content/uploads/2015/11/Health care-outcomes-index-2014.pdf.

(2017) Value-based health care in Korea. A pioneer in Asia. https://perspectives.eiu.com/sites/default/files/Value-based-health care-in-Korea_1.pdf.

Ellis, R. and McGuire, T. (1990) Optimal payment systems for health services. *Journal of Health Economics*, *9*(4), 375–396.

Ellis, R. P., Alam, M. and Gupta, I. (2000) Health insurance in India: Prognosis and prospectus. *Economic and Political Weekly*, *35*(4), 207–217.

Employees State Insurance Corporation (ESIC) (2016) *Annual Report*. New Delhi: ESIC.

(2018) *Employees State Insurance Scheme: Annual Report*. New Delhi: ESIC.

Ensor, T. and Langenbrunner, J. (2002) Allocating resources and paying providers. In M. McKee, J. Healey and J. Falkingham (Eds.), *Health Care in Central Asia*. Buckingham, England: Open University Press.

Esping-Andersen, G. (Ed.). (1996) *Welfare States in Transition: National Adaptations in Global Economies*. London: Sage.

Evans, T. G., Chowdhury, A. M. R., Evans, D. B., Fidler, A., Lindelow, M., Mills, A. and Scheil-Adlung, X. (2012) Thailand's universal coverage scheme: Achievements and challenges. *An Independent Assessment of the First*, *10*, 2001–2010.

Fan, Shenggen, Kanbur, Ravi, and Zhang, Xiaobo (2011) China's regional disparities: Experience and policy. *Review of Development Finance*, *1* (1), 47–56. doi: 10.1016/j.rdf.2010.10.001.

Fang, Hai (undated). The Chinese Health Care System. International Health Systems Profiles. London: Commonwealth Secretariat. https://international.commonwealthfund.org/countries/china/.

Flynn, Sean M. (2016) Want to ditch ObamaCare? Let's copy Singapore's health care miracle. *Fox News*, November 29. www.foxnews.com/opinion/want-to-ditch-obamacare-lets-copy-singapores-health-care-miracle.

Foo, Patricia K., Lee, Robin S. and Fong, Kyna (2017) Physician prices, hospital prices, and treatment choice in labor and delivery. *American Journal of Health Economics*, *3*(3), 422–453.

Food and Health Bureau (2008) Your health your life: Report on first stage public consultation on health care reform. www.fhb.gov.hk/beStrong/files/consultation/HCR_Report_eng.pdf. Retrieved 11 April 2011.

(2010) *My health my choice: Health care reform second stage consultation document*. My Health My Choice. www.myhealthmychoice.gov.hk/pdf/consultation_full_eng.pdf. Retrieved 3 May 2011.

(2015) Press Release: Steering Committee on Review of HA releases report (with photo/video). Food and Health Bureau, HKSAR Government. Retrieved from www.fhb.gov.hk/en/press_and_publications/press/2015/press150714.htm.

(2017) Hong Kong's Domestic Health Accounts- Estimates of Health Expenditure, 1989/90 – 2016/17 (SHA 2011). Food and Health Bureau, HKSAR Government. Retrieved from www.fhb.gov.hk/statistics/download/dha/en/table2_1617.pdf.

(2018) Hong Kong's Domestic Health Accounts. Estimates of Health Expenditure, 1989/90 – 2017/18 (SHA 2011). Retrieved from www.fhb.gov.hk/statistics/download/dha/en/table2_1718.pdf.

(n.d.) Statistics-Domestic Health Accounts. Food and Health Bureau, HKSAR Government. Retrieved from www.fhb.gov.hk/statistics/en/dha/dha_summary_report.htm.

(2019) About VHIS:Voluntary Health Insurance Scheme. Retrieved from www.vhis.gov.hk/en/about_us/scheme.html.

Fox, A. M. and Reich, M. R. (2013) Political economy of reform. In Alexander S. Preker, Marianne E. Lindner, Dov Chernichovsky, and Onno P. Schellekens (Eds.), *Scaling Up Affordable Health Insurance* (pp. 395–433). Washington, DC: World Bank.

Fox, A. M. and Reich, M. R. (2015) The politics of universal health coverage in low and middle-income countries: A framework for evaluation and action. *Journal of Health Politics, Policy and Law*, *40*, 1023–1060.

Fuchs, V. R. (1978) The supply of surgeons and the demand for operations. Working paper #236. National Bureau of Economic Research. Retrieved from www.nber.org/papers/w0236.

(2011) *Who Shall Live?: Health, Economics and Social Choice*. Singapore: World Scientific.

Gauld, R. and Gould, D. (2002) *The Hong Kong Health Sector: Development and Change*. Hong Kong: The Chinese University Press.

Glied, S. (2000) Managed care. In *Handbook of Health Economics* (Vol. 1, pp. 707–753). Amsterdam: Elsevier.

Global Change Data Lab (n.d.) Coronavirus Pandemic (COVID-19)-the data. Our World in Data-University of Oxford. Retrieved on 25 December 2020, from https://ourworldindata.org/coronavirus-data.

Global Health Observatory (2020) World Health Organization. Retrieved from https://apps.who.int/gho/data/view.main.UHCFINANCIALPROTECTION01v.

Goh, Timothy and Chong, Clara (2020) Experts say S'pore has one of the lowest ICU admission and fatality rates, thanks to effective strategy. *Straits Times*, 31 July 2020. www.straitstimes.com/singapore/health/just-03-of-cases-here-admitted-to-icu.

Government of Hong Kong SAR (2018) Press Release. September 20. www.info.gov.hk/gia/general/201809/20/P2018092000625.htm.

(n.d.) Vital statistics – Center for Health Protection, Government of HKSAR. (n.d.). Retrieved 28 July 2020, from www.chp.gov.hk/en/static/24025.html.

Government of India (2012) *Rural Health Statistics in India in 2012*. New Delhi: Ministry of Health.

(2015) *Rural Health Statistics in India in 2015*. New Delhi: Ministry of Health.

(2018) *Rural Health Statistics in India in 2018*. New Delhi: Ministry of Health.

GovTech (2020) Medisave accounts and balances, Annual. Government of Singapore. Retrieved on 25 December 2020, from https://data.gov.sg/dataset/medisave-accounts-and-balances-annual?view_id=ec34d730–053b-4088-8905-d9fa2b450276&resource_id=f17cb073-d086–4435-b3a7-b430e593433e.

Goyen, M. and Debatin, J. F. (2009) Healthcare costs for new technologies. *European Journal of Nuclear Medicine and Molecular Imaging*, *36*, 139–143. doi: 10.1007/s00259–008-0975-y.

Greener, I. (2002) Understanding NHS reform: The policy-transfer, social learning, and path-dependency perspectives. *Governance*, *15*(2), 161–183.

Gupta, I. and Chowdhury, S. (2014) Public financing for health coverage in India: Who spends, who benefits, and at what cost? *Economic and Political Weekly*, *XLIX*, 59–63.

Haeder, S. F. (2012) Beyond path dependence: Explaining healthcare reform and its consequences. *Policy Studies Journal*, *40*, 65–86.

Haggard, S. and Kaufman, R. (2008) *Development, Democracy, and Welfare States: Latin America, East Asia, and Eastern Europe*. Princeton; Oxford: Princeton University Press. doi: 10.2307/j.ctv10crf4b.

Hai Fang (2020) International Health Care System Profiles: China. Commonwealth Fund. www.commonwealthfund.org/international-health-policy-center/countries/china#:~:text=In%202018%2C%2076.7%20million%20people,funds%20for%20direct%20health%20expenses.

Hannah, A. (2020) Evaluating the role of bricolage in US Health Care policy reform. *Policy & Politics*, *48*(3), 485–502.

Hanvoravongchai, P. (2013) *Health Financing Reform in Thailand: Toward Universal Coverage under Fiscal Constraints,* World Bank UNICO Series #20. Washington, DC: World Bank.

Harris, R. A. and Milkis, S. M. (1989) *The Politics of Regulatory Change: A Tale of Two Agencies*. USA: Oxford University Press.

Hartley, Kris and Jarvis, Darryl S. L. (2020) Policymaking in a low-trust state: Legitimacy, state capacity, and responses to COVID-19 in Hong Kong. *Policy and Society*, *39*(3), 403–423. doi: 10.1080/14494035.2020.1783791.

Harvard Team (1999) *Improving Hong Kong's Health Care System: Why and For Whom?*. Hong Kong: HKSAR Government Printing Department.

Haseltine, William A. (2013) *Affordable Excellence: The Singapore Health Care Story*. Singapore: NUS Press. https://nuspress.nus.edu.sg/products/affordable-excellence-the-singapore-health care-story.

He, Alex Jingwei (2018) Public satisfaction with the health system and popular support for state involvement in an East Asian Welfare Regime: Health Policy Legitimacy of Hong Kong. *Social Policy & Administration*, *52*(3), 750–770. doi: 10.1111/spol.12274.

Health and Welfare Bureau (2000) *Lifelong Investment in Health: Consultation Document on Health Care Reform*. Hong Kong: Printing Department of the Government of the Hong Kong Special Administrative Region.

Health Insurance Task Force (2016) Managing the Cost of Health Insurance in Singapore. Life Insurance Association, Singapore. Retrieved from www.lia.org.sg/media/1521/managingsingaporehealthinsurancecost_hitf_20161013.pdf.

"Healthy life expectancy (HALE)" (n.d.). Retrieved 11 August 2020, from https://photius.com/rankings/healthy_life_table2.html.

Heller, Peter S. (1999) Ageing in Asia: Challenges for fiscal policy. *Journal of Asian Economics*, *10*(1), 37–63.

Himmelstein, D. U., Thorne, D., Warren, E. and Woolhandler, S. (2009) Medical bankruptcy in the United States, 2007: Results of a national study. *American Journal of Medicine*, *122*(8), 741–746.

Ho Kyeong, Jang (2018) Mooncare: South Korea's proposed health insurance reform in full. *Korea Expose*, May 31, 2018. www.koreaexpose.com/mooncare-south-korea-proposed-health-insurance-reform/.

Hollaway, K. (2012) *Drug Policy and the Use of Pharmaceuticals in Health Care Delivery in Thailand*. New Delhi: World Health Organization Regional Office for Southeast Asia. www.searo.who.int/entity/medicines/thailand_situational_analysis.pdf.

Hood, C. (1983) *The Tools of Government*. London: Macmillan.

(2007) Intellectual obsolescence and intellectual makeovers: Reflections on the tools of government after two decades. *Governance*, *20*(1), 127–144.

Hood, C. and Dixon, R. (2015) What we have to show for 30 years of new public management: Higher costs, more complaints. *Governance*, *28*(3), 265–267.

Howlett, M. (2019) *Designing Public Policies: Principles and Instruments*. London: Routledge.

Howlett, M. and Del Rio, P. (2015) The parameters of policy portfolios: Verticality and horizontality in design spaces and their consequences for policy mix formulation. *Environment and Planning C: Government and Policy*, *33*(5), 1233–1245.

Howlett, M. and Lejano, R. P. (2013) Tales from the crypt: The rise and fall (and rebirth?) of policy design. *Administration & Society*, *45*(3), 357–381.

Howlett, M. and Rayner, J. (2013) Patching vs packaging in policy formulation: Assessing policy portfolio design. *Politics and Governance*, *1*(2), 170–182.

Howlett, M. and Tosun, J. (Eds.). (2018) *Policy Styles and Policy-Making: Exploring the Linkages*. Routledge.

Howlett, M., Mukherjee, I. and Rayner, J. (2017) The elements of effective program design: A two-level analysis. In *Handbook of Policy Formulation*. Cheltenham: Edward Elgar Publishing.

Howlett, M., Mukherjee, I. and Woo, J. J. (2015) From tools to toolkits in policy design studies: The new design orientation towards policy formulation research. *Policy & Politics*, *43*(2), 291–311.

Howlett, M., Ramesh, M. and Perl, A. (2009) *Studying Public Policy: Policy Cycles and Policy Subsystems* (Vol. 3). Oxford: Oxford University Press.

(2020) *Studying Public Policy: Policy Cycles and Policy Subsystems* (Vol. 4). Oxford: Oxford University Press.

Hsiao, W. C. and Shaw, P. W. (2007) *Social Health Insurance for Developing Nations*. Washington, DC: World Bank Publications.

Hsiao, W. C. and Shaw, R. P. (Eds.). (2007) *Social Health Insurance for Developing Nations*. Washington, DC: The World Bank.

Hu, G., Chen, Y., Liu, Q. et al. (2019) Patient experience of hospital care in China: Major findings from the Chinese Patient Experience Questionnaire Survey (2016–2018). *BMJ Open*, *9*(9), e031615.

Hu, G., Ding, H., Liu, S., Wang, Z., Hu, G. and Liu, Y. (2020) Influence of patient and hospital characteristics on inpatient satisfaction in China's tertiary hospitals: A cross-sectional study. *Health Expectations*, *23*, 115–124. doi: 10.1111/hex.12974.

Huntington, D. and Hort, K. (2015) Public Hospital Governance in Asia and the Pacific, Vol. 1, Number 1. Asia Pacific Observatory on Health Systems and Policies.

Institute for Health Metrics and Evaluation (2010) *GBD Profile: India. Global burden of diseases, injuries, and risk factors study 2010*. Institute for Health Metrics and Evaluation. Retrieved from www.healthdata.org/sites/default/files/files/country_profiles/GBD/ihme_gbd_country_report_india.pdf.

Institute for Health Policy and Systems Research (2015) Submission on the consultation document on Voluntary Health Insurance Scheme (VHIS). 15 May 2015.

Jacobs, D. (2000) Low Inequality with Low Redistribution? An Analysis of Income Distribution in Japan, South Korea, and Taiwan Compared to Britain, *Casepaper* 33, London: Centre for Analysis of Social Exclusion, London School of Economics.

Jeon, B. and Kwon, S. (2013) Effect of private health insurance on health care utilization in a universal public insurance system: A case of South Korea. *Health Policy*, *113*(1–2), 69–76.

Jeong, H. S. and Niki, R. (2012) Divergence in the development of public health insurance in Japan and the Republic of Korea: A multiple-payer versus a single payer system. *International Social Security Review*, *65*(2), 51–73.

Jochim, A. E. and May, P. J. (2010) Beyond subsystems: Policy regimes and governance. *Policy Studies Journal*, *38*(2), 303–327.

Jongudomsuk, P., Srithamrongsawat, S., Patcharanarumol, W. et al. (2015) *The Kingdom of Thailand Health System Review*. Health Systems in Transition, Vol. 5. Manila: WHO Regional Office for the Western Pacific.

Jose, Annmary (2019) India's regional disparity and its policy responses. *Journal of Public Affairs*. doi: 10.1002/pa.1933.

Kapoor, G., Sriram, A., Joshi, J., Nandi, A. and Laxminarayan, R. (2020) COVID-19 in India: State-wise estimates of current hospital beds, intensive care unit (ICU) beds and ventilators. The Center For Disease Dynamics, Economics & Policy. Retrieved from https://cddep.org/wp-content/uploads/2020/04/State-wise-estimates-of-current-beds-and-ventilators_24Apr2020.pdf.

Karan, A., Selvaraj, S. and Mahal, A. (2014) Moving to universal coverage? Trends in the burden of out-of-pocket payments for health care across social groups in India, 1999–2000 to 2011–12. *PLoS One*, *9*(8), e105162. doi: 10.1371/journal.pone.0105162.

Kathleen, F. (2019) Does Singapore need a new approach to health care? Comparing SDP's health care policy with 3M health care scheme, *The Online Citizen*, 15 July 2019, www.theonlinecitizen.com/2019/07/15/does-singapore-need-a-new-approach-to-health care-comparing-sdps-health care-policy-with-3m-health care-scheme/.

Keane, M. and Thakur, R. (2018) Health care spending and hidden poverty in India. *Research in Economics*, 72(4), 435–451.

Kekez, Anka, Howlett, Michael and Ramesh, M. (2018) Varieties of collaboration in public service delivery. *Policy Design and Practice*, *1*(4), 243–252. doi: 10.1080/25741292.2018.1532026.

Khaw, B. W. (2005) The best health care that Singaporeans can afford. Speech in Parliament. Retrieved 8 March 2005, from www.moh.gov.sg/corp/about/newsroom/speeches/details.do?id=30507042.

Khoo, H. S., Lim, Y. W. and Vrijhoef, H. J. (2014) Primary health care system and practice characteristics in Singapore. *Asia Pacific Family Medicine*, *13*(1), 8. doi: 10.1186/s12930-014-0008-x.

Kim, S. J., Kim, S. J., Han, K.-T. and Park, E.-C. (2017) Medical costs, Cesarean delivery rates, and length of stay in specialty hospitals vs. non-specialty hospitals in South Korea. *PLoS One*, *12*(11), e0188612.

Kim, A. M., Cho, S., Kim, H. J., Jung, H. et al. (2018) Primary care patients' preference for hospitals over clinics in Korea. *International Journal of Environment Research*, *15*(6), 1119.

Kongsawat, S. (2000) *The Evaluation of the Sixth Round Implementation of Low-Income Card for the Poor Households*. Nonthaburi: Ministry of Public Health.

Konisky, David and Teodoro, Manuel P. (2015) When governments regulate governments (July 1, 2015). *American Journal of Political Science*. doi: 10.1111/ajps.12221.

Kozhimannil, K. B., Graves, A. J., Ecklund, A. M., Shah, N., Aggarwal, R. and Snowden, J. M. (2018) Cesarean delivery rates and costs of child-birth in a state Medicaid program after implementation of a blended payment policy. *Medical Care*, *56*(8), 658–664.

Kumar, N. and Rani, R. (2019) Regional disparities in social development: Evidence from states and Union Territories of India. *South Asian Survey*, 26(1), 1–27. doi: 10.1177/0971523118825388.

Kumar, R. and Pal, R. (2018) India achieves WHO recommended doctor population ratio: A call for paradigm shift in public health discourse!. *Journal of Family Medicine and Primary Care*, 7(5), 841.

Kumar, S. G., Adithan, C., Harish, B. N., Sujatha, S., Roy, G. and Malini, A. (2013) Antimicrobial resistance in India: A review. *Journal of Natural Science, Biology, and Medicine*, 4(2), 286.

Kwon, Soonman (2009) Thirty years of national health insurance in South Korea: Lessons for achieving universal health care coverage. Health Policy and Planning, 24, 63–71. doi: 10.1093/heapol/czn037.

(2018) *Advancing Universal Health Coverage: What Developing Countries Can Learn from the Korean Experience?* Universal Health Coverage Study Series No. 33. Washington, DC: The World Bank.

(2020) *Covid-19: Lessons from South Korea*. Health Systems Global. Available online at https://healthsystemsglobal.org/news/covid-19-lessons-from-south-korea/.

Kyu, H. H., Abate, D., Abate, K. H., Abay, S. M., Abbafati, C., Abbasi, N. and Abdollahpour, I. (2018) Global, regional, and national disability-adjusted life-years (DALYs) for 359 diseases and injuries and healthy life expectancy (HALE) for 195 countries and territories, 1990–2017: A systematic analysis for the Global Burden of Disease Study 2017. *The Lancet*, *392*(10159), 1859–1922.

La Forgia, G. and Nagpal, S. (2012) *Government-Sponsored Health Insurance In India: Are You Covered?* Washington, DC: World Bank.

Laffont, Jean-Jacques and Tirole, Jean (1993) *A Theory of Incentives in Procurement and Regulation*. Cambridge, MA, MIT Press.

Lai, Wei Lin. (2016) Paying for Health care. In Lee Chien Earn and K. Satku (Eds.), *Singapore's Health Care System: What 50 Years Have Achieved*. Singapore: World Scientific.

Lam, Calvin (2015) Is there a doctor in the house?. *Harbour Times*, October 13, 2015. http://harbourtimes.com/2015/10/13/is-there-a-doctor-in-the-house/.

Landy, Marc K., Roberts, Marc J. and Thomas, Stephen R. (1990) *The Environmental Protection Agency: Asking the Wrong Questions*. Oxford: Oxford University Press.

Langenbrunner, J. C. and Liu, X. (2005) How to pay? Understanding and using payment incentives. In Alexander Preker and John Langenbrunner (eds). *Spending Wisely* (pp. 89–106). Washington, DC: World Bank.

Langenbrunner, J. C. and Somanathan, A. (2011) *Financing Health Care in East Asia and the Pacific: Best Practices and Remaining Challenges*. Washington, DC: The World Bank.

Langenbrunner, J. C. and Wiley, M. M. (2002) Hospital payment mechanisms: Theory and practice in transition countries. In Martin McKee and Judith Healy (Eds.), *Hospitals in a Changing Europe* (pp. 150–176). Buckingham: Open University Press.

Langenbrunner, J. C., O'Duagherty, S. and Cashin, C. S. (Eds.). (2009) *Designing and Implementing Health Care Provider Payment Systems: " How-to" Manuals*. Washington, DC: The World Bank.

Law, C. K. and Yip, P. S. F. (2003) Healthy life expectancy in Hong Kong Special. *Bulletin of the World Health Organization*, *81*(1), 43–47.

Lee, V. W. Y., Cheng, F. W. T., Fong, F. Y. H. et al. (2018) Branded versus generic drug use in chronic disease management in Hong Kong – Perspectives of health care professionals and the general public. *Hong Kong Medical Journal*, *24*(6), 554–560. doi: 10.12809/hkmj177087.

Lee, Sabinne, Hwang, Changho and Jae Moon, M. (2020) Policy learning and crisis policy-making: Quadruple-loop learning and COVID-19 responses in South Korea. *Policy and Society*, *39*(3), 363–381. doi: 10.1080/14494035.2020.1785195.

Legislative Council (2005) Panel on Health Services. Background Paper Prepared by the Legislative Council Secretariat for the Meeting on 19 July 2005 – Health Care Reform; accessed at www.legco.gov.hk/yr04-05/english/panels/hs/papers/hs0719cb2–2252-1e.pdf.

(2014) Panel on Health Services- Updated background brief prepared by the Legislative Council Secretariat for the meeting on 20 January 2014. Resources allocation among hospital clusters by the Hospital Authority. LC Paper No. CB(2)671/13-14(05). Legislative Council, HKSAR Government. Retrieved from www.legco.gov.hk/yr13-14/english/panels/hs/papers/hs0120cb2-671-5-e.pdf.

Explainer: Why Singapore's Covid-19 death rate is the world's lowest, Today, 17 September 2020. www.todayonline.com/singapore/why-singapores-covid-19-death-rate-worlds-lowest-0.

Lagasse, Jeff (2020) COVID-19 impact on hospitals worse than previously estimated. Health Care Finance, 1 June 2020. www.healthcarefinancenews.com/news/covid-19-impact-hospitals-worse-previously-estimated#node-140145:~:text=Modeling%20estimates%20based%20on%20Chinese%20data,had%20been%20predicted.%20By%20April%209.

Li, X. (2017) The primary health-care system in China. *Lancet*, *390*(10112), 2584–2594. doi: 10.1016/S0140-6736(17)33109-4.

Li, Shantong and Xu, Zhaoyuan (2008) *The Trend of Regional Income Disparity in the People's Republic of China*. ADBI Discussion Paper 85. Tokyo: Asian Development Bank Institute. www.adb.org/sites/default/files/publication/156724/adbi-dp85.pdf.

Limwattananon, S., Neelsen, S., O'Donnell, O., Prakongsai, P., Tangcharoensathien, V. and van Doorslaer, E. (2013) *Universal Coverage on a Budget: Impacts on Health Care Utilization and Out-of-Pocket Expenditures in Thailand*. Tinbergen Institute Discussion Paper (TI 2013-067/V).

Limwattananon, S., Tangcharoensathien, V. and Prakongsai, P. (2005) Equity in Financing Health care: Impact of universal access to health care in Thailand. Equitap Project Working Paper # 16.

List of countries by government spending as percentage of GDP (n.d.) Wikipedia. https://en.wikipedia.org/wiki/List_of_countries_by_government_spending_as_percentage_of_GDP.

Liu, C. and Haseltine, W. (2015) 'The Singaporean Health Care System'. Duke-NUS Graduate Medical School and ACCESS Health International. http://international.commonwealthfund.org/countries/singapore/.

Liu, Gordon G., Samantha, A. V. and Xuezhi, H. (2017) China's health reform update. *Annual Review of Public Health*, *38*, 431–448.

Liu, J. (2009) How to achieve universal coverage in China: Study on the impact factors of development of social medical insurance system. *Comparative Economic and Social Systems*, *2*, 115–122.

Liu, X. and Hsiao, W. C. L. (1995) The cost escalation of social health insurance plans in China: Its implication for public policy. *Social Science & Medicine*, *41*(8), 1095–1101.

Liu, X. and Yi, Y. (2004) *The Health Sector in China: Policy and Institutional Review*. Washington, DC: World Bank Background. http://siteresources.worldbank.org/INTEAPREG-TOPHEANUT/.

Liu, X., Liu, Y. and Chen, N. (2000) The Chinese experience of hospital price regulation. *Health Policy Planning*, *15*(2), 157–163.

Ma, X., Wang, H., Yang, L., Shi, L. and Liu, X. (2019) Realigning the incentive system for China's primary health care providers. *BMJ*, *365*, l2406.

Malhotra, S., Patnaik, I., Roy, S. and Shah, A. (2018) Fair play in Indian health insurance. Available at SSRN 3179354.

Marmor, R., Freeman, R. and Okma, K. (2005) Comparative perspectives and policy learning in the World of Health Care. *Journal of Comparative Policy Analysis*, 7(4), 331–448.

Mathauer, Inke and Xu, Ke (2009) *An Analysis of the Health Financing System of the Republic of Korea and Options to Strengthen Health Financing Performance*. Geneva: World Health Organization. www.who.int/health_financing/documents/hsfr_e_09-korea.pdf.

Maurya, D. S. (2015) Inter-organizational relations in public-private partnerships: National health insurance in India, Doctoral thesis. National University of Singapore.

Maurya, D. and Mintrom, M. (2020). Policy entrepreneurs as catalysts of broad system change: The case of social health insurance adoption in India. *Journal of Asian Public Policy*, 13(1), 18–34.

Maurya, D. and Ramesh, M. (2019) Program design, implementation and performance: The case of social health insurance in India. *Health Economics, Policy and Law*, *14*(4), 487–508.

May, P. J., Sapotichne, J. and Workman, S. (2006) Policy coherence and policy domains. *Policy Studies Journal*, *34*(3), 381–403.

McGuire, T. G. (2000) *Physician agency. In Handbook of Health Economics* (Vol. 1, pp. 461–536). Amsterdam: Elsevier.

McIntyre, D. (2007) *Learning from Experience: Health Care Financing in Low and Middle Income Countries*. Geneva: Global Forum for Health Research.

McKee, M., Balabanova, D., Basu, S., Ricciardi, W. and Stuckler, D. (2013) Universal health coverage: A quest for all countries but under threat in some. *Value in Health*, *16*(1), S39–S45.

McPherson, K., Strong, P. M., Epstein, A. and Jones, L. (1981) Regional variation in the use of common surgical procedures. *Social Sciences and Medicine*, *15A*, 273–288.

Mei, Ciqi (2020) Policy style, consistency and the effectiveness of the policy mix in China's fight against COVID-19. *Policy and Society*, *39*(3), 309–325. doi: 10.1080/14494035.2020.1787627.

Meng, Q., Li, L. and Eggleston, K. (2008) Health service delivery in China: A literature review. *Health Economics*, *17*(2), 149–165.

Mercurio, Bryan (2019) *Drugs Patents and Policy: A Contextual Study of Hong Kong*. Cambridge: Cambridge University Press.

Miller, L. and Lu, W. (2018, September 19) These are the economies with the most (and Least) efficient health care. Retrieved 28 July 2020, from www.bloomberg.com/news/articles/2018-09-19/u-s-near-bottom-of-health-index-hong-kong-and-singapore-at-top.

Milliman Limited (2010a) Feasibility study on the key features of the health protection scheme. In Study Report- Information Centre- Health Care Reform Second Stage Public Consultation. Retrieved 5 May 2011 from www.myhealthmychoice.gov.hk/pdf/studyreport/hps_feasibility_study.pdf.

(2010b) Review of literatures on the theoretical framework on analyzing the role of private health insurance in a health care system. Retrieved 5 May 2011 from www.myhealthmychoice.gov.hk/pdf/studyreport/insurance_background_research.pdf.

Ministry of Health (2018) Fee Benchmarks Advisory Committee Report. Retrieved from www.moh.gov.sg/docs/librariesprovider5/pressroom/press-releases/fee-benchmarks-advisory-committee-report.pdf.

Ministry of Statistics and Programme Implementation (MOSPI) (2006) *Morbidity, Health Care and the Condition of the Aged 2004*. New Delhi: Government of India. http://mospi.nic.in/sites/default/files/publication_reports/507_final.pdf.

Mishra, S. and Mohanty, S. K. (2019) Out-of-pocket expenditure and distress financing on institutional delivery in India. *International Journal for Equity in Health*, *18*(1), 99.

Mitnick, B. M. (1981) The strategic uses of regulation – And deregulation. *Business Horizons*, *24*(2), 71–83.

MOPH (2014) Annual Report, 2014. Bangkok, Thailand.

(2017) Annual Report, 2017. Bangkok, Thailand.

(2019) Annual Report, 2019. Bangkok, Thailand.

MOH Singapore (1993) Affordable Health Care: A White Paper. Ministry of Health, Singapore. Retrieved from www.moh.gov.sg/resources-statistics/information-paper/affordable-health-care-(1993-white-paper).

(2015) Survey Of Patient Satisfaction With Public Health care Institutions. News Highlights 13 June 2016. www.moh.gov.sg/news-highlights/details/2015-survey-of-patient-satisfaction-with-public-health care-institutions.

(2016) *Lfe Expectancy in Singapore*. Singapore: Ministry of Health. Retrieved from www.moh.gov.sg/content/dam/moh_web/PressRoom/Highlights/2016/cos/Life%20Expectancy%20in%20Singapore%20Information%20Note_120416.pdf.

MOH (n.d.) Hospital admission rate by age and sex. Ministry of Health, Singapore. Retrieved from https://data.gov.sg/dataset/hospital-admission-rate-by-age-and-sex?view_id=50d36d2c-0099-471e-8445-db8cc03135de&resource_id=9f9ca3a2–0e0f-4e3c-bc23-a9b4533dabd2.

Mohanan, M., Hay, K. and Mor, N. (2016) Quality of health care in India: Challenges, priorities, and the road ahead. *Health Affairs*, *35*(10), 1753–1758.

MOHW (n.d.) South Korea – Medical Devices & Diagnostics: Global Health Technology Assessment Road Map. Retrieved 23 July 2020, from https://tools.ispor.org/HTARoadMaps/S-KoreaMD.asp.

Mok, K. H. and Lau, M. (2002) Changing government role for socio-economic development in Hong Kong in the twenty-first century. *Policy Studies*, *23*(2), 107–124.

Mor, N. and Yip, Winnie (2015) What role should governments play in healthcare? www.weforum.org/agenda/2015/12/what-role-should-governments-play-in-healthcare/.

Moran, M. (2000) Understanding the welfare state: The case of health care. *The British Journal of Politics and International Relations*, 2(2), 135–160.

Mounier-Jack, S., Griffiths, U. K., Closser, S., Burchett, H. and Marchal, B. (2014) Measuring the health systems impact of disease control programmes: A critical reflection on the WHO building blocks framework. *BMC Public Health*, *14*, 278. doi: 10.1186/1471-2458-14-278.

Mukherjee, I. and Bali, A. (2018) Capacities and customization in policy design. In M. Howlett and Ishani Mukherjee (Eds.), *Routledge Handbook of Policy Design*. London: Routledge.

Murray, C. J., Barber, R. M., Foreman, K. J. et al. (2015) Global, regional, and national disability-adjusted life years (DALYs) for 306 diseases and injuries and healthy life expectancy (HALE) for 188 countries, 1990–2013: Quantifying the epidemiological transition. *The Lancet*, *386*(10009), 2145–2191.

Musgrave, P. and World Bank (1996) *Public and Private Roles in Health: Theory and Financing Patterns*. Washington, DC: World Bank.

Na, S. and Kwon, S. (2015) *Building Systems for Universal Health Coverage in South Korea*. Health, Nutrition and Population (HNP) Discussion Paper. Washington, DC: World Bank.

Nagral, S. and Nundy, S. (2017) We need to end "cut" practice in Indian health care. *The BMJ Opinion*. Retrieved from https://blogs.bmj.com/bmj/2017/07/07/we-need-to-end-cut-practice-in-indian-health care/.

New, William (2019) Progress on Universal Health Coverage Lagging: WHO Report Released Ahead of High Level UHC Meeting. Health Policy Watch, 22/09/2019. https://healthpolicy-watch.news/un-members-to-decide-on-faster-progress-better-financial-protections-for-universal-health-coverage/.

Ng, Irene Y. H. (2018) Definitions and Measurements of Poverty. Published in SSR-Snippet, Cutting Edge Applied Research. Issue 2. Retrieved from www.fas.nus.edu.sg/ssr/PDF%20Format/SSR%20Snippets/Vol%202/Snippet%202_final.pdf.

Nguyen, H. T. H., Bales, S., Wagstaff, A. and Dao, H. (2017) Getting incentives right? The impact of hospital capitation payment in Vietnam. *Health Economics*, *26*(2), 263–272.

NHSO (2014) Annual Fiscal Report, 2014. Bangkok, Thailand.

Nolte, E. and McKee, M. (2004) Does healthcare save lives? Avoidable mortality revisited. Research report. Nuffield Trust. www.nuffieldtrust.org.uk/files/2017-01/does-healthcare-save-lives-web-final.pdf.

OECD (2012) Health Care Quality Review: Korea. Assessment and Recommendations. Health Division, Directorate for Employment, Labour and Social Affairs. www.oecd.org/health/qualityreviews.

(2016) Health policy in Korea. www.oecd.org/korea/Health-Policy-in-Korea-April-2016.pdf.

(2018) *Health at a Glance-2018*. Paris: OECD. Retrieved from www.oecd-ilibrary.org/social-issues-migration-health/health-at-a-glance-asia-pacific-2018_health_glance_ap-2018-en;jsessionid=-BSIbNoGWYO32jQs04GbN6xP.ip-10-240-5-102.

(2019) *Health at a Glance-2019*. Paris: OECD. Retrieved from: doi: 10.1787/888934017747.

OECD/WHO (2018) *Health at a Glance: Asia/Pacific 2018: Measuring Progress towards Universal Health Coverage*. Paris: OECD Publishing. doi: 10.1787/health_glance_ap-2018-en.

Ooi, E. W. (2005) *The World Bank's Assistance to China's Health Sector*. Washington, DC: The World Bank Operations Evaluation Department.

Osental, Duffie (2018) Private health insurance still booming in South Korea. Insurance Business Asia, 8 December 2018. www.insurancebusinessmag.com/asia/news/breaking-news/private-health-insurance-still-booming-in-south-korea-118259.aspx.

Oster, E. and McClelland, Spencer (2019) Why the C-Section Rate Is So High? The Atlantic. 17 October 2019. www.theatlantic.com/ideas/archive/2019/10/c-section-rate-high/600172/.

Pacific Prime 2018 Cost of International Health Insurance (2018) www.pacificprime.com/blog/cost-of-health-insurance-2018-report-release.html.

Panagariya, A. (2008a) *India: The Emerging Giant*. New Delhi: Oxford University Press.

(2008b) India: The crisis in rural health care. *The Brookings Institution*, *24*.

Patcharanarumol, W., Panichkriangkrai, W., Sommanuttaweechai, A., Hanson, K., Wanwong, Y. and Tangcharoensathien, V. (2018) Strategic purchasing and health system efficiency: A comparison of two financing schemes in Thailand. *PLoS One*, *13*(4), e0195179.

Patnaik, I., Shah, A. and Roy, S. (2018) *The Rise of Government-Funded Health Insurance in India*. Chicago: National Institute of Public Finance and Policy.

Paul, P., Tong, Y. T. and Narayanan, S. (2015) *Caring for Our People: 50 Years of Health Care in Singapore*. Singapore: MOH Holdings for the Ministry of Health.

Pauly, M. V. (2009) *Doctors and Their Workshops: Economic Models of Physician Behavior*. Chicago: University of Chicago Press.

Pauly, M. V., McGuire, T. G. and Barros, P. P. (Eds.). (2012) *Handbook of Health Economics* (Vol. 2). Amsterdam: Elsevier.

Perneger, Thomas V., Peytremann-Bridevaux, Isabelle and Combescure, Christophe (2020) Patient satisfaction and survey response in 717 hospital surveys in Switzerland: A cross-sectional study. *BMC Health Services Research*, *20*(1), 1–8.

Peters, B. G. (2004) Back to the centre? Rebuilding the state. *The Political Quarterly*, *75*, 130–140.

(2015) *Pursuing Horizontal Management. The Politics of Public Sector Coordination*. Kansas: University Press of Kansas.

(2018a) The challenge of policy coordination. *Policy Design and Practice*, *1*(1), 1–11.

(2018b) *Policy Problems and Policy Design*. Cheltenham: Edward Elgar Publishing.

Phua, Kai Hong (1990) *Privatization and Restructuring of Health Services in Singapore*. Singapore: Institute of Policy Studies.

Pierson, P. (2000) Increasing returns, path dependence, and the study of politics. *American Political Science Review*, 251–267.

Pitayarangsarit, S. and Tangcharoensathien, V. (2009) Sustaining capacity in health policy and systems research in Thailand. *Bulletin of the World Health Organization*, *87*(1), 72–74.

Poverty in China (2020) Retrieved 10 August 2020, from https://en.wikipedia.org/wiki/Poverty_in_China.

Poverty in South Korea (2020) https://en.wikipedia.org/wiki/Poverty_in_South_Korea.

Poverty Situation – Overview: Census and Statistics Department (2019, December 13). Retrieved 10 August 2020, from www.censtatd.gov.hk/hkstat/sub/so461.jsp.

Preker, A. S. and Harding, A. (2003) A conceptual framework for the organizational reforms of hospitals. In A. Harding and A. S. Preker (Eds.), *Innovations in Health Service Delivery: The Corporatization of Public Hospitals* (pp. 23–78). Washington, DC: World Bank.

Preker, A. S., Lindner, M. E., Chernichovsky, D. and Schellekens, O. P. (Eds.). (2013) *Scaling Up Affordable Health Insurance: Staying the Course*. Washington, DC: The World Bank.

Preker, Alexander S. and Langebrunner, John C. (Eds.). (2005) *Spending Wisely: Buying Health Services for the Poor* (pp. 89–106). Washington DC: World Bank.

Preker, Alexander S and Langebrunner, John C. (2005) *Spending Wisely: Buying Health Services for the Poor*. Washington, DC: World Bank.

Primary Care Networks (undated). Ministry of Health, Singapore. Retrieved from www.moh.gov.sg/our-health care-system/health care-services-and-facilities/primary-care-networks.

Prinja, S., Bahuguna, P., Gupta, I., Chowdhury, S. and Trivedi, M. (2019) Role of insurance in determining utilization of health care and financial risk protection in India. *PLoS One*, *14*(2), e0211793.

The Print (2020) 99% of Mumbai ICU beds are occupied, BMC data shows. The Print, 29 May, 2020. Retrieved from https://theprint.in/india/99-of-mumbai-icu-beds-are-occupied-bmc-data-shows/431621/.

PTI (2018) India 145th among 195 countries in health care access, quality: Lancet. *Times of India*. Retrieved from https://timesofindia.indiatimes.com/india/india-145th-among-195-countries-in-health care-access-quality-lancet/articleshow/64283179.cms?utm_source=contentofinterest&utm_medium=text&utm_campaign=cppstPTI (2019, July 04). India has one doctor for every 1,457 citizens: Govt. Retrieved 6 August 2020, from www.business-standard.com/article/pti-stories/india-has-one-doctor-for-every-1-457-citizens-govt-119070401127_1.html.

(2019, September 30) 111 hospitals listed on AB-PMJAY website to 'name and shame' them: Vardhan. Retrieved 6 August 2020, from https://health.economictimes.indiatimes.com/news/hospitals/111-hospitals-listed-on-ab-pmjay-website-to-name-and-shame-them-vardhan/71380014.

Pwee, K. H. (2009) Health technology assessment in Singapore. *International Journal of Technology Assessment in Health Care*, 25(Supplement 1), 234–240.

Qian, J. (2017) Reforming Public Hospitals In China. EAI Background Brief No. 1236. 13 April 2017.

Qian, Jiwei (2015) Reallocating authority in the Chinese health system: An institutional perspective. *Journal of Asian Public Policy*, *2015*. doi: 10.1080/17516234.2014.1003454.

Quah, Elizabeth and Siong, Neo Boon (2016) Evolving the governance of public health care institutions – A continuous improvement journey. In C. E. Lee and K. Satku (Eds.), *Singapore's Health Care System: What 50 Years Have Achieved* (pp. 57–74). Singapore: World Scientific.

Radwan, I. (2005) *India – Private Health Services For the Poor, Report, Health, Nutrition and Population Discussion Paper*. Washington, DC: World Bank.

Raghavan, P., Barnagarwala, T. and Ghosh, A. (2020) COVID fight: Govt system in front, private hospitals do the distancing. The Indian Express. April 30, 2020.

Ramesh, M. (2004) *Social Policy in East and Southeast Asia: Education, Health, Housing and Income Maintenance*. London: RoutledgeCurzon, ISBN: 0415332559.

(2008a) Reasserting the role of the state in the health care sector: Lessons from Asia. *Policy and Society*, 27(2), 129–136.

Ramesh, M. (2008b) Competition and control in public hospital reforms in Singapore. *American Review of Public Administration*, *38*(1), 62–79.

(2008c) Autonomy and control in public hospital reforms in Singapore. *American Review of Public Administration*, *38*(1), 62–79.

(2009) Economic crisis and its social impacts lessons from the 1997 Asian economic crisis. *Global Social Policy*, 9(1 suppl), 79–99. doi: 10.1177/1468018109106886.

Ramesh, M. (2012) Health care reform in Hong Kong: The politics of liberal non-democracy. *The Pacific Review*, *25*(4), 455–471.

(2013) Health care reform in Vietnam: Chasing shadows. *Journal of Contemporary Asia*, *43*(3), 399–412.

Ramesh, M. and Araral, E. (2010) *Reasserting the Public in Public Services: New Public Management Reforms*. London: Routledge.

Ramesh, M. and Asher, M. G. (2000) *Welfare Capitalism in Southeast Asia: Social Security, Health and Education Policies*. London: Macmillan.

Ramesh, M. and Wu, Xun (2009) Health policy reform in China: Lessons from Asia. *Social Science and Medicine*, *68*(12), 2256–2262.

Ramesh, M., Howlett, M. P. and Saguin, K. (2016) Measuring individual-level analytical, managerial and political policy capacity: A survey instrument.

Ramesh, M., Wu, Xun and He, Jingwei (2013) Health governance and health care reforms in China. *Health Policy and Planning*. doi: 10.1093/heapol/czs109.

Ramesh, M., Wu, X. and He, A. J. (2014) Health governance and healthcare reforms in China. *Health Policy and Planning*, *29*(6), 663–672.

Ramesh, M., Wu, X. and Howlett, M. (2015) Second best governance? Governments and governance in the imperfect world of health care delivery in China, India and Thailand in comparative perspective. *Journal of Comparative Policy Analysis: Research and Practice*, *17*(4), 342–358.

Rana, R. H., Alam, K. and Gow, J. (2020) Selection of private or public hospital care: Examining the care-seeking behaviour of patients with private health insurance. *BMC Health Services Research*, *20*, 1–17.

Rao, M. G. (2019) Redesigning the fiscal transfer system in India. *Economic & Political Weekly*, *54*(31), 52–60.

Rao, M. G. and Choudhury, M. (2012). *Health Care Financing Reforms in India*. New Delhi: National Institute of Public Finance and Policy.

Rao, S. (2015) Inter-State Comparisons on Health Outcomes in Major States and a Framework for Resource Devolution for Health, Background Study for the 14th Finance Commission, Government of India, New Delhi, India.

(2017) *Do We Care? India's Health Sysytem*. New Delhi: Oxford University Press.

Roberts, M., Hsiao, W., Berman, P. and Reich, M. (2003) *Getting Health Reform Right: A Guide to Improving Performance and Equity*. Oxford: Oxford University Press.

Roberts, M. J., Hsiao, W. C., Berman, P. and Reich, M. R. (2003) *Getting Health Reform Right*. New York: Oxford University Press.

Russell, S. (2004) The economic burden of illness for households in developing countries: A review of studies focusing on malaria, tuberculosis, and human immunodeficiency virus/acquired immunodeficiency syndrome. *American Journal of Tropical Medicine and Hygiene*, *71*(Supp. 2), 147–155.

Sabel, Charles F. and Zeitlin, Jonathan (2012) Experimentalist governance. In David, Levi-Faur (Ed.), *Oxford Handbook of Governance*. Oxford: Oxford University Press.

Salgado, R. (2017) Asia: At risk of growing old before becoming rich?, Regional Economic Outlook, April 2017. Accessed 15 August 2020.

Samans, R., Blanke, J., Drzeniek, M. and Corrigan, G. (2018) The Inclusive Development Index 2018 Summary and Data Highlights. Geneva, Switzerland: World Economic Forum. In World Economic Forum. URL: www3.weforum. org/docs/WEF_Forum_IncGrwth_2018.pdf (дата обращения: 11.07. 2018).

Sangar, S., Dutt, V. and Thakur, R. (2019) Distress financing of out-of-pocket health expenditure in India. *Review of Development Economics*, *23*(1), 314–330.

Schaffrin, A., Sewerin, S. and Seubert, S. (2014) The innovativeness of national policy portfolios–climate policy change in Austria, Germany, and the UK. *Environmental Politics*, *23*(5), 860–883.

Sengupta, A. and Nundy, S. (2005) The private health sector in India. *British Medical Journal*, *331*(7526), 1157–1158.

Shan, L., Wu, Q., Liu, C. et al. (2017) Perceived challenges to achieving universal health coverage: A cross-sectional survey of social health insurance managers/administrators in China. *BMJ Open*, 7, e014425. doi: 10.1136/bmjopen-2016-014425.

Shen, Y. C., Eggleston, K., Lau, J. and Schmid, C. H. (2007) Hospital ownership and financial performance: what explains the different findings in the empirical literature? *INQUIRY: The Journal of Health Care Organization, Provision, and Financing*, *44*(1), 41–68.

Singapore Business Review (2019) Chart of the Day: Public hospitals snapped up three-fourths of tertiary admissions in 2006–2018. https://sbr.com.sg/health care/news/chart-day-public-hospitals-snapped-three-fourths-tertiary-admissions-in-2006-2018.

Singh, P., Ravi, S. and Chakraborty, S. (2020) *Is India's Health Infrastructure Equipped to Handle an Epidemic*. Washington, DC: Brookings Institution Press. Retrieved from www.brookings.edu/blog/up-front/2020/03/24/is-indias-health-infrastructure-equipped-to-handle-an-epidemic/.

SingHealth (2018) Overview 2017–2018. SingHealth Duke NUS Academic Medical Center. Retrieved from www.singhealth.com.sg/about-singhealth/newsroom/Documents/SingHealth-AR17-Insert.pdf.

Smullen, Amanda and Hong, Phua Kai (2015) Comparing the health care systems of high-performing Asian Countries. *Asia & the Pacific Policy Studies*, *2*(2), 347–355. doi: 10.1002/app5.76.

Solt, Frederick (2019) The Standardized World Income Inequality Database, Version 8. https://doi.org/10.7910/DVN/LM4OWF, Harvard Dataverse, V4.

Sparkes, S. P., Bump, J. B., Özçelik, E. A., Kutzin, J. and Reich, M. R. (2019) Political economy analysis for health financing reform. *Health Systems & Reform*, *5*(3), 183–194.

Sparkes, Susan P., Kutzin, Joseph, Soucat, Agnès, Bump, Jesse B. and Reich, Michael R. (2019) Introduction to special issue on political economy of health financing reform. *Health System & Reform*, *3*, 179–182.

Statista (2018a) Apollo hospital average length of patient stay in India 2013–2018. Retrieved from www.statista.com/statistics/976534/alos-apollo-hospitals-india/.

(2018b) Average length of stay in hospitals in China 2018, by region (in days). Retrieved from www.statista.com/statistics/225527/average-length-of-stay-in-hospitals-in-china-by-region/.

Straits Times (2017) Govt spending on health care to rise sharply in next 3–5 years: Heng Swee Keat, 7 December. www.straitstimes.com/singapore/health/govt-spending-on-health care-to-rise-sharply-in-next-3-5-years-heng-swee-keat.

Tan, H., Lim, J. C. W., Yun, S. T. S. et al. (2016) Healthcare regulation. In Lee Chien Earn and K. Satku (Eds.), *Singapore's Health Care System: What 50 Years Have Achieved*. Singapore: World Scientific.

Tan, K. B. and Earn Lee, C. (2019) Integration of primary care with hospital services for sustainable universal health coverage in Singapore. *Health Systems & Reform*, *5*(1), 18–23.

Tan, Kenneth Paul (2016) *Governing Global-City Singapore: Legacies and Futures after Lee Kuan Yew*. London: Routledge.

Tang, Kwong-Leung (1998) *Colonial State and Social Policy: Social Welfare Development in Hong Kong 1842–1997*. Lanham: University Press of America.

(2000) *Social Welfare Development in East Asia*. New York: St. Martin's Press.

Tangcharoensathien, V., Limwattananon, S., Patcharanarumol, W., Thammatacharee, J., Jongudomsuk, P. and Sirilak, S. (2007) Achieving universal coverage in Thailand: What lessons do we learn. Working Paper, International Health Policy Program, Thailand.

Tangcharoensathien, V., Supachutikul, A., Patcharanarumol, W. et al. (2015) Public Hospital Governance in Thailand: A case study on universal health coverage, health care accreditation and the emerging medico-legal environment. In D. Huntington and K. Hort (Eds.), *Public Hospital Governance in Asia and the Pacific*. Asia Pacific Observatory on Health Systems and Policies.

Tangcharoensathien, V., Limwattananon, S., Patcharanarumol, W. and Thammatacharee, J. (2014) Monitoring and evaluating progress towards universal health coverage in Thailand. *PLoS Medicine*, *11*(9), e1001726.

(2015) Achieving universal health coverage goals in Thailand: The vital role of strategic purchasing. *Health Policy and Planning*, *30*(9), 1152–1161.

Tangcharoensathien, V., Patcharanarumol, W., Kulthanmanusorn, A., Saengruang, N. and Kosiyaporn, H. (2019) The political economy of UHC reform in Thailand: Lessons for low-and middle-income countries. *Health Systems & Reform*, *5*(3), 195–208.

Tangcharoensathien, V., Tisayaticom, K., Suphanchaimat, R., Vongmongkol, V., Viriyathorn, S. and Limwattananon, S. (2020) Financial risk protection of Thailand's Universal Health Coverage: Results from series of national household surveys between 1996 and 2015.

Tangcharoensathien, V., Wibulpholprasert, S. and Nitayaramphong, S. (2004) Knowledge-based changes to health systems: The Thai experience in policy development. *Bulletin of the World Health Organization, 82*, 750–756.

Teerawattananon, Y., Tangcharoensathien, V., Tantivess, S. and Mills, A. (2003). Health sector regulation in Thailand: Recent progress and the future agenda. *Health Policy, 63*(3), 323–338.

Tejativaddhana, Phudit et al. (2018) Developing primary health care in Thailand Innovation in the use of socio-economic determinants, Sustainable Development Goals and the district health strategy. *Public Administration and Policy, 21*(1).

Thailand's Poverty on the Rise Amid Slowing Economic Growth (2020, March 5). Retrieved 10 August 2020, from www.worldbank.org/en/news/press-release/2020/03/03/thailands-poverty-on-the-rise-amid-slowing-economic-growth.

Thiagarajan, K. (2020) Covid-19 exposes the high cost of India's reliance on private healthcare. *BMJ, 2020*, 370. doi: 10.1136/bmj.m3506.

Thit, W. M., Thanapak, W., Nareenuch, N., Leelakittisin, B., Khongmun, P., Ponthongmak, W. and Looareesuwan, P. (2016) eHealth in Thailand: Interoperability and Health Information Standards. In B. Kijsanayotin (Ed.), Thai Health Information Standards Development Center, Health System Research Institution.

Thomas, Joan Sara et al. (2016) A brief history of public health in Singapore. In Lee Chin Earn and K. Satku (Eds.), *Singapore's Health Care System: What 50 Years Have Achieved*. Singapore: World Scientific.

Tollen, L., Keating, E. and Weil, Alan (2020) How administrative spending contributes to excess US Health Spending. *Health Affairs*. doi: 10.1377/hblog20200218.375060.

Travis, P., Egger, D., Davies, P. and Mechbal, A. (2002) *Towards Better Stewardship: Concepts and Critical Issues*. Geneva, Switzerland: World Health Organization.

Trivedi, M., Saxena, A., Mavalankar, D. and Sharma, M. (2020) Improving hospital-based processes for effective implementation of pmjay: Evidence from early implementation. National Health Authority Working Paper. Retrieved from https://pmjay.gov.in/sites/default/files/2020-02/WP_IIPH_study_2.pdf.

Tsang, Donald (2006) Big market, small government. 18 September 2006. www.ceo.gov.hk/eng/press/oped.htm. Visited 16 April 2011.

Tucci, John (2010) The Singapore health system – Achieving positive health outcomes with low expenditure. https://web.archive.org/web/20100419231253/http://www.watsonwyatt.com/europe/pubs/health care/render2.asp?ID=13850.

Tuohy, C. H. (2018) *Remaking Policy: Scale, Pace, and Political Strategy in Health Care Reform* (Vol. 54). Toronto: University of Toronto Press.

Tuohy, C. H. and Glied, Sherry (2011) The Political Economy of Healthcare in Sherry Glied and Peter Smith (Eds.), *The Oxford Handbook of Health Economics*. doi: 10.1093/oxfordhb/9780199238828.013.0004.

Vaidya, K. (2018, January 17) Draft Act ready to prevent 'cut' practice among Maharashtra docs. Retrieved 6 August 2020, from www.dnaindia.com/mumbai/report-draft-act-ready-to-prevent-cut-practice-among-maharashtra-docs-2576065.

Van Doorslaer, Eddy et al. (2007) Catastrophic payments for health care in Asia. *Health Economics*, *16*, 1159–1184. doi: 10.1002/hec.1209.

Verhoest, K., Van Thiel, S., Bouckaert, G. and Lægreid, P. (2012) *Government Agencies: Practices and Lessons from 30 Countries*. London: Palgrave Macmillan.

Virani, A. (2019) Notions of policy effectiveness and implications for policy design: Insights from public-private partnerships in India. *Policy Design and Practice*, 2(2), 198–214.

Virani, A. and Ramesh, M. (2019) How neglecting policy mechanisms can lead to policy failure: Insights from public–private partnerships in India's health sector. In G. Capano, M. Howlett, M. Ramesh and A. Virani (Eds.), *Making Policies Work: First- and Second-Order Mechanisms in Policy Design* (pp. 76–95). Cheltenham: Edward Elgar.

Wagstaff, A. et al. (2018) Progress on catastrophic health spending in 133 countries: A retrospective observational study. *The Lancet: Global Health*, 6(2), e169–e179. doi: 10.1016/s2214-109x(17)30429-1.

Wagstaff, A. and Lindelow, M. (2008) Can insurance increase financial risk?: The curious case of health insurance in China. *Journal of Health Economics*, 27(4), 990–1005.

Wagstaff, A., Yip,W., Lindelow, M. and Hsiao,W. C. (2009) China's health system and its reform: A review of recent studies. *Health Economics*, *18* (S2), S7–S23.

Wagstaff, Adam (2005) Health Systems in East Asia : What Can Developing Countries Learn from Japan and the Asian Tigers. Policy Research Working Paper WPS3790, The World Bank. www-wds.worldbank.org/servlet/WDSContentServer/WDSP/IB/2005/12/07/000016406_20051207105757/Rendered/PDF/wps3790.pdf.

Wang, S. (2004) China's health system: From crisis to opportunity. *Yale-China Health Journal*, *3*, 5–49.

Wang, Y. and Wang, W. (2007) Escalation in health expenditure and difficult access to care in China. In J. Chen and Y. Wang (Eds.), *Development of Social Security in China: Health Services and Protection in Transformation*. Beijing: Social Science Academic Press (in Chinese).

Wei, F., Shuli, Z., Yuhui, Z., Peipei, C. and John, G. (2018) Research in health policy making in China: Out-of-pocket payments in Healthy China 2030. *BMJ*, *2018*, 360. doi: 10.1136/bmj.k234.

Wendt, C. (2009) Mapping European healthcare systems: A comparative analysis of financing, service provision and access to healthcare. *Journal of European Social Policy*, *19*(5), 432–445.

White, W. D. (2014) Professional self-regulation in medicine. *Virtual Mentor*, *16*(4), 275–278. doi: 10.1001/virtualmentor.2014.16.4.hlaw1-1404.

WHO (2007) *Everybody's Business. Strengthening Health Systems to Improve Health Outcomes: WHO's Framework for Action*. Geneva: World Health Organization.

(2014) *Health Systems Governance for Universal Health Coverage: Action Plan*. Geneva: WHO, Department of Health Systems Governance and Financing.

(2020a) Questions and Answers on Universal Health Coverage. 12 September 2020 www.who.int/healthsystems/topics/financing/uhc_qa/en/.

(2020b) What is universal health coverage? 21 January 2020, www.who.int/westernpacific/news/q-a-detail/what-is-universal-health-coverage#:~:text=The%20goal%20of%20universal%20health,hardship%20when%20paying%20for%20them.&text=detecting%20health%20conditions%20early%3B,helping%20patients%20with%20rehabilitation.

(undated) "Governance" Health Systems Strengthening Glossary. www.who.int/healthsystems/hss_glossary/en/index5.html.

(n.d.) Global Health. Observatory data repository. https://apps.who.int/gho/data/view.main.UHCFINANCIALPROTECTION01v. Accessed 15 August 2020.

WHO 2000, World Health Report (2000) *Health Systems: Improving Performance*. Geneva: World Health Organization. www.who.int/whr/2000/en/whr00_en.pdf.

WHO 2010, World health report (2010) *Health Systems Financing: The Path to Universal Coverage*. Geneva: World Health Organization. www.who.int/whr/2010/whr10_en.pdf.

WHO Europe (2008) Stewardship/Governance of health systems in the WHO European Region. www.euro.who.int/__data/assets/pdf_file/0016/70180/RC58_edoc09.pdf?ua=1.

Wibulpolprasert, S., Sirilak, S., Ekachampaka, P. and Wattanamano, N. (2011) Thailand health profile 2008–2010. https://bps.moph.go.th/new_bps/thailand_health_profile/Thailand%20Health%20Profile%20Report%202008-2010.pdf.

Willis, K. and Lewis, S. (2016) Which are better, public or private hospitals?. The Conversation, 18 March 2016. https://theconversation.com/which-are-better-public-or-private-hospitals-54338.

Wilsford, D. (1994) Path dependency, or why history makes it difficult but not impossible to reform health care systems in a big way. *Journal of Public Policy*, 251–283.

Witter, S. et al. (2019) Health system strengthening – Reflections on its meaning, assessment, and our state of knowledge. *International Journal of Health Planning and Management*, *34*, e1980–e1989. doi: 10.1002/hpm.2882.

Wong, Cheuk Nam and Tin, Keith YK (2015) Universal Health Coverage Assessment: Hong Kong. Global Network for Health Equity (GNHE). http://gnhe.org/blog/wp-content/uploads/2015/05/GNHE-UHC-assessment_Hong-Kong-1.pdf Accessed 16 January 2020.

Wong, J. (2006) *Healthy Democracies: Welfare Politics in Taiwan and South Korea*. Ithaca: Cornell University Press.

Woo, L. H. (2006) *Health care financing and health care services for the elderly*. The Voice, 34. Retrieved 3 May 2011, from www.thevoice.org.hk/v0348/00002.htm.

Woo, J. J. (2020) Policy capacity and Singapore's response to the COVID-19 pandemic. *Policy and Society*, *39*(3), 345–362, doi: 10.1080/14494035.2020.1783789.

World Bank (2020a) *World Development Indicators*. Washington, DC: World Bank. https://databank.worldbank.org/source/world-development-indicators.

(2020b) UHC service coverage index. Retrieved 11 August 2020, from https://data.worldbank.org/indicator/SH.UHC.SRVS.CV.XD.

World Bank and World Health Organization (2019) *Healthy China: Deepening Health Reform in China: Building High-Quality and Value-Based Service Delivery*. Washington, DC: World Bank.

World Health Organization (2017) Primary health care systems (PRIMASYS). Case study from Thailand.

Wu, Xun and Ramesh, M. (2009) Health care reforms in developing countries: Potentials and realities. *Development and Change*, *40*(3), 531–549.

Wu, X. and Ramesh, M. (2014) Market imperfections, government imperfections, and policy mixes: Policy innovations in Singapore. *Policy Sciences*, *47*(3), 305–320.

Wu, X., Ramesh, M. and Howlett, M. (2015) Policy capacity: A conceptual framework for understanding policy competences and capabilities. *Policy and Society*, *34*(3), 165–171.

Xu, H., He, J., Zheng, K. and Kwan, H. (2010) The Chinese health care system. In J. Johnson and C. Stoskopf (Eds.), *Comparative Health Systems: Global Perspective*. Boston: Jones & Bartlett.

Xu, Jin, Weiyan, Jian, Kun, Zhu, Soonman, Kwon and Hai, Fang (2019) Reforming public hospital financing in China: Progress and challenges. *BMJ*, *365*, l4015.

Xu, K., Evans, D. B., Kawabata, K., Zeramdini, R., Klavus, J. and Murray, C. J. (2003) Household catastrophic health expenditure: A multicountry analysis. *The Lancet*, *362*(9378), 111–117.

Yeo, Y. H., Lee, K. H. and Kim, H. J. (2016) Closure of a local public hospital in Korea: Focusing on the organizational life cycle. *Journal of Health Care Leadership*, *8*, 95–105. doi: 10.2147/JHL.S113070.

Yin, Jason Dean-Chen and He, Alex Jingwei (2018) Health insurance reforms in Singapore and Hong Kong: How the two ageing asian tigers respond to health financing challenges?. *Health Policy*, *122*(7), 693–697.

Yip, P. S. F. et al. (2014) What makes the hospitalisation system more efficient? An application of the decomposition method to Hong Kong morbidity data. *BMJ Open*, *4*, e003903. doi: 10.1136/bmjopen-2013-003903.

Yip, W. C. and Hsiao, W. C. (2003) Autonomizing a hospital system: Corporate control by central authorities in Hong Kong. In A. S. Preker and A. Harding (Eds.), *Innovations in Health Service Delivery: The Corporatization of Public Hospitals* (pp. 391–424). Washington, DC: World Bank.

Yip, W. and Hsiao, W. (2009) China's health care reform: A tentative assessment. *China Economic Review*, *20*, 613–619.

Yip, W. et al. (2019) 10 years of health-care reform in China: Progress and gaps in Universal Health Coverage. *The Lancet*, 394(10204), 1192–1204.

Yip, W. and Mahal, A. (2008) The health care systems of China and India: Performance and future challenges. *Health Affairs*, *27*(4), 921–932.

Yip, Winnie (1998) Physicians response to medicare fee reductions: Changes in the volume and intensity of supply of coronary artery bypass graft (CABG) procedures for the medicare and provider sectors. *Journal of Health Economics*, *17*(6), 675–699.

Yoo, S. L., Kim, D. J., Lee, S. M., Kang, W. G., Kim, S. Y., Lee, J. H. and Suh, D. C. (2019) Improving patient access to new drugs in South Korea: Evaluation of the national drug formulary system. *International Journal of Environmental Research and Public Health*, *16*(2), 288.

Yonhap (2019) Korea to up health insurance premiums 3.5% in 2019. *Korea Herald*. November 6. www.koreaherald.com/view.php?ud=20181106000130.

Yong, Lee Jin et al. (2012) Health care organizations' attitudes toward pay-for-performance in Korea. *Health Policy*, *108*(2–3), 277–285. doi: 10.1016/j.healthpol.2012.09.002.

Youngkong, S., Baltussen, R., Tantivess, S., Mohara, A. and Teerawattananon, Y. (2012). Multicriteria decision analysis for including health interventions in the universal health coverage benefit package in Thailand. *Value in Health*, *15*(6), 961–970.

Yulchon, L. L. C. (2018) Impacts and Implications of President Health Care Initiative – Moon Jae-In Care at a Glance. March 16. www.lexology.com/library/detail.aspx?g=bbbd2d0a-fff8-463c-9a0b-0e754b5733d3.

Zhong, J. (2001) Influencing factors on medical quality and measures for management. *Chinese Journal of Hospital Management*, *17*(9), 536.

Zweifel, P. and Manning, W. G. (2000) *Moral hazard and consumer incentives in health care. In Handbook of Health Economics* (Vol. 1, pp. 409–459). Amsterdam: Elsevier.

Index